Methods in Molecular Biology

Series Editor
John M. Walker
School of Life and Medical Sciences
University of Hertfordshire
Hatfield, Hertfordshire, AL10 9AB, UK

For further volumes:
http://www.springer.com/series/7651

Proteostasis

Methods and Protocols

Edited by

Rune Matthiesen

Computational and Experimental Biology Group, Department of Health Promotion and Chronic Diseases, National Health Institute Dr. Ricardo Jorge, INSA, I.P., Lisboa, Portugal

Humana Press

Editor
Rune Matthiesen
Computational and Experimental Biology Group
Department of Health Promotion and Chronic Diseases
National Health Institute Dr. Ricardo Jorge, INSA, I.P.
Lisboa, Portugal

ISSN 1064-3745 ISSN 1940-6029 (electronic)
Methods in Molecular Biology
ISBN 978-1-4939-3754-7 ISBN 978-1-4939-3756-1 (eBook)
DOI 10.1007/978-1-4939-3756-1

Library of Congress Control Number: 2016939242

Printed on acid-free paper

This Humana Press imprint is published by Springer Nature
The registered company is Springer Science+Business Media LLC New York

Preface

Proteostasis or protein homeostasis is the process by which cells control the abundance and folding of the proteome. Proteostasis appears to be involved in many diseases and aging. For example, retinal dystrophies, neurodegenerative diseases, inflammatory diseases, infectious diseases, and cancer are broad categories of diseases already linked to proteostasis. For instance, several genes related to the ubiquitin pathway are implicated in retinal dystrophies. Retinal dystrophies are a group of rare diseases that affect individuals worldwide. Protein misfolding, aggregation, and accumulation are a common hallmark in various neurodegenerative diseases. The autophagy-lysosomal pathway and the ubiquitin-proteasome system, the two main intracellular degradation machineries, are essential for cell survival under stress conditions, for clearance of intracellular pathogens, for the maintenance of cellular homeostasis and play an essential role in cancer survival upon drug treatment. In addition, proteasome inhibitors, which typically target protease subunits of the proteasome, have been shown to reverse liver cancers in xenograft models and prolong time of survival of patients with certain blood cancers (e.g., multiple myeloma and multiple cell lymphoma).

The importance of proteostasis in diseases has fostered the development of a large number of technologies to obtain deep insight into the underlining mechanistic events. The technologies are based on fluorescence/confocal microscopy, expression arrays, mass spectrometry, and diverse range of transfection models combined with biochemical assays. The methodologies target the proteins in a stationary quantitative way but also protein turnover rate can be estimated in vivo by pulsed labeling and novel technologies like bioorthogonal click chemistry. Protein homeostasis is regulated by a broad range of posttranslational modifications of which ubiquitin and SUMO are the most frequently studied. Posttranslational modifications of human proteins do not exclusively influence human health. For example, proteins from pathogens can also be heavily ubiquitinated and thereby targeted for degradation. Furthermore, interaction between the adenoviral capsid protein VI and Nedd4.2, a cellular ubiquitin ligase, is essential for virus infection highlighting the role of proteostasis in host–pathogen interactions.

This book highlights the role of proteostasis in human health and associated disease model systems. It provides state-of-the-art protocols to study and target proteostasis for therapeutics. This book is designed and written mainly by proteostasis experts with the ambitious aim to become the future reference book on proteostasis in human health.

I acknowledge the great enthusiasm of the Proteostasis COST action members in supporting the realization of the book. Chapters were delivered on time and it has been a true pleasure to work with the authors. Special thanks go to Dr. R. Menezes, Dr. Gemma Marfany, and Dr. Gustavo J. Gutierrez who provided assistance in reviewing chapters.

The left part of the cover image depicts the ubiquitin coating of Salmonella bacteria inside an infected epithelial cell (as part of the autophagy process). Bacteria are in red, ubiquitin in green, nuclei of infected cells in blue (image kindly provided by Professor Rudi Beyaert). Center image displays the structure of ubiquitin (kindly provided by Simona Polo).

Top right depicts Human ARPE-19 cells transfected with the human deubiquitinating enzyme ATXN3. Green: ATXN3, Red: Actin fillaments stained with phalloidin, Cyan: acetylated alpha-tubulin. Nuclei are stained with DAPI (Photo by Vasileios Toulis from Gemma Marfany's group).

Lisboa, Portugal ***Rune Matthiesen***

Contents

Contributors

CARMEN AGUADO • *Intracellular Protein Degradation and Rare Diseases Laboratory, Centro de Investigación Príncipe Felipe, Valencia, Spain; Centro de Investigación Biomédica en Red de Enfermedades Raras (CIBERER), Valencia, Spain*
FABIENNE AILLET • *Ubiquitylation and Cancer Molecular Biology, Inbiomed, San Sebastian, Spain*
FRANCISCO AMADO • *Department of Chemistry, Mass Spectrometry Centre, QOPNA, University of Aveiro, Aveiro, Portugal*
SISLEY AUSTIN • *Intracellular Transport of Viral Structures, MFP CNRS UMR 5234, University of Bordeaux, Bordeaux, France*
MIKEL AZKARGORTA • *Proteomics Platform CICbioGUNE, CIBERehd, ProteoRed-ISCIII, Parque Tecnologico de Bizkaia, Derio, Spain*
ROSA BARRIO • *CIC bioGUNE, Derio, Spain*
RAMON BARTRONS • *Departament de Ciències Fisiològiques II, IDIBELL, Campus de Bellvitge, L'Hospitalet de Llobregat, Universitat de Barcelona, Barcelona, Spain*
MICHEL BECUWE • *Institut Jacques Monod, UMR 7592 CNRS/Univ. Paris Diderot, Sorbonne Paris Cité, Paris, France; Department of Genetics and Complex Diseases, Harvard School of Public Health, Boston, MA, USA*
BORJA BELDA-PALAZÓN • *Instituto de Biología Molecular y Celular de Plantas, CSIC-Universidad Politécnica de Valencia, Valencia, Spain*
RUDI BEYAERT • *Unit of Molecular Signal Transduction in Inflammation, Inflammation Research Center, VIB, Ghent, Belgium; Department of Biomedical Molecular Biology, Ghent University, Ghent, Belgium*
EWA BLASZCZAK • *Department of Genetics and Cell Physiology, Institute of Experimental Biology, University of Wroclaw, Wroclaw, Poland*
GUILLAUME BOSSIS • *Equipe Labellisée par la Ligue contre le Cancer, Institut de Génétique Moléculaire de Montpellier, CNRS UMR 5535, Université de Montpellier, CNRS, Montpellier Cedex 5, France*
FRÉDÉRIQUE BROCKLY • *Equipe Labellisée par la Ligue contre le Cancer, Institut de Génétique Moléculaire de Montpellier, CNRS UMR 5535, Université de Montpellier, CNRS, Montpellier Cedex 5, France*
ANA SOFIA CARVALHO • *Computational and Experimental Biology Group, Department of Health Promotion and Chronic Diseases, National Health Institute Dr. Ricardo Jorge, INSA, I.P., Lisboa, Portugal*
CESYEN CEDEÑO • *Department of Structural Biology, Vlaams Instituut voor Biotechnologie (VIB), Vrije Universiteit Brussel (VUB), Brussels, Belgium*
DIONYSIOS V. CHARTOUMPEKIS • *Department of Pharmacology and Chemical Biology, University of Pittsburgh, Pittsburgh, PA, USA*
ERIC CHEVET • *Inserm ERL440 "Oncogenesis Stress Signaling", Université Rennes 1, Rennes, France; Centre de Lutte Contre le Cancer Eugène Marquis, Rennes, France*
NIKI CHONDROGIANNI • *Institute of Biology, Medicinal Chemistry and Biotechnology, National Hellenic Research Foundation, Athens, Greece*

AARON CRANSTON • *MISSION Therapeutics Limited, Moneta, Babraham Research Campus, Cambridge, UK*
MONICA CUBILLOS-ROJAS • *Departament de Ciències Fisiològiques II, IDIBELL, Campus de Bellvitge, L'Hospitalet de Llobregat, Universitat de Barcelona, Barcelona, Spain*
MICHAŁ DADLEZ • *Department of Biophysics, Institute of Biochemistry and Biophysics, Polish Academy of Sciences, Warszawa, Poland*
MIEKE DELVAEYE • *Unit of Molecular Signal Transduction in Inflammation, Inflammation Research Center, VIB, Ghent, Belgium; Department of Biomedical Molecular Biology, Ghent University, Ghent, Belgium*
C. NUNES DO SANTOS • *Molecular Nutrition and Health, Instituto de Biologia Experimental Tecnológica, Oeiras, Portugal; Instituto de Tecnologia Química e Biológica António Xavier, Oeiras, Portugal*
PEDRO DOMINGUES • *Department of Chemistry, Mass Spectrometry Centre, QOPNA, University of Aveiro, Aveiro, Portugal*
KRZYSZTOF DRABIKOWSKI • *Department of Biophysics, Institute of Biochemistry and Biophysics, Polish Academy of Sciences, Warszawa, Poland*
FELIX ELORTZA • *Proteomics Platform CICbioGUNE, CIBERehd, ProteoRed-ISCIII, Parque Tecnologico de Bizkaia, Derio, Spain*
RICARDO ESCALANTE • *Instituto de Investigaciones Biomédicas Alberto Sols, Madrid, Spain*
IRAIDE ESCOBES • *Proteomics Platform CICbioGUNE, CIBERehd, ProteoRed-ISCIII, Parque Tecnologico de Bizkaia, Derio, Spain*
MARA ESPOSITO • *Laboratory of Pathophysiological Cell Signaling (PACS), Department of Biology, Faculty of Sciences and Bioengineering Sciences, Vrije Universiteit Brussel (VUB), Brussels, Belgium*
ROSA FARRÀS • *Centro de Investigación Príncipe Felipe, Valencia, Spain*
ALEJANDRO FERRANDO • *Instituto de Biología Molecular y Celular de Plantas, CSIC-Universidad Politécnica de Valencia, Valencia, Spain*
RITA FERREIRA • *Department of Chemistry, Mass Spectrometry Centre, QOPNA, University of Aveiro, Aveiro, Portugal*
CASSIE FORDER • *MISSION Therapeutics Limited, Moneta, Babraham Research Campus, Cambridge, UK*
ALEJANDRO GARANTO • *Departament de Genètica, Facultat de Biologia, Universitat de Barcelona, Barcelona, Spain; Department of Human Genetics, and Donders Center for Neurosciences, Radboud University Medical Centre, Nijmegen, The Netherlands*
G. GARCIA • *Molecular Nutrition and Health, Instituto de Biologia Experimental Tecnológica, Oeiras, Portugal; Instituto de Tecnologia Química e Biológica António Xavier, Oeiras, Portugal*
MICHAEL H. GLICKMAN • *Department of Biology, Technion-Israel Institute of Technology, Haifa, Israel*
AMNON GOLAN • *Department of Biology and Environment, University of Haifa at Oranim, Tivon, Israel; Protein Purification Center, Ramat Yohanan, Israel*
AMANDA GONÇALVES • *Bio Imaging Core, Inflammation Research Center, VIB, Ghent, Belgium*
GUSTAVO J. GUTIERREZ • *Laboratory of Pathophysiological Cell Signaling (PACS), Department of Biology, Faculty of Sciences and Bioengineering Sciences, Vrije Universiteit Brussel (VUB), Brussels, Belgium*
JEANINE HARRIGAN • *MISSION Therapeutics Limited, Moneta, Babraham Research Campus, Cambridge, UK*

Rasmus Hartmann-Petersen • *Department of Biology, Section for Biomolecular Sciences, University of Copenhagen, Copenhagen N, Denmark*
Carina I. Holmberg • *Research Programs Unit, Translational Cancer Biology, University of Helsinki, Helsinki, Finland*
Junie Hovsepian • *Institut Jacques Monod, UMR 7592 CNRS/Univ. Paris Diderot, Sorbonne Paris Cité, Paris, France*
Xavier Jacq • *MISSION Therapeutics Limited, Moneta, Babraham Research Campus, Cambridge, UK*
Sweta Jha • *Research Programs Unit, Translational Cancer Biology, University of Helsinki, Helsinki, Finland*
Jens V. Johansen • *Biotech Research and Innovation Centre, University of Copenhagen, Copenhagen N, Denmark*
Oded Kleifeld • *Department of Biochemistry and Molecular Biology, Monash University, Clayton, VIC, Australia*
Erwin Knecht • *Intracellular Protein Degradation and Rare Diseases Laboratory, Centro de Investigación Príncipe Felipe, Valencia, Spain; Centro de Investigación Biomédica en Red de Enfermedades Raras (CIBERER), Valencia, Spain*
Varun Gopala Krishna • *School of Life Sciences, Queen's Medical Centre, University of Nottingham, Nottingham, UK*
Marcos Lahuerta • *Intracellular Protein Degradation and Rare Diseases Laboratory, Centro de Investigación Príncipe Felipe, Valencia, Spain*
Rob Layfield • *School of Life Sciences, Queen's Medical Centre, University of Nottingham, Nottingham, UK*
Sébastien Léon • *Institut Jacques Monod, UMR 7592 CNRS/Univ. Paris Diderot, Sorbonne Paris Cité, Paris, France; Membrane Trafficking, Ubiquitin and Signaling Lab, Institut Jacques Monod, UMR7592 CNRS/Université Paris-Diderot, Paris Cedex, France*
Catherine Lindon • *Department of Genetics, University of Cambridge, Cambridge, UK*
Marie Lork • *Unit of Molecular Signal Transduction in Inflammation, Inflammation Research Center, VIB, Ghent, Belgium; Department of Biomedical Molecular Biology, Ghent University, Ghent, Belgium*
Diana Macedo • *Instituto de Tecnologia Química e Biológica, Universidade Nova de Lisboa, Oeiras, Portugal*
Gemma Marfany • *Departament de Genètica, Facultat de Biologia, Universitat de Barcelona, Barcelona, Spain; Centro de Investigación Biomédica en Red de Enfermedades Raras (CIBERER), Instituto de Salud Carlos III, Barcelona, Spain; Institut de Biomedicina de la Universitat de Barcelona (IBUB), Barcelona, Spain*
Zrinka Marijanovic • *Instituto de Medicina Molecular, Lisbon, Portugal*
Elena Maspero • *IFOM, Fondazione Istituto FIRC di Oncologia Molecolare, Milan, Italy*
Lydia Mata-Cantero • *Ubiquitylation and Cancer Molecular Biology, Inbiomed, San Sebastian, Spain*
Olli Matilainen • *Research Programs Unit, Translational Cancer Biology, University of Helsinki, Helsinki, Finland*
Rune Matthiesen • *Computational and Experimental Biology Group, Department of Health Promotion and Chronic Diseases, National Health Institute Dr. Ricardo Jorge, INSA, I.P., Lisboa, Portugal*

Ugo Mayor • *Biokimika eta Biologia Molekularra Saila, Zientzia eta Teknologia Fakultatea, University of the Basque Country (UPV/EHU), Leioa, Spain; Ikerbasque, Basque Foundation for Science, Bilbao, Spain*

Lauren McLellan • *MISSION Therapeutics Limited, Moneta, Babraham Research Campus, Cambridge, UK*

R. Menezes • *Molecular Nutrition and Health, Instituto de Biologia Experimental Tecnológica, Oeiras, Portugal; Instituto de Tecnologia Química e Biológica António Xavier, Oeiras, Portugal*

Mingwei Min • *Department of Genetics, University of Cambridge, Cambridge, UK; Department of Cell Biology, Harvard Medical School, Boston, MA, USA*

Esther La Monaca • *Laboratory of Pathophysiological Cell Signaling (PACS), Department of Biology, Faculty of Sciences and Bioengineering Sciences, Vrije Universiteit Brussel (VUB), Brussels, Belgium*

Sandra Muñoz-Braceras • *Instituto de Investigaciones Biomédicas Alberto Sols, Madrid, Spain*

Sofie V. Nielsen • *Department of Biology, Section for Biomolecular Sciences, University of Copenhagen, Copenhagen N, Denmark*

Neil J. Oldham • *School of Chemistry, University of Nottingham, Nottingham, UK*

Tiago Fleming Outeiro • *Faculdade de Ciências Médicas, CEDOC—Chronic Diseases Research Center, NOVA Medical School, Universidade Nova de Lisboa, Lisboa, Portugal; Instituto de Fisiologia, Faculdade de Medicina da Universidade de Lisboa, Lisboa, Portugal; Department of NeuroDegeneration and Restorative Research, Center for Nanoscale Microscopy and Molecular Physiology of the Brain, University Medical Center Göttingen, Göttingen, Germany*

Nikoletta Papaevgeniou • *Institute of Biology, Medicinal Chemistry and Biotechnology, National Hellenic Research Foundation, Athens, Greece*

Eva Pérez-Jiménez • *Intracellular Protein Degradation and Rare Diseases Laboratory, Centro de Investigación Príncipe Felipe, Valencia, Spain*

Elah Pick • *Department of Biology and Environment, University of Haifa at Oranim, Tivon, Israel*

Marc Piechaczyk • *Equipe Labellisée par la Ligue contre le Cancer, Institut de Génétique Moléculaire de Montpellier, CNRS UMR 5535, Université de Montpellier, CNRS, Montpellier Cedex 5, France*

Elin J. Pietras • *Biotech Research and Innovation Centre, University of Copenhagen, Copenhagen N, Denmark*

Simona Polo • *IFOM, Fondazione Istituto FIRC di Oncologia Molecolare, Milan, Italy; Dipartimento di Oncologia ed Emato-oncologia, Università degli Studi di Milano, Milan, Italy*

Esben G. Poulsen • *Department of Biology, Section for Biomolecular Sciences, University of Copenhagen, Copenhagen N, Denmark*

Claude Prigent • *Centre National de la Recherche Scientifique, UMR 6290, Rennes, France; Institut de Génétique et Développement de Rennes, Université de Rennes 1, Rennes, France*

Gwenaël Rabut • *Centre National de la Recherche Scientifique, UMR 6290, Rennes, France; Institut de Génétique et Développement de Rennes, Université de Rennes 1, Rennes, France; Institute of Genetics and Development of Rennes, UMR 6290 CNRS-UR1, Rennes, France*

JUANMA RAMIREZ • *Biokimika eta Biologia Molekularra Saila, Zientzia eta Teknologia Fakultatea, University of the Basque Country (UPV/EHU), Leioa, Spain*
FABIENNE RAYNE • *Intracellular Transport of Viral Structures, MFP CNRS UMR 5234, University of Bordeaux, Bordeaux, France*
MARKO RISTIC • *Equipe Labellisée par la Ligue contre le Cancer, Institut de Génétique Moléculaire de Montpellier, CNRS UMR 5535, Université de Montpellier, CNRS, Montpellier Cedex 5, France*
MANUEL S. RODRÍGUEZ • *IPBS, Université de Toulouse, CNRS, UPS and ITAV, Université de Toulouse, CNRS, UPS, Toulouse, France*
JOSE LUIS ROSA • *Departament de Ciències Fisiològiques II, IDIBELL, Campus de Bellvitge, L'Hospitalet de Llobregat, Universitat de Barcelona, Barcelona, Spain*
SUSANA SÁNCHEZ-TENA • *Departament de Ciències Fisiològiques II, IDIBELL, Campus de Bellvitge, L'Hospitalet de Llobregat, Universitat de Barcelona, Barcelona, Spain*
TAIANE SCHNEIDER • *Departament de Ciències Fisiològiques II, IDIBELL, Campus de Bellvitge, L'Hospitalet de Llobregat, Universitat de Barcelona, Barcelona, Spain*
DANIEL SCOTT • *School of Life Sciences, Queen's Medical Centre, University of Nottingham, Nottingham, UK*
MARK S. SEARLE • *School of Chemistry, University of Nottingham, Nottingham, UK; Centre for Biomolecular Sciences, University of Nottingham, Nottingham, UK*
BARRY SHAW • *School of Life Sciences, Queen's Medical Centre, University of Nottingham, Nottingham, UK*
SARA SIGISMUND • *IFOM, Fondazione Istituto FIRC di Oncologia Molecolare, Milan, Italy*
CORNELIA STEINHAUER • *Biotech Research and Innovation Centre, University of Copenhagen, Copenhagen N, Denmark*
JO STRACHAN • *Institute of Cell Biology, University of Edinburgh, Edinburgh, UK*
GERASIMOS P. SYKIOTIS • *Service of Endocrinology, Diabetology, and Metabolism, Lausanne University Hospital, Lausanne, Switzerland; Faculty of Biology and Medicine, University of Lausanne, Lausanne, Switzerland*
SAÏD TAOUJI • *BMY Screen, Bordeaux, France*
SANDRA TENREIRO • *Instituto de Medicina Molecular, Lisboa, Portugal; Faculdade de Ciências Médicas, CEDOC—Chronic Diseases Research Center, NOVA Medical School, Universidade Nova de Lisboa, Lisboa, Portugal*
DAVID J. TOOTH • *School of Life Sciences, Queen's Medical Centre, University of Nottingham, Nottingham, UK*
VASILEIOS TOULIS • *Departament de Genètica, Facultat de Biologia, Universitat de Barcelona, Barcelona, Spain*
EVELIEN VAN HAMME • *Bio Imaging Core, Inflammation Research Center, VIB, Ghent, Belgium*
FRANCESC VENTURA • *Departament de Ciències Fisiològiques II, IDIBELL, Campus de Bellvitge, L'Hospitalet de Llobregat, Universitat de Barcelona, Barcelona, Spain*
RUI VITORINO • *Department of Physiology and Cardiothoracic Surgery, Faculty of Medicine, University of Porto, Porto, Portugal; Department of Medical Sciences, iBiMED-Institute for Biomedicine, University of Aveiro, Aveiro, Portugal; PortugalQOPNA, Centro de Espectrometria de Massa, Departamento de Química, Universidade de Aveiro, Aveiro, Portugal*
NING WEI • *Department of Molecular, Cellular and Developmental Biology, Yale University, New Haven, CT, USA*

HARALD WODRICH • *Intracellular Transport of Viral Structures, MFP CNRS UMR 5234, University of Bordeaux, Bordeaux, France*
WENDY XOLALPA • *Proteomics Platform CICbioGUNE, CIBERehd, ProteoRed-ISCIII, Parque Tecnologico de Bizkaia, Derio, Spain*
PANOS G. ZIROS • *Service of Endocrinology, Diabetology, and Metabolism, Lausanne University Hospital, Lausanne, Switzerland*

Chapter 1

UPS Activation in the Battle Against Aging and Aggregation-Related Diseases: An Extended Review

Nikoletta Papaevgeniou and Niki Chondrogianni

Abstract

Aging is a biological process accompanied by gradual increase of damage in all cellular macromolecules, i.e., nucleic acids, lipids, and proteins. When the proteostasis network (chaperones and proteolytic systems) cannot reverse the damage load due to its excess as compared to cellular repair/regeneration capacity, failure of homeostasis is established. This failure is a major hallmark of aging and/or aggregation-related diseases. Dysfunction of the major cellular proteolytic machineries, namely the proteasome and the lysosome, has been reported during the progression of aging and aggregation-prone diseases. Therefore, activation of these pathways is considered as a possible preventive or therapeutic approach against the progression of these processes. This chapter focuses on UPS activation studies in cellular and organismal models and the effects of such activation on aging, longevity and disease prevention or reversal.

Key words Ubiquitin-proteasome system, Aging, Longevity, Aggregation-related diseases, Proteostasis, Proteasome activation

1 Aging and Aggregation-Related Diseases

1.1 Aging/Models of Aging

Aging is a multifactorial, natural process leading to gradual functional deterioration, continuing decline of self-defensive mechanisms, reduced homeostatic capacity of all tissues and an exponential accumulation of damage (in nucleic acids, proteins, and lipids) that leads to increased death incidence. The progression of aging is dynamically affected by both genetic and environmental factors. As long as equilibrium between cellular insults (mediated by stressors both from the micro- but also the macro-environment) and cellular repair/regeneration capacity is conserved, the cell/organism overcomes the damage that is produced without any fatal alterations in its phenotype and its physiology. However, once this balance is disturbed, the damaged molecules accumulate fast and multiple vicious circles of additional insults commence. As a result, an irreversible failure of homeostasis with compromised molecular pathways occurs. This failure eventually leads to aging and increased

Rune Matthiesen (ed.), *Proteostasis: Methods and Protocols*, Methods in Molecular Biology, vol. 1449,
DOI 10.1007/978-1-4939-3756-1_1, © Springer Science+Business Media New York 2016

rates of morbidity and mortality [1, 2]. Given the effects of aging on a pleiad of key pathways, it is logical that it constitutes a major risk factor for several pathologies including aggregation-related disorders [3, 4].

The establishment of several short-lived model organisms, such as yeast, nematode worms, flies and rodents along with the use of primary mammalian cell cultures as well as the use of isolated tissues from donors of different ages are the main tools to investigate the aging process and to decipher its regulation. More specifically, the cellular and organismal models that are most commonly used in aging studies are:

The **replicative senescence model** is until now the most accepted cellular model to study human aging. The model is based on the notion that normal human fibroblasts may undergo a limited number of divisions in culture before they gradually reach a state of irreversible growth arrest. This process is termed as replicative senescence or Hayflick limit and it is believed to recapitulate most of the human aging features [5].

Saccharomyces cerevisiae (*S. cerevisiae*) is often used in the study of various molecular pathways that govern the aging progression. There are two types of life-span that can be dissected in this model, namely the replicative and the chronological. The replicative (mitotic) life-span is defined by the number of daughter cells that a single mother yeast cell produces, whereas chronological life-span or stationary phase (post-mitotic) is defined by the time period during which the nondividing yeast cells can remain viable. Given those two types of life-span, it is suggested that *S. cerevisiae* is an attractive model to study the life-span of various human cell types, and thus mitotically active types but also post-mitotic types [6].

The soil nematode ***Caenorhabditis elegans*** (*C. elegans*) is a post-mitotic multicellular eukaryotic model organism that due to its advantages is heavily used to study aging. *C. elegans* shares many fundamental cellular/molecular structures and biological properties with more advanced organisms (including humans with which *C. elegans* shares 40 % homology), characteristics that nominate the nematode as an ideal model organism. Moreover, it is the first multicellular organism with known cell lineage and completely sequenced genome.

The fruit fly ***Drosophila melanogaster*** (*D. melanogaster*) has been used as a model organism for nearly a century. It is mostly composed of post-mitotic cells, it has a short life cycle/span and shows gradual aging. There is a 60 % conservation of genes between flies and humans [7] while 77 % of all known human disease genes have fly homologues [8]. Consequently, this insect is frequently used as a model organism in aging studies.

Rodents are frequently used in animal testing with mice and rats being the most used ones. The high degree of gene conservation between rodents and humans (i.e., humans share over 90 %

homology with mice into corresponding regions of conserved synteny; [9]), the possibilities of genetic manipulation of their genomes but also their relative short life expectancy are few of the advantages in using those animals as models to study aging. On top of that, the so far obtained results from studies on caloric restriction (CR) and pharmacological anti-aging/prolongevity treatments that have revealed increased relevance to humans further advocate for the use of those animals in aging studies [10].

Using the abovementioned models, numerous genes, proteins, and functional networks have been identified so far, thus permitting to establish the current known hallmarks of aging [2].

1.2 Aggregation-Related Diseases

In general, most of the misfolded and/or aggregated proteins are subjected to degradation by the cellular proteolytic machineries. However, there are few proteins (native and mutant) that are resistant to the degradation systems due to their tendency to form β-sheet-enriched oligomers that are finally packed into inclusion bodies or extracellular plaques. This characteristic accumulation of protease-resistant aggregated proteins is a common feature in protein misfolding disorders, including Alzheimer's disease (AD), Parkinson's disease (PD), Huntington's disease (HD), amyotrophic lateral sclerosis (ALS), and prion diseases (PrD).

1.2.1 Alzheimer's Disease

AD is the most known and common cause of dementia worldwide representing 65–75 % of all dementia cases [11]. It is a polygenic disorder that is characterized by loss of synaptic connections, extensive neurodegeneration and brain atrophy. AD patients can have an early onset mainly due to genetic mutations or a late onset, the latter being the most common case. The key hallmarks of AD are the deposition of intracellular, filamentous aggregates that consist of hyper-phosphorylated Tau protein (intracellular neurofibrillary tangles; NFTs) and amyloid-β (Aβ) extracellular plaques [12–14]. Aβ is produced through the presenilin-mediated cleavage of a transmembrane protein that normally regulates the synaptic function, namely Amyloid Precursor Protein (APP). Early onset of AD is characterized by the expression of both mutant APP (mAPP) and presenilin 1 and 2, which are required for the active function of γ-secretase to produce the Aβ peptide through APP breakdown [15–17]. Late onset of AD is induced by genetic and environmental factors with aging being one of the main risk factors. Mutation in the apolipoprotein E ε4 allele represents one pivotal genetic factor involved in this sporadic AD form [18]. Various other genes have been implicated to the sporadic late onset of AD as CLU, CR1, and PICALM [19]. The consecutive neurodegenerative alterations lead to a gradual decline in cognitive functions, especially in memory and visual-spatial orientation ending up to the individual's incapability to live functionally.

Most of the therapeutic approaches have focused so far on Aβ production, degradation, and prevention of its toxicity, on Tau formation and on general neuroprotection [20]. Various in vitro and in vivo models of the disease like neuroblastoma cell lines, mammals, *Aplysia*, zebra fish, fruit fly, and nematode mutant strains expressing the human Aβ peptide have been exploited [21, 22]. Here we summarize data regarding UPS activation as a promising therapeutic approach against AD.

1.2.2 Parkinson's Disease

PD is the second most common neurodegenerative disease characterized by muscular rigidity, bradykinesia, and uncontrollable tremor that worsen gradually in severity. The main pathologoanatomical feature of PD is the loss of a large portion of substantia nigra dopaminergic neurons [23, 24]. The gradual accumulation of inclusion bodies in the neuronal cytoplasm that consists of α-synuclein, parkin, UHC-L1, ubiquitin, and neurofilaments, namely Lewy bodies leads to irreversible neurodegeneration.

α-Synuclein is a 14 kDa protein that normally regulates vesicle trafficking during neurotransmission signaling through a chaperone-like activity [24]. Oligomeric and fibrillar conformations of α-synuclein (that polymerizes into fibrils in vitro) induce toxicity through (a) impairment of the function of several organelles, (b) alterations of the proper signal transmission through synapses, and (c) inhibition of the proteostasis mechanisms [24].

Parkin is the second important protein that exerts a distinct role on PD pathology while it is also responsible for autosomal recessive juvenile parkinsonism. It is a RING-domain E3 ligase that under normal conditions regulates the degradation of synaptic transmission-associated proteins and prevents the creation of aggregates while it is also essential for the regulation of mitophagy and mitochondrial equilibrium [25, 26]. Parkin mutations may lead to substrate recognition impairment and prevent the interaction with E2 enzymes. Lewy body inclusions in turn affect the normal function of Parkin by interfering to its normal ability to regulate degradation, thus leading to high toxicity [27].

Other molecules that have been identified to play a critical role in PD onset and progression are UCH-L1, PINK1, and DJ-1. UCH-L1 is a deubiquitinase, PINK1 is a serine/threonine kinase that acts protectively under conditions of proteasome inhibition, while DJ-1 has been shown to exert chaperone activity and protease activity both resulting in prevention of α-synuclein accumulation and aggregation [28]. It is obvious that the gene products targeted in familial PD are somehow associated to the UPS; either as UPS substrates (α-synuclein, parkin, synphilin-1, mutated DJ-1) or as components of the degradation pathway (parkin, ubiquitin, C-terminal hydrolase L1; [29]).

1.2.3 Huntington's Disease

HD is an autosomal dominant neurodegenerative disorder which is characterized by gradual degeneration of striatum neurons, affects muscle coordination, and causes mental decline and psychopathological problems [30]. Huntingtin (HTT) is the key protein involved in HD pathogenesis. More specifically, wild type (wt) huntingtin gene (htt) bears 6–35 CAG repeats in the N-terminus producing a polyglutamine (polyQ) tract. In contrast, in mutated htt gene the CAG triplet repeat stretch overpasses 36 repeats promoting a toxic gain of function, a feature that coincides with the onset of HD pathology [31]. The onset, progression, as well as severity of the disease are directly affected by the polyQ length. HD is a proteinopathy mainly characterized by intracellular inclusions bodies (IBs) formed by mutant HTT (mHTT) aggregates [32]. These IBs are gradually increasing in number and size thus impeding the normal function of neurons. Several studies have suggested that mHTT is cleaved to produce a shorter N-terminal fragment containing the polyQ expansion that eventually induces the protein fragment to misfold and form aggregates. Neurotoxicity has been linked to either the soluble and/or the aggregated form of the misfolded protein as well as to the aggregation process itself. The various forms of mHTT protein have been suggested to affect transcriptional regulation through the interaction with various transcription cofactors (activators or repressors), to promote apoptosis, to enhance the intracellular production of reactive oxygen species, to affect caspase activation, and to inhibit proteasome function.

1.2.4 Amyotrophic Lateral Sclerosis

ALS is a motor neuron degenerative disorder with severe symptoms and an expeditious progress from symptoms onset, ending to muscular atrophy, weakness, and eventually death due to degeneration of the respiratory muscles. The main cells that are affected are the pyramidal Betz cells in the motor cortex, the large anterior horn cells of the spinal cord, and the lower cranial motor nuclei of the brainstem [33]. ALS is mainly a sporadic disease but 10% of ALS cases are familial [34]. The pathologoanatomical signature of the disease is the accumulation of insoluble proteins that form intracellular aggregates (Skein-like inclusions, SLIs) as found in samples from human patients and animal models of ALS [35, 36].

Superoxide dismutase 1 (SOD1) missense mutations play a distinct role to most cases of the familial onset of the disease [37]. Toxic gain of function is believed to occur while increased levels of intracellular protein aggregates of mutant SOD1 (mSOD1) that disturb the unfolded protein response (UPR) and mitochondrial functionality are also revealed [38]. Several other proteins have been also implicated to ALS, including ALSIN, TDP-43, nuclear protein FUS, ubiquilin 2, p62, optineurin, and valosin-containing protein [39]. The causes are basically unknown in the absence of family history (sporadic ALS). C9ORF72 is one of the locuses on chromosome 9p identified to be involved in the sporadic ALS onset together with UNC13A, a presynaptic protein that normally acts in the neurotransmission signaling procedure [34]. It was recently pointed that most of the involved proteins in both sporadic and familial ALS share aggregation-prone properties that may ultimately act toxically and inhibitory to the proteostasis network.

1.2.5 Prion Diseases

PrDs, also known as transmissible spongiform encephalopathies, are infectious neurodegenerative disorders with acute and severe symptoms including memory and movement control problems, visual dysfunction and cognitive inability [40, 41]. Severe neuronal loss in prion-affected sections leads to the development of a "spongy" architecture which is the main anatomical characteristic of the disease. The most known PrDs are divided into three groups: the sporadic group including Jakob-Creutzfeldt disease (JCD); the genetic group including genetic JCD, Gerstmann-Sträussler-Sneaker disease, and fatal familial insomnia; and the infectious group including Kuru, variant JCD, and iatrogenic JCD.

All known mammalian PrDs are caused by the scrapie prion protein (PrP^{Sc}) an abnormal form of the naturally occurring protein PrP^{C}, a cell surface membrane [42]. The role of PrP^{C} is not yet fully elucidated. PrP knockout mice exhibit only minor abnormalities but more recently, it was shown that that neuronal expression and regulated proteolysis of PrP^{C} are essential for myelin maintenance [43]. Moreover, mice devoid of PrP^{C} exhibit an altered hippocampal long-term potentiation [44] while it was also suggested that PrP^{c} is necessary for the self-renewal of long-term hematopoietic stem cells [45].

PrP^{Sc} is a β-sheet-enriched isoform [46] able to self-propagate and fold in a variety of distinct ways [47]. This self-replication mechanism leads to the formation of spontaneous extracellular aggregates (prion deposits; [48]). Prions are at least partially protease-resistant proteins and therefore they tend to constantly accumulate. Moreover, PrP^{Sc} has the ability to interact with PrP^{C} and change its conformation into the infectious isoform, thus initiating a vicious cycle that potentiates the disease progression. Even a small quantity of PrP^{Sc} is enough to trigger the conversion of PrP^{C} to PrP^{Sc} as shown in vitro [49] but also in vivo [50].

Apart from the PrP, additional proteins have been shown to share prion-like domains. These domains endow the proteins with the self-replicating ability that is necessary for the formation of amyloid-like deposits. For example, it has been shown that TDP-43 mutations facilitate the conversion of misfolded proteins to aggregation-prone prion-like conformation, resulting in the ALS-related aggregates found in many familial ALS cases [51]. The latter case is the so-called prion paradigm, where otherwise harmless proteins can be converted to a pathogenic form by a small number of misfolded, nucleating proteins [52]. Nevertheless, cautiousness should be attributed since with the exception of PrP, the rest of the aggregation-prone proteins are not infectious agents.

1.3 Proteostasis in Normal Aging and Aggregation-Related Diseases

The proteome is challenged constantly and proteome integrity (proteostasis) is one of the nodal points that needs to be preserved in order to maintain organismal homeostasis. Therefore, it is not surprising that a group of specific molecules is dedicated to preserve the cellular protein load and therefore the cellular proteostasis. A complicated surveillance network of cellular mechanisms that inspect every aspect of protein biology from synthesis and folding to trafficking and clearance is set as responsible for proteostasis [53]. One primary arsenal of this network is constituted by chaperones that assure the correct folding/function of proteins and their maintenance in a correctly folded/functional mode. If however this arm of the proteostasis network fails, the secondary arsenal takes over to degrade the damaged, unfolded, aggregated and in general unwanted proteins. This arm includes the ubiquitin-proteasome system (UPS; which is the theme of this chapter) and the autophagy-lysosome system (for a recent review refer to [54, 55]). Upon failure of all surveillance systems, failure of proteostasis occurs with detrimental effects on the cellular physiology and life. It is not thus astonishing that the loss of proteostasis is considered as one of the hallmarks of aging [2] and that this loss is strongly related to the onset and progression of aging and aggregation-related diseases.

2 Introduction: The Ubiquitin System

Ubiquitin is a highly conserved protein that covalently modifies proteins through the ubiquitination process. There are three main steps that are gradually followed in order for an ubiquitin moiety to be added on a protein. These three steps are characterized by the action of three different types of ligases, namely E1 (ubiquitin-activating enzymes), E2 (ubiquitin-conjugating enzymes), and E3 (ubiquitin-ligase enzymes). The cycles of ubiquitination for a given protein can occur once thus leading to mono-ubiquitination or can be repeated several times on the same lysine thus leading to polyubiquitination. Depending on the moieties of ubiquitin added on a protein along with the lysine residues used for this binding, the localization/intracellular trafficking, activity, protein-protein interactions, participation in different signaling pathways, and degradation either by the 26S proteasome or by autophagy-lysosome system can be signaled [56, 57]. Polyubiquitin chains with at least four moieties constitute the signal for the 26S proteasome-mediated recognition and degradation of the protein substrate with the most frequent signal being the K48-linked ubiquitin chain [58]. To prevent energy loss, once the tagged substrate is recognized by the proteasome for degradation, specific deubiquitinases (DUBs) remove the polyubiquitin chains; those ubiquitin molecules can be reused [59]. The abovementioned proteins constitute the UPS (Fig. 1).

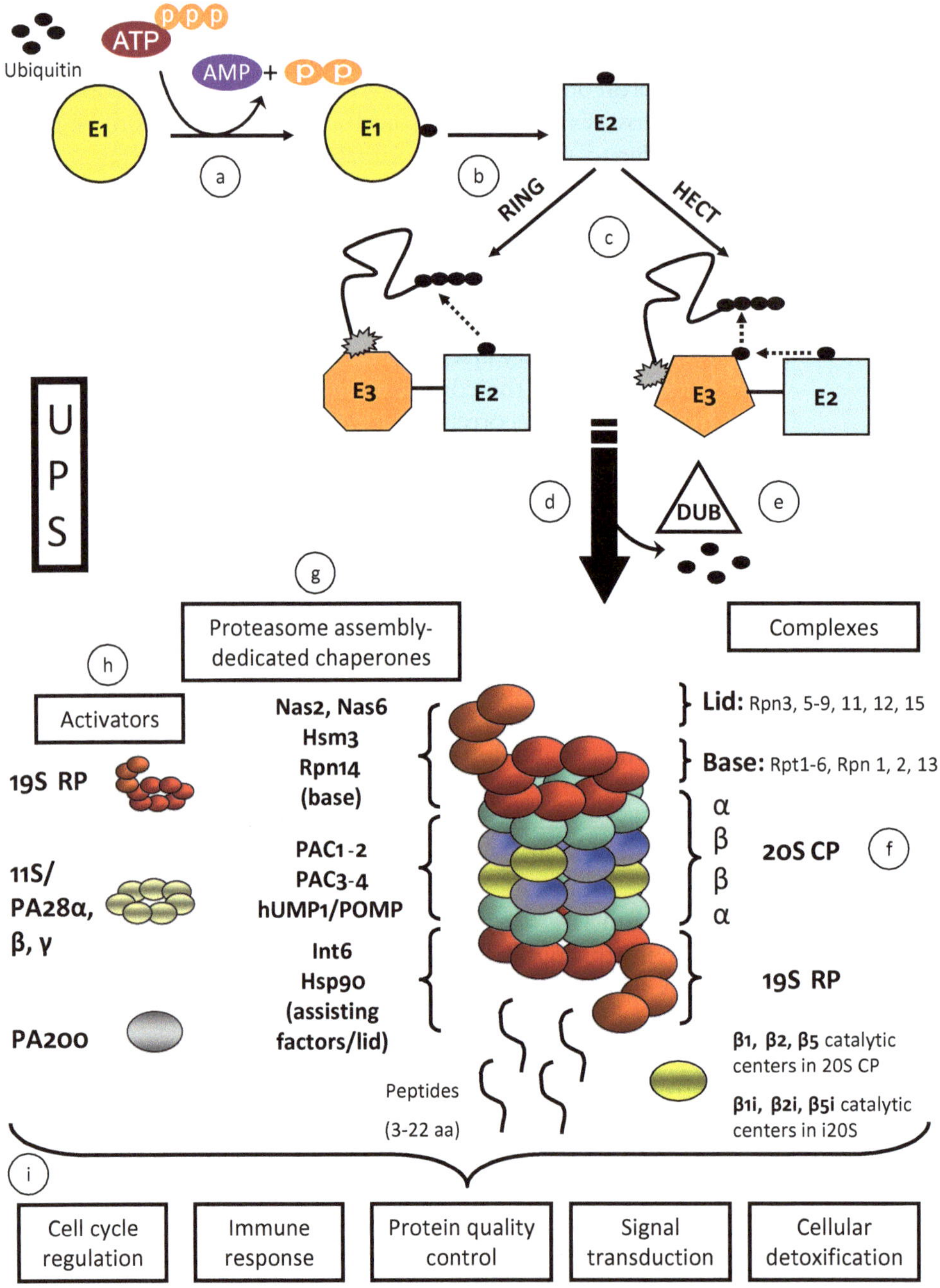

Fig. 1 The ubiquitin-proteasome system (UPS). **(a)** Ubiquitin activation through ubiquitin-activating enzyme (E1). **(b)** Activated ubiquitin is transferred to ubiquitin-conjugating enzyme (E2). **(c)** RING domain ligase: the ubiquitin-charged E2 binds to the E3 ligase that carries the substrate for degradation and ubiquitin is directly transferred to the substrate. HECT domain ligase: ubiquitin is firstly transferred from E2 to the E3 ligase that carries the substrate for degradation and then to the substrate. All three steps are repeated to result in substrate polyubiquitination. **(d)** Ubiquitinated protein is recognized by the proteasome, captured and processed for degradation. Short peptides (3–22 aa) are released at the end of the process. **(e)** Following substrate recognition, polyubiquitin chain is cleaved off through deubiquitinases (DUB) and free ubiquitin is released in order to be reused. **(f)** 26S proteasome structure; the constituent subunits appear for each subcomplex. In the case of 20S proteasome, β1, β2, and β5 subunits are the catalytic centers of the complex, whereas in the case of i20S these subunits are *de novo* substituted by β1i, β2i, and β5i subunits. **(g)** Proteasome assembly-dedicated chaperones (or assisting factors in the case of lid assembly). **(h)** Major proteasome activators that can be located on the top of 20S complex. **(i)** The various proteasome complexes are involved in multiple cellular pathways/processes

2.1 E1, E2, E3 Enzymes

The first two-step reaction in the ubiquitination process is catalyzed by the ubiquitin-activating enzymes (E1), in an ATP-dependent process that results in an activated ubiquitin molecule. More specifically, the E1 enzyme binds ATP and ubiquitin and catalyzes ubiquitin C-terminal acyl-adenylation. Ubiquitin is then transferred to the catalytic cysteine of the E1 enzyme producing a high-energy thioester bond and forming ubiquitin–E1 complex [60]. There are two human genes that have been so far identified to produce E1s, namely Uba1 and Uba6 [61, 62]. As expected, E1s can collaborate with multiple E2s.

The ubiquitin-conjugating enzymes (E2) catalyze the transfer of the activated ubiquitin from E1 to its own catalytic cysteine residue where a thioester bond is formed. So far, 35 E2 enzymes have been identified in humans while in other eukaryotes the number ranges between 16 and 35 [63]. Each E2 can activate a palette of E3 ligases in an hierarchical manner thus producing multiple different but specific E2–E3 combinations.

The final step of ubiquitination is catalyzed by E3 ligases forming an isopeptide bond between the C-terminal glycine of ubiquitin and a lysine of the target protein. The two main classes of E3 ligases (classified according to the domain that they possess) are the homologous to the E6-AP carboxyl terminus (HECT) domain proteins and the really interesting new gene (RING) domain proteins where one can find monomeric and multisubunit RING finger ligases [64]. The RING group of E3s along with the RING-related E3s, such as members of the U-box family, the plant homeodomain (PHD), and leukemia-associated protein (LAP) finger proteins, is the largest group of E3 ligases [65]. HECT-domain E3s firstly accept through a thioester linkage the ubiquitin moiety and then they transfer it to the protein substrate, whereas RING-domains E3s bind the cooperating E2 and they mediate the direct transfer of ubiquitin from E2 to the target protein [64] (Fig. 1). More than 600 E3 ligases have been annotated in humans [66] from which ~30 are HECT-domain E3 ligases. Most of the multisubunit RING E3s belong to the cullin RING ligase (CRL) superfamily [67] with SCF complex (consisting of S-phase phase kinase-associated protein 1/Skp1, cullin, and F-box protein) and anaphase-promoting complex (APC/C) being the most known complexes. Both complexes assure the correct cell cycle progression [68].

2.2 Deubiquitinases

Ubiquitination can be reversed through the act of specific proteases, namely deubiquitinating enzymes (DUBs; also known as deubiquitinases, deubiquitinating peptidases, ubiquitin isopeptidases, deubiquitinating isopeptidases, ubiquitin proteases, and ubiquitin hydrolases; [69]). DUBs cleave ubiquitin from protein substrates and other molecules and thus they act antagonistically to the ubiquitination process. Apart from their role in protein degradation they have been also implicated in several other pathways including cell growth and differentiation, membrane protein trafficking, development, neuronal diseases, and transcriptional regulation while they are also responsible for ubiquitin activation and recycling [70, 71]. Approximately 100 DUBs have been annotated in humans, grouped into two classes: cysteine proteases and zinc-dependent metalloproteases. Cysteine proteases include ubiquitin-specific proteases (USPs), ovarian-tumor (OTU) domain proteases, ubiquitin C-terminal hydrolases (UCHs), and Machado-Josephin domain proteases (MJDs) while metalloproteases contain a Jab1/MPN metalloenzyme (JAMM) domain [69, 70].

3 Introduction: The Proteasome System

The proteasome is a large multisubunit enzyme complex hosting multiple catalytic centers and is responsible for the clearance of short-lived normal, regulatory proteins but also for the elimination of unwanted (misfolded, damaged, or in any way abnormal) proteins [72, 73]. The 20S core proteasome (CP) is the main complex that hosts the catalytic activities of the multienzyme while various regulators can be attached in either one or both ends of the 20S, giving rise to supra-proteasome complexes with 19S regulatory particle (RP) being the most common. The various proteasome complexes are thus engaged in the regulation of numerous biological processes including signal transduction, cell cycle control, cell differentiation, stress response, quality control, antigen presentation, and cellular detoxification [74].

3.1 20S Core Proteasome: Structure, Assembly, and Localization

3.1.1 Structure

The 20S CP is a barrel-like structure composed of 28 subunits (14 α-type and 14 β-type) arranged in four seven-membered rings with a molecular weight of 700 kDa and a diameter of 120–160 Å (Fig. 1). The α-type subunits form the two external rings and create an aperture of 10–15 Å through which the protein substrate enters to reach the three catalytic centers of the CP that are located in the inner β-rings. More specifically, β1, β2, and β5 subunits possess the caspase-like (C-L or PGPH), the trypsin-like (T-L), and chymotrypsin-like (CT-L) activities, respectively. The α-subunits also offer the matrix for the binding of the various regulators that modify the specific activity of the CP [74].

3.1.2 Assembly

The assembly of the eukaryotic 20S CP is highly orchestrated, assisted by several proteasome-dedicated chaperones. This assembly initiates with the α-ring formation that it then serves as a template for the incorporation of the β-subunits. Up to now, four different proteasome assembling chaperons (PACs), namely PAC1-PAC4 (Pba1-4 in yeast; [75]) and the proteasome maturation factor POMP (Ump1 in yeast; [76–78]) have been isolated. PAC1-PAC2 heterodimer is responsible for the α-ring formation as well as for the prevention of incorrect dimerization. PAC3–PAC4 heterodimer assists the incorporation of pro-β2 subunit that is followed by the incorporation of β3, β4, pro-β5, pro-β6, pro-β1, and pro-β7 subunits. PAC3-PAC4 gets displaced once β4 and hUMP1/POMP join the complex. hUMP1/POMP then assists the serial incorporation of the rest β subunits [79]. The two half-CP are dimerized with the help of Hsc73 which is then released, the β-propeptides are self-cleaved, and UMP1/POMP is the first substrate of the newly assembled CP [80]. CP maturation induces an affinity switch mechanism that reduces its affinity for PAC1-PAC2 and thus enables the RP to dislocate the dimer and to get attached on the CP [81].

3.1.3 Localization

Intracellular proteasomes localize in the cytoplasm, the nucleus and the ER and can constitute approximately up to 5 % of the total cellular protein content depending on the cell type [82]. However, the 20S core proteasome has been identified to get attached to the plasma membrane thus suggesting its potential release in the extracellular space, e.g., in the alveolar lining fluid, epididymal fluid and possibly during the acrosome reaction. Moreover, active (reported as circulating) proteasomes have been detected in normal human plasma but also in plasma from patients suffering from various forms of malignancies, autoimmune diseases, sepsis, and trauma [83]. It was lately shown that activated immune cells can export assembled proteasomes (fully functional) as microparticles, thus possibly revealing the mode of extracellular proteasomes generation. Moreover, 19S particles as well as the PA28 activator were also detected in these microparticles [84].

3.2 26S Proteasome: Structure and Assembly

3.2.1 Structure

One or two RP may bind in the CP ends; the RP-CP configuration is termed as 26S complex whereas the RP-CP-RP configuration is termed as 30S complex. The RP is responsible for the substrate recognition, unfolding, deubiquitination, and translocation. It is subdivided into two smaller complexes, namely the base and the lid [85, 86]. The base is composed of six AAA-ATPases (Rpt1-6) along with three non-ATPases namely Rpn1, Rpn2, and Rpn13. The ATPases are responsible for the unfolding of the protein substrate, the opening of the α-gated channel on the CP, and the translocation of the unfolded protein towards the inner proteolytic cavity of the proteasome. Both Rpn1-Rpn2 and the ATPases are necessary for substrate translocation and gating of the proteolytic channel [87], while Rpn13 together with Rpn10 act as integral ubiquitin receptors thus recognizing the tagged substrates [88, 89]. Moreover, Rpn10 acts as a "bridge" subunit that connects the base and the lid. The lid is composed by 9 Rpn subunits namely Rpn3, 5–9, 11, 12, and 15. Rpn11 serves as a deubiquitinating enzyme [90] while it stabilizes the otherwise weak interaction between the CP and the RP [91].

3.2.2 Assembly

The incorporation of the base subunits is the first step in the RP assembly. Rpn14, Nas6, Nas2, and Hsm3 (PAAF1, gankyrin/p28, p27, and S5b in human, respectively) are the yeast 19S-specific assembly factors assisting the RP assembly and not found on the mature 26S proteasome [92, 93]. These four factors can be also found named as RAC (RP assembling chaperones) 1, 2, 3, and 4, respectively [94]. Three intermediates are produced, namely RPN1-RPT2-RPT1-Hsm3, Nas6–RPT3–RPT6–RPN14, and Nas2–RPT5–RPT4. These intermediates form the base complex and Rpn2 and Rpn13 are finally added to give rise to the final base complex that will be bound to the lid through Rpn10. Following Rpn10 binding, the chaperones are detached from the base.

The lid assembly is not fully elucidated. Recent studies suggest that Rpn5, 6, 8, and 11 form an initial stable module where Rpn3, 7, and 15 then bind and the full lid is formed through the addition of Rpn12 [95]. Hsp90 [96] and Yin6 (ortholog of the mammalian Int6) [97] are two assisting factors identified in the lid formation in yeast.

3.3 Various Proteasome Forms

3.3.1 Immuno-proteasome

Upon interferon γ (IFNγ) stimulation, the constitutively expressed catalytic subunits are de novo replaced by their cytokine inducible counterparts, namely β1i (LPM2 or PSMB9), β2i (MECL-1 or PSMB10), and β5i (LPM7 or PSMB8), thus giving rise to the immunoproteasome or i20S [98]. Immunoproteasomes exhibit increased CT-L activity and decreased C-L activity, thus facilitating antigen presentation due to the generation of antigenic peptides with increased affinity for MHC class I clefts. Mice lacking immunoproteasomes display major alterations in antigen presentation [99]. Despite this particular role, increasing number of studies implicate immunoproteasomes in processes irrelevant to antigen presentation like the adaptive response of the cells to oxidative stressors in order to preserve homeostasis [100], aging [101, 102], and longevity [103].

The activities of the immunoproteasome can be altered through the binding of various activators like the RP but also the 11S complex (also known as PA28/REG/PA26), a heptameric IFNγ-inducible protein that induces the degradation of short peptides in an ATP-independent manner [104]. There are three 11S isoforms in higher eukaryotes, namely PA28α, β, and γ (or REGα, β, and γ; [105]).

3.3.2 Hybrid Proteasomes

Upon binding of an RP in one end of the CP and an 11S in the other end, hybrid proteasomes are produced [106]. It is believed that the RP serves at the substrate recognition while the 11S complex alters the proteolytic potential of the CP.

3.3.3 Thymo-proteasomes

A specific catalytic β5 subunit has been isolated in mouse cortical thymic epithelial cells, namely β5t [107]. A similar subunit with thymus-specific expression was then revealed in humans as well

[108]. More specifically, β5i subunit is substituted by the proteolytic active subunit β5t in the relative tissue, thus giving rise to the thymoproteasomes. Thymoproteasomes contain β1i and β2i along with β5t, but notably not the constitutive β subunits [108]. In contrast to β5i incorporation, β5t insertion leads to markedly decreased CT-L activity, a feature that was shown to be necessary for the positive selection of developing thymocytes [107, 109].

3.3.4 Other Forms of Proteasomes

An additional tissue-specific subunit has been identified in *D. melanogaster* where Prosalpha6 subunit is replaced by the testis-specific subunit Prosalpha6T. It is suggested that this substitution is necessary for spermatogenesis [110, 111].

Finally, PA200/Blm10 (human/*S. cerevisiae*) is another activator that similarly to the 11S induces peptides degradation by the CP in an ATP-independent manner [112]. This activator has been implicated so far in various processes ranging from proteasome assembly [112] and inhibition [113], to DNA repair [114] and mitochondrial checkpoint regulation [115].

3.4 Regulation of the Proteasome Expression and Function

3.4.1 Transcriptional Regulation

Although the proteasome structure and function is extensively studied, the transcriptional regulation of the proteasome genes is still not fully elucidated. Rpn4 is a yeast transcription factor controlling the expression levels of the proteasomal genes bearing the proteasome-associated control element (PACE) in their promoters [116]. Rpn4 controls proteasome expression under both normal and stress conditions including proteasome inhibition and DNA damage [117]. Recently, a minimal hexamer "PACE-core" sequence that is responsive to Rpn4 was identified. These PACE-cores are present in many genes related to proteasome function (including the proteasome assembly chaperones), although they cannot substitute for the known PACE of the subunits [118]. Nevertheless, no human homologue of RPN4 has been identified thus far.

Nuclear factor (erythroid-derived 2)-like 2 (Nrf2) is a transcription factor that has been implicated in the regulation of proteasome genes in mammals. Nrf2 is the main responsible for the expression of various antioxidant enzymes [119], including several components of the proteostasis network namely, chaperons and proteasome subunits under specific conditions [120]. Nrf2 belongs to the family of Cap'n'collar (Cnc) transcription factors. It is responsible for the cellular transcriptional response to oxidative stressors and electrophilic xenobiotics thus being nominated as the central mediator of a prominent antioxidant response system. Kelch-like ECH-associated protein 1 (Keap1) is the main regulator that keeps Nrf2 in the cytoplasm and mediates its proteasomal degradation [121]. Upon a stimulus, Nrf2 may become phosphorylated and/or Keap1 may be modified, resulting in the disruption of the Keap1-Nrf2 complex and the nuclear translocation of Nrf2 [122]. In the nucleus, Nrf2 heterodimerizes with small musculo-aponeurotic fibrosarcoma (Maf) proteins and recognizes a cis-acting DNA element namely antioxidant response element (ARE) or electrophile responsive element (EpRE; 5′-TGA[C/T]NNNGC-3′) on its target genes, thereby conducting their transcription [123–125]. Several studies have reported the Nrf2-mediated proteasome induction as it will be discussed in various sections below. The nematode ortholog, SKN-1 has been also implicated in the regulation of proteasome genes. More specifically, depending on the redox conditions, proteasomal genes have been shown to be regulated by SKN-1 [126]. SKN-1 has been shown to exert pivotal role in longevity [126, 127] and resistance to oxidative stress. Moreover, it has been shown that proteasome deregulation/inhibition imposes SKN-1 translocation to the nucleus and promotes proteasome subunits upregulation [128–131]. We have also found that proteasome activation through the overexpression of *pbs-5* proteasome subunit and the consequent life-span extension is at least partially SKN-1-dependent [132].

Nuclear factor erythroid-derived 2-related factor 1 (Nrf1, also known as NFE2L1/LCRF1/TCF11) is also a member of the CNC family [133]. *NFE2L1* gene encodes two main isoforms [134]: Nrf1 (a short isoform) and TCF11 (a long isoform). TCF11 was shown to regulate the induction of proteasome genes, rather than Nrf2, after proteasome inhibition via an ERAD-dependent feedback loop [135, 136]. It was further elucidated that in normal conditions, proteasomes are active and they degrade Nrf1. In contrast, when there is a partial proteasome inhibition, proteasomes proceed to limited proteolysis thus releasing the processed Nrf1 (lacking its N-terminal region) from the ER which is also the active Nrf1 form that promotes gene expression [137]. Interestingly, if Nrf1 expression is lost in the brain, various proteasome subunits get downregulated and it was suggested that Nrf1 perturbations may be at least partially responsible for neurodegenerative diseases progression [59]. Given the interplay between Nrf1 and the proteasome, such possibility could also implicate the proteasome in this Nrf1-dependent process.

Finally, it was recently shown that the expression of β catalytic subunits and especially β5 subunit in mammals is regulated by constitutively activated signal transducer and activator of transcription 3 (STAT3; [138]). There is more available data for the transcription factors of the immunosubunits. More specifically, interferon regulatory factor-1 (IRF-1) has been suggested to be the master regulator for the concerted expression of immunoproteasome subunits [139, 140]. More recently, the transcription factor PU.1 was shown to bind and transactivate PSMB8, PSMB9 and PSMB10 (immunosubunits) promoters. Furthermore, PU.1-dependent transactivation and PU.1 expression were shown to be repressed by PML/RARα [141].

3.4.2 Posttranslational Modifications

The various proteasome regulators (RP/19S, PA28/11S and PA200/Blm10) that have been described above alter drastically the proteasome activities. Apart from this kind of proteasome activity regulation, several posttranslational modifications (PTMs) such as oxidation, phosphorylation, ubiquitination, O-linked addition of N-acetylglucosamine, glycosylation, N-acetylation, and lipid peroxidation may also have an impact on proteasome function.

Rpt3 and Rpt5 are two subunits that have been shown to be carbonylated (oxidized) in human end-stage heart failure and experimental myocardial ischemia [142, 143]. In both cases, this oxidation leads to proteasome activities compromise.

Phosphorylation is one of the most frequent PTMs that have been detected in several CP and RP subunits. Two CP subunits namely, α7 and α3 subunits were initially identified to be phosphorylated and proteasomes with α7 phosphorylated subunit have elevated activity levels [144]. It was additionally found that α7 phosphorylation stabilizes 26S proteasomes and upon IFNγ treatment, 26S proteasomes are destabilized due to α7 dephosphorylation [145]. Caseine kinase II was identified to be the kinase responsible for this phosphorylation [146]. Calcium/calmodulin-dependent protein kinase II (CaMKII) and polo-like kinase (Plk) were also identified as proteasome-phosphorylating kinases. More specifically, CaMKII phosphorylates Rpt6 both in vitro and in vivo and consequently stimulates proteasome activity and plays a regulatory role in remodeling of synaptic connections [147]. Plk was found to interact with all α subunits but α2 and β1, β2, β3, β5, and β7 subunits, to phosphorylate α3 and α7 subunits in vivo and to enhance proteasome activities [148]. Using MS/MS, Kikuchi et al. [149] identified 33 Ser/Thr phosphorylation sites in 15 subunits of the yeast proteasome and showed that dephosphorylation of the 19S RP results in a 30% decrease in ATPase activity. Other groups have found additional subunits subjected to phosphorylation (α1, α2, α7, and β6 in mammalian proteasomes [150, 151]. In contrast to the abovementioned activating properties of the phosphorylation of proteasome subunits, diminished 26S activity in failing human hearts is suggested to be related to the impaired docking of the RP to the CP as a result of decreased Rpt subunit ATPase activity and α7 phosphorylation [152]. DNA damage induces phosphorylation of several α-subunits (α5, α6, α7), thus probably affecting protein-protein interactions and gate opening due to the increased net negative charge given by the phosphate groups [153].

Ubiquitinated forms of α5, α6, α7, and β5 have been identified following doxorubicin treatment. Ubiquitination of proteasome subunits inhibits CT-L and C-L activities in vitro while in vivo doxorubicin treatment enhances proteasome activities in parallel to the decreased levels of ubiquitination thus suggesting that the proteasome activities upon DNA damage are regulated by ubiquitination [153].

O-Linked addition of the monosaccharide N-acetylglucosamine (O-GlcNAc) has been shown to inhibit the 26S proteolytic activities but not the 20S activities. It was further shown that the ATPase activity is inhibited and Rpt2 is identified as a substrate for this kind of PMT [154]. It was also suggested that the O-GlcNAc system may participate in neurodegeneration and this is at least partially linked with the inhibition of the proteasome [155]. In addition, O-GlcNAc-sites have been identified in CP subunits, namely α1 (Ser5), α4 (Ser130), α5 (Ser198), and α6 (Ser110) and the β subunit β6 (Ser57 and Ser208; [156]).

Subunits α1, α2, α3, β4, β5, and β6 of the murine cardiac 20S proteasome were identified to be glycosylated without however revealing whether this has a positive or a negative effect on proteasome activities [150].

N-Acetylation was also shown to affect proteasome subunits. More specifically, all α-type subunits and β3 and β4 subunits were found acetylated in yeast and CT-L activity was shown to be elevated in a mutant that cannot perform N-acetylation [157]. Rpt4, Rpt5, Rpt6, Rpn2, Rpn3, Rpn5, Rpn6, Rpn8, Rpt3, and Rpn11 were also found acetylated in yeast but nevertheless, the activities were not altered [158] whereas Rpt3 and 6 and Rpn1, 5 and 6 were found acetylated in murine proteasomes [150].

Proteasome subunits can also be subjected to modification by the lipid peroxidation product 4-hydroxy-2-nonenal (HNE). HNE modification of α1, α2, and α4 subunits during cardiac ischemia/reperfusion results in reduced peptidase activities [159, 160]. A similar decrease was also found in epidermis samples from old donors and HNE-modification of certain α-subunits was involved in the age-related decline of the proteasome function [161]. Accordingly, HNE modification promoted proteasome activity decline in neural PC6 cells [162].

Finally, other types of PMTs like N-myristoylation [158], S-glutathionylation [163], and nitrosylation [150] have also been identified. N-myristoylation of Rpt2 does not alter proteasome activities but it controls proteasome localization [164]. S-glutathionylation of α5 subunit promotes gate opening and therefore stimulation of 20S activity while the effects of nitrosylation are not yet elucidated. Despite the abovementioned changes of proteasome activity, several PTMs have been shown to indirectly alter the function of the proteasome through alterations in the ability of various RP subunits to directly interact with protein substrates (e.g., autoubiquitination of Rpn13; [165], monoubiquitination of Rpn10; [166], in situ ubiquitination of Rpn10 (S5a), Rpt5, and Uch37DUB; [167]).

3.5 Elimination of Proteasomes

Proteasomes are degraded through the lysosomal machinery. They have been found in autophagic vacuoles, thus suggesting that they follow the pathway of nonselective autophagy. Nevertheless, under starvation conditions, they follow the heat-shock cognate protein of 73 kDa (hsc73)-mediated transport [168].

4 Proteasome Status During Aging/Senescence in Cellular and Organismal Models

4.1 Cellular Senescence

Proteasome activities diminish upon progression of senescence of human fibroblasts [169]. Moreover, partial inhibition of the proteasome by 50% in young cells (in levels analogous to the levels normally found in senescent cells) elicits a premature senescence phenotype [170] in a p53-dependent process [171]. Elaborate analysis of the expression of the various proteasome subunits during the senescence progression has revealed the critical role of the β-catalytic subunits that have been suggested to act as the rate-limiting factors in the proteasome assembly pathway [169]. Additionally, senescent cells exhibit a reduced response to IFNγ, thus resulting in lower expression levels of immunosubunits [172]. Apart from expression and assembly alterations during senescence, the proteasomal function is also affected by the accumulation of damaged, aggregated, and cross-linked proteins as shown by the negative effect of lipofuscin on proteasome activities [173].

In contrast to senescent fibroblasts, fibroblasts derived from healthy centenarians exhibit proteasome activities similar to the ones exhibited by cells derived from younger donors. Both of these cultures differ significantly in terms of proteasome potential with the cultures derived from older donors that are not centenarians [174]. These results further advocate for the pivotal role of proteasome in cellular senescence and aging.

4.2 Model Organisms

4.2.1 *Saccharomyces cerevisiae*

Proteasomal function has been reported to deteriorate during stationary phase conditions [175] and the decreased proteolysis has been correlated with increased rates of 26S proteasomes disassembly [176]. Recently, the important role of Cdc48-Vms1 complex in the preservation of the 26S proteasome assembly was revealed [177]. Upon starvation, a relocalization of the proteasome subunits from the nucleus into cytoplasmic structures termed as proteasome storage granules (PSGs) occurs [178]. The nuclear-to-cytosolic proteasome relocalization upon starvation is affected by chronological aging since young cells efficiently relocalize the proteasomes and form PSGs in contrast to the old cells. This process is dependent on two of the three N-acetylation complexes [179]. PSG formation requires fully assembled 26S proteasomes and Rpn11 proteasome subunit is crucial for both PSGs formation and cell survival during stationary phase [180]. Finally, 20S core sequestration into PSGs is mediated by Blm10 whereas upon resumption of cell growth Blm10 facilitates nuclear import of the 20S particles [181].

4.2.2 *Caenorhabditis elegans*

Cell-specific photoconvertible reporters assaying proteasome activity in the nematodes have revealed an impaired UPS function in the dorsorectal neurons of 7-day-old worms as compared to the one found in young adults. In contrast, no alterations are scored in body-wall muscle cells thus suggesting a cell type-specific decline of the proteasome in nematodes [182]. The pivotal role of the proteasome in the aging procedure of the animal is exhibited by the fact that deletion/knockdown of various 19S and 20S proteasome subunits elicited premature aging and shortened life-span [132, 183, 184]. Finally, increased ROS levels (that are strongly related to chronological age) are linked with impaired UPS activity and this in turn may potentiate disease progression [185].

4.2.3 *Drosophila melanogaster*

During the progression of aging, the proteasome function becomes gradually impaired in *D. melanogaster* fruits flies. More specifically, the 26S proteasome assembly has been shown to be impaired during aging. This impairment is also accompanied by a significant reduction of the endogenous ATP levels. In bright contrast, the 20S proteasome function is slightly increased, thus suggesting a possible compensatory mechanism in response to loss of 26S integrity [186]. Other studies have shown that the proteasome function is decreased in the somatic tissues upon the aging progression but however elevated proteasome activities are maintained in the gonads and the eggs of the aged flies [187].

4.2.4 *Rodents*

Proteasome activity and/or expression are compromised in various tissues in mice and rats including adipose [188], retina [189, 190], liver [191–193], lung [191], muscle [194], brain [192], spinal cord [162], heart [195, 196], hippocampus, and cortex [162]. On the other hand, increased levels of immunosubunit-containing proteasomes are usually detected in aging tissues [194, 197]. Nevertheless, several controversial results have been reported suggesting that proteasomal activity alterations in brain differ between species and brain regions [162, 198, 199]. In bright contrast, enhanced proteasome activity levels were found in the longest-living rodents, namely the naked mole rats [200].

A transgenic β5t-overexpressing mouse has decreased CT-L activity and eventually exhibits a premature senescent phenotype that leads to shortened life-span. Moreover, the animals accumulate polyubiquitinated and oxidized proteins while they are more prone to age-associated metabolic disorders [201]. Similar results were obtained in LMP2 (β1i) knockout mice [202]. Accordingly, PA28γ-deficient mice age prematurely [203]. Finally, CT-L proteasome activity is lower in the senescence-accelerated mouse prone 8 (SAMP8) as compared to the relative control SAMR1 that exhibits normal aging phenotype [204].

4.2.5 *Homo sapiens*

Decreased levels of proteasome expression and/or function has been revealed during the progression of aging in several human tissues including lymphocytes [205, 206], lens [207], skeletal muscle [208], and epidermis [161, 209], with controversial results for few tissues [210–212]. Additionally, compositional but not functional alterations have been also suggested for tissues like liver [213]. In bright contrast, proteasome function is maintained in fibroblasts derived from healthy centenarians [174]. The effects of aging and cellular senescence on the various levels of proteasome regulation are summarized in Fig. 2.

5 Proteasome Impairment During Aggregation-Related Diseases

5.1 Alzheimer's Disease

The link between UPS and the AD onset and progression was initially suggested when senile plaques were stained positively for ubiquitin [214] while elevated levels of UBB^{+1} (a mutated ubiquitin form) were detected in sporadic and familial AD [215]. When proteasome activities of different parts of the brains of AD patients were tested, diminished levels were detected, thus verifying the link between dysfunctional UPS and AD [216, 217]. A vicious circle exists since Aβ, paired helical filament-tau and the UBB^{+1} are all identified as inhibitors of the proteasome function [218–220] and this inhibition further leads to β-amyloid precursor protein (AβPP), Aβ, and tau accumulation [219, 221]. ER stress is induced in activated astrocytes from AD brains and autophagy is increased [222]. Nevertheless, marked inhibition of proteasome activities and impairment in the autophagic flux is monitored in cells overexpressing AβPP mutant isoform thus suggesting that the whole proteolysis network is affected during AD [223]. Finally, several E3 ligases such as parkin [224], HRD1 [225], and UCHL-1 [226] are downregulated in AD while E2-25K, a nontraditional ubiquitin-conjugating enzyme is accumulated in AD samples [227].

5.2 Parkinson's Disease

Aggregated and monomeric α-synuclein is deleterious for neurons viability due to its inhibitory role on both 20S and 26S proteasome activities. It was additionally shown that aggregated α-synuclein directly interacts with Rpt5 subunit [228]. In mutant α-synuclein transgenic mice a remarkable downregulation of proteasome activity is recorded [229]. Nevertheless, it was suggested that α-synuclein expression levels per se do not significantly affect proteasome activities, subunit expression, assembly, and function but additional mechanisms contributing to α-synuclein aggregation are central players in the deterioration of the UPS during PD [230].

Parkin has been identified as an interacting protein of various proteasome subunits such as α4 [231], Rpt6 [232], and Rpn10 [233]. Wild-type parkin has been shown to activate the 26S proteasome (*see* in Subheading 6.2.2) in contrast to PD-linked parkin mutants that lose this ability, thus impairing the 26S proteasome assembly [234]. In accordance, parkin knockout mice and flies exhibit reduced proteasome activity [234].

The 20S [235] but also the 26S [236] proteasome activities are diminished in the substantia nigra of PD patients while reduced levels of α-subunits, RP and 11S complexes have been revealed in

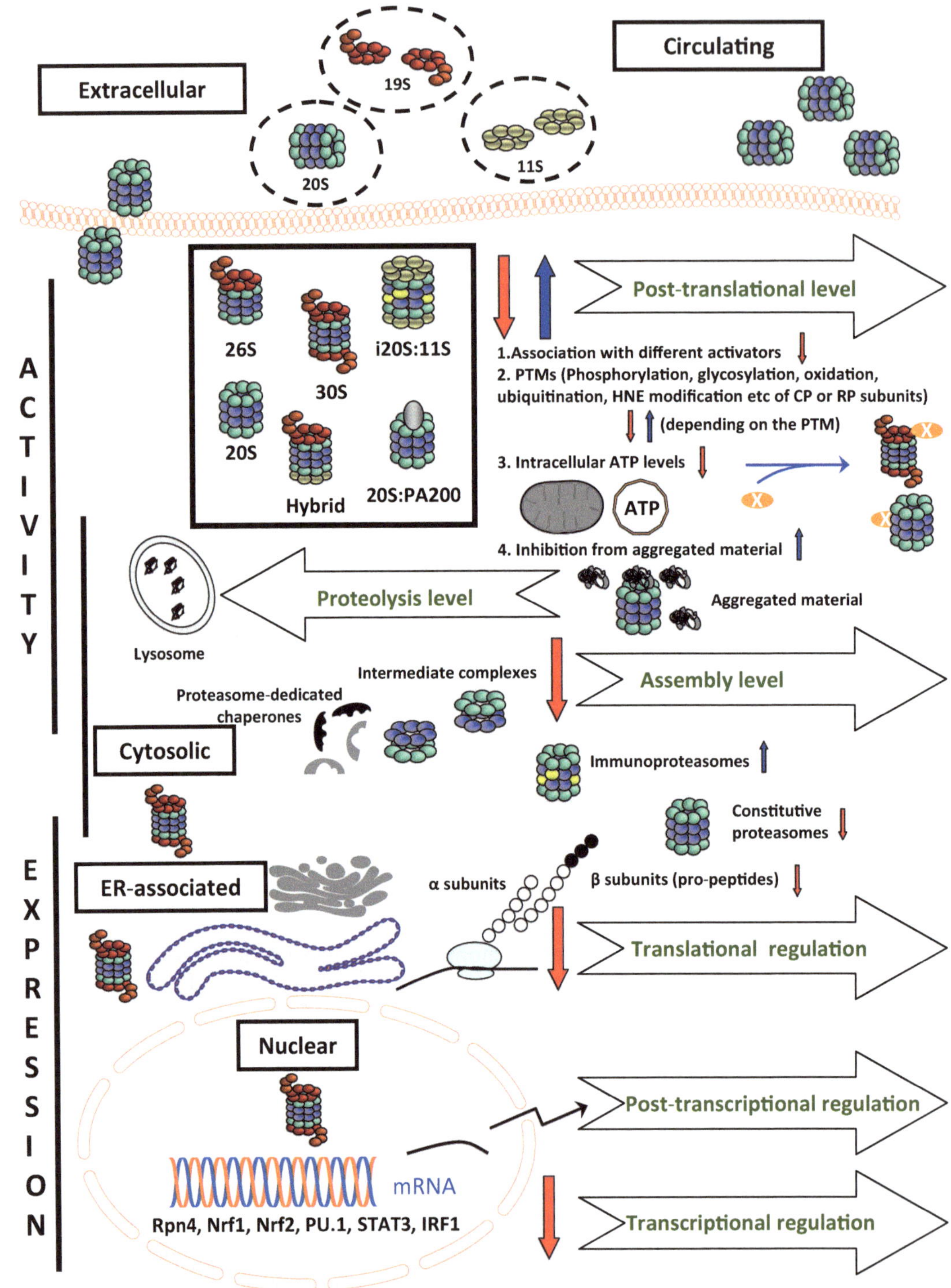

Fig. 2 The effects of aging on the various levels of proteasome regulation. Proteasomes can be found in the nucleus (nuclear), in the cytosol (cytosolic), attached to the endoplasmic reticulum mediating the ERAD (ER-associated) as well as in the extracellular space (named as extracellular or circulating). 30S complex appears in the various compartments in the figure for the scope of presentation but various complexes have been detected in the different compartments in vivo. However, so far only circulating 20S complexes have been isolated although 19S and 11S complexes have been also detected in various somatic fluids. The supra-complexes that constitute the proteasome potential include the constitutive proteasome (20S) and the immunoproteasome (i20S), the 26S and 30S complexes, the hybrid proteasomes, the i20S:11S as well as the 20S:PA200 complexes as shown in the

brain samples from sporadic PD patients [237, 238]. Finally, reduced proteasome activities are also detected in peripheral blood lymphocytes of patients with PD thus paving the way to the development of a potential peripheral biomarker of PD [239].

5.3 Huntington's Disease

As in all proteinopathies, the accumulation of aggregated proteins suggest a failure of the proteostasis network per se. PolyQ aggregates and ubiquitin co-localize in brain samples from HD patients [32], while mutant ubiquitin (UBB^{+1}) has been also detected in IBs [240]. Numerous studies have shown that mHTT inhibits proteasome function in cellular models as well as in vivo in animal models or patients thus suggesting choking or clogging of the proteasome by mHTT aggregates [241–244]. In an attempt to find the biochemical cause of proteasome inhibition, polyQ-containing proteins were shown to get kinetically trapped within proteasomes, thus inhibiting them [245, 246]. A selective inhibition of 26S proteasome but not 20S complex is also suggested and this is related to the interaction of HTT filaments with the 19S particles [247] as well as to ATP depletion due to the HD-induced dysfunction of mitochondria [248]. However, an indirect proteasome inhibition has been also suggested [249], while efficient degradation of expanded polyQ sequences without inhibitory effects on the proteasome has also been shown [250]. The abovementioned studies verify the contradictory results regarding UPS function and HD onset and progression. Furthermore, studies in HD mouse models challenge the concept of proteasome impairment during HD. Bett et al. [251] have revealed that overall proteasome function is not impaired by trapped mutant polyQ in R6/2 HD mice, while Maynard et al. [252] reported that although expression of N-mHtt caused a general UPS inhibition in PC12 cells, no inhibition was detected in the brains of R6/2 and R6/1 mice. Finally, dynamic recruitment of fully active proteasomes into IBs has been also suggested [253].

Fig. 2 (continued) inserted square. As potentially all proteins, the proteasome expression and function may be regulated in the following levels: transcriptional, posttranscriptional, translational, and posttranslational level. The two additional levels that appear in the figure, namely the assembly and the proteolysis level, constitute parts of post-regulation but given their importance in proteasome biology, we have included them here as additional regulatory levels. Multiple studies have already revealed an effect on proteasome expression and/or function/activity in several of those levels [e.g. identified transcription factors that regulate proteasomal RNA expression, regulative conditions for the shift between the expression of constitutive proteasome subunits or immunosubunits, chaperones that regulate its assembly, various PTMs (*X* in the figure represents the various groups that can be added or altered on the various proteasome subunits), association with different activators, alterations by aggregated material or alterations due to the energetic status of the cell]. During the progression of aging and senescence in organisms and cell cultures respectively, several of these regulatory levels are affected. The *red arrows* indicate decrease/downregulation and the *blue arrows* indicate increase/upregulation of pathways that have been shown to eventually affect the proteasome content and/or function during aging and senescence. Some of these regulatory levels affect mainly the proteasome content, some affect the proteasome activity without altering the content and some affect both as shown by the lines on the left of the figure. For more details, please refer to the text

5.4 Amyotrophic Lateral Sclerosis

The detection of ubiquitin and ubiquitin ligases within the ALS-related protein aggregates in ALS mutant mice [254] and in samples from ALS patients [255–257] indicates the possible involvement of UPS in ALS pathophysiology. Deposition of TDP-43 protein aggregates leads to proteasome inhibition [258]. Nevertheless inclusion bodies have been suggested to exert a possible neuroprotective role, given that monomeric and oligomeric misfolded ALS proteins are the actual toxic molecules in motor neurons [259]. Motor neuron-specific knockout mice for Rpt3 19S subunit possess inclusions with ALS-related proteins such as optineurin, ubiquilin 2, FUS, and TDP-43, thus indicating that decreased proteasome activity may result in ALS phenotype [260]. Accordingly, cells from rat spinal cords treated with lactacystin possess reduced proteasome activity and accumulate neurofilaments [261]. In line with these results, the UPS is found inhibited in terms of activity and/or expression in neuronal cell lines overexpressing human mSOD1 [262], in $SOD1^{G93A}$ transgenic mice [263, 264], as well as in samples from ALS patients [265–267]. Finally, upregulation of immunosubunits [268], PA28γ [268], and PA28αβ [269] occur in the motor neurons of $SOD1^{G93A}$ transgenic mice.

5.5 Prion Diseases

Accumulation and aggregation of prion and prion-like proteins in intracellular inclusions and extracellular plaques have been reported to impair protein homeostasis and to provoke cellular stress [270]. Prions cause severe ER stress [271] accompanied by the consequent downregulation of protein translation through chronic eIF2α phosphorylation [272] and impairment of ER protein translocation [273]. The above observations may link the proteolytic pathways to PrD. In addition, abnormal levels of ubiquitin and ubiquitinated proteins have been detected in intracellular inclusions located in the brain tissue, while PrP^{Sc} specifically inhibits the β-type proteasome subunits in two different neuronal cell lines and prion-infected mouse brain. Immunoblot analysis revealed no loss of subunits, while oligomeric inhibitory PrP species directly inhibit the activities of the 20S particle without affecting the 26S assembly. Collectively, the loss of proteolytic activity results from an inhibitory effect on the proteasome [274]. More recently, it was suggested that PrP aggregates inhibit the proteasome by stabilizing the closed conformation of the 20S proteasome and therefore obstruct the entry of the substrate [275]. Table 1 summarizes the proteostasis factors that have been found to be affected upon the progression of aggregation-related diseases.

Table 1
Proteostasis factors in neurodegenerative diseases

Neurodegenerative disease	(a) Impaired proteostasis factor (related to proteolysis)	(b) Manipulated proteostasis factor
Alzheimer's disease	1. High levels of UBB^{+1} 2. Decreased proteasome activities 3. Induced ER stress 4. Increased autophagy 5. Downregulation of E3 ligases	1. Overexpression of *pbs-5*, *aip-1* and key E3 ligases 2. Downregulation of VHL-1 ligase 3. Inhibition of USP14 DUB 4. Compound-mediated proteasome activation (CNB-001, apomorphine, quercetin, resveratrol, rasagiline, thioflavin T, methylene blue, geldamycin, polysaccharide PS5, ganoderic acid DM, lithium)
Parkinson's disease	1. Proteasome inhibition by aggregated and monomeric α-synuclein 2. α-Synuclein and RPT-5 interaction 3. Decreased proteasome activities 4. Proteasome assembly impairment by mutant parkin	1. Acceleration of 19S assembly and proteasome activation via wt parkin 2. Increased mitochondrial activity 3. Increased levels of K48-linked polyubiquitin 4. Upregulation of heat-shock proteins 5. Compound-mediated rescue of the PD-induced proteasome impairment (apomorphine, pramipexole, D3 receptor-preferring agonist D-264, rasagiline, coenzyme Q10) 6. Compound-mediated proteasome stimulation (sepiapterin, pueranin, n-butylidenephthalide, acetylcorynoline, rasagiline)
Huntington's disease	1. PolyQ aggregates and ubiquitin interaction 2. UBB^{+1} detection 3. Proteasome inhibition (chocking or clogging by polyQ aggregates, interaction with HTT filaments) 4. ATP depletion 5. Recruitment of fully active proteasomes into inclusion bodies	1. Overexpression of *pbs-5*, *rpn-6.1*, *rpn-11*, PA28γ, key E3 ligases and USP14 2. Downregulation of VHL-1 ligase 3. Activation of PKA, Akt, and CKB kinases 4. ROCKs inhibition 5. Overexpression of ubiquilins 6. Compound-mediated rescue of the HD-induced proteasome impairment (CGS21680 agonist, benzamil, baclofen, scyllo-inositol, sulforaphane, EGb 761 extract)
Amyotrophic lateral sclerosis	1. Ubiquitin and ubiquitin ligases interaction with aggregates 2. Proteasome inhibition 3. Decreased proteasome activities 4. Upregulation of immunoproteasome subunits	1. Overexpression of key E3 ligases 2. Overexpression of TorsinA, Derlin-1 and p62 3. Compound-mediated proteasome activation (PAP1, pyrazolone, bee venom, melittin, methyl pyruvate) 4. Nrf2 activation via CDDO-TFEA
Prion diseases	1. ER stress 2. Abnormal levels of ubiquitinated proteins 3. Inhibition of β-type subunits 4. Decreased 20S activities 5. Stabilization of 20S closed conformation	1. Congo red derivatives-mediated proteasome activation

Overview of the proteostasis factors (related to proteolysis) that have been (a) found altered in various neurodegenerative diseases and, (b) subjected to manipulation in the context of a potential therapeutic strategy

6 Proteasome Activation During Aging

Manipulation of several UPS-related factors in various cellular and organismal models results in an increase of the proteasome/UPS function with various effects in the cellular/animals life-span and stress resistance. The so far revealed factors include 20S and 19S proteasome subunits, other proteasome activators, E2 and E3 ligases and deubiquitinases. Moreover, the proteasome has been shown to get activated under various conditions and through several molecular pathways while there are also few compounds that have been shown to promote its activation. These factors/conditions/pathways in the various cellular and animal models that ultimately affect aging, longevity, and stress resistance are summarized below.

6.1 *Saccharomyces cerevisiae*

The yeast orthologs for α- and β-type proteasome subunits are PRE5/6/8/9/10, PUP2, SCL1 and PRE1/2/3/4/7, PUP1/3, respectively. Accordingly, the yeast 19S complex ATPases and non-ATPases are termed RPT1-6 and RPN1-12, respectively (Table 2).

6.1.1 20S and 19S Proteasome Subunits and Other Proteasome Activators

20S proteasome activity gets elevated upon α5 subunit S-glutathionylation and the consequent gate opening which results in increased ability of the yeast cells to degrade oxidized proteins [163].

Blm10 is an alternative proteasome activator identified in *S. cerevisiae* [276]. Enhanced degradation of peptide substrates is scored upon binding of Blm10 on the 20S core particle through a gate opening strategy [277].

6.1.2 E1, E2, and E3 Ligases

The Mub1/Ubr2 ubiquitin ligase complex is responsible for Rpn4 (the yeast transcription factor controlling the expression of proteasome genes) tagging for proteasomal degradation [278, 279]. Loss of *UBR2* and *MUB1* results in stabilization and increase of Rpn4 levels and a consequent induction of 20S and 26S subunits expression levels. The elevated protein levels are accompanied by enhanced activity levels that eventually lead to life-span extension. This extension is exclusively related to the increased proteasome function and the downstream degradation of unstable proteins [280].

6.1.3 Deubiquitinases

Low ubiquitin levels in yeast are sensed and trigger the expression of Ubp6, a proteasome-associated DUB. As a consequence, increased numbers of proteasomes loaded with Ubp6 are monitored with parallel alterations in proteasome function and ultimately, the restoration of the ubiquitin pool [281]. More recently, it was shown that the ubiquitin chain of ubiquitinated proteins is bound to the 26S-associated DUB, Ubp6, and this interaction promotes ATP hydrolysis and enhancement of their own degradation [282].

Table 2
Proteasome subunit nomenclature in different organisms

		Gene name				
Protein type definition		***S. cerevisiae***	***C. elegans***	***D. melanogaster***	***M. musculus***	***H. sapiens***
Reference	**Subunit**	**[451]**	**[452]**	**[7]**	**[9]**	**[453]**
20S proteasome subunits						
Proteasome subunit, alpha type, 1	α6	PRE5	pas-6	Prosalpha6	Psma1	PSMA1
Proteasome subunit, alpha type, 2	α2	PRE8	pas-2	Prosalpha2	Psma2	PSMA2
Proteasome subunit, alpha type, 3	α7	PRE10	pas-7	Prosalpha7	Psma3	PSMA3
Proteasome subunit, alpha type, 4	α3	PRE9	pas-3	Prosalpha3T	Psma4	PSMA4
Proteasome subunit, alpha type, 5	α5	PUP2	pas-5	Prosalpha5	Psma5	PSMA5
Proteasome subunit, alpha type, 6	α1	SCL1	pas-1	Prosalpha1	Psma6	PSMA6
Proteasome subunit, alpha type, 7	α4	PRE6	pas-4	Prosalpha4T1	Psma7	PSMA7
Proteasome subunit, beta type, 1	β6	PRE7	pbs-6	Prosbeta6	Psmb1	PSMB1
Proteasome subunit, beta type, 2	β4	PRE1	pbs-4	Prosbeta4R1	Psmb2	PSMB2
Proteasome subunit, beta type, 3	β3	PUP3	pbs-3	Prosbeta3	Psmb3	PSMB3
Proteasome subunit, beta type, 4	β7	PRE4	pbs-7	Prosbeta7	Psmb4	PSMB4
Proteasome subunit, beta type, 5	β5	PRE2	pbs-5	Prosbeta5	Psmb5	PSMB5
Proteasome subunit, beta type, 6	β1	PRE3	pbs-1	Prosbeta1	Psmb6	PSMB6
Proteasome subunit, beta type, 7	β2	PUP1	pbs-2	Prosbeta2R2	Psmb7	PSMB7

(continued)

Table 2
(continued)

		Gene name				
Protein type definition		***S. cerevisiae***	***C. elegans***	***D. melanogaster***	***M. musculus***	***H. sapiens***
Reference	**Subunit**	**[451]**	**[452]**	**[7]**	**[9]**	**[453]**
19S proteasome subunits						
Proteasome 26S subunit, ATPase, 1	rpt2	rpt2	rpt-2	rpt2	Psmc1	PSMC1
Proteasome 26S subunit, ATPase, 2	rpt1	rpt1	rpt-1	rpt1	Psmc2	PSMC2
Proteasome 26S subunit, ATPase, 3	rpt5	rpt5	rpt-5	rpt5	Psmc3	PSMC3
Proteasome 26S subunit, ATPase, 4	rpt3	rpt3	rpt-3	rpt3	Psmc4	PSMC4
Proteasome 26S subunit, ATPase, 5	rpt6	rpt6	rpt-6	rpt6	Psmc5	PSMC5
Proteasome 26S subunit, ATPase, 6	rpt4	rpt4	rpt-4	rpt4	Psmc6	PSMC6
Proteasome 26S subunit, non-ATPase, 1	rpn2	rpn2	rpn-2	rpn2	Psmd1	PSMD1
Proteasome 26S subunit, non-ATPase, 2	rpn1	rpn1	rpn-1	rpn1	Psmd2	PSMD2
Proteasome 26S subunit, non-ATPase, 3	rpn3	rpn3	rpn-3	rpn3	Psmd3	PSMD3
Proteasome 26S subunit, non-ATPase, 4	rpn10	rpn10	rpn-10	rpn10	Psmd4	PSMD4
Proteasome 26S subunit, non-ATPase, 6	rpn7	rpn7	rpn-7	rpn7	Psmd6	PSMD6
Proteasome 26S subunit, non-ATPase, 7	rpn8	rpn8	rpn-8	rpn8	Psmd7	PSMD7
proteasome 26S subunit, non-ATPase, 8	rpn12	rpn12	rpn-12	rpn12	Psmd8	PSMD8
Proteasome 26S subunit, non-ATPase, 11	rpn6	rpn6	rpn-6.1	rpn6	Psmd11	PSMD11
Proteasome 26S subunit, non-ATPase, 12	rpn5	rpn5	rpn-5	rpn5	Psmd12	PSMD12
Proteasome 26S subunit, non-ATPase, 13	rpn9	rpn9	rpn-9	rpn9	Psmd13	PSMD13
Proteasome 26S subunit, non-ATPase, 14	rpn11	rpn11	rpn-11	rpn11	Psmd14	PSMD14

Ubp3 is a conserved DUB that suppresses accelerated replicative aging and heat-stress sensitivity through the induction of proteasome-mediated degradation of cytotoxic proteins or (depending on the stage at which the damaged protein is committed for destruction) through their rescue from destruction [283].

6.1.4 Other Conditions and Compounds

Ump1 is a proteasome-dedicated assembly chaperone in yeast. Upon its overexpression, yeast cells exhibit increased resistance to various oxidative stressors, enhanced degradation rates of oxidized proteins, and elongated chronological life-span. All those effects are positively correlated with the elevated levels of CT-L activity exhibited by the overexpressors [229]. In accordance, deletion of *UMP1* gene results in increased levels of protein oxidation and reduced survival during stationary phase [175].

Overexpression of the heat-shock protein Hsp104 drives to elevated levels of disaggregase activity resulting in lower levels of protein aggregates and importantly in restored levels of UPS activity in aged yeast cells. Nevertheless, under those conditions the proteasome levels are unaffected and the cellular life-span is not altered [284].

PAP1 peptide (proteasome-activating peptide 1) activates the 20S proteasome activity through α-gate opening. Yeast cells are then able to effectuate more sufficient clearance of the oxidized proteins and therefore to exhibit an increased resistance to oxidative stress [285].

CR extends the replicative and chronological life-span [286, 287]. Increased levels of CT-L activity are scored in CR yeast cells accompanied by decreased levels of oxidized/ carbonylated proteins. Young CR yeast cells carry lower amounts of ubiquitinated proteins as compared to the control cells while CR preserves the ubiquitinylating ability of aged yeast cells, thus resulting in increased viability of the CR cells [288].

Adc17 is a newly identified chaperone that has been suggested to adjust proteasome assembly upon increased demand. It interacts with Rpt6 subunit (without being part of the proteasome) to assist an early step during proteasome assembly in yeast and it is induced upon conditions of proteasomes deficiency. As a result, Adc17 is important for biogenesis of adequate proteasome levels during stress and consequently for cell viability [289].

Finally, it was recently shown that the proteasome-mediated life-span extension is partially correlated to the deregulation of the AMPK signaling pathway. More specifically, increased proteasome activity is linked to premature activation of respiration that induces a mitohormetic response with beneficial impact on yeast life-span [290].

6.2 *Caenorhabditis elegans*

The nematode orthologs for α- and β-type proteasome subunits are PAS1-7 and PBS1-7, respectively. Accordingly, the nematode 19S complex ATPases and non-ATPases are termed RPT1-6 and RPN1-12, respectively (Table 2).

6.2.1 20S and 19S Proteasome Subunits

We have recently shown that overexpression of *pbs-5* catalytic subunit results in proteasome activation in terms of both content and activity. As a result, *pbs-5*-overexpressing animals exhibit extended life-span and ameliorated healthspan while they are more resistant to oxidative stress [132]. A similar phenotype has been achieved through the overexpression of a 19S subunit, namely *rpn-6*. Transgenic nematodes possess elevated proteasome activities that lead to increased survival to oxidative and mild heat stress and ameliorated response to proteotoxicity [291]. This particular subunit has been correlated with the increased proteasome activity that is detected in the long-lived *glp-1* mutants. Interestingly, *pbs-5* subunit is the only other proteasome subunit that is moderately increased in those animals [291]. A similar induction is also observed for the ortholog of *rpn-6*, namely PSMD11, and the ortholog of *pbs-5*, namely β5, in human embryonic stem cells [292] and in human embryonic fibroblasts [293], respectively, as described in Subheading 6.5.1.

AIP-1 (homologue of mammalian AIRAP) is a non-constitutive 19S proteasome subunit that is induced following exposure to arsenite. Upon *aip-1* overexpression, the nematodes conduct a more effective degradation of damaged proteins in stress response conditions, e.g., following arsenic treatment, and fumarylacetoacetate or maleylacetoacetate treatment [294, 295]. In contrast, silencing of *aip-1* results in shorter life-span [184]. As described in Subheading 6.5.1, its mammalian homologue, AIRAP, promotes 20S proteasome activation that enables the cells to cope with proteotoxic stress induced by an environmental toxin like arsenite [296].

6.2.2 E1, E2, and E3 Ligases

Modulation of several E3 ligases has been shown to result in life-span extension mainly through the enhanced degradation of key components for longevity and stress resistance. For example, the conserved insulin/IGF-1 signaling (IIS) pathway is a major pathway that governs the nematodes growth and differentiation [297] with DAF-16 (transcription factor of the FOXO family that is the downstream regulator of the IIS pathway) being the central player [298]. The results regarding DAF-16 effects on proteasome activities per se are controversial. The wt form of the main IIS receptor, DAF-2, has been shown to positively affect the activities of the proteasome since *daf-2* mutants (where *daf-16* expression is elevated), possess lower proteasomal activity [299]. In contrast, Vilchez et al. [291] have suggested that in *glp-1* mutants CT-L proteasome activity is increased through DAF-16 activation while we have also revealed a DAF-16 positive dependence in the *pbs-5*-overexpressing nematodes [132]. A similar positive dependence was also suggested by Holmberg's group using an in vivo reporter system for UPS activity [300].

EGF pathway has been also implicated with the UPS. A positive regulation of the UPS activity via Ras-MAPK pathway and the EOR-1 and EOR-2 transcription factors has been suggested [301].

This increase is correlated with SKR-5, a Skp-1-like protein, upon the loss of which, no UPS activation is observed while shorter life-span is monitored [301].

Proteasome subunit expression is induced through SKN-1 upon proteasome deregulation or inhibition [128–131]. H_2O_2 pretreatment of nematodes leads to SKN-1-mediated 20S proteasome activity elevation but notably not to alteration of the 26S activity [302, 303]. Moreover, it has been shown that IIS affects proteasome activity in a SKN-1-dependent manner [304]. Loss of a WD40 repeat protein, namely WDR-23, is accompanied by accumulation of SKN-1 in the nucleus and subsequent extension of life-span and increased resistance to stress. WDR-23 interacts with CUL4/DDB-1 ubiquitin ligase in order to target SKN-1 for degradation [128]. It is however noteworthy that UPS-independent regulation of SKN-1 through WDR-23 has been also suggested [305].

Elevated levels of proteasome activity accompanied by increased levels of various proteasome subunits have been also revealed in various dietary restriction (DR) nematode models [291, 306]. WWP-1 is a HECT E3 ligase that has been shown to be indispensable for the DR-mediated life-span extension [307]. Moreover, its overexpression in ad libitum-fed nematodes promotes a moderate but still significant 20% life-span extension in a FOXA transcription factor *pha-4*-dependent way. Ubiquitination of specific substrates that are pivotal for DR-related longevity has been suggested as the mode of action of WWP-1 and the crucial E2 ligase that collaborates with WWP-1, namely UBC-18 has been also identified [307]. In agreement, overexpression of the human WWP1 delays the progression of cellular senescence in human fibroblasts, while irreversible premature senescence is established upon its knockdown [308] (*see* Subheading 6.5.1).

6.2.3 Deubiquitinases

Modulation of UBH-4 DUB in *C. elegans* has been implicated with alterations in proteasome activities and with notable effects in stress/proteotoxicity resistance and longevity. More specifically, *ubh-4* silencing results in proteasome activity induction without alterations of the relative expression levels. *Ubh-4* was identified as a DAF-16 target gene that may slightly affect life-span of wt animals with no effects on animals with suppressed IIS pathway [300]. Accordingly, when *uchl5*, the human ortholog of *ubh-4*, is knocked down, increased UPS activity is monitored [309] (*see* Subheading 6.5.1).

6.2.4 Other Conditions and Compounds

Stress adaptation has been shown to occur in nematodes following repeated exposure to mild heat shock or mild doses of oxidants and this hormetic effect has been linked to enhanced longevity [310, 311]. More recently, it was revealed that the mild adaptive stress induced by exposure to H_2O_2 results in elevated proteasome activity [302].

Exposure of nematodes to UV increases UPS function via the activation of the innate immune system with a consequent increased proteostasis and systemic stress resistance [312].

Protein aggregation has been also shown to affect proteasome function and activity. Increased RNA expression levels of key UPS-relevant genes (i.e., *pdr-1*, *ubc-7*, *pas-5*, *pbs-4*, *rpt-2*, and *psmd9*) are detected in transgenic animals overexpressing A53T human synuclein, an aggregation-prone protein found in cellular inclusions in PD, Lewy body dementia, and multiple system atrophy [313].

Although several compounds have been described to promote proteasome activation in cells in vitro [314], only few of them have been examined for their proteasome-activating properties in *C. elegans* and their downstream effects in life-span. Quercetin, a known polyphenolic compound, induces proteasome activation and consequently inhibits $A\beta_{1-42}$-induced paralysis in nematodes [315]. Given that quercetin is a life-span-extending compound [316], one cannot rule out the possibility that this is also related to the induced proteasome activation. Several plant extracts were recently tested in *C. elegans* subjected to high glucose levels for reversal of the glucose-induced survival reduction. Extracts from hibiscus, elderberries, jiaogulan, and blackberries leaves have been identified as potent rescuers while they also promote proteasome activation thus suggesting an efficient degradation of glucose-impaired proteins [317]. Additionally, quercetin prevents glucose-induced reduction of survival through SIR-2.1, DAF-12, and MDT-15 that activate UPR and proteasomal degradation [318]. More recently, a catechin-enriched green tea extract was shown to completely reverse the glucose-induced decrease of life-span. Furthermore, it was shown that the recorded survival extension was dependent on *sir-2.1* and most importantly on *uba-1* that encodes for the unique E1-ubiquitin-activating enzyme in *C. elegans.* This extract stimulates the proteasome activities and thus reverses the glucose-mediated damage through the activation of adaptive responses that include proteasomal degradation [319]. Enhanced activity accompanied by elevated levels of *rpn-5* is monitored following treatment with acetylcorynoline, a Chinese herb-derived alkaloid component [320]. We have also recently shown that feeding of wt *C. elegans* with 18α-glycyrrhetinic acid, a triterpenoid, promotes life-span extension that is dependent on proteasome activation [321]. Finally, osmotic stress caused by NaCl treatment, leads to elevated levels of proteasome degradation as a protective action against stress-induced accumulation of damaged proteins [322].

6.3 *Drosophila melanogaster*

The fly orthologs for α- and β-type proteasome subunits are Prosalpha1-7 and Prosbeta1-7, respectively. Accordingly, the *Drosophila* 19S complex ATPases and non-ATPases are termed RPT1-6 and RPN1-12, respectively (Table 2).

6.3.1 20S and 19S Proteasome Subunits

Ectopic overexpression of *Rpn11* 19S complex subunit attenuates the age-related decline of proteasome activities. As a consequence, the flies exhibit an elongated life-span [323].

6.3.2 E1, E2, and E3 Ligases

Loss-of-function mutations of the *Drosophila* Ubiquitin Activating Enzyme, Uba1 results in reduced life-span and in severe motility defects. Even loss of one of the two alleles results in a significant life-span reduction [324]. Parkin is an E3 ubiquitin ligase that dictates the degradation of various proteins via the UPS [325] while *parkin* mutations are involved in autosomal-recessive PD [326]. Overexpression of *parkin* in flies is accompanied by increased levels of proteasome activity [234], in accordance with in vitro results [231, 234, 327]. This parkin-mediated proteasome activation is independent of parkin's E3 ligase activity. The proteasome function enhancement is related to parkin-mediated enhanced interactions between the 19S complex subunits. In accordance, parkin-null *Drosophila* exhibit decreased proteasome activity [234]. A more recent study has revealed that both ubiquitous and neuron-specific *parkin* overexpression results in elongated mean as well as maximum life-span. Moreover, those long-lived flies also exhibit decreased protein aggregation levels during the progression of aging [328].

6.3.3 Deubiquitinases

The DUB Leon/USP5 is essential for viability and tissue maintenance during *Drosophila* development. Leon mutants exhibit abnormal ubiquitin homeostasis, characterized by increased tissue disorder and augmented death incidents. Notably in those mutants, protein expression levels of proteasome subunits along with the relative enzymatic activities are elevated as a compensation mechanism in response to aberrant ubiquitin homeostasis [329]. Nevertheless, impaired degradation levels of ubiquitinated substrates are monitored.

USP2 DUB prevents uncontrollable activation of the fly immune response in unchallenged conditions by controlling the proteasomal degradation of Imd, an NF-κB-like *Drosophila* factor. Apart from the obvious action of USP2 related to the K48-ubiquitin chain cleavage from Imd, a synergistic binding of USP2 and Imd on the proteasome further alters proteasome-mediated Imd degradation [330].

6.3.4 Other Conditions and Compounds

DmPI31 is the *Drosophila* homolog of the mammalian PI31, a known inhibitor of the 20S proteasome [331, 332]. As opposed to the mammalian homolog, DmPI31 functions as an activator of 26S proteasomes in vitro but also in vivo, since its overexpression in flies suppresses the phenotypes that are caused by dominant temperature-sensitive proteasome alleles (rough eye phenotype; [333]).

Basic leucine zipper protein CncC has been shown to be a transcriptional regulator of the *Drosophila* 26S proteasome [334]. Impaired proteasome function triggers a CncC-mediated upregulation of the proteasome subunits. Conversely, induction of CncC leads to elevated proteasome expression and activity. Nevertheless, prolonged CncC overexpression results in shorter life-span [335]. Exposure of female flies to low H_2O_2 doses promotes increase of proteasome activity and 20S proteasome expression in a CncC-dependent manner [302].

Finally, several proteasome subunits have been shown to be induced upon exposure of flies to low doses of γ-irradiation and to lead to life-span extension [336, 337].

6.4 Rodents

The rodent orthologs for α- and β-type proteasome subunits are Psma1-7 and Psmb1-7, respectively. Accordingly, the rodent 19S complex ATPases and non-ATPases are termed Psmc1-6 and Psmd1-14, respectively (Table 2).

6.4.1 20S and 19S Proteasome Subunits and Other Proteasome Activators and Components

PA28α is the only proteasome component that has been so far manipulated. More specifically, transgenic mice with cardiomyocyte-restricted PA28α overexpression exhibit diminished aberrant protein aggregation in their hearts. This results in decreased levels of cardiac hypertrophy and consequently, in increased life-span. Therefore, PA28α overexpression may promote protection from cardiac proteinopathy following ischemia [338].

6.4.2 Other Conditions and Compounds

The naked mole rat (*Heterocephalus glaber*) is a nice model of exceptional life-span since it is the longest-living rodent known (~31 years maximum life-span). The proteasomal activities of this rodent are 1.5-fold higher than the ones exhibited by the "normal" mice while they are also maintained in high levels upon the progression of aging. Moreover, they exhibit attenuated age-dependent accumulation of ubiquitinated proteins and cysteine oxidation [200]. In the liver of these animals, more active 20S and 26S proteasomes accompanied by an enhanced proportion of immunoproteasomes are scored [103]. A cytosolic protein factor was shown to interact with the proteasome and to stimulate its activity. Heat shock proteins 72 and 40 were identified as some of the constituents of the unknown factor which however is still not totally characterized. Upon exposure of proteasomes isolated from yeast, mouse and human samples to the cytosolic proteasome-depleted fractions from the naked-mole rat, induction of proteasome activity occurs, thus suggesting a conserved action of this factor across species [339]. A theory that long-lived species may have superior mechanisms to ensure protein quality has been also suggested recently following analysis of protein quality control players in rodents, marsupials and bats [340].

High levels of 20S and 26S proteasome activities are scored in the frontal cortex of transgenic mice overproducing IGF-1 with PI3-kinase/mTOR signaling being involved. The same stimulation is also detected in cell cultures upon IGF-1 stimulation [341].

Late-onset DR in mice and rats is beneficial since it promotes restoration of proteasome activation and reduction of oxidative damage [342]. In other tissues like the rat spleen, DR does not induce proteasome activity but nevertheless, in the same samples, DR leads to decreased levels of ubiquitinated proteins [343]. Lifelong CR induces T-L proteasome activity but not CT-L and this increase is suggested to be related to the elevated levels of Hsp90 that are revealed in CR animals [344]. Mild CR counteracts the age-related decrease of proteasome activity in rats liver [345] while increased proteasome biogenesis occurs in the same tissue in response to DR [192]. Short-term food deprivation induces UPS function through induced expression of E3 ubiquitin ligases, muscle RING-finger protein-1 (*Murf1*), and muscle atrophy F-box protein or Atrogin-1 (*Fbxo32*) [346]. Treatment of rats with T3 induces the expression of Atrogin-1 and MuRF1 and enhances the proteasome activities by ~40 % whereas the UPS remains activated during extended periods of untreated hyperthyroidism [347]. Finally, gene expression analysis in mice subjected to DR revealed the induction of Psmc3 19S subunit and PA28α [348].

Sulforaphane and 3H-1,2-dithiole-3-thione (D3T) are natural compounds that are capable of activating genes that bear the antioxidant response element (ARE) in their promoters through Nrf2 induction [123, 349]. Nineteen proteasome subunits are upregulated by D3T in wt mice as opposed to *nrf2*-disrupted mice. This upregulation is followed by increased proteasome activities [120] and is tissue-specific [350]. 26S/20S proteasome subunits, including PSMB5, the subunit that is responsible for the CT-L proteasome activity are identified among the gene clusters that are under the Nrf2-mediated regulation [351]. Several additional compounds have been shown to alter proteasome activities in mouse models for various diseases. These compounds will be presented in the relative sections. Finally, proteolysis-inducing factor (PIF) is a glycoprotein firstly identified in cancer patients that acts as an enhancer of the proteasome subunits expression and activities in skeletal muscle in vivo [352].

6.5 Mammalian Cells

6.5.1 20S and 19S Proteasome Subunits and Other Proteasome Activators and Components

Since β5 catalytic subunit is the catalytic center for the CT-L activity, many groups have attempted its overexpression in several cell lines. In the stable transfectants, enhanced proteasome activities and/or expression and/or assembly are monitored. Furthermore, β5 overexpression (a) in WI-38/T and IMR90 human fibroblast cell lines and in HL-60 human promyelocytic leukemia cells endows cells with an increased capacity to cope with various oxidants (EtOH, tBHP, H_2O_2, and $FeCl_3$) while human primary cells overexpressing β5 subunit exhibit a ~15–20 % life-span extension [293], (b) in dermal fibroblasts from elderly donors results in diminished levels of aging markers such as oxidized and ubiquitinated proteins, SA-β-galactosidase activity and p21 content [353], (c) in lens epithelial cells leads to increased capacity to cope with oxidative stress [354], (d) in human bone marrow stromal cells restores their capacity for growth while they remain pluripotent for longer [355], and (e) in murine neuroblastoma leads to increased resistance against H_2O_2 toxicity and protein oxidation [350]. Similar results were also obtained upon overexpression of β1 subunit which is the catalytic center for C-L activity [293, 353], while β1 overexpression in human bronchial epithelial cells promotes a protection from cigarette smoke-induced ER stress through enhanced proteasome activities [356]. Accordingly, β5i immunosubunit overexpression in lymphoblasts and HeLa cells leads to elevated CT-L and T-L activities [357], while T-L activity is induced following overexpression of the β1i immunosubunit [358].

With regard to 19S proteasome subunits, human embryonic stem cells (hESCs) overexpressing the 19S PSMD11 subunit have more 26S proteasomes with potential effects in their pluripotency and differentiation capacity [291]. Overexpression of AIRAP, an inducible 19S subunit, promotes proteasome activation upon exposure to an environmental toxic factor, namely arsenite and confers protection in primary mouse embryonic fibroblasts (MEFs) and primary cells of the murine proximal tubule epithelia [359]. AIRAP association on the 19S cap promotes changes in the assembly of the various proteasome complexes favoring the stability of hybrid proteasomes [296]. A similar protection is observed in nematodes by overexpression of its homologue, AIP-1, as described in Subheading 6.2.1.

The association of PA28 activator with the proteasome has been shown to play a role in antigen presentation. Nevertheless, it was recently shown that PA28α overexpression in rat cardiomyocytes results in stabilization and increase of 11S proteasomes that leads to increased resistance to oxidative stress [338].

Finally, proteasome activation has been achieved in human fibroblasts through overexpression of hUMP1/POMP proteasome assembly chaperone. More specifically, overexpression of hUMP1/POMP in WI-38/T fibroblasts leads to enhanced proteasome activities and assembly that ultimately lead to resistance to oxidative stressors [360].

6.5.2 E1, E2, and E3 Ligases

Several E3 ligases have been modulated in various cell lines and have been shown to exert pro-longevity effects, mainly through the induced degradation of their target proteins that inhibit cell growth. Nevertheless, there are no reports showing a simultaneous modulation of proteasome activity. We will just report here the overexpression of two ubiquitin ligases that have been correlated with aging and proteasome degradation: WWP1 and CHIP ligase. We refer to the human WW domain-containing E3 ubiquitin protein ligase 1 (WWP1) as (a) it is implicated in cellular senescence [308], (b) its nematode ortholog has been shown to be essential for the DR-mediated life-span extension [307], and (c) DR has been shown to induce proteasome expression and activities [344, 348]. Therefore, there is a potential link between WWP1 with the proteasome activities that is however still unrevealed. WWP1 overexpression delays cellular senescence in human diploid fibroblasts through the enhanced degradation of p27(Kip1) while its knockdown leads to premature senescence [308].

The ubiquitin ligase CHIP (carboxyl terminus of HSP70-interacting protein) has been shown to regulate protein quality control and to affect longevity. More specifically, it has been shown that CHIP-deficient mice possess lower levels of proteasome activities and increased levels of oligomerized proteins that eventually lead to reduced life-span and premature aging phenotypes [361]. It was recently revealed that CHIP saves SirT6 (a lysine deacetylase/ADR ribosylase, member of the sirtuin family) from degradation through noncanonical ubiquitination. CHIP overexpression leads to SirT6 stabilization that endows cells with resistance to cellular stress and elevated DNA repair capacity [362]. CHIP overexpression remains to be shown if it may induce proteasome activities/function.

6.5.3 Deubiquitinases

Knockdown of UCHL5 (UCH37) promotes the clearance of aggregation-prone proteins in human U-2OS osteosarcoma cells through increased UPS function similarly to its nematode ortholog UBH-4 [300]. A similar effect is observed upon silencing of UCHL5 in HeLa cells where increased degradation rates of ubiquitinated proteins are scored but notably not enhanced hydrolytic proteasome capacity [309]. Given that silencing of its ortholog in nematodes promotes life-span extension, it would be interesting to see whether modulation of UCHL5 has the same effects in cells and higher eukaryotes.

USP14, another DUB, inhibits the degradation of ubiquitinated proteins both in vitro and in vivo. In agreement with the results from UCHL5 silencing, treatment of MEFs with a selective and reversible inhibitor of USP14, namely IU1, accelerates the degradation of ubiquitinated or oxidized proteins through proteasome activation [363].

It was recently shown that occupancy of Usp14 (a DUB reversibly associated with 26S proteasomes; [309]) or Uch37 (a constitutive DUB of the 26S proteasomes; [309]) by the polyubiquitin chains of tagged proteins leads to enhanced degradation of these substrates through stimulation of ATP hydrolysis [282].

6.5.4 Other Conditions and Compounds

Several natural or synthetic compounds have been shown to stimulate proteasome activities and function in mammalian cell cultures. Oleuropein, the most abundant constituent found in *Olea europea* leaves, olives, and olive oil, has been shown to stimulate the proteasome activities and function in various human embryonic fibroblasts. This induction is accompanied by reduced levels of oxidized proteins, while long-term treatment promotes cellular life-span extension [364]. Various phenolic and flavonoid constituents of the bee pollen induce CT-L proteasome activity in HFL-1 human fibroblasts [365]. Curcumin is a natural phenol that positively alters proteasome activities in human keratinocytes [366]. An algae extract protects human keratinocytes from the UV-mediated proteasome inactivation [195]. More recently, the synthetic peptide, PAP1 was shown to stimulate CT-L activity in fibroblasts and consequently to protect from oxidative damage and protein aggregation [367].

D3T activates Nrf2 and leads to induction of proteasome subunit protein levels and activity in wt MEFs. This induction is lost upon Nrf2 knockout [120]. A similar enhancement was revealed upon treatment of murine neuroblastoma cells with sulforaphane, a bioactive molecule within the isothiocyanate group of organosulfur compounds [350] as well as in HeLa cells [368]. We have also identified a proteasome-activating compound, namely the triterpenoid 18α-glycyrrhetinic acid [369]. Long-term treatment of human fibroblasts with this compound results in stimulation of the proteasome activities/assembly and function and ultimately in cellular life-span extension and increased resistance to oxidative stress. A similar phenotype was revealed upon chronic treatment of human fibroblasts with the flavonoid quercetin [370]. Although we have not checked whether this proteasome activation is Nrf2-dependent, this possibility cannot be excluded given that quercetin is a known Nrf2 activator [371]. Proteasome activation has been also achieved in Hepa1c1c7 mouse hepatocytes by zerumbone (a sesquiterpene isolated from the plant *Zingiber zerumbet*; [372]). Finally, Nrf2 and proteasome have been shown to be key mediators of human embryonic stem cells (hESCs) physiology. Nrf2 expression decreases upon differentiation while Nrf2 activation delays it through regulation of the proteasome activity. Accordingly, treatment of hESCs with t-BHQ or sulforaphane results in Nrf2-dependent increase of proteasome activities and in delayed differentiation and preservation of cellular pluripotency for longer [373].

Various cardiovascular diseases are characterized by proteasome functional insufficiency and protein control failure. Elevated levels of cGMP along with the downstream activation of cGMP-dependent protein kinase (PKG) have been demonstrated to prevent and reverse already existing hypertrophy and to inhibit the pathways related to hypertrophy [374]. Therefore, while seeking for a potential link between PKG and the UPS pathway, it was shown that overexpression of the protein kinase G (PKG) in rat ventricular myocytes induces proteasome activities resulting in enhanced clearance of misfolded proteins, thus protecting from cardiac proteinopathies [375].

20S levels and activity are augmented upon calpain-mediated processing of the 26S subunit Rpn10. More specifically, upon mitochondrial impairment, Rpn10 is cleaved by calpain, thus resulting in 26S disassembly with a concurrent increase of 20S levels [376].

Treatment of cells with IGF-1 results in elevated levels of CT-L activity in rat glioblastoma cells and WI38 human fibroblasts with an initial peak at 15 min of stimulation. Activities remain elevated for 24 h following IGF-1 addition but no quantitative alterations are observed with the exception of a slight increase of β5 expression. This induction is abolished in knockout cells for IGF-1 receptor. Accordingly, the Akt/PI3-kinase/mTOR cascade signaling is also involved given that in the presence of the relative inhibitors, proteasome activation by IGF-1 is significantly reduced [341]. Finally, PIF was also shown to enhance the proteasome potential in murine myoblasts in vitro through the induction of NF-κB [352, 377].

6.6 *Homo sapiens*

The human orthologs for α- and β-type proteasome subunits are PSMA1-7 and PSMB1-7, respectively. Accordingly, the human 19S complex ATPases and non-ATPases are termed PSMC1-6 and PSMD1-14, respectively (Table 2).

6.6.1 Other Conditions and Compounds

There are so far no population studies examining the possibility of proteasome activation. The only report comes from a study where volunteers were supplemented with zinc. More specifically, zinc supplementation for 7 weeks promoted the stimulation of both CT-L proteasome activity and MSR (methionine sulfoxide reductase) activity. In accordance, zinc supplemented donors exhibited reduced levels of oxidized protein thus suggesting the possible role of proteasome activation as an anti-aging strategy in vivo [378]. The various means of proteasome activation in cellular and organismal models are summarized in Fig. 3.

7 Proteasome Activation During Aggregation-Related Diseases

Proteasome activation has been attempted in several cellular and organismal models of aggregation-related diseases. Table 1 summarizes the proteostasis factors that have been subjected to various types of manipulation in the context of a potential therapeutic strategy.

7.1 Alzheimer's Disease (AD)

Given that UPS regulates the presynaptic protein turnover in the nervous system [379], it is not surprising that proteasome inhibition severely affects AD progression and normal synaptic function [22]. It is not additionally unexpected to attempt UPS activation as a therapeutic approach for AD.

7.1.1 20S and 19S Proteasome Subunits

Using a temperature-inducible *C. elegans* strain that expresses human $A\beta_{1-42}$ in muscle cells and that eventually is driven to paralysis [380], we have shown that *pbs-5* overexpression results in proteasome-mediated decreased levels of total but also oligomeric Aβ. This decrease is accompanied by significantly lower paralysis rates [132]. In the same nematode AD model, AIP-1 overexpression (an inducible 19S subunit) results in reduced Aβ levels, aggregation, and toxicity [295].

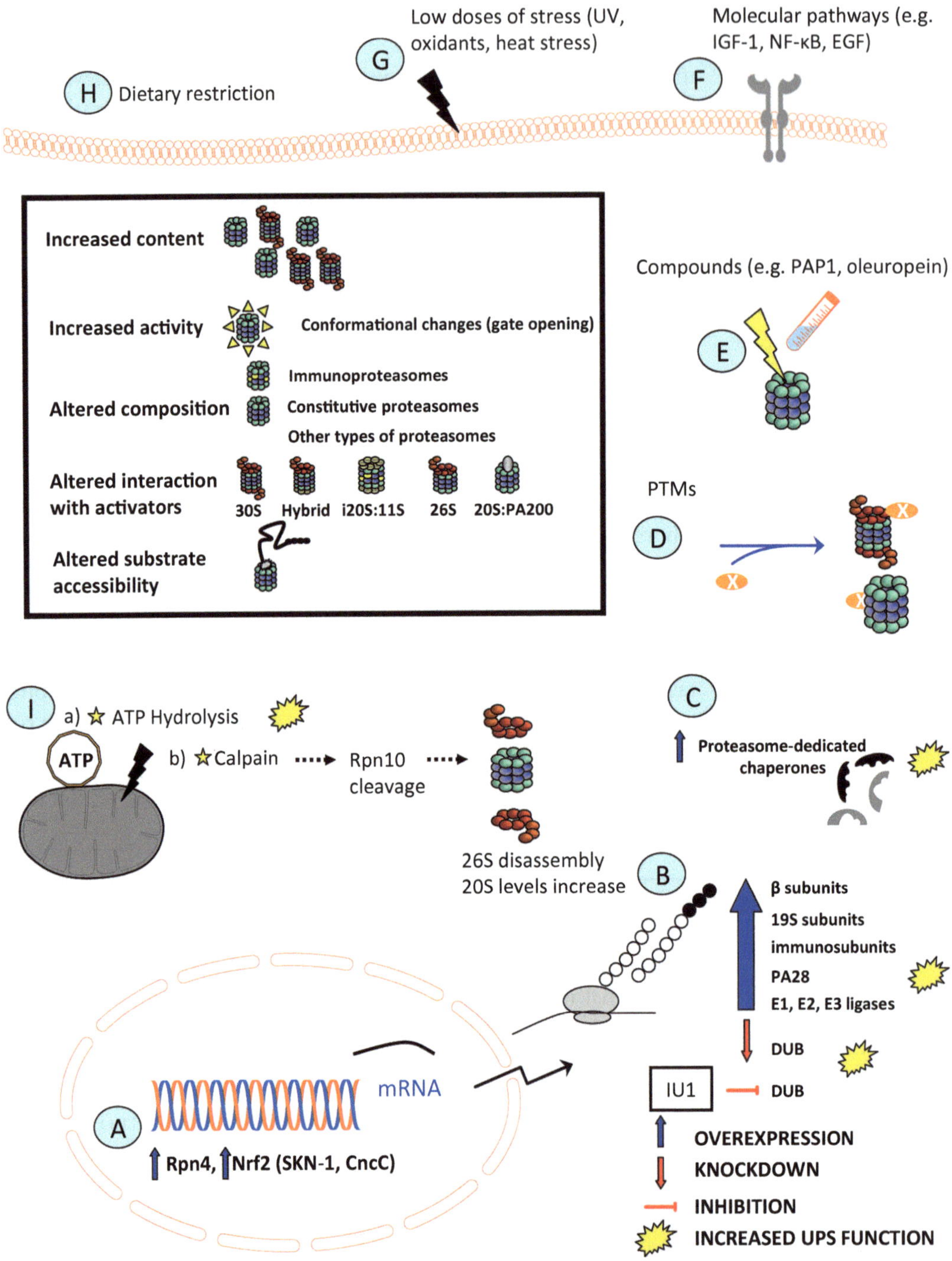

Fig. 3 Means of UPS activation. UPS has been shown to be enhanced through the manipulation of various constituents. UPS activation may refer to increased proteasome content, increased activity (with or without quantitative alterations) that may occur following conformational changes of the complexes that result in α-gate opening, altered ratios of proteasome types (constitutive proteasomes, immunoproteasomes, or other proteasome types), altered interactions with the various activators and altered substrate accessibility as shown in the inserted *square*. The end result of such alterations is the enhancement of UPS activity/function. The so far investigated means of UPS activation include: (*A*) Activation of proteasome-related transcription factors such as Rpn4 in yeast and Nrf2/SKN-1/CncC in mammals/nematodes/flies, respectively. This activation may occur following treatment with a specific compound or following genetic manipulation of these factors. (*B*) Overexpression of UPS components

7.1.2 E1, E2, and E3 Ligases

UPS activation during AD has been attempted through the induction/overexpression of key ligases. More specifically, co-transfection of APP and HRD1 (an ubiquitin ligase that induces proteasome-mediated degradation of ubiquitinated APP) results in reduced Aβ levels and aggregation in HEK293 cells [381]. Overexpression of Fbx2 (an SCF(Fbx2)-E3 ligase) in the primary cortical and hippocampal neurons of transgenic mice overexpressing mAPP (Tg2576), reduces the levels of BACE1, the β-secretase that induces β-amyloidogenesis, and consequently the Aβ levels and ameliorates the synaptic function in vivo [382]. CHIP E3 ligase also drives BACE1 to proteasome-dependent degradation and in parallel regulates p53-mediated trans-repression of BACE1 at both transcriptional and posttranslational level. As a result, reduced Aβ levels are monitored [383]. In an AD model of *C. elegans*, loss of VHL-1 (Hippel-Lindau tumor-suppressor homolog; an E3 ligase for HIF-1 transcription factor) results in delayed paralysis rates and resistance to Aβ proteotoxicity [384].

7.1.3 Deubiquitinases

Inhibition of USP14 DUB by the specific inhibitor IU1, an active-site-directed thiol protease inhibitor, leads to enhanced tau degradation via increased proteasome activities [363].

7.1.4 Other Conditions and Compounds

CNB-001 is a 5-lipoxygenase (5-LOX) inhibitor. Treatment of APP/PS1 AD transgenic mice with CNB-001 activates the eIF2α/ATF4 arm of the UPR that eventually activates both proteasome and autophagic flux and eventually promotes increased rates of Aβ clearance, thus resulting in ameliorated memory function [385].

Treatment of AD transgenic mice with a dopamine receptor agonist, namely apomorphine results in stimulation of proteasome activities and enhanced removal of Aβ and hyper-phosphorylated tau. As a consequence, ameliorated memory function is observed [386].

Quercetin is a proteasome activator and treatment of a transgenic nematode AD model with this polyphenol results in lower levels of Aβ aggregates and to decelerated paralysis rates [315]. Another polyphenol, namely resveratrol also reduces Aβ levels

Fig. 3 (continued) such as β-type 20S subunits, 19S subunits, immunosubunits, PA28 activator, various E1, E2, and E3 ligases or knockdown (or compound-mediated inhibition) of UPS components such as DUBs. (*C*) Enhancement of proteasome assembly through manipulation of proteasome-dedicated chaperones. (*D*) Enhancement of proteasome activity through various PTMs (*X* in the figure represents the various groups that can be added or altered on the various proteasome subunits; please refer to the text for details). (*E*) Direct allosteric alterations of the proteasome structure through the direct binding/interaction of specific natural or chemical compounds. (*F*) Activation of specific pathways that ultimately affect UPS content and/or function such as IGF-1, NF-κB or EGF. (*G*) Exposure to low doses of stress such as UV, oxidants, or heat stress that promote hormetic response that may finally promote UPS activation. (*H*) Effects of dietary protocols such as dietary restriction. (*I*) Cellular energy alterations that ultimately affect: (*a*) ATP hydrolysis and thus proteasome activity or, (*b*) proteases that are responsive to energy alterations and may regulate proteasome assembly/activity/function

[387]. Given that red grapes and red wine are characterized by increased resveratrol concentrations, these results coincide with epidemiological studies suggesting a reverse correlation between red wine intake and AD incidence [388]. More recently, we have also shown that constant feeding of various AD nematode models with the triterpenoid 18α-glycyrrhetinic acid (a previously identified proteasome activator) confers lower paralysis rates accompanied by decreased Aβ deposits, thus ultimately leading to deceleration of the AD phenotype progression. More importantly, similar positive outcomes were also scored in human and murine cells of nervous origin that were subjected to 18α-glycyrrhetinic acid treatment [321].

Rasagiline is an inhibitor of cholinesterase and MAO-A and B that has been shown to stimulate the proteasome activities. Its derivative, namely TV3326 was shown to be neuroprotective and anti-apoptotic in SH-SY5Y and PC-12 cells treated with exogenous Aβ peptide. One cannot rule out the possibility of a link between these positive outcomes and proteasome stimulation [389, 390]. Thioflavin T (ThT) has been shown to reduce Aβ aggregation in vivo in nematodes and this anti-aggregation activity was related to alterations in proteasome function, autophagy and molecular chaperones [391]. Methylene blue, a member of phenothiazines family enhances CT-L and T-L proteasome activities in the brain. This increased proteasome function was linked to the reduced Aβ levels in transgenic mice under chronic methylene blue supplementation and the downstream improved learning and memory functions [392]. Treatment of cells expressing the double truncated $Tau_{151-391}$ with geldamycin, a natural inhibitor of HSP90, results in decreased $Tau_{151-391}$ half-life due to enhanced proteasome degradation [393]. Finally, cellular treatment with polysaccharide PS5 derived from *Rubia cordifolia* and the organic compound ganoderic acid DM leads to an enhanced proteasome-mediated clearance of the intracellular Aβ aggregates [394].

Acetylcholinesterase (AChE) is an enzyme that inactivates acetylcholine at synapses and neuromuscular junctions and is downregulated in AD brains but notably it is still present and activated in amyloid plaques and tangle formations. Cell treatment with lithium results in rapid enhancement of synaptic AChE proteasome-mediated degradation [395].

7.2 Parkinson's Disease (PD)

Given the link between all the PD-associated genes with the UPS in one or the other way, UPS activation may serve as a potential anti-PD approach.

7.2.1 20S and 19S Proteasome Subunits

Overexpression of either 20S or 19S proteasome subunits has not been investigated so far in relation to PD progression. However, the importance of the proper proteasome function in PD was exhibited when upon conditional overexpression of mutated Rpt2 subunit in mice 26S proteasome malfunction occurs and ultimately formation of Lewy-like inclusions and neurodegeneration are established [396].

7.2.2 E1, E2, and E3 Ligases and Ubiquitin

Wt parkin (a key E3 ligase in PD) has been shown to activate the 26S proteasome in an E3 ligase activity-independent manner with an N-terminal ubiquitin-like domain within parkin being critical for this activation through enhancement of the interaction between 19S proteasomal subunits. As a result, wt parkin accelerates the assembly of the 19S RP and thus proteasome activity [234]. In accordance, parkin overexpression in neuroepithelioma cells has been shown to enhance proteasome activity [327] while 26S proteasome activity is upregulated in transgenic flies overexpressing wt parkin or any other form of parkin that possesses the N-terminal parkin fragment containing the necessary for activation UBL domain [234]. Upregulation of wt parkin extends the flies' life-span through decreased levels of protein aggregates, increased levels of K48-linked polyubiquitin and increased turnover of mitofusin (a mitochondrial fusion-promoting factor) followed by changes in mitochondrial morphology and an increase in mitochondrial activity [328].

In a PD *Drosophila* model that overexpresses wt α-synuclein in the eye [397], co-expression of wt Ub protects against α-synuclein-induced toxicity (eye degeneration, locomotor dysfunction, and dopaminergic neurodegeneration) in a K48-polyubiquitin linkage-dependent manner [398], thus suggesting that UPS upregulation might be an attractive anti-PD strategy.

7.2.3 Other Conditions and Compounds

Upregulation of heat-shock proteins protects neuroblastoma cells from the 1-methyl-4-phenylpyridinium ion (MPP^{+})-induced neurotoxicity through inhibition of α-synuclein expression and UPS activation in terms of both ubiquitination rates and proteasome activities [399]. In an attempt to elucidate the molecular mechanism of MPP+ toxicity, Shang et al. [400] revealed that overexpression of neuronal nitric oxide synthase (nNOS) significantly enhances proteasome activity with a consequent reduction of apoptosis rates. In the same study, sepiapterin treatment resulted to nNOS activity restoration (that is negatively affected upon MPP+-induced oxidative stress) with the downstream inhibition of superoxide formation, the enhancement of proteasome activity accompanied by decreased levels of ubiquitinated proteins and the attenuation of apoptosis in MPP+-treated cells [400]. Using the same PD model, pretreatment with pueranin results in attenuation of the MPP(+)-induced dysfunction of the proteasome with a consequent delay of apoptosis [401].

Using a PD *C. elegans* model (α-synuclein overexpression in muscle cells), Fu et al. [320, 402] have shown that treatment with n-butylidenephthalide (a naturally occurring component derived from the chloroform extract of *Angelica sinensis*; [402]) or treatment with acetylcorynoline (the major alkaloid component derived from the traditional Chinese medical herb *Corydalis bungeana*; [320]) decreases 6-hydroxydopamine-mediated dopaminergic neuron degeneration, prevents α-synuclein aggregation, recovers lipid content, restores food-sensing behavior and dopamine levels, and prolongs life-span. In both treatments, proteasome activity enhancement is suggested through the upregulation of *rpn-6* [402] and *rpn-5* [320], respectively. Rasagiline, the inhibitor of monoamine oxidase MAO-B, is a phase 3 anti-PD drug that has been shown to improve pathology symptoms like motor dysfunction [403]. Rasagiline enhances proteasome activity levels in neuroblastoma cells, thus leading to high anti-apoptotic activity [390]. DNA array studies indicate that rasagiline increases the expression of the genes coding mitochondrial energy synthesis, inhibitors of apoptosis, transcription factors, kinases and UPS, sequentially in a time-dependent way [404].

Systemic administration of proteasome inhibitors in the brain of rats or mice results in progressive PD development and has been suggested to constitute an appropriate model of the PD onset and progression [405, 406]). Apomorphine has been found to ameliorate motor activity probably through rescuing proteasome-mediated degradation in mice treated with lactacystin [235]. Pramipexole alleviates lactacystin-mediated proteasome dysfunction resulting in attenuation of the dopaminergic neuronal death in lactacystin-treated mice [407]. Pretreatment with the D3 receptor-preferring agonist D-264 totally blocks the proteasome inhibition and microglial activation in the substantia nigra thus improving behavioral performance and attenuating both MPTP- and lactacystin-induced DA neuron loss [408]. A similar protection was also revealed for rasagiline [409] and for coenzyme Q10 (that protects against proteasome impairment through induction of ATP production and therefore through enhancement of UPS function; [410]).

7.3 Huntington's Disease (HD)

In a conditional mouse model of HD, reversal of neuropathology and motor dysfunction was exhibited with a disappearance of inclusions upon blockade of the constant influx of the mHTT [411]. Therefore, HD pathology might be reversible and a link with the proteostasis network is revealed suggesting that UPS activation could be a potential anti-HD approach.

7.3.1 20S and 19S Proteasome Subunits

Upregulation of *pbs-5* subunit in *C. elegans* leads to enhanced proteasome activities and in turn to reduced polyQ toxicity and improved motility in transgenic worms expressing Q35 in body wall muscle cells or Q40 in neurons [132]. Accordingly, overexpression of *rpn-6.1* 19S subunit results in reduced polyQ toxicity and aggregates levels [291]. Overexpression of PA28γ in HD cells results in recovered proteasome function and in improved cell viability. However, overexpression of *rpn-10* did not result in either proteasome activation or neuroprotection [412]. Ectopic overexpression of a 19S complex subunit, namely *Rpn11*, was shown to attenuate the age-related decline of the proteasome activity in *Drosophila*. As a consequence, the flies exhibit an elongated life-span. Accordingly, *Rpn11* overexpression leads to decreased polyQ-induced toxicity and neurodegeneration [323].

7.3.2 E1, E2, and E3 Ligases

E6-AP E3 ubiquitin ligase promotes the degradation of misfolded polyQ proteins resulting in a suppression of aggregate formation and cell death in cellular HD model [413]. CHIP overexpression suppresses the formation of insoluble aggregates by mutant polyQ proteins in differentiated neuronal cells as well as in an HD zebrafish model [414]. Hrd1 is an endoplasmic reticulum (ER) membrane-E3 ligase with its catalytic active RING finger facing the cytosol that is upregulated in cells overexpressing the N-terminal fragment of htt containing an expanded polyQ tract (httN). Enhanced expression of Hrd1 results in increased degradation of httN and in decreased levels of httN-induced cell death [415]. Similar results are obtained upon overexpression of Parkin [232]. In an HD *C. elegans* model, loss of VHL-1 (a HIF E3 ligase) results in elevated resistance to polyQ toxicity with concomitant decreased paralysis rates [384]. Accordingly, increased resistance to proteotoxic stress is also observed upon loss of Mub1/Ubr2 ubiquitin ligase complex that results in Rpn4 stabilization [280].

7.3.3 Deubiquitinases

Overexpression of the USP14 DUB in mHTT-expressing cells leads to diminished levels of cellular aggregates mainly via the UPS. Specifically, the serine-threonine kinase IRE1 is an ER stress-associated protein that is activated during mHTT toxicity. USP14 overexpression counteracts the IRE1 activation thus leading to reduced rates of cell degeneration [416].

7.3.4 Other Conditions and Compounds

Activation of protein kinase A (PKA) confers Rpt6 phosphorylation that in turn results in increased proteasome activity, reduced mHTT aggregates and improved motor capacity of an HD mouse model [203]. Proteasome impairment through HTT aggregates has been shown to be alleviated by Akt kinase [417] as well as by brain-type creatine kinase (CKB) [203]. Inhibition of Rho-associated kinases (ROCKs) in cellular models of HD reduces the aggregation levels of mHTT via activation of the UPS and macroautophagy [418].

Ubiquilins are proteins that are speculated to function as shuttle factors to transfer misfolded proteins to the proteasome since they have the ability to bind ubiquitin moieties conjugated onto proteins via their UBA domain and subunits of the proteasome via their UBL domain [419]. Overexpression of ubiquilin-1 suppresses polyQ toxicity in cell culture and *C. elegans* models of HD [420], as well as in an HD mouse model where extension of life-span, delayed htt inclusions formation and attenuated ER stress in the hippocampus are scored. Nevertheless, motor defects are not ameliorated [421]. Overexpression of NUB1, a negative regulator of ubiquitin-like protein 1 results in elevated degradation rates of mHTT and thus in lower levels of aggregates and neuronal survival [422].

Various compounds have been identified to alleviate the mHTT-related proteasome impairment like an agonist of the A(2A) adenosine receptor (A(2A) receptor), namely CGS21680

[423], benzamil, an amiloride derivative [424], baclofen, a $GABA_B$ receptor agonist [425], and scyllo-inositol [426]. Finally, sulforaphane, a natural compound derived from broccoli and other vegetables, is a potent activator of both proteasome and autophagy in mice. Sulforaphane treatment enhances the proteasomal degradation of mHTT and induces cell survival in HD cell models [427]. A similar increase in proteasome activity accompanied by more efficient degradation of pathologic polyQ variants is also exerted by the antioxidant *Ginkgo biloba* extract EGb 761 [428].

7.4 Amyotrophic Lateral Sclerosis (ALS)

Given the link between UPS and ALS onset and progression, proteasome activation could be an ALS-targeted therapeutic strategy.

7.4.1 E1, E2, and E3 Ligases

Overexpression of various E3 ligases that target mSOD1 for degradation has shown promising results as potential targets for ALS therapy. More specifically, overexpression of dorphin (identified to promote the proteasome-mediated degradation of mSOD1 and to prevent neurotoxicity; [429]) ameliorates the ALS phenotype in the relevant transgenic mice [430]. Accordingly, overexpression of the ERAD E3 ubiquitin ligase Gp78 targets mSOD1 for ERAD resulting in increased cell viability and reduced SOD1 aggregation levels [431]. Finally, a mitochondrial ubiquitin ligase, namely MITOL, interacts with ubiquitinated mSOD1 but not wt SOD1 and its overexpression results in the enhanced clearance of mSOD1 and in the suppression of mitochondrial accumulation of mSOD1 [432].

7.4.2 Other Conditions and Compounds

Activation of UPR has been shown to be beneficial in conditions of ALS pathology. TorsinA is an AAA+ family member with molecular chaperone-like activity. TorsinA overexpression rescues an ALS *C. elegans* model from the mSOD1-specific ER stress increase and restores normal neuronal function. These positive effects are mediated through enhanced mSOD1 targeting for proteasome degradation [433]. Accordingly, overexpression of the ER-resident factor Derlin-1 results in suppression of the activation of ER stress and in increased proteasomal and autophagosomal turnover of mSOD1 [434].

Overexpression of p62 (sequestosome 1), an adaptor protein for the autophagy pathway, reduces TDP-43 aggregates through enhanced proteasome and autophagy function [435].

Treatment of human neuroblastoma $SOD1^{G93A}$ cells with the synthetic peptide PAP1 leads to decreased levels of mSOD1 aggregates and enhanced cytoprotection through the enhanced proteasome activities mediated via conformational alterations of the proteasome gate [367]. Two proteasome subunits (PSMC1 and PSMC4) have been identified as target proteins of pyrazolone (a five-membered-ring lactam). Treatment of PC12-$SOD1^{G93A}$ cells with pyrazolone results in proteasome activation and the downstream delay of ALS progression [436].

Bee venom and its anti-inflammatory component, melittin, alleviate proteasome activity impairment in human $SOD1^{G85R}$-expressing NSC34 motor neuron cells and in human ALS $SOD1^{G93A}$ mouse model, respectively [437, 438]. Similar results are obtained following treatment of mouse N2A cells overexpressing mutant SIGMAR1 (a gene involved in familial ALS; [439]) with methyl pyruvate, a mitochondrial TCA cycle substrate. The proteasome activity is restored, mitochondrial ATP production is enhanced and aggregation-prone TDP-43 mislocalization is prevented [440].

Malfunction of Nrf2 pathway has been revealed in few ALS patients [441] and in cultures of $SOD1^{G93A}$ motor neurons from the relevant transgenic mice [442]. Use of an Nrf2 activator, namely CDDO trifluoroethylamide (CDDO-TFEA), results in activation of Nrf2 and in deceleration of neurodegeneration [443]. Given that proteasome genes are Nrf2 target genes [120], one cannot rule out the possibility that proteasome activation might also occur and contribute to this neuroprotection.

7.5 Prion Diseases

Prion clearance and the relative proteolytic pathways may constitute a potential therapeutic target for PrDs given that (a) the pathogenesis of the disease is directly related to constant PrP^{Sc} aggregation [444] and (b) diminished PrP^{Sc} levels result in reversal of cognitive deficits and neurophysiological dysfunction of prion-infected mice [445, 446]. The so far collected data suggest that both lysosomal and proteasomal degradation may play significant roles in prion degradation [447]. Nevertheless, scarce data exist regarding the modulation of UPS as an anti-prion therapeutic approach.

7.5.1 E1, E2, and E3 Ligases

The responsible E3 ligase for the unglycosylated PrP (ugPrP) has been identified: Hrd1-Hrd3 in yeast [448] and Gp78 in mammalian cells [449]. Although overexpression of either of those ligases has not been attempted in relation to PrD progression, potential positive results may be expected similarly to what has been shown in ALS with Gp78 (see above).

7.5.2 Other Conditions and Compounds

Congo red derivatives WSP774 and WSP677 have been shown to enhance the proteasome-mediated degradation of PrP^{Sc} in infected cells and thus to alleviate the inhibitory effect of PrP^{Sc} on proteasome function [450]. The efficacy of other proteasome activating compounds like sulforaphane, quercetin, the DUB inhibitor IU1 and all the other molecules that have been so far investigated in various aggregation-prone diseases as mentioned above, remain untested in relation to PrD. Therefore, one cannot rule out the possibility that they could be potential anti-prion candidates. The effects of aggregation-related diseases on the proteasome and the outcome of UPS activation are summarized in Fig. 4.

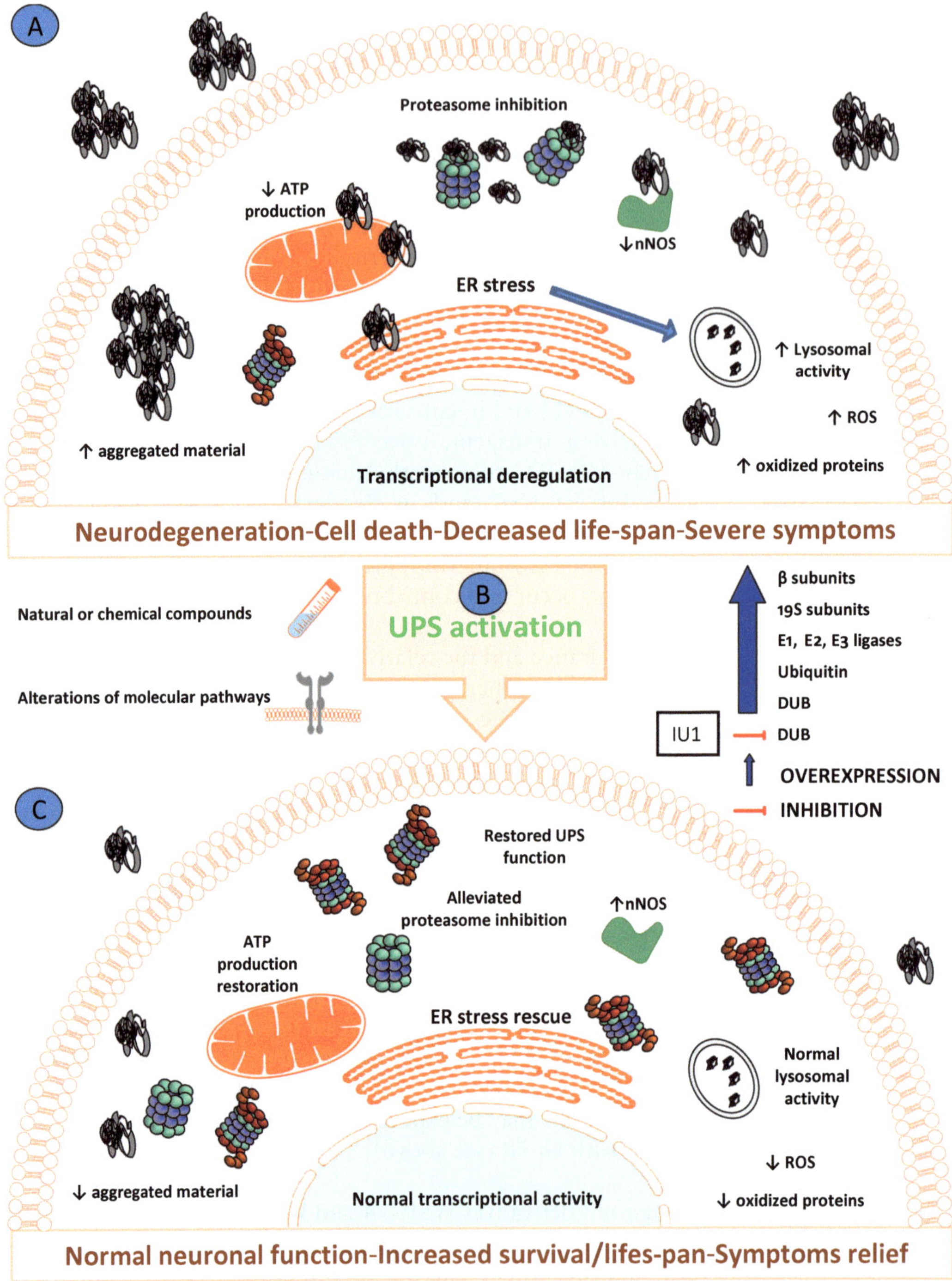

Fig. 4 The effects of aggregation-related diseases on the proteasome and the outcome of UPS activation. Aggregation-related diseases are characterized by increased amount of protease-resistant misfolded/aggregated proteins. (*A*) In neuronal cells from organisms suffering from an aggregation-related disease, aggregated material induces (among others) transcriptional deregulation, inhibition of several key enzymes including nNOS and the proteasome, defects in mitochondria that lead to decreased ATP production that further affects UPS function and ER stress due to the enhanced aggregate load and the inhibition of the normal proteasome function. These defects initiate a vicious circle of constant accumulation of aggregates, additional proteasome inhibition, and constant oxidative stress. The autophagy-lysosome system is induced to compensate for the reduced UPS activity but the end result includes neurodegeneration, cell death and decreased life-span. (*B*) UPS

8 Concluding Remarks

Life is linked to conditions of increased stress (e.g., oxidative stress due to respiration, UV stress by sun exposure). Nevertheless, excessive stress is not compatible with survival. Therefore, proteostasis mechanisms (with proteolytic modules forming the ultimate arsenal) have been evolved to assure the balance between the inevitable stress conditions and cellular/organismal homeostasis. Upon malfunction of these mechanisms due to intrinsic (e.g., mutations, loss- or gain-of-function alterations) or extrinsic (e.g., environmental stress factors) causes, this balance is destroyed. Therefore, the preservation or even the enhancement of proteostasis mechanisms function seems to be beneficial for cellular survival. This is further supported by the fact that most of the pro-longevity factors and pathways enhance the function of protcolysis modules leading to extended life-span, ameliorated response to stress and alleviation of aggregation-related disease phenotypes. Most of these studies have been performed in lower eukaryotes. Therefore additional studies in higher eukaryotes followed by human population studies (wherever possible) are necessary. These studies will finally validate the correlation between aging/aggregation-related diseases and enhanced proteostasis mechanisms.

In the case of the UPS and its potential enhancement, further studies are needed to fully elucidate the regulatory mechanisms behind such activation. For example, although genetic and compound-mediated UPS activation has been successful, the molecular mechanisms behind such modulation are not fully investigated. Questions that remain to be elucidated include regulation of transcription, assembly, trafficking, and elimination as well as posttranslational regulation of the various UPS components. The same mechanisms should be then thoroughly examined in the context of aging or a given aggregation-related disease as age- or disease-specific alterations might be expected. Given that overactivation might also prove to be detrimental, highly orchestrated UPS activation is necessary in order to be able to suggest a manipulation as an anti-aging/anti-aggregation preventive/therapeutic strategy. In the case of activating compounds, one should be very cautious with the translation of results, considering the possible (but still uncovered) side targets of a given molecule. Identification of activating molecules that are constituents of human regular diet should be also explored since they provide extra advantages;

Fig. 4 (continued) activation has been achieved in cellular and organismal models of aggregation-related diseases through the genetic manipulation of several UPS constituents, through treatment with natural or chemical compounds as well as through alterations of specific molecular pathways (please refer to text for details). (*C*) The abovementioned manipulations result in alleviated proteasome inhibition, ER stress rescue and restoration of (a) ATP production, (b) transcriptional activity, (c) lysosomal activity, and (d) UPS activity, among others. The cellular aggregate load decreases promoting normal neuronal function and increased survival/life-span

beneficial UPS activation or preservation should probably commence before heavily aggregated proteins get established in the cellular milieu thus in a young age before we can even detect such alterations. Therefore, diet constituents might be ideal for such approach. Addressing these questions will further pave the way to the establishment of therapeutic but also preventive strategies in the battle against aging and age-related diseases.

Acknowledgements

Cited work has been funded by Research Funding Program: Thales "GenAge" [ΘΑΛΗΣ ΑΠ:10479/3.7.12 MIS380228] and Thales "MAESTRO" co-financed by the European Union (European Social Fund—ESF) and Greek national funds through the Operational Program "Education and Lifelong Learning" of the National Strategic Reference Framework (NSRF), a KRIPIS project "STHENOS," an IKYDA 2012 fellowship, a Scientific Project funded by John S. Latsis Public Benefit Foundation and a grant from Empirikion foundation to N.C. The authors would like to acknowledge networking support by the Proteostasis COST Action (BM1307).

References

1. Chondrogianni N, Sakellari M, Lefaki M, Papaevgeniou N, Gonos ES (2014) Proteasome activation delays aging in vitro and in vivo. Free Radic Biol Med 71:303–320
2. Lopez-Otin C, Blasco MA, Partridge L, Serrano M, Kroemer G (2013) The hallmarks of aging. Cell 153(6):1194–1217
3. Campisi J, Andersen JK, Kapahi P, Melov S (2011) Cellular senescence: a link between cancer and age-related degenerative disease? Semin Cancer Biol 21(6):354–359
4. Niccoli T, Partridge L (2012) Ageing as a risk factor for disease. Curr Biol 22(17):R741–R752
5. Hayflick L (1965) The limited in vitro lifetime of human diploid cell strains. Exp Cell Res 37:614–636
6. Vachova L, Cap M, Palkova Z (2012) Yeast colonies: a model for studies of aging, environmental adaptation, and longevity. Oxid Med Cell Longev 2012:601836
7. Adams MD, Celniker SE, Holt RA, Evans CA, Gocayne JD, Amanatides PG, Scherer SE, Li PW, Hoskins RA, Galle RF, George RA, Lewis SE, Richards S, Ashburner M, Henderson SN, Sutton GG, Wortman JR, Yandell MD, Zhang Q, Chen LX, Brandon RC, Rogers YH, Blazej RG, Champe M, Pfeiffer BD, Wan KH, Doyle C, Baxter EG, Helt G, Nelson CR, Gabor GL, Abril JF, Agbayani A, An HJ, Andrews-Pfannkoch C, Baldwin D, Ballew RM, Basu A, Baxendale J, Bayraktaroglu L, Beasley EM, Beeson KY, Benos PV, Berman BP, Bhandari D, Bolshakov S, Borkova D, Botchan MR, Bouck J, Brokstein P, Brottier P, Burtis KC, Busam DA, Butler H, Cadieu E, Center A, Chandra I, Cherry JM, Cawley S, Dahlke C, Davenport LB, Davies P, de Pablos B, Delcher A, Deng Z, Mays AD, Dew I, Dietz SM, Dodson K, Doup LE, Downes M, Dugan-Rocha S, Dunkov BC, Dunn P, Durbin KJ, Evangelista CC, Ferraz C, Ferriera S, Fleischmann W, Fosler C, Gabrielian AE, Garg NS, Gelbart WM, Glasser K, Glodek A, Gong F, Gorrell JH, Gu Z, Guan P, Harris M, Harris NL, Harvey D, Heiman TJ, Hernandez JR, Houck J, Hostin D, Houston KA, Howland TJ, Wei MH, Ibegwam C, Jalali M, Kalush F, Karpen GH, Ke Z, Kennison JA, Ketchum KA, Kimmel BE, Kodira CD, Kraft C, Kravitz S, Kulp D, Lai Z, Lasko P, Lei Y, Levitsky AA, Li J, Li Z, Liang Y, Lin X, Liu X, Mattei B, McIntosh TC, McLeod MP, McPherson D, Merkulov G, Milshina NV, Mobarry C, Morris J, Moshrefi A, Mount SM, Moy M, Murphy B, Murphy L, Muzny DM, Nelson DL, Nelson

DR, Nelson KA, Nixon K, Nusskern DR, Pacleb JM, Palazzolo M, Pittman GS, Pan S, Pollard J, Puri V, Reese MG, Reinert K, Remington K, Saunders RD, Scheeler F, Shen H, Shue BC, Siden-Kiamos I, Simpson M, Skupski MP, Smith T, Spier E, Spradling AC, Stapleton M, Strong R, Sun E, Svirskas R, Tector C, Turner R, Venter E, Wang AH, Wang X, Wang ZY, Wassarman DA, Weinstock GM, Weissenbach J, Williams SM, WoodageT, Worley KC, Wu D, Yang S, Yao QA, Ye J, Yeh RF, Zaveri JS, Zhan M, Zhang G, Zhao Q, Zheng L, Zheng XH, Zhong FN, Zhong W, Zhou X, Zhu S, Zhu X, Smith HO, Gibbs RA, Myers EW, Rubin GM, Venter JC (2000) The genome sequence of Drosophila melanogaster. *Science* 287(5461):2185–2195.

8. Reiter LT, Potocki L, Chien S, Gribskov M, Bier E (2001) A systematic analysis of human disease-associated gene sequences in Drosophila melanogaster. Genome Res 11(6):1114–1125
9. Mouse Genome Sequencing C, Waterston RH, Lindblad-Toh K, Birney E, Rogers J, Abril JF, Agarwal P, Agarwala R, Ainscough R, Alexandersson M, An P, Antonarakis SE, Attwood J, Baertsch R, Bailey J, Barlow K, Beck S, Berry E, Birren B, Bloom T, Bork P, Botcherby M, Bray N, Brent MR, Brown DG, Brown SD, Bult C, Burton J, Butler J, Campbell RD, Carninci P, Cawley S, Chiaromonte F, Chinwalla AT, Church DM, Clamp M, Clee C, Collins FS, Cook LL, Copley RR, Coulson A, Couronne O, Cuff J, Curwen V, Cutts T, Daly M, David R, Davies J, Delehaunty KD, Deri J, Dermitzakis ET, Dewey C, Dickens NJ, Diekhans M, Dodge S, Dubchak I, Dunn DM, Eddy SR, Elnitski L, Emes RD, Eswara P, Eyras E, Felsenfeld A, Fewell GA, Flicek P, Foley K, Frankel WN, Fulton LA, Fulton RS, Furey TS, Gage D, Gibbs RA, Glusman G, Gnerre S, Goldman N, Goodstadt L, Grafham D, Graves TA, Green ED, Gregory S, Guigo R, Guyer M, Hardison RC, Haussler D, Hayashizaki Y, Hillier LW, Hinrichs A, Hlavina W, Holzer T, Hsu F, Hua A, Hubbard T, Hunt A, Jackson I, Jaffe DB, Johnson LS, Jones M, Jones TA, Joy A, Kamal M, Karlsson EK, Karolchik D, Kasprzyk A, Kawai J, Keibler E, Kells C, Kent WJ, Kirby A, Kolbe DL, Korf I, Kucherlapati RS, Kulbokas EJ, Kulp D, Landers T, Leger JP, Leonard S, Letunic I, Levine R, Li J, Li M, Lloyd C, Lucas S, Ma B, Maglott DR, Mardis ER, Matthews L, Mauceli E, Mayer JH, McCarthy M, McCombie WR, McLaren S, McLay K, McPherson JD, Meldrim J, Meredith B, Mesirov JP, Miller W, Miner TL, Mongin E, Montgomery KT, Morgan M, Mott R, Mullikin JC, Muzny DM, Nash WE, Nelson JO, Nhan MN, Nicol R, Ning Z, Nusbaum C, O'Connor MJ, Okazaki Y, Oliver K, Overton-Larty E, Pachter L, Parra G, Pepin KH, Peterson J, Pevzner P, Plumb R, Pohl CS, Poliakov A, Ponce TC, Ponting CP, Potter S, Quail M, Reymond A, Roe BA, Roskin KM, Rubin EM, Rust AG, Santos R, Sapojnikov V, Schultz B, Schultz J, Schwartz MS, Schwartz S, Scott C, Seaman S, Searle S, Sharpe T, Sheridan A, Shownkeen R, Sims S, Singer JB, Slater G, Smit A, Smith DR, Spencer B, Stabenau A, Stange-Thomann N, Sugnet C, Suyama M, Tesler G, Thompson J, Torrents D, Trevaskis E, Tromp J, Ucla C, Ureta-Vidal A, Vinson JP, Von Niederhausern AC, Wade CM, Wall M, Weber RJ, Weiss RB, Wendl MC, West AP, Wetterstrand K, Wheeler R, Whelan S, Wierzbowski J, Willey D, Williams S, Wilson RK, Winter E, Worley KC, Wyman D, Yang S, Yang SP, Zdobnov EM, Zody MC, Lander ES (2002) Initial sequencing and comparative analysis of the mouse genome. *Nature* 420(6915):520–562.
10. Vanhooren V, Libert C (2013) The mouse as a model organism in aging research: usefulness, pitfalls and possibilities. Ageing Res Rev 12(1):8–21
11. Brookmeyer R, Evans DA, Hebert L, Langa KM, Heeringa SG, Plassman BL, Kukull WA (2011) National estimates of the prevalence of Alzheimer's disease in the United States. Alzheimers Dement 7(1):61–73
12. Oddo S (2008) The ubiquitin-proteasome system in Alzheimer's disease. J Cell Mol Med 12(2):363–373
13. Wischik CM, Novak M, Thogersen HC, Edwards PC, Runswick MJ, Jakes R, Walker JE, Milstein C, Roth M, Klug A (1988) Isolation of a fragment of tau derived from the core of the paired helical filament of Alzheimer disease. Proc Natl Acad Sci U S A 85(12):4506–4510
14. Glenner GG, Wong CW, Quaranta V, Eanes ED (1984) The amyloid deposits in Alzheimer's disease: their nature and pathogenesis. Appl Pathol 2(6):357–369
15. Goate A, Chartier-Harlin MC, Mullan M, Brown J, Crawford F, Fidani L, Giuffra L, Haynes A, Irving N, James L et al (1991) Segregation of a missense mutation in the amyloid precursor protein gene with familial Alzheimer's disease. Nature 349(6311):704–706
16. Levy-Lahad E, Wijsman EM, Nemens E, Anderson L, Goddard KA, Weber JL, Bird TD, Schellenberg GD (1995) A familial Alzheimer's disease locus on chromosome 1. Science 269(5226):970–973

17. Sherrington R, Rogaev EI, Liang Y, Rogaeva EA, Levesque G, Ikeda M, Chi H, Lin C, Li G, Holman K, Tsuda T, Mar L, Foncin JF, Bruni AC, Montesi MP, Sorbi S, Rainero I, Pinessi L, Nee L, Chumakov I, Pollen D, Brookes A, Sanseau P, Polinsky RJ, Wasco W, Da Silva HA, Haines JL, Perkicak-Vance MA, Tanzi RE, Roses AD, Fraser PE, Rommens JM, St George-Hyslop PH (1995) Cloning of a gene bearing missense mutations in early-onset familial Alzheimer's disease. Nature 375(6534):754–760
18. Strittmatter WJ, Saunders AM, Schmechel D, Pericak-Vance M, Enghild J, Salvesen GS, Roses AD (1993) Apolipoprotein E: high-avidity binding to beta-amyloid and increased frequency of type 4 allele in late-onset familial Alzheimer disease. Proc Natl Acad Sci U S A 90(5):1977–1981
19. Ferrari R, Moreno JH, Minhajuddin AT, O'Bryant SE, Reisch JS, Barber RC, Momeni P (2012) Implication of common and disease specific variants in CLU, CR1, and PICALM. Neurobiol Aging 33(8):1846.e1847–1818
20. Han SH, Mook-Jung I (2014) Diverse molecular targets for therapeutic strategies in Alzheimer's disease. J Korean Med Sci 29(7):893–902
21. Ribeiro FM, Camargos ER, de Souza LC, Teixeira AL (2013) Animal models of neurodegenerative diseases. Rev Bras Psiquiatr 35(Suppl 2):S82–S91
22. Upadhya SC, Hegde AN (2007) Role of the ubiquitin proteasome system in Alzheimer's disease. BMC Biochem 8(Suppl 1):S12
23. Recasens A, Dehay B (2014) Alpha-synuclein spreading in Parkinson's disease. Front Neuroanat 8:159
24. Recchia A, Debetto P, Negro A, Guidolin D, Skaper SD, Giusti P (2004) Alpha-synuclein and Parkinson's disease. FASEB J 18(6):617–626
25. Chan NC, Chan DC (2011) Parkin uses the UPS to ship off dysfunctional mitochondria. Autophagy 7(7):771–772
26. Tanaka K, Suzuki T, Hattori N, Mizuno Y (2004) Ubiquitin, proteasome and parkin. Biochim Biophys Acta 1695(1-3):235–247
27. Giasson BI, Lee VM (2001) Parkin and the molecular pathways of Parkinson's disease. Neuron 31(6):885–888
28. Cookson MR (2004) Roles of the proteasome in neurodegenerative disease: refining the hypothesis. Ann Neurol 56(3):315–316
29. Kruger R, Eberhardt O, Riess O, Schulz JB (2002) Parkinson's disease: one biochemical pathway to fit all genes? Trends Mol Med 8(5):236–240
30. Bano D, Zanetti F, Mende Y, Nicotera P (2011) Neurodegenerative processes in Huntington's disease. Cell Death Dis 2, e228
31. Andrew S, Theilmann J, Almqvist E, Norremolle A, Lucotte G, Anvret M, Sorensen SA, Turpin JC, Hayden MR (1993) DNA analysis of distinct populations suggests multiple origins for the mutation causing Huntington disease. Clin Genet 43(6):286–294
32. DiFiglia M, Sapp E, Chase KO, Davies SW, Bates GP, Vonsattel JP, Aronin N (1997) Aggregation of huntingtin in neuronal intranuclear inclusions and dystrophic neurites in brain. Science 277(5334):1990–1993
33. Ajroud-Driss S, Siddique T (2014) Sporadic and hereditary amyotrophic lateral sclerosis (ALS). Biochim Biophys Acta
34. Ajroud-Driss S, Siddique T (2015) Sporadic and hereditary amyotrophic lateral sclerosis (ALS). Biochim Biophys Acta 1852(4):679–684
35. Kato S (2008) Amyotrophic lateral sclerosis models and human neuropathology: similarities and differences. Acta Neuropathol 115(1):97–114
36. Strong MJ, Kesavapany S, Pant HC (2005) The pathobiology of amyotrophic lateral sclerosis: a proteinopathy? J Neuropathol Exp Neurol 64(8):649–664
37. Siddique T, Pericak-Vance MA, Brooks BR, Roos RP, Hung WY, Antel JP, Munsat TL, Phillips K, Warner K, Speer M et al (1989) Linkage analysis in familial amyotrophic lateral sclerosis. Neurology 39(7):919–925
38. Deng HX, Shi Y, Furukawa Y, Zhai H, Fu R, Liu E, Gorrie GH, Khan MS, Hung WY, Bigio EH, Lukas T, Dal Canto MC, O'Halloran TV, Siddique T (2006) Conversion to the amyotrophic lateral sclerosis phenotype is associated with intermolecular linked insoluble aggregates of SOD1 in mitochondria. Proc Natl Acad Sci U S A 103(18):7142–7147
39. Blokhuis AM, Groen EJ, Koppers M, van den Berg LH, Pasterkamp RJ (2013) Protein aggregation in amyotrophic lateral sclerosis. Acta Neuropathol 125(6):777–794
40. Morinet F (2014) Prions: a model of conformational disease? Pathol Biol (Paris) 62(2):96–99
41. Rabinowitz J, Slyuzberg M, Ritsner M, Mark M, Popper M, Ginath Y (1994) Changes in diagnosis in a 9-year national longitudinal sample. Compr Psychiatry 35(5):361–365
42. Hegde RS, Mastrianni JA, Scott MR, DeFea KA, Tremblay P, Torchia M, DeArmond SJ, Prusiner SB, Lingappa VR (1998) A transmembrane form of the prion protein in neurodegenerative disease. Science 279(5352):827–834

43. Bremer J, Baumann F, Tiberi C, Wessig C, Fischer H, Schwarz P, Steele AD, Toyka KV, Nave KA, Weis J, Aguzzi A (2010) Axonal prion protein is required for peripheral myelin maintenance. Nat Neurosci 13(3):310–318
44. Maglio LE, Perez MF, Martins VR, Brentani RR, Ramirez OA (2004) Hippocampal synaptic plasticity in mice devoid of cellular prion protein. Brain Res Mol Brain Res 131(1-2):58–64
45. Zhang CC, Steele AD, Lindquist S, Lodish HF (2006) Prion protein is expressed on long-term repopulating hematopoietic stem cells and is important for their self-renewal. Proc Natl Acad Sci U S A 103(7):2184–2189
46. Pan KM, Baldwin M, Nguyen J, Gasset M, Serban A, Groth D, Mehlhorn I, Huang Z, Fletterick RJ, Cohen FE et al (1993) Conversion of alpha-helices into beta-sheets features in the formation of the scrapie prion proteins. Proc Natl Acad Sci U S A 90(23):10962–10966
47. Prusiner SB, Scott MR, DeArmond SJ, Cohen FE (1998) Prion protein biology. Cell 93(3):337–348
48. Masel J, Jansen VA (1999) The kinetics of proteinase K digestion of linear prion polymers. Proc Biol Sci 266(1431):1927–1931
49. Saborio GP, Permanne B, Soto C (2001) Sensitive detection of pathological prion protein by cyclic amplification of protein misfolding. Nature 411(6839):810–813
50. Wang F, Wang X, Yuan CG, Ma J (2010) Generating a prion with bacterially expressed recombinant prion protein. Science 327(5969):1132–1135
51. Kim HJ, Kim NC, Wang YD, Scarborough EA, Moore J, Diaz Z, MacLea KS, Freibaum B, Li S, Molliex A, Kanagaraj AP, Carter R, Boylan KB, Wojtas AM, Rademakers R, Pinkus JL, Greenberg SA, Trojanowski JQ, Traynor BJ, Smith BN, Topp S, Gkazi AS, Miller J, Shaw CE, Kottlors M, Kirschner J, Pestronk A, Li YR, Ford AF, Gitler AD, Benatar M, King OD, Kimonis VE, Ross ED, Weihl CC, Shorter J, Taylor JP (2013) Mutations in prion-like domains in hnRNPA2B1 and hnRNPA1 cause multisystem proteinopathy and ALS. Nature 495(7442):467–473
52. Jucker M, Walker LC (2013) Self-propagation of pathogenic protein aggregates in neurodegenerative diseases. Nature 501(7465):45–51
53. Calamini B, Morimoto RI (2012) Protein homeostasis as a therapeutic target for diseases of protein conformation. Curr Top Med Chem 12(22):2623–2640
54. Cuervo AM, Wong E (2014) Chaperone-mediated autophagy: roles in disease and aging. Cell Res 24(1):92–104
55. Vilchez D, Saez I, Dillin A (2014) The role of protein clearance mechanisms in organismal ageing and age-related diseases. Nat Commun 5:5659
56. Glickman MH, Ciechanover A (2002) The ubiquitin-proteasome proteolytic pathway: destruction for the sake of construction. Physiol Rev 82(2):373–428
57. Mukhopadhyay D, Riezman H (2007) Proteasome-independent functions of ubiquitin in endocytosis and signaling. Science 315(5809):201–205
58. Ciechanover A, Stanhill A (2014) The complexity of recognition of ubiquitinated substrates by the 26S proteasome. Biochim Biophys Acta 1843(1):86–96
59. Lee CS, Lee C, Hu T, Nguyen JM, Zhang J, Martin MV, Vawter MP, Huang EJ, Chan JY (2011) Loss of nuclear factor E2-related factor 1 in the brain leads to dysregulation of proteasome gene expression and neurodegeneration. Proc Natl Acad Sci U S A 108(20):8408–8413
60. Pickart CM, Eddins MJ (2004) Ubiquitin: structures, functions, mechanisms. Biochim Biophys Acta 1695(1-3):55–72
61. Chiu YH, Sun Q, Chen ZJ (2007) E1-L2 activates both ubiquitin and FAT10. Mol Cell 27(6):1014–1023
62. Kudo M, Sugasawa K, Hori T, Enomoto T, Hanaoka F, Ui M (1991) Human ubiquitin-activating enzyme (E1): compensation for heat-labile mouse E1 and its gene localization on the X chromosome. Exp Cell Res 192(1):110–117
63. van Wijk SJ, de Vries SJ, Kemmeren P, Huang A, Boelens R, Bonvin AM, Timmers HT (2009) A comprehensive framework of E2-RING E3 interactions of the human ubiquitin-proteasome system. Mol Syst Biol 5:295
64. Metzger MB, Hristova VA, Weissman AM (2012) HECT and RING finger families of E3 ubiquitin ligases at a glance. J Cell Sci 125(Pt 3):531–537
65. Deshaies RJ, Joazeiro CA (2009) RING domain E3 ubiquitin ligases. Annu Rev Biochem 78:399–434
66. Li W, Bengtson MH, Ulbrich A, Matsuda A, Reddy VA, Orth A, Chanda SK, Batalov S, Joazeiro CA (2008) Genome-wide and functional annotation of human E3 ubiquitin ligases identifies MULAN, a mitochondrial E3 that regulates the organelle's dynamics and signaling. PLoS One 3(1), e1487

67. Petroski MD, Deshaies RJ (2005) Function and regulation of cullin-RING ubiquitin ligases. Nat Rev Mol Cell Biol 6(1):9–20
68. Mocciaro A, Rape M (2012) Emerging regulatory mechanisms in ubiquitin-dependent cell cycle control. J Cell Sci 125(Pt 2):255–263
69. Wilkinson KD (2009) DUBs at a glance. J Cell Sci 122(Pt 14):2325–2329
70. Amerik AY, Hochstrasser M (2004) Mechanism and function of deubiquitinating enzymes. Biochim Biophys Acta 1695(1-3):189–207
71. Reyes-Turcu FE, Ventii KH, Wilkinson KD (2009) Regulation and cellular roles of ubiquitin-specific deubiquitinating enzymes. Annu Rev Biochem 78:363–397
72. Ciechanover A, Kwon YT (2015) Degradation of misfolded proteins in neurodegenerative diseases: therapeutic targets and strategies. Exp Mol Med 47, e147
73. Schmidt M, Finley D (2014) Regulation of proteasome activity in health and disease. Biochim Biophys Acta 1843(1):13–25
74. Goldberg AL (2007) Functions of the proteasome: from protein degradation and immune surveillance to cancer therapy. Biochem Soc Trans 35(Pt 1):12–17
75. Hirano Y, Hendil KB, Yashiroda H, Iemura S, Nagane R, Hioki Y, Natsume T, Tanaka K, Murata S (2005) A heterodimeric complex that promotes the assembly of mammalian 20S proteasomes. Nature 437(7063):1381–1385
76. Burri L, Hockendorff J, Boehm U, Klamp T, Dohmen RJ, Levy F (2000) Identification and characterization of a mammalian protein interacting with 20S proteasome precursors. Proc Natl Acad Sci U S A 97(19):10348–10353
77. Griffin TA, Slack JP, McCluskey TS, Monaco JJ, Colbert RA (2000) Identification of proteassemblin, a mammalian homologue of the yeast protein, Ump1p, that is required for normal proteasome assembly. Mol Cell Biol Res Commun 3(4):212–217
78. Witt E, Zantopf D, Schmidt M, Kraft R, Kloetzel PM, Kruger E (2000) Characterisation of the newly identified human Ump1 homologue POMP and analysis of LMP7(beta 5i) incorporation into 20S proteasomes. J Mol Biol 301(1):1–9
79. Hirano Y, Kaneko T, Okamoto K, Bai M, Yashiroda H, Furuyama K, Kato K, Tanaka K, Murata S (2008) Dissecting beta-ring assembly pathway of the mammalian 20S proteasome. EMBO J 27(16):2204–2213
80. Gu ZC, Enenkel C (2014) Proteasome assembly. Cell Mol Life Sci 71(24):4729–4745
81. Wani PS, Rowland MA, Ondracek A, Deeds EJ, Roelofs J (2015) Maturation of the proteasome core particle induces an affinity switch that controls regulatory particle association. Nat Commun 6:6384
82. Marguerat S, Schmidt A, Codlin S, Chen W, Aebersold R, Bahler J (2012) Quantitative analysis of fission yeast transcriptomes and proteomes in proliferating and quiescent cells. Cell 151(3):671–683
83. Sixt SU, Dahlmann B (2008) Extracellular, circulating proteasomes and ubiquitin – incidence and relevance. Biochim Biophys Acta 1782(12):817–823
84. Bochmann I, Ebstein F, Lehmann A, Wohlschlaeger J, Sixt SU, Kloetzel PM, Dahlmann B (2014) T Lymphocytes export proteasomes by way of microparticles: a possible mechanism for generation of extracellular proteasomes. J Cell Mol Med 18(1):59–68
85. Takada LT, Geschwind MD (2013) Prion diseases. Semin Neurol 33(4):348–356
86. Tomko RJ Jr, Hochstrasser M (2013) Molecular architecture and assembly of the eukaryotic proteasome. Annu Rev Biochem 82:415–445
87. Rosenzweig R, Osmulski PA, Gaczynska M, Glickman MH (2008) The central unit within the 19S regulatory particle of the proteasome. Nat Struct Mol Biol 15(6):573–580
88. Groll M, Bochtler M, Brandstetter H, Clausen T, Huber R (2005) Molecular machines for protein degradation. Chembiochem 6(2):222–256
89. Nickell S, Beck F, Scheres SH, Korinek A, Forster F, Lasker K, Mihalache O, Sun N, Nagy I, Sali A, Plitzko JM, Carazo JM, Mann M, Baumeister W (2009) Insights into the molecular architecture of the 26S proteasome. Proc Natl Acad Sci U S A 106(29):11943–11947
90. Verma R, Aravind L, Oania R, McDonald WH, Yates JR 3rd, Koonin EV, Deshaies RJ (2002) Role of Rpn11 metalloprotease in deubiquitination and degradation by the 26S proteasome. Science 298(5593):611–615
91. Pathare GR, Nagy I, Bohn S, Unverdorben P, Hubert A, Korner R, Nickell S, Lasker K, Sali A, Tamura T, Nishioka T, Forster F, Baumeister W, Bracher A (2012) The proteasomal subunit Rpn6 is a molecular clamp holding the core and regulatory subcomplexes together. Proc Natl Acad Sci U S A 109(1):149–154

92. Kim S, Saeki Y, Fukunaga K, Suzuki A, Takagi K, Yamane T, Tanaka K, Mizushima T, Kato K (2010) Crystal structure of yeast rpn14, a chaperone of the 19S regulatory particle of the proteasome. J Biol Chem 285(20):15159–15166
93. Roelofs J, Park S, Haas W, Tian G, McAllister FE, Huo Y, Lee BH, Zhang F, Shi Y, Gygi SP, Finley D (2009) Chaperone-mediated pathway of proteasome regulatory particle assembly. Nature 459(7248):861–865
94. Tomko RJ Jr, Hochstrasser M (2011) Order of the proteasomal ATPases and eukaryotic proteasome assembly. Cell Biochem Biophys 60(1-2):13–20
95. Fukunaga K, Kudo T, Toh-e A, Tanaka K, Saeki Y (2010) Dissection of the assembly pathway of the proteasome lid in Saccharomyces cerevisiae. Biochem Biophys Res Commun 396(4):1048–1053
96. Imai J, Maruya M, Yashiroda H, Yahara I, Tanaka K (2003) The molecular chaperone Hsp90 plays a role in the assembly and maintenance of the 26S proteasome. EMBO J 22(14):3557–3567
97. Sha Z, Yen HC, Scheel H, Suo J, Hofmann K, Chang EC (2007) Isolation of the Schizosaccharomyces pombe proteasome subunit Rpn7 and a structure-function study of the proteasome-COP9-initiation factor domain. J Biol Chem 282(44):32414–32423
98. Tanaka K (2013) The proteasome: from basic mechanisms to emerging roles. Keio J Med 62(1):1–12
99. Kincaid EZ, Che JW, York I, Escobar H, Reyes-Vargas E, Delgado JC, Welsh RM, Karow ML, Murphy AJ, Valenzuela DM, Yancopoulos GD, Rock KL (2012) Mice completely lacking immunoproteasomes show major changes in antigen presentation. Nat Immunol 13(2):129–135
100. Pickering AM, Koop AL, Teoh CY, Ermak G, Grune T, Davies KJ (2010) The immunoproteasome, the 20S proteasome and the PA28alphabeta proteasome regulator are oxidative-stress-adaptive proteolytic complexes. Biochem J 432(3):585–594
101. Gavilan MP, Castano A, Torres M, Portavella M, Caballero C, Jimenez S, Garcia-Martinez A, Parrado J, Vitorica J, Ruano D (2009) Age-related increase in the immunoproteasome content in rat hippocampus: molecular and functional aspects. J Neurochem 108(1):260–272
102. Mishto M, Bellavista E, Santoro A, Stolzing A, Ligorio C, Nacmias B, Spazzafumo L, Chiappelli M, Licastro F, Sorbi S, Pession A, Ohm T, Grune T, Franceschi C (2006) Immunoproteasome and LMP2 polymorphism in aged and Alzheimer's disease brains. Neurobiol Aging 27(1):54–66
103. Rodriguez KA, Edrey YH, Osmulski P, Gaczynska M, Buffenstein R (2012) Altered composition of liver proteasome assemblies contributes to enhanced proteasome activity in the exceptionally long-lived naked mole-rat. PLoS One 7(5), e35890
104. Groettrup M, Kirk CJ, Basler M (2010) Proteasomes in immune cells: more than peptide producers? Nat Rev Immunol 10(1):73–78
105. Rechsteiner M, Hill CP (2005) Mobilizing the proteolytic machine: cell biological roles of proteasome activators and inhibitors. Trends Cell Biol 15(1):27–33
106. Tanahashi H, Kito K, Ito T, Yoshioka K (2010) MafB protein stability is regulated by the JNK and ubiquitin-proteasome pathways. Arch Biochem Biophys 494(1):94–100
107. Murata S, Sasaki K, Kishimoto T, Niwa S, Hayashi H, Takahama Y, Tanaka K (2007) Regulation of CD8+ T cell development by thymus-specific proteasomes. Science 316(5829):1349–1353
108. Tomaru U, Ishizu A, Murata S, Miyatake Y, Suzuki S, Takahashi S, Kazamaki T, Ohara J, Baba T, Iwasaki S, Fugo K, Otsuka N, Tanaka K, Kasahara M (2009) Exclusive expression of proteasome subunit {beta}5t in the human thymic cortex. Blood 113(21):5186–5191
109. Xing Y, Jameson SC, Hogquist KA (2013) Thymoproteasome subunit-beta5T generates peptide-MHC complexes specialized for positive selection. Proc Natl Acad Sci U S A 110(17):6979–6984
110. Yuan X, Miller M, Belote JM (1996) Duplicated proteasome subunit genes in Drosophila melanogaster encoding testes-specific isoforms. Genetics 144(1):147–157
111. Zhong L, Belote JM (2007) The testis-specific proteasome subunit Prosalpha6T of D. melanogaster is required for individualization and nuclear maturation during spermatogenesis. Development 134(19):3517–3525
112. Fehlker M, Wendler P, Lehmann A, Enenkel C (2003) Blm3 is part of nascent proteasomes and is involved in a late stage of nuclear proteasome assembly. EMBO Rep 4(10):959–963
113. Lehmann A, Jechow K, Enenkel C (2008) Blm10 binds to pre-activated proteasome core particles with open gate conformation. EMBO Rep 9(12):1237–1243
114. Ustrell V, Hoffman L, Pratt G, Rechsteiner M (2002) PA200, a nuclear proteasome activator involved in DNA repair. EMBO J 21(13):3516–3525

115. Sadre-Bazzaz K, Whitby FG, Robinson H, Formosa T, Hill CP (2010) Structure of a Blm10 complex reveals common mechanisms for proteasome binding and gate opening. Mol Cell 37(5):728–735
116. Mannhaupt G, Schnall R, Karpov V, Vetter I, Feldmann H (1999) Rpn4p acts as a transcription factor by binding to PACE, a nonamer box found upstream of 26S proteasomal and other genes in yeast. FEBS Lett 450(1-2):27–34
117. Ju D, Xu H, Wang X, Xie Y (2007) Ubiquitin-mediated degradation of Rpn4 is controlled by a phosphorylation-dependent ubiquitylation signal. Biochim Biophys Acta 1773(11):1672–1680
118. Shirozu R, Yashiroda H, Murata S (2015) Identification of minimum Rpn4-responsive elements in genes related to proteasome functions. FEBS Lett 589(8):933–940
119. Nguyen T, Yang CS, Pickett CB (2004) The pathways and molecular mechanisms regulating Nrf2 activation in response to chemical stress. Free Radic Biol Med 37(4):433–441
120. Kwak MK, Wakabayashi N, Greenlaw JL, Yamamoto M, Kensler TW (2003) Antioxidants enhance mammalian proteasome expression through the Keap1-Nrf2 signaling pathway. Mol Cell Biol 23(23):8786–8794
121. Furukawa M, Xiong Y (2005) BTB protein Keap1 targets antioxidant transcription factor Nrf2 for ubiquitination by the Cullin 3-Roc1 ligase. Mol Cell Biol 25(1):162–171
122. Surh YJ, Kundu JK, Na HK (2008) Nrf2 as a master redox switch in turning on the cellular signaling involved in the induction of cytoprotective genes by some chemopreventive phytochemicals. Planta Med 74(13):1526–1539
123. Itoh K, Chiba T, Takahashi S, Ishii T, Igarashi K, Katoh Y, Oyake T, Hayashi N, Satoh K, Hatayama I, Yamamoto M, Nabeshima Y (1997) An Nrf2/small Maf heterodimer mediates the induction of phase II detoxifying enzyme genes through antioxidant response elements. Biochem Biophys Res Commun 236(2):313–322
124. Venugopal R, Jaiswal AK (1998) Nrf2 and Nrf1 in association with Jun proteins regulate antioxidant response element-mediated expression and coordinated induction of genes encoding detoxifying enzymes. Oncogene 17(24):3145–3156
125. Wasserman WW, Fahl WE (1997) Comprehensive analysis of proteins which interact with the antioxidant responsive element: correlation of ARE-BP-1 with the chemoprotective induction response. Arch Biochem Biophys 344(2):387–396
126. Tullet JM, Hertweck M, An JH, Baker J, Hwang JY, Liu S, Oliveira RP, Baumeister R, Blackwell TK (2008) Direct inhibition of the longevity-promoting factor SKN-1 by insulin-like signaling in C. elegans. Cell 132(6):1025–1038
127. Bishop NA, Guarente L (2007) Genetic links between diet and lifespan: shared mechanisms from yeast to humans. Nat Rev Genet 8(11):835–844
128. Choe KP, Przybysz AJ, Strange K (2009) The WD40 repeat protein WDR-23 functions with the CUL4/DDB1 ubiquitin ligase to regulate nuclear abundance and activity of SKN-1 in Caenorhabditis elegans. Mol Cell Biol 29(10):2704–2715
129. Kahn Nate W, Rea Shane L, Moyle S, Kell A, Johnson Thomas E (2008) Proteasomal dysfunction activates the transcription factor SKN-1 and produces a selective oxidative-stress response in Caenorhabditis elegans. Biochem J 409(1):205
130. Oliveira RP, Porter Abate J, Dilks K, Landis J, Ashraf J, Murphy CT, Blackwell TK (2009) Condition-adapted stress and longevity gene regulation by Caenorhabditis elegans SKN-1/Nrf. Aging Cell 8(5):524–541
131. Park SK, Tedesco PM, Johnson TE (2009) Oxidative stress and longevity in Caenorhabditis elegans as mediated by SKN-1. Aging Cell 8(3):258–269
132. Chondrogianni N, Georgila K, Kourtis N, Tavernarakis N, Gonos ES (2015) 20S proteasome activation promotes life span extension and resistance to proteotoxicity in Caenorhabditis elegans. FASEB J 29(2):611–622
133. Chan JY, Han XL, Kan YW (1993) Cloning of Nrf1, an NF-E2-related transcription factor, by genetic selection in yeast. Proc Natl Acad Sci U S A 90(23):11371–11375
134. Luna L, Johnsen O, Skartlien AH, Pedeutour F, Turc-Carel C, Prydz H, Kolsto AB (1994) Molecular cloning of a putative novel human bZIP transcription factor on chromosome 17q22. Genomics 22(3):553–562
135. Radhakrishnan SK, Lee CS, Young P, Beskow A, Chan JY, Deshaies RJ (2010) Transcription factor Nrf1 mediates the proteasome recovery pathway after proteasome inhibition in mammalian cells. Mol Cell 38(1):17–28
136. Steffen J, Seeger M, Koch A, Kruger E (2010) Proteasomal degradation is transcriptionally controlled by TCF11 via an ERAD-dependent feedback loop. Mol Cell 40(1):147–158
137. Sha Z, Goldberg AL (2014) Proteasome-mediated processing of Nrf1 is essential for coordinate induction of all proteasome subunits and p97. Curr Biol 24(14):1573–1583

138. Vangala JR, Dudem S, Jain N, Kalivendi SV (2014) Regulation of PSMB5 protein and beta subunits of mammalian proteasome by constitutively activated signal transducer and activator of transcription 3 (STAT3): potential role in bortezomib-mediated anticancer therapy. J Biol Chem 289(18):12612–12622
139. Foss GS, Prydz H (1999) Interferon regulatory factor 1 mediates the interferon-gamma induction of the human immunoproteasome subunit multicatalytic endopeptidase complex-like 1. J Biol Chem 274(49):35196–35202
140. Namiki S, Nakamura T, Oshima S, Yamazaki M, Sekine Y, Tsuchiya K, Okamoto R, Kanai T, Watanabe M (2005) IRF-1 mediates upregulation of LMP7 by IFN-gamma and concerted expression of immunosubunits of the proteasome. FEBS Lett 579(13):2781–2787
141. Yang XW, Wang P, Liu JQ, Zhang H, Xi WD, Jia XH, Wang KK (2014) Coordinated regulation of the immunoproteasome subunits by PML/RARalpha and PU.1 in acute promyelocytic leukemia. Oncogene 33(21):2700–2708
142. Divald A, Kivity S, Wang P, Hochhauser E, Roberts B, Teichberg S, Gomes AV, Powell SR (2010) Myocardial ischemic preconditioning preserves postischemic function of the 26S proteasome through diminished oxidative damage to 19S regulatory particle subunits. Circ Res 106(12):1829–1838
143. Predmore JM, Wang P, Davis F, Bartolone S, Westfall MV, Dyke DB, Pagani F, Powell SR, Day SM (2010) Ubiquitin proteasome dysfunction in human hypertrophic and dilated cardiomyopathies. Circulation 121(8):997–1004
144. Mason GG, Hendil KB, Rivett AJ (1996) Phosphorylation of proteasomes in mammalian cells. Identification of two phosphorylated subunits and the effect of phosphorylation on activity. Eur J Biochem 238(2):453–462
145. Bose S, Stratford FL, Broadfoot KI, Mason GG, Rivett AJ (2004) Phosphorylation of 20S proteasome alpha subunit C8 (alpha7) stabilizes the 26S proteasome and plays a role in the regulation of proteasome complexes by gamma-interferon. Biochem J 378(Pt 1):177–184
146. Castano JG, Mahillo E, Arizti P, Arribas J (1996) Phosphorylation of C8 and C9 subunits of the multicatalytic proteinase by casein kinase II and identification of the C8 phosphorylation sites by direct mutagenesis. Biochemistry 35(12):3782–3789
147. Djakovic SN, Schwarz LA, Barylko B, DeMartino GN, Patrick GN (2009) Regulation of the proteasome by neuronal activity and calcium/calmodulin-dependent protein kinase II. J Biol Chem 284(39):26655–26665
148. Feng Y, Longo DL, Ferris DK (2001) Polo-like kinase interacts with proteasomes and regulates their activity. Cell Growth Differ 12(1):29–37
149. Kikuchi J, Iwafune Y, Akiyama T, Okayama A, Nakamura H, Arakawa N, Kimura Y, Hirano H (2010) Co- and post-translational modifications of the 26S proteasome in yeast. Proteomics 10(15):2769–2779
150. Gomes AV, Zong C, Edmondson RD, Li X, Stefani E, Zhang J, Jones RC, Thyparambil S, Wang GW, Qiao X, Bardag-Gorce F, Ping P (2006) Mapping the murine cardiac 26S proteasome complexes. Circ Res 99(4):362–371
151. Zong C, Gomes AV, Drews O, Li X, Young GW, Berhane B, Qiao X, French SW, Bardag-Gorce F, Ping P (2006) Regulation of murine cardiac 20S proteasomes: role of associating partners. Circ Res 99(4):372–380
152. Day SM, Divald A, Wang P, Davis F, Bartolone S, Jones R, Powell SR (2013) Impaired assembly and post-translational regulation of 26S proteasome in human end-stage heart failure. Circ Heart Fail 6(3):544–549
153. Moiseeva TN, Bottrill A, Melino G, Barlev NA (2013) DNA damage-induced ubiquitylation of proteasome controls its proteolytic activity. Oncotarget 4(9):1338–1348
154. Zhang F, Su K, Yang X, Bowe DB, Paterson AJ, Kudlow JE (2003) O-GlcNAc modification is an endogenous inhibitor of the proteasome. Cell 115(6):715–725
155. Liu K, Paterson AJ, Zhang F, McAndrew J, Fukuchi K, Wyss JM, Peng L, Hu Y, Kudlow JE (2004) Accumulation of protein O-GlcNAc modification inhibits proteasomes in the brain and coincides with neuronal apoptosis in brain areas with high O-GlcNAc metabolism. J Neurochem 89(4):1044–1055
156. Overath T, Kuckelkorn U, Henklein P, Strehl B, Bonar D, Kloss A, Siele D, Kloetzel PM, Janek K (2012) Mapping of O-GlcNAc sites of 20S proteasome subunits and Hsp90 by a novel biotin-cystamine tag. Mol Cell Proteomics 11(8):467–477
157. Kimura Y, Takaoka M, Tanaka S, Sassa H, Tanaka K, Polevoda B, Sherman F, Hirano H (2000) N(Alpha)-acetylation and proteolytic activity of the yeast 20S proteasome. J Biol Chem 275(7):4635–4639
158. Kimura Y, Saeki Y, Yokosawa H, Polevoda B, Sherman F, Hirano H (2003) N-Terminal modifications of the 19S regulatory particle subunits of the yeast proteasome. Arch Biochem Biophys 409(2):341–348

159. Bulteau AL, Ikeda-Saito M, Szweda LI (2003) Redox-dependent modulation of aconitase activity in intact mitochondria. Biochemistry 42(50):14846–14855
160. Farout L, Mary J, Vinh J, Szweda LI, Friguet B (2006) Inactivation of the proteasome by 4-hydroxy-2-nonenal is site specific and dependant on 20S proteasome subtypes. Arch Biochem Biophys 453(1):135–142
161. Bulteau AL, Petropoulos I, Friguet B (2000) Age-related alterations of proteasome structure and function in aging epidermis. Exp Gerontol 35(6-7):767–777
162. Keller JN, Hanni KB, Markesbery WR (2000) Possible involvement of proteasome inhibition in aging: implications for oxidative stress. Mech Ageing Dev 113(1):61–70
163. Demasi M, Hand A, Ohara E, Oliveira CL, Bicev RN, Bertoncini CA, Netto LE (2014) 20S proteasome activity is modified via S-glutathionylation based on intracellular redox status of the yeast Saccharomyces cerevisiae: implications for the degradation of oxidized proteins. Arch Biochem Biophys 557:65–71
164. Kimura A, Kato Y, Hirano H (2012) N-myristoylation of the Rpt2 subunit regulates intracellular localization of the yeast 26S proteasome. Biochemistry 51(44):8856–8866
165. Besche HC, Sha Z, Kukushkin NV, Peth A, Hock EM, Kim W, Gygi S, Gutierrez JA, Liao H, Dick L, Goldberg AL (2014) Autoubiquitination of the 26S proteasome on Rpn13 regulates breakdown of ubiquitin conjugates. EMBO J 33(10):1159–1176
166. Isasa M, Katz EJ, Kim W, Yugo V, Gonzalez S, Kirkpatrick DS, Thomson TM, Finley D, Gygi SP, Crosas B (2010) Monoubiquitination of RPN10 regulates substrate recruitment to the proteasome. Mol Cell 38(5):733–745
167. Jacobson AD, MacFadden A, Wu Z, Peng J, Liu CW (2014) Autoregulation of the 26S proteasome by in situ ubiquitination. Mol Biol Cell 25(12):1824–1835
168. Cuervo AM, Palmer A, Rivett AJ, Knecht E (1995) Degradation of proteasomes by lysosomes in rat liver. Eur J Biochem 227(3):792–800
169. Chondrogianni N, Stratford FL, Trougakos IP, Friguet B, Rivett AJ, Gonos ES (2003) Central role of the proteasome in senescence and survival of human fibroblasts: induction of a senescence-like phenotype upon its inhibition and resistance to stress upon its activation. J Biol Chem 278(30):28026–28037
170. Chondrogianni N, Gonos ES (2004) Proteasome inhibition induces a senescence-like phenotype in primary human fibroblasts cultures. Biogerontology 5(1):55–61
171. Chondrogianni N, Trougakos IP, Kletsas D, Chen QM, Gonos ES (2008) Partial proteasome inhibition in human fibroblasts triggers accelerated M1 senescence or M2 crisis depending on p53 and Rb status. Aging Cell 7(5):717–732
172. Stratford FL, Chondrogianni N, Trougakos IP, Gonos ES, Rivett AJ (2006) Proteasome response to interferon-gamma is altered in senescent human fibroblasts. FEBS Lett 580(16):3989–3994
173. Grune T, Jung T, Merker K, Davies KJ (2004) Decreased proteolysis caused by protein aggregates, inclusion bodies, plaques, lipofuscin, ceroid, and 'aggresomes' during oxidative stress, aging, and disease. Int J Biochem Cell Biol 36(12):2519–2530
174. Chondrogianni N, Petropoulos I, Franceschi C, Friguet B, Gonos ES (2000) Fibroblast cultures from healthy centenarians have an active proteasome. Exp Gerontol 35(6-7):721–728
175. Chen Q, Thorpe J, Ding Q, El-Amouri IS, Keller JN (2004) Proteasome synthesis and assembly are required for survival during stationary phase. Free Radic Biol Med 37(6):859–868
176. Bajorek M, Finley D, Glickman MH (2003) Proteasome disassembly and downregulation is correlated with viability during stationary phase. Curr Biol 13(13):1140–1144
177. Tran JR, Brodsky JL (2014) The Cdc48-Vms1 complex maintains 26S proteasome architecture. Biochem J 458(3):459–467
178. Laporte D, Salin B, Daignan-Fornier B, Sagot I (2008) Reversible cytoplasmic localization of the proteasome in quiescent yeast cells. J Cell Biol 181(5):737–745
179. van Deventer S, Menendez-Benito V, van Leeuwen F, Neefjes J (2015) N-terminal acetylation and replicative age affect proteasome localization and cell fitness during aging. J Cell Sci 128(1):109–117
180. Saunier R, Esposito M, Dassa EP, Delahodde A (2013) Integrity of the Saccharomyces cerevisiae Rpn11 protein is critical for formation of proteasome storage granules (PSG) and survival in stationary phase. PLoS One 8(8), e70357
181. Weberruss MH, Savulescu AF, Jando J, Bissinger T, Harel A, Glickman MH, Enenkel C (2013) Blm10 facilitates nuclear import of proteasome core particles. EMBO J 32(20):2697–2707
182. Hamer G, Matilainen O, Holmberg CI (2010) A photoconvertible reporter of the ubiquitin-proteasome system in vivo. Nat Methods 7(6):473–478

183. Ghazi A, Henis-Korenblit S, Kenyon C (2007) Regulation of Caenorhabditis elegans lifespan by a proteasomal E3 ligase complex. Proc Natl Acad Sci U S A 104(14):5947–5952
184. Yun C, Stanhill A, Yang Y, Zhang Y, Haynes CM, Xu CF, Neubert TA, Mor A, Philips MR, Ron D (2008) Proteasomal adaptation to environmental stress links resistance to proteotoxicity with longevity in Caenorhabditis elegans. Proc Natl Acad Sci U S A 105(19):7094–7099
185. Segref A, Kevei E, Pokrzywa W, Schmeisser K, Mansfeld J, Livnat-Levanon N, Ensenauer R, Glickman MH, Ristow M, Hoppe T (2014) Pathogenesis of human mitochondrial diseases is modulated by reduced activity of the ubiquitin/proteasome system. Cell Metab 19(4):642–652
186. Vernace VA, Arnaud L, Schmidt-Glenewinkel T, Figueiredo-Pereira ME (2007) Aging perturbs 26S proteasome assembly in Drosophila melanogaster. FASEB J 21(11):2672–2682
187. Fredriksson A, Johansson Krogh E, Hernebring M, Pettersson E, Javadi A, Almstedt A, Nystrom T (2012) Effects of aging and reproduction on protein quality control in soma and gametes of Drosophila melanogaster. Aging Cell 11(4):634–643
188. Dasuri K, Zhang L, Ebenezer P, Fernandez-Kim SO, Bruce-Keller AJ, Szweda LI, Keller JN (2011) Proteasome alterations during adipose differentiation and aging: links to impaired adipocyte differentiation and development of oxidative stress. Free Radic Biol Med 51(9):1727–1735
189. Louie JL, Kapphahn RJ, Ferrington DA (2002) Proteasome function and protein oxidation in the aged retina. Exp Eye Res 75(3):271–284
190. Kapphahn RJ, Bigelow EJ, Ferrington DA (2007) Age-dependent inhibition of proteasome chymotrypsin-like activity in the retina. Exp Eye Res 84(4):646–654
191. Breusing N, Arndt J, Voss P, Bresgen N, Wiswedel I, Gardemann A, Siems W, Grune T (2009) Inverse correlation of protein oxidation and proteasome activity in liver and lung. Mech Ageing Dev 130(11-12):748–753
192. Dasuri K, Zhang L, Ebenezer P, Liu Y, Fernandez-Kim SO, Keller JN (2009) Aging and dietary restriction alter proteasome biogenesis and composition in the brain and liver. Mech Ageing Dev 130(11-12):777–783
193. Hayashi T, Goto S (1998) Age-related changes in the 20S and 26S proteasome activities in the liver of male F344 rats. Mech Ageing Dev 102(1):55–66
194. Ferrington DA, Husom AD, Thompson LV (2005) Altered proteasome structure, function, and oxidation in aged muscle. FASEB J 19(6):644–646
195. Bulteau AL, Szweda LI, Friguet B (2002) Age-dependent declines in proteasome activity in the heart. Arch Biochem Biophys 397(2):298–304
196. Li F, Zhang L, Craddock J, Bruce-Keller AJ, Dasuri K, Nguyen A, Keller JN (2008) Aging and dietary restriction effects on ubiquitination, sumoylation, and the proteasome in the heart. Mech Ageing Dev 129(9):515–521
197. Gohlke S, Mishto M, Textoris-Taube K, Keller C, Giannini C, Vasuri F, Capizzi E, D'Errico-Grigioni A, Kloetzel PM, Dahlmann B (2014) Molecular alterations in proteasomes of rat liver during aging result in altered proteolytic activities. Age (Dordr) 36(1):57–72
198. Abd El Mohsen MM, Iravani MM, Spencer JP, Rose S, Fahim AT, Motawi TM, Ismail NA, Jenner P (2005) Age-associated changes in protein oxidation and proteasome activities in rat brain: modulation by antioxidants. Biochem Biophys Res Commun 336(2):386–391
199. Zeng BY, Medhurst AD, Jackson M, Rose S, Jenner P (2005) Proteasomal activity in brain differs between species and brain regions and changes with age. Mech Ageing Dev 126(6-7):760–766
200. Perez VI, Buffenstein R, Masamsetti V, Leonard S, Salmon AB, Mele J, Andziak B, Yang T, Edrey Y, Friguet B, Ward W, Richardson A, Chaudhuri A (2009) Protein stability and resistance to oxidative stress are determinants of longevity in the longest-living rodent, the naked mole-rat. Proc Natl Acad Sci U S A 106(9):3059–3064
201. Tomaru U, Takahashi S, Ishizu A, Miyatake Y, Gohda A, Suzuki S, Ono A, Ohara J, Baba T, Murata S, Tanaka K, Kasahara M (2012) Decreased proteasomal activity causes age-related phenotypes and promotes the development of metabolic abnormalities. Am J Pathol 180(3):963–972
202. Ding Q, Martin S, Dimayuga E, Bruce-Keller AJ, Keller JN (2006) LMP2 knock-out mice have reduced proteasome activities and increased levels of oxidatively damaged proteins. Antioxid Redox Signal 8(1-2):130–135
203. Lin YS, Cheng TH, Chang CP, Chen HM, Chern Y (2013) Enhancement of brain-type creatine kinase activity ameliorates neuronal deficits in Huntington's disease. Biochim Biophys Acta 1832(6):742–753

204. Bayram B, Nikolai S, Huebbe P, Ozcelik B, Grimm S, Grune T, Frank J, Rimbach G (2013) Biomarkers of oxidative stress, antioxidant defence and inflammation are altered in the senescence-accelerated mouse prone 8. Age (Dordr) 35(4):1205–1217
205. Carrard G, Dieu M, Raes M, Toussaint O, Friguet B (2003) Impact of ageing on proteasome structure and function in human lymphocytes. Int J Biochem Cell Biol 35(5):728–739
206. Ponnappan U, Zhong M, Trebilcock GU (1999) Decreased proteasome-mediated degradation in T cells from the elderly: a role in immune senescence. Cell Immunol 192(2):167–174
207. Viteri G, Carrard G, Birlouez-Aragon I, Silva E, Friguet B (2004) Age-dependent protein modifications and declining proteasome activity in the human lens. Arch Biochem Biophys 427(2):197–203
208. Strucksberg KH, Tangavelou K, Schroder R, Clemen CS (2010) Proteasomal activity in skeletal muscle: a matter of assay design, muscle type, and age. Anal Biochem 399(2):225–229
209. Petropoulos I, Conconi M, Wang X, Hoenel B, Bregegere F, Milner Y, Friguet B (2000) Increase of oxidatively modified protein is associated with a decrease of proteasome activity and content in aging epidermal cells. J Gerontol A Biol Sci Med Sci 55(5):B220–B227
210. Bossola M, Muscaritoli M, Valenza V, Panocchia N, Tazza L, Cascino A, Laviano A, Liberatori M, Lodovica Moussier M, Rossi Fanelli F, Luciani G (2004) Anorexia and serum leptin levels in hemodialysis patients. Nephron Clin Pract 97(3):c76–c82
211. Fry CS, Drummond MJ, Glynn EL, Dickinson JM, Gundermann DM, Timmerman KL, Walker DK, Volpi E, Rasmussen BB (2013) Skeletal muscle autophagy and protein breakdown following resistance exercise are similar in younger and older adults. J Gerontol A Biol Sci Med Sci 68(5):599–607
212. Zetterberg M, Petersen A, Sjostrand J, Karlsson J (2003) Proteasome activity in human lens nuclei and correlation with age, gender and severity of cataract. Curr Eye Res 27(1):45–53
213. Bellavista E, Martucci M, Vasuri F, Santoro A, Mishto M, Kloss A, Capizzi E, Degiovanni A, Lanzarini C, Remondini D, Dazzi A, Pellegrini S, Cescon M, Capri M, Salvioli S, D'Errico-Grigioni A, Dahlmann B, Grazi GL, Franceschi C (2014) Lifelong maintenance of composition, function and cellular/subcellular distribution of proteasomes in human liver. Mech Ageing Dev 141–142:26–34
214. Perry G, Friedman R, Shaw G, Chau V (1987) Ubiquitin is detected in neurofibrillary tangles and senile plaque neurites of Alzheimer disease brains. Proc Natl Acad Sci U S A 84(9):3033–3036
215. van Leeuwen FW, van Tijn P, Sonnemans MA, Hobo B, Mann DM, Van Broeckhoven C, Kumar-Singh S, Cras P, Leuba G, Savioz A, Maat-Schieman ML, Yamaguchi H, Kros JM, Kamphorst W, Hol EM, de Vos RA, Fischer DF (2006) Frameshift proteins in autosomal dominant forms of Alzheimer disease and other tauopathies. Neurology 66(2 Suppl 1):S86–S92
216. Keller JN, Hanni KB, Markesbery WR (2000) Impaired proteasome function in Alzheimer's disease. J Neurochem 75(1):436–439
217. Lopez Salon M, Morelli L, Castano EM, Soto EF, Pasquini JM (2000) Defective ubiquitination of cerebral proteins in Alzheimer's disease. J Neurosci Res 62(2):302–310
218. Keck S, Nitsch R, Grune T, Ullrich O (2003) Proteasome inhibition by paired helical filament-tau in brains of patients with Alzheimer's disease. J Neurochem 85(1):115–122
219. Tseng BP, Green KN, Chan JL, Blurton-Jones M, LaFerla FM (2008) Abeta inhibits the proteasome and enhances amyloid and tau accumulation. Neurobiol Aging 29(11):1607–1618
220. Lindsten K, de Vrij FM, Verhoef LG, Fischer DF, van Leeuwen FW, Hol EM, Masucci MG, Dantuma NP (2002) Mutant ubiquitin found in neurodegenerative disorders is a ubiquitin fusion degradation substrate that blocks proteasomal degradation. J Cell Biol 157(3):417–427
221. Kumar P, Ambasta RK, Veereshwarayya V, Rosen KM, Kosik KS, Band H, Mestril R, Patterson C, Querfurth HW (2007) CHIP and HSPs interact with beta-APP in a proteasome-dependent manner and influence Abeta metabolism. Hum Mol Genet 16(7):848–864
222. Nijholt DA, de Graaf TR, van Haastert ES, Oliveira AO, Berkers CR, Zwart R, Ovaa H, Baas F, Hoozemans JJ, Scheper W (2011) Endoplasmic reticulum stress activates autophagy but not the proteasome in neuronal cells: implications for Alzheimer's disease. Cell Death Differ 18(6):1071–1081
223. Cecarini V, Bonfili L, Cuccioloni M, Mozzicafreddo M, Rossi G, Buizza L, Uberti D, Angeletti M, Eleuteri AM (2012) Crosstalk between the ubiquitin-proteasome system and autophagy in a human cellular model of Alzheimer's disease. Biochim Biophys Acta 1822(11):1741–1751

224. Lonskaya I, Desforges NM, Hebron ML, Moussa CE (2013) Ubiquitination increases parkin activity to promote autophagic alpha-synuclein clearance. PLoS One 8(12), e83914
225. Kaneko M, Okuma Y, Nomura Y (2012) Molecular approaches to the treatment, prophylaxis, and diagnosis of Alzheimer's disease: possible involvement of HRD1, a novel molecule related to endoplasmic reticulum stress, in Alzheimer's disease. J Pharmacol Sci 118(3):325–330
226. Choi HD, Seo PJ, Son BW, Kang BW (2004) Synthesis of 2-(4-hydroxyphenyl)benzofurans and their application to beta-amyloid aggregation inhibitor. Arch Pharm Res 27(1):19–24
227. Riederer BM, Leuba G, Vernay A, Riederer IM (2011) The role of the ubiquitin proteasome system in Alzheimer's disease. Exp Biol Med (Maywood) 236(3):268–276
228. Snyder H, Mensah K, Theisler C, Lee J, Matouschek A, Wolozin B (2003) Aggregated and monomeric alpha-synuclein bind to the S6' proteasomal protein and inhibit proteasomal function. J Biol Chem 278(14):11753–11759
229. Chen L, Thiruchelvam MJ, Madura K, Richfield EK (2006) Proteasome dysfunction in aged human alpha-synuclein transgenic mice. Neurobiol Dis 23(1):120–126
230. Martin-Clemente B, Alvarez-Castelao B, Mayo I, Sierra AB, Diaz V, Milan M, Farinas I, Gomez-Isla T, Ferrer I, Castano JG (2004) Alpha-Synuclein expression levels do not significantly affect proteasome function and expression in mice and stably transfected PC12 cell lines. J Biol Chem 279(51):52984–52990
231. Dachsel JC, Lucking CB, Deeg S, Schultz E, Lalowski M, Casademunt E, Corti O, Hampe C, Patenge N, Vaupel K, Yamamoto A, Dichgans M, Brice A, Wanker EE, Kahle PJ, Gasser T (2005) Parkin interacts with the proteasome subunit alpha4. FEBS Lett 579(18):3913–3919
232. Tsai YC, Fishman PS, Thakor NV, Oyler GA (2003) Parkin facilitates the elimination of expanded polyglutamine proteins and leads to preservation of proteasome function. J Biol Chem 278(24):22044–22055
233. Sakata E, Yamaguchi Y, Kurimoto E, Kikuchi J, Yokoyama S, Yamada S, Kawahara H, Yokosawa H, Hattori N, Mizuno Y, Tanaka K, Kato K (2003) Parkin binds the Rpn10 subunit of 26S proteasomes through its ubiquitin-like domain. EMBO Rep 4(3):301–306
234. Um JW, Im E, Lee HJ, Min B, Yoo L, Yoo J, Lubbert H, Stichel-Gunkel C, Cho HS, Yoon JB, Chung KC (2010) Parkin directly modulates 26S proteasome activity. J Neurosci 30(35):11805–11814
235. McNaught KS, Jenner P (2001) Proteasomal function is impaired in substantia nigra in Parkinson's disease. Neurosci Lett 297(3):191–194
236. Jenner P, Olanow CW (1998) Understanding cell death in Parkinson's disease. Ann Neurol 44(3 Suppl 1):S72–S84
237. McNaught KS, Belizaire R, Isacson O, Jenner P, Olanow CW (2003) Altered proteasomal function in sporadic Parkinson's disease. Exp Neurol 179(1):38–46
238. McNaught KS, Jnobaptiste R, Jackson T, Jengelley TA (2010) The pattern of neuronal loss and survival may reflect differential expression of proteasome activators in Parkinson's disease. Synapse 64(3):241–250
239. Blandini F, Sinforiani E, Pacchetti C, Samuele A, Bazzini E, Zangaglia R, Nappi G, Martignoni E (2006) Peripheral proteasome and caspase activity in Parkinson disease and Alzheimer disease. Neurology 66(4):529–534
240. van Leeuwen FW, Verwer RW, Spence H, Evans DA, Burbach JP (1998) The magnocellular neurons of the hypothalamo-neurohypophyseal system display remarkable neuropeptidergic phenotypes leading to novel insights in neuronal cell biology. Prog Brain Res 119:115–126
241. Bennett EJ, Shaler TA, Woodman B, Ryu KY, Zaitseva TS, Becker CH, Bates GP, Schulman H, Kopito RR (2007) Global changes to the ubiquitin system in Huntington's disease. Nature 448(7154):704–708
242. Hunter JM, Lesort M, Johnson GV (2007) Ubiquitin-proteasome system alterations in a striatal cell model of Huntington's disease. J Neurosci Res 85(8):1774–1788
243. Seo H, Sonntag KC, Isacson O (2004) Generalized brain and skin proteasome inhibition in Huntington's disease. Ann Neurol 56(3):319–328
244. Wang CE, Tydlacka S, Orr AL, Yang SH, Graham RK, Hayden MR, Li S, Chan AW, Li XJ (2008) Accumulation of N-terminal mutant huntingtin in mouse and monkey models implicated as a pathogenic mechanism in Huntington's disease. Hum Mol Genet 17(17):2738–2751
245. Holmberg CI, Staniszewski KE, Mensah KN, Matouschek A, Morimoto RI (2004) Inefficient degradation of truncated polyglutamine proteins by the proteasome. EMBO J 23(21):4307–4318
246. Venkatraman P, Wetzel R, Tanaka M, Nukina N, Goldberg AL (2004) Eukaryotic proteasomes cannot digest polyglutamine sequences and release them during degradation of polyglutamine-containing proteins. Mol Cell 14(1):95–104

247. Diaz-Hernandez M, Valera AG, Moran MA, Gomez-Ramos P, Alvarez-Castelao B, Castano JG, Hernandez F, Lucas JJ (2006) Inhibition of 26S proteasome activity by huntingtin filaments but not inclusion bodies isolated from mouse and human brain. J Neurochem 98(5):1585–1596
248. Orr AL, Li S, Wang CE, Li H, Wang J, Rong J, Xu X, Mastroberardino PG, Greenamyre JT, Li XJ (2008) N-terminal mutant huntingtin associates with mitochondria and impairs mitochondrial trafficking. J Neurosci 28(11):2783–2792
249. Hipp MS, Patel CN, Bersuker K, Riley BE, Kaiser SE, Shaler TA, Brandeis M, Kopito RR (2012) Indirect inhibition of 26S proteasome activity in a cellular model of Huntington's disease. J Cell Biol 196(5):573–587
250. Juenemann K, Schipper-Krom S, Wiemhoefer A, Kloss A, Sanz Sanz A, Reits EA (2013) Expanded polyglutamine-containing N-terminal huntingtin fragments are entirely degraded by mammalian proteasomes. J Biol Chem 288(38):27068–27084
251. Bett JS, Goellner GM, Woodman B, Pratt G, Rechsteiner M, Bates GP (2006) Proteasome impairment does not contribute to pathogenesis in R6/2 Huntington's disease mice: exclusion of proteasome activator REGgamma as a therapeutic target. Hum Mol Genet 15(1):33–44
252. Maynard CJ, Bottcher C, Ortega Z, Smith R, Florea BI, Diaz-Hernandez M, Brundin P, Overkleeft HS, Li JY, Lucas JJ, Dantuma NP (2009) Accumulation of ubiquitin conjugates in a polyglutamine disease model occurs without global ubiquitin/proteasome system impairment. Proc Natl Acad Sci U S A 106(33):13986–13991
253. Schipper-Krom S, Juenemann K, Jansen AH, Wiemhoefer A, van den Nieuwendijk R, Smith DL, Hink MA, Bates GP, Overkleeft H, Ovaa H, Reits E (2014) Dynamic recruitment of active proteasomes into polyglutamine initiated inclusion bodies. FEBS Lett 588(1):151–159
254. Bruijn LI, Becher MW, Lee MK, Anderson KL, Jenkins NA, Copeland NG, Sisodia SS, Rothstein JD, Borchelt DR, Price DL, Cleveland DW (1997) ALS-linked SOD1 mutant G85R mediates damage to astrocytes and promotes rapidly progressive disease with SOD1-containing inclusions. Neuron 18(2):327–338
255. Leigh PN, Whitwell H, Garofalo O, Buller J, Swash M, Martin JE, Gallo JM, Weller RO, Anderton BH (1991) Ubiquitin-immunoreactive intraneuronal inclusions in amyotrophic lateral sclerosis. Morphology, distribution, and specificity. Brain 114(Pt 2):775–788
256. Matsumoto S, Goto S, Kusaka H, Imai T, Murakami N, Hashizume Y, Okazaki H, Hirano A (1993) Ubiquitin-positive inclusion in anterior horn cells in subgroups of motor neuron diseases: a comparative study of adult-onset amyotrophic lateral sclerosis, juvenile amyotrophic lateral sclerosis and Werdnig-Hoffmann disease. J Neurol Sci 115(2):208–213
257. Mendonca DM, Chimelli L, Martinez AM (2006) Expression of ubiquitin and proteasome in motorneurons and astrocytes of spinal cords from patients with amyotrophic lateral sclerosis. Neurosci Lett 404(3):315–319
258. Giordana MT, Piccinini M, Grifoni S, De Marco G, Vercellino M, Magistrello M, Pellerino A, Buccinna B, Lupino E, Rinaudo MT (2010) TDP-43 redistribution is an early event in sporadic amyotrophic lateral sclerosis. Brain Pathol 20(2):351–360
259. Guo Y, Li C, Wu D, Wu S, Yang C, Liu Y, Wu H, Li Z (2010) Ultrastructural diversity of inclusions and aggregations in the lumbar spinal cord of SOD1-G93A transgenic mice. Brain Res 1353:234–244
260. Tashiro Y, Urushitani M, Inoue H, Koike M, Uchiyama Y, Komatsu M, Tanaka K, Yamazaki M, Abe M, Misawa H, Sakimura K, Ito H, Takahashi R (2012) Motor neuron-specific disruption of proteasomes, but not autophagy, replicates amyotrophic lateral sclerosis. J Biol Chem 287(51):42984–42994
261. Tsuji S, Kikuchi S, Shinpo K, Tashiro J, Kishimoto R, Yabe I, Yamagishi S, Takeuchi M, Sasaki H (2005) Proteasome inhibition induces selective motor neuron death in organotypic slice cultures. J Neurosci Res 82(4):443–451
262. Urushitani M, Kurisu J, Tsukita K, Takahashi R (2002) Proteasomal inhibition by misfolded mutant superoxide dismutase 1 induces selective motor neuron death in familial amyotrophic lateral sclerosis. J Neurochem 83(5):1030–1042
263. Basso M, Massignan T, Samengo G, Cheroni C, De Biasi S, Salmona M, Bendotti C, Bonetto V (2006) Insoluble mutant SOD1 is partly oligoubiquitinated in amyotrophic lateral sclerosis mice. J Biol Chem 281(44):33325–33335
264. Cheroni C, Peviani M, Cascio P, Debiasi S, Monti C, Bendotti C (2005) Accumulation of human SOD1 and ubiquitinated deposits in the spinal cord of SOD1G93A mice during motor neuron disease progression correlates with a decrease of proteasome. Neurobiol Dis 18(3):509–522
265. Dangond F, Hwang D, Camelo S, Pasinelli P, Frosch MP, Stephanopoulos G, Stephanopoulos G, Brown RH Jr, Gullans SR

(2004) Molecular signature of late-stage human ALS revealed by expression profiling of postmortem spinal cord gray matter. Physiol Genomics 16(2):229–239

266. Marino M, Papa S, Crippa V, Nardo G, Peviani M, Cheroni C, Trolese MC, Lauranzano E, Bonetto V, Poletti A, DeBiasi S, Ferraiuolo L, Shaw PJ, Bendotti C (2014) Differences in protein quality control correlate with phenotype variability in 2 mouse models of familial amyotrophic lateral sclerosis. Neurobiol Aging

267. Kabashi E, Agar JN, Strong MJ, Durham HD (2012) Impaired proteasome function in sporadic amyotrophic lateral sclerosis. Amyotroph Lateral Scler 13(4):367–371

268. Cheroni C, Marino M, Tortarolo M, Veglianese P, De Biasi S, Fontana E, Zuccarello LV, Maynard CJ, Dantuma NP, Bendotti C (2009) Functional alterations of the ubiquitin-proteasome system in motor neurons of a mouse model of familial amyotrophic lateral sclerosis. Hum Mol Genet 18(1):82–96

269. Bendotti C, Marino M, Cheroni C, Fontana E, Crippa V, Poletti A, De Biasi S (2012) Dysfunction of constitutive and inducible ubiquitin-proteasome system in amyotrophic lateral sclerosis: implication for protein aggregation and immune response. Prog Neurobiol 97(2):101–126

270. Hetz C, Mollereau B (2014) Disturbance of endoplasmic reticulum proteostasis in neurodegenerative diseases. Nat Rev Neurosci 15(4):233–249

271. Hetz C, Chevet E, Harding HP (2013) Targeting the unfolded protein response in disease. Nat Rev Drug Discov 12(9):703–719

272. Moreno JA, Radford H, Peretti D, Steinert JR, Verity N, Martin MG, Halliday M, Morgan J, Dinsdale D, Ortori CA, Barrett DA, Tsaytler P, Bertolotti A, Willis AE, Bushell M, Mallucci GR (2012) Sustained translational repression by eIF2alpha-P mediates prion neurodegeneration. Nature 485(7399):507–511

273. Rane NS, Kang SW, Chakrabarti O, Feigenbaum L, Hegde RS (2008) Reduced translocation of nascent prion protein during ER stress contributes to neurodegeneration. Dev Cell 15(3):359–370

274. Kristiansen M, Deriziotis P, Dimcheff DE, Jackson GS, Ovaa H, Naumann H, Clarke AR, van Leeuwen FW, Menendez-Benito V, Dantuma NP, Portis JL, Collinge J, Tabrizi SJ (2007) Disease-associated prion protein oligomers inhibit the 26S proteasome. Mol Cell 26(2):175–188

275. Deriziotis P, Andre R, Smith DM, Goold R, Kinghorn KJ, Kristiansen M, Nathan JA, Rosenzweig R, Krutauz D, Glickman MH, Collinge J, Goldberg AL, Tabrizi SJ (2011) Misfolded PrP impairs the UPS by interaction with the 20S proteasome and inhibition of substrate entry. EMBO J 30(15):3065–3077

276. Schmidt M, Haas W, Crosas B, Santamaria PG, Gygi SP, Walz T, Finley D (2005) The HEAT repeat protein Blm10 regulates the yeast proteasome by capping the core particle. Nat Struct Mol Biol 12(4):294–303

277. Dange T, Smith D, Noy T, Rommel PC, Jurzitza L, Cordero RJ, Legendre A, Finley D, Goldberg AL, Schmidt M (2011) Blm10 protein promotes proteasomal substrate turnover by an active gating mechanism. J Biol Chem 286(50):42830–42839

278. Ju D, Wang X, Xu H, Xie Y (2008) Genome-wide analysis identifies MYND-domain protein Mub1 as an essential factor for Rpn4 ubiquitylation. Mol Cell Biol 28(4):1404–1412

279. Wang L, Mao X, Ju D, Xie Y (2004) Rpn4 is a physiological substrate of the Ubr2 ubiquitin ligase. J Biol Chem 279(53):55218–55223

280. Kruegel U, Robison B, Dange T, Kahlert G, Delaney JR, Kotireddy S, Tsuchiya M, Tsuchiyama S, Murakami CJ, Schleit J, Sutphin G, Carr D, Tar K, Dittmar G, Kaeberlein M, Kennedy BK, Schmidt M (2011) Elevated proteasome capacity extends replicative lifespan in Saccharomyces cerevisiae. PLoS Genet 7(9), e1002253

281. Hanna J, Meides A, Zhang DP, Finley D (2007) A ubiquitin stress response induces altered proteasome composition. Cell 129(4):747–759

282. Peth A, Nathan JA, Goldberg AL (2013) The ATP costs and time required to degrade ubiquitinated proteins by the 26S proteasome. J Biol Chem 288(40):29215–29222

283. Oling D, Eisele F, Kvint K, Nystrom T (2014) Opposing roles of Ubp3-dependent deubiquitination regulate replicative life span and heat resistance. EMBO J 33(7):747–761

284. Andersson V, Hanzen S, Liu B, Molin M, Nystrom T (2013) Enhancing protein disaggregation restores proteasome activity in aged cells. Aging 5(11):802–812

285. Dal Vechio FH, Cerqueira F, Augusto O, Lopes R, Demasi M (2013) Peptides that activate the 20S proteasome by gate opening increased oxidized protein removal and reduced protein aggregation. Free Radic Biol Med 67C:304–313

286. Barros MH, Bandy B, Tahara EB, Kowaltowski AJ (2004) Higher respiratory activity decreases mitochondrial reactive oxygen release and increases life span in Saccharomyces cerevisiae. J Biol Chem 279(48):49883–49888
287. Fabrizio P, Gattazzo C, Battistella L, Wei M, Cheng C, McGrew K, Longo VD (2005) Sir2 blocks extreme life-span extension. Cell 123(4):655–667
288. da Cunha FM, Demasi M, Kowaltowski AJ (2011) Aging and calorie restriction modulate yeast redox state, oxidized protein removal, and the ubiquitin-proteasome system. Free Radic Biol Med 51(3):664–670
289. Hanssum A, Zhong Z, Rousseau A, Krzyzosiak A, Sigurdardottir A, Bertolotti A (2014) An inducible chaperone adapts proteasome assembly to stress. Mol Cell 55(4):566–577
290. Yao Y, Tsuchiyama S, Yang C, Bulteau AL, He C, Robison B, Tsuchiya M, Miller D, Briones V, Tar K, Potrero A, Friguet B, Kennedy BK, Schmidt M (2015) Proteasomes, Sir2, and Hxk2 form an interconnected aging network that impinges on the AMPK/Snf1-regulated transcriptional repressor Mig1. PLoS Genet 11(1), e1004968
291. Vilchez D, Morantte I, Liu Z, Douglas PM, Merkwirth C, Rodrigues AP, Manning G, Dillin A (2012) RPN-6 determines C. elegans longevity under proteotoxic stress conditions. Nature 489(7415):263–268
292. Vilchez D, Boyer L, Morantte I, Lutz M, Merkwirth C, Joyce D, Spencer B, Page L, Masliah E, Berggren WT, Gage FH, Dillin A (2012) Increased proteasome activity in human embryonic stem cells is regulated by PSMD11. Nature 489(7415):304–308
293. Chondrogianni N, Tzavelas C, Pemberton AJ, Nezis IP, Rivett AJ, Gonos ES (2005) Overexpression of proteasome beta5 assembled subunit increases the amount of proteasome and confers ameliorated response to oxidative stress and higher survival rates. J Biol Chem 280(12):11840–11850
294. Ferguson AA, Springer MG, Fisher AL (2010) Skn-1-dependent and -independent regulation of aip-1 expression following metabolic stress in Caenorhabditis elegans. Mol Cell Biol 30(11):2651–2667
295. Hassan WM, Merin DA, Fonte V, Link CD (2009) AIP-1 ameliorates beta-amyloid peptide toxicity in a Caenorhabditis elegans Alzheimer's disease model. Hum Mol Genet 18(15):2739–2747
296. Stanhill A, Haynes CM, Zhang Y, Min G, Steele MC, Kalinina J, Martinez E, Pickart CM, Kong XP, Ron D (2006) An arsenite-inducible 19S regulatory particle-associated protein adapts proteasomes to proteotoxicity. Mol Cell 23(6):875–885
297. Kenyon C (2005) The plasticity of aging: insights from long-lived mutants. Cell 120(4):449–460
298. Lin K, Dorman JB, Rodan A, Kenyon C (1997) daf- 16: an HNF-3/forkhead family member that can function to double the life span of Caenorhabditis elegans. Science 278:1319–1322
299. Stout GJ, Stigter EC, Essers PB, Mulder KW, Kolkman A, Snijders DS, van den Broek NJ, Betist MC, Korswagen HC, Macinnes AW, Brenkman AB (2013) Insulin/IGF-1-mediated longevity is marked by reduced protein metabolism. Mol Syst Biol 9:679
300. Matilainen O, Arpalahti L, Rantanen V, Hautaniemi S, Holmberg CI (2013) Insulin/IGF-1 signaling regulates proteasome activity through the deubiquitinating enzyme UBH-4. Cell Rep 3(6):1980–1995
301. Liu G, Rogers J, Murphy CT, Rongo C (2011) EGF signalling activates the ubiquitin proteasome system to modulate C. elegans lifespan. EMBO J 30(15):2990–3003
302. Pickering AM, Staab TA, Tower J, Sieburth D, Davies KJ (2013) A conserved role for the 20S proteasome and Nrf2 transcription factor in oxidative stress adaptation in mammals, Caenorhabditis elegans and Drosophila melanogaster. J Exp Biol 216(Pt 4):543–553
303. Przybysz AJ, Choe KP, Roberts LJ, Strange K (2009) Increased age reduces DAF-16 and SKN-1 signaling and the hormetic response of Caenorhabditis elegans to the xenobiotic juglone. Mech Ageing Dev 130(6):357–369
304. Li X, Matilainen O, Jin C, Glover-Cutter KM, Holmberg CI, Blackwell TK (2011) Specific SKN-1/Nrf stress responses to perturbations in translation elongation and proteasome activity. PLoS Genet 7(6), e1002119
305. Leung CK, Hasegawa K, Wang Y, Deonarine A, Tang L, Miwa J, Choe KP (2014) Direct interaction between the WD40 repeat protein WDR-23 and SKN-1/Nrf inhibits binding to target DNA. Mol Cell Biol 34(16):3156–3167
306. Depuydt G, Xie F, Petyuk VA, Shanmugam N, Smolders A, Dhondt I, Brewer HM, Camp DG, Smith RD, Braeckman BP (2013) Reduced insulin/insulin-like growth factor-1 signaling and dietary restriction inhibit translation but preserve muscle mass in Caenorhabditis elegans. Mol Cell Proteomics 12(12):3624–3639

307. Carrano AC, Liu Z, Dillin A, Hunter T (2009) A conserved ubiquitination pathway determines longevity in response to diet restriction. Nature 460(7253):396–399
308. Cao X, Xue L, Han L, Ma L, Chen T, Tong T (2011) WW domain-containing E3 ubiquitin protein ligase 1 (WWP1) delays cellular senescence by promoting p27(Kip1) degradation in human diploid fibroblasts. J Biol Chem 286(38):33447–33456
309. Koulich E, Li X, DeMartino GN (2008) Relative structural and functional roles of multiple deubiquitylating proteins associated with mammalian 26S proteasome. Mol Biol Cell 19(3):1072–1082
310. Cypser JR, Johnson TE (2002) Multiple stressors in Caenorhabditis elegans induce stress hormesis and extended longevity. J Gerontol A Biol Sci Med Sci 57(3):B109–B114
311. Lithgow GJ, White TM, Melov S, Johnson TE (1995) Thermotolerance and extended life-span conferred by single-gene mutations and induced by thermal stress. Proc Natl Acad Sci U S A 92(16):7540–7544
312. Ermolaeva MA, Segref A, Dakhovnik A, Ou HL, Schneider JI, Utermohlen O, Hoppe T, Schumacher B (2013) DNA damage in germ cells induces an innate immune response that triggers systemic stress resistance. Nature 501(7467):416–420
313. Vartiainen S, Pehkonen P, Lakso M, Nass R, Wong G (2006) Identification of gene expression changes in transgenic C. elegans overexpressing human alpha-synuclein. Neurobiol Dis 22(3):477–486
314. Chondrogianni N, Gonos ES (2012) Structure and function of the ubiquitin-proteasome system: modulation of components. Prog Mol Biol Transl Sci 109:41–74
315. Regitz C, Marie Dussling L, Wenzel U (2014) Amyloid-beta (Abeta1-42)-induced paralysis in Caenorhabditis elegans is inhibited by the polyphenol quercetin through activation of protein degradation pathways. Mol Nutr Food Res 58(10):1931–1940
316. Kampkotter A, Timpel C, Zurawski RF, Ruhl S, Chovolou Y, Proksch P, Watjen W (2008) Increase of stress resistance and lifespan of Caenorhabditis elegans by quercetin. Comp Biochem Physiol B Biochem Mol Biol 149(2):314–323
317. Fitzenberger E, Deusing DJ, Wittkop A, Kler A, Kriesl E, Bonnlander B, Wenzel U (2014) Effects of plant extracts on the reversal of glucose-induced impairment of stress-resistance in Caenorhabditis elegans. Plant Foods Hum Nutr 69(1):78–84
318. Fitzenberger E, Deusing DJ, Marx C, Boll M, Luersen K, Wenzel U (2014) The polyphenol quercetin protects the mev-1 mutant of Caenorhabditis elegans from glucose-induced reduction of survival under heat-stress depending on SIR-2.1, DAF-12, and proteasomal activity. Mol Nutr Food Res 58(5):984–994
319. Deusing DJ, Winter S, Kler A, Kriesl E, Bonnlander B, Wenzel U, Fitzenberger E (2015) A catechin-enriched green tea extract prevents glucose-induced survival reduction in Caenorhabditis elegans through sir-2.1 and uba-1 dependent hormesis. Fitoterapia 102:163–170
320. Fu RH, Wang YC, Chen CS, Tsai RT, Liu SP, Chang WL, Lin HL, Lu CH, Lu CH, Wei JR, Wang ZW, Shyu WC, Lin SZ (2014) Acetylcorynoline attenuates dopaminergic neuron degeneration and alpha-synuclein aggregation in animal models of Parkinson's disease. Neuropharmacology 82:108–120
321. Papaevgeniou N, Sakellari M, Jha S, Tavernarakis N, Holmberg CI, Gonos ES, Chondrogianni N (2016) 18α-Glycyrrhetinic acid proteasome activator decelerates aging and alzheimer's disease progression in C. elegans and Neuronal cultures. *Antioxid Redox Signal in press*
322. Burkewitz K, Choe KP, Lee EC, Deonarine A, Strange K (2012) Characterization of the proteostasis roles of glycerol accumulation, protein degradation and protein synthesis during osmotic stress in C. elegans. PLoS One 7(3), e34153
323. Tonoki A, Kuranaga E, Tomioka T, Hamazaki J, Murata S, Tanaka K, Miura M (2009) Genetic evidence linking age-dependent attenuation of the 26S proteasome with the aging process. Mol Cell Biol 29(4):1095–1106
324. Liu HY, Pfleger CM (2013) Mutation in E1, the ubiquitin activating enzyme, reduces Drosophila lifespan and results in motor impairment. PLoS One 8(1), e32835
325. Shimura H, Hattori N, Kubo S, Mizuno Y, Asakawa S, Minoshima S, Shimizu N, Iwai K, Chiba T, Tanaka K, Suzuki T (2000) Familial Parkinson disease gene product, parkin, is a ubiquitin-protein ligase. Nat Genet 25(3):302–305
326. Kitada T, Asakawa S, Hattori N, Matsumine H, Yamamura Y, Minoshima S, Yokochi M, Mizuno Y, Shimizu N (1998) Mutations in the parkin gene cause autosomal recessive juvenile parkinsonism. Nature 392(6676): 605–608
327. Hyun DH, Lee M, Hattori N, Kubo S, Mizuno Y, Halliwell B, Jenner P (2002)

Effect of wild-type or mutant Parkin on oxidative damage, nitric oxide, antioxidant defenses, and the proteasome. J Biol Chem 277(32):28572–28577

328. Rana A, Rera M, Walker DW (2013) Parkin overexpression during aging reduces proteotoxicity, alters mitochondrial dynamics, and extends lifespan. Proc Natl Acad Sci U S A 110(21):8638–8643

329. Wang CH, Chen GC, Chien CT (2014) The deubiquitinase Leon/USP5 regulates ubiquitin homeostasis during Drosophila development. Biochem Biophys Res Commun 452(3):369–375

330. Engel E, Viargues P, Mortier M, Taillebourg E, Coute Y, Thevenon D, Fauvarque MO (2014) Identifying USPs regulating immune signals in Drosophila: USP2 deubiquitinates Imd and promotes its degradation by interacting with the proteasome. Cell Commun Signal 12:41

331. Chu-Ping M, Slaughter CA, DeMartino GN (1992) Purification and characterization of a protein inhibitor of the 20S proteasome (macropain). Biochim Biophys Acta 1119(3):303–311

332. McCutchen-Maloney SL, Matsuda K, Shimbara N, Binns DD, Tanaka K, Slaughter CA, DeMartino GN (2000) cDNA cloning, expression, and functional characterization of PI31, a proline-rich inhibitor of the proteasome. J Biol Chem 275(24):18557–18565

333. Bader M, Benjamin S, Wapinski OL, Smith DM, Goldberg AL, Steller H (2011) A conserved F box regulatory complex controls proteasome activity in Drosophila. Cell 145(3):371–382

334. Grimberg KB, Beskow A, Lundin D, Davis MM, Young P (2011) Basic leucine zipper protein Cnc-C is a substrate and transcriptional regulator of the Drosophila 26S proteasome. Mol Cell Biol 31(4):897–909

335. Tsakiri EN, Sykiotis GP, Papassideri IS, Terpos E, Dimopoulos MA, Gorgoulis VG, Bohmann D, Trougakos IP (2013) Proteasome dysfunction in Drosophila signals to an Nrf2-dependent regulatory circuit aiming to restore proteostasis and prevent premature aging. Aging Cell 12(5):802–813

336. Moskalev A, Shaposhnikov M, Turysheva E (2009) Life span alteration after irradiation in Drosophila melanogaster strains with mutations of Hsf and Hsps. Biogerontology 10(1):3–11

337. Moskalev AA, Pliusnina EN, Zainullin VG (2007) The influence of low doze gamma-irradiation on life span of Drosophila mutants with defects of DNA damage sensation and repair. Radiats Biol Radioecol 47(5):571–573

338. Li J, Horak KM, Su H, Sanbe A, Robbins J, Wang X (2011) Enhancement of proteasomal function protects against cardiac proteinopathy and ischemia/reperfusion injury in mice. J Clin Invest 121(9):3689–3700

339. Rodriguez KA, Osmulski PA, Pierce A, Weintraub ST, Gaczynska M, Buffenstein R (2014) A cytosolic protein factor from the naked mole-rat activates proteasomes of other species and protects these from inhibition. Biochim Biophys Acta 1842(11):2060–2072

340. Pride H, Yu Z, Sunchu B, Mochnick J, Coles A, Zhang Y, Buffenstein R, Hornsby PJ, Austad SN, Perez VI (2015) Long-lived species have improved proteostasis compared to phylogenetically-related shorter-lived species. Biochem Biophys Res Commun 457(4): 669–675

341. Crowe E, Sell C, Thomas JD, Johannes GJ, Torres C (2009) Activation of proteasome by insulin-like growth factor-I may enhance clearance of oxidized proteins in the brain. Mech Ageing Dev 130(11-12):793–800

342. Goto S, Takahashi R, Araki S, Nakamoto H (2002) Dietary restriction initiated in late adulthood can reverse age-related alterations of protein and protein metabolism. Ann N Y Acad Sci 959:50–56

343. Zhang L, Li F, Dimayuga E, Craddock J, Keller JN (2007) Effects of aging and dietary restriction on ubiquitination, sumoylation, and the proteasome in the spleen. FEBS Lett 581(28):5543–5547

344. Selsby JT, Judge AR, Yimlamai T, Leeuwenburgh C, Dodd SL (2005) Life long calorie restriction increases heat shock proteins and proteasome activity in soleus muscles of Fisher 344 rats. Exp Gerontol 40(1-2):37–42

345. Bonelli MA, Desenzani S, Cavallini G, Donati A, Romani AA, Bergamini E, Borghetti AF (2008) Low-level caloric restriction rescues proteasome activity and Hsc70 level in liver of aged rats. Biogerontology 9(1):1–10

346. Shavlakadze T, Soffe Z, Anwari T, Cozens G, Grounds MD (2013) Short-term feed deprivation rapidly induces the protein degradation pathway in skeletal muscles of young mice. J Nutr 143(4):403–409

347. O'Neal P, Alamdari N, Smith I, Poylin V, Menconi M, Hasselgren PO (2009) Experimental hyperthyroidism in rats increases the expression of the ubiquitin ligases atrogin-1 and MuRF1 and stimulates multiple proteolytic pathways in skeletal muscle. J Cell Biochem 108(4):963–973

348. Lee CK, Klopp RG, Weindruch R, Prolla TA (1999) Gene expression profile of aging and

its retardation by caloric restriction. Science 285(5432):1390–1393

349. Kwak MK, Itoh K, Yamamoto M, Sutter TR, Kensler TW (2001) Role of transcription factor Nrf2 in the induction of hepatic phase 2 and antioxidative enzymes in vivo by the cancer chemoprotective agent, 3H-1, 2-dimethiole-3-thione. Mol Med 7(2):135–145
350. Kwak MK, Huang B, Chang H, Kim JA, Kensler TW (2007) Tissue specific increase of the catalytic subunits of the 26S proteasome by indirect antioxidant dithiolethione in mice: enhanced activity for degradation of abnormal protein. Life Sci 80(26):2411–2420
351. Kwak MK, Wakabayashi N, Itoh K, Motohashi H, Yamamoto M, Kensler TW (2003) Modulation of gene expression by cancer chemopreventive dithiolethiones through the Keap1-Nrf2 pathway. Identification of novel gene clusters for cell survival. J Biol Chem 278(10):8135–8145
352. Lorite MJ, Smith HJ, Arnold JA, Morris A, Thompson MG, Tisdale MJ (2001) Activation of ATP-ubiquitin-dependent proteolysis in skeletal muscle in vivo and murine myoblasts in vitro by a proteolysis-inducing factor (PIF). Br J Cancer 85(2):297–302
353. Hwang JS, Hwang JS, Chang I, Kim S (2007) Age-associated decrease in proteasome content and activities in human dermal fibroblasts: restoration of normal level of proteasome subunits reduces aging markers in fibroblasts from elderly persons. J Gerontol A Biol Sci Med Sci 62(5):490–499
354. Liu Y, Liu X, Zhang T, Luna C, Liton PB, Gonzalez P (2007) Cytoprotective effects of proteasome beta5 subunit overexpression in lens epithelial cells. Mol Vis 13:31–38
355. Lu L, Song HF, Wei JL, Liu XQ, Song WH, Yan BY, Yang GJ, Li A, Yang WL (2014) Ameliorating replicative senescence of human bone marrow stromal cells by PSMB5 overexpression. Biochem Biophys Res Commun 443(4):1182–1188
356. Malhotra D, Thimmulappa R, Vij N, Navas-Acien A, Sussan T, Merali S, Zhang L, Kelsen SG, Myers A, Wise R, Tuder R, Biswal S (2009) Heightened endoplasmic reticulum stress in the lungs of patients with chronic obstructive pulmonary disease: the role of Nrf2-regulated proteasomal activity. Am J Respir Crit Care Med 180(12):1196–1207
357. Gaczynska M, Rock KL, Spies T, Goldberg AL (1994) Peptidase activities of proteasomes are differentially regulated by the major histocompatibility complex-encoded genes for LMP2 and LMP7. Proc Natl Acad Sci U S A 91(20):9213–9217
358. Gaczynska M, Goldberg AL, Tanaka K, Hendil KB, Rock KL (1996) Proteasome subunits X and Y alter peptidase activities in opposite ways to the interferon-gamma-induced subunits LMP2 and LMP7. J Biol Chem 271(29):17275–17280
359. Sok J, Calfon M, Lu J, Lichtlen P, Clark SG, Ron D (2001) Arsenite-inducible RNA-associated protein (AIRAP) protects cells from arsenite toxicity. Cell Stress Chaperones 6(1):6–15
360. Chondrogianni N, Gonos ES (2007) Overexpression of hUMP1/POMP proteasome accessory protein enhances proteasome-mediated antioxidant defence. Exp Gerontol 42(9):899–903
361. Min JN, Whaley RA, Sharpless NE, Lockyer P, Portbury AL, Patterson C (2008) CHIP deficiency decreases longevity, with accelerated aging phenotypes accompanied by altered protein quality control. Mol Cell Biol 28(12):4018–4025
362. Ronnebaum SM, Wu Y, McDonough H, Patterson C (2013) The ubiquitin ligase CHIP prevents SirT6 degradation through noncanonical ubiquitination. Mol Cell Biol 33(22):4461–4472
363. Lee BH, Lee MJ, Park S, Oh DC, Elsasser S, Chen PC, Gartner C, Dimova N, Hanna J, Gygi SP, Wilson SM, King RW, Finley D (2010) Enhancement of proteasome activity by a small-molecule inhibitor of USP14. Nature 467(7312):179–184
364. Katsiki M, Chondrogianni N, Chinou I, Rivett AJ, Gonos ES (2007) The olive constituent oleuropein exhibits proteasome stimulatory properties in vitro and confers life span extension of human embryonic fibroblasts. Rejuvenation Res 10(2):157–172
365. Graikou K, Kapeta S, Aligiannis N, Sotiroudis G, Chondrogianni N, Gonos E, Chinou I (2011) Chemical analysis of Greek pollen – antioxidant, antimicrobial and proteasome activation properties. Chem Cent J 5(1):33
366. Ali RE, Rattan SI (2006) Curcumin's biphasic hormetic response on proteasome activity and heat-shock protein synthesis in human keratinocytes. Ann N Y Acad Sci 1067:394–399
367. Dal Vechio FH, Cerqueira F, Augusto O, Lopes R, Demasi M (2014) Peptides that activate the 20S proteasome by gate opening increased oxidized protein removal and reduced protein aggregation. Free Radic Biol Med 67:304–313

368. Gan N, Wu YC, Brunet M, Garrido C, Chung FL, Dai C, Mi L (2010) Sulforaphane activates heat shock response and enhances proteasome activity through up-regulation of Hsp27. J Biol Chem 285(46):35528–35536
369. Kapeta S, Chondrogianni N, Gonos ES (2010) Nuclear erythroid factor 2-mediated proteasome activation delays senescence in human fibroblasts. J Biol Chem 285(11):8171–8184
370. Chondrogianni N, Kapeta S, Chinou I, Vassilatou K, Papassideri I, Gonos ES (2010) Anti-ageing and rejuvenating effects of quercetin. Exp Gerontol 45(10):763–771
371. Tanigawa S, Fujii M, Hou DX (2007) Action of Nrf2 and Keap1 in ARE-mediated NQO1 expression by quercetin. Free Radic Biol Med 42(11):1690–1703
372. Ohnishi K, Nakahata E, Irie K, Murakami A (2013) Zerumbone, an electrophilic sesquiterpene, induces cellular proteo-stress leading to activation of ubiquitin-proteasome system and autophagy. Biochem Biophys Res Commun 430(2):616–622
373. Jang J, Wang Y, Kim HS, Lalli MA, Kosik KS (2014) Nrf2, a regulator of the proteasome, controls self-renewal and pluripotency in human embryonic stem cells. Stem Cells 32(10):2616–2625
374. Takimoto E, Champion HC, Li M, Belardi D, Ren S, Rodriguez ER, Bedja D, Gabrielson KL, Wang Y, Kass DA (2005) Chronic inhibition of cyclic GMP phosphodiesterase 5A prevents and reverses cardiac hypertrophy. Nat Med 11(2):214–222
375. Ranek MJ, Terpstra EJ, Li J, Kass DA, Wang X (2013) Protein kinase g positively regulates proteasome-mediated degradation of misfolded proteins. Circulation 128(4):365–376
376. Huang Q, Wang H, Perry SW, Figueiredo-Pereira ME (2013) Negative regulation of 26S proteasome stability via calpain-mediated cleavage of Rpn10 subunit upon mitochondrial dysfunction in neurons. J Biol Chem 288(17):12161–12174
377. Whitehouse AS, Tisdale MJ (2003) Increased expression of the ubiquitin-proteasome pathway in murine myotubes by proteolysis-inducing factor (PIF) is associated with activation of the transcription factor NF-kappaB. Br J Cancer 89(6):1116–1122
378. Cabreiro F, Perichon M, Jatje J, Malavolta M, Mocchegiani E, Friguet B, Petropoulos I (2008) Zinc supplementation in the elderly subjects: effect on oxidized protein degradation and repair systems in peripheral blood lymphocytes. Exp Gerontol 43(5):483–487
379. Speese SD, Trotta N, Rodesch CK, Aravamudan B, Broadie K (2003) The ubiquitin proteasome system acutely regulates presynaptic protein turnover and synaptic efficacy. Curr Biol 13(11):899–910
380. Link CD, Taft A, Kapulkin V, Duke K, Kim S, Fei Q, Wood DE, Sahagan BG (2003) Gene expression analysis in a transgenic Caenorhabditis elegans Alzheimer's disease model. Neurobiol Aging 24(3):397–413
381. Kaneko M, Koike H, Saito R, Kitamura Y, Okuma Y, Nomura Y (2010) Loss of HRD1-mediated protein degradation causes amyloid precursor protein accumulation and amyloid-beta generation. J Neurosci 30(11):3924–3932
382. Gong B, Chen F, Pan Y, Arrieta-Cruz I, Yoshida Y, Haroutunian V, Pasinetti GM (2010) SCFFbx2-E3-ligase-mediated degradation of BACE1 attenuates Alzheimer's disease amyloidosis and improves synaptic function. Aging Cell 9(6):1018–1031
383. Singh AK, Pati U (2015) CHIP stabilizes amyloid precursor protein via proteasomal degradation and p53-mediated trans-repression of beta-secretase. Aging Cell
384. Mehta R, Steinkraus KA, Sutphin GL, Ramos FJ, Shamieh LS, Huh A, Davis C, Chandler-Brown D, Kaeberlein M (2009) Proteasomal regulation of the hypoxic response modulates aging in C elegans. Science 324(5931):1196–1198
385. Valera E, Dargusch R, Maher PA, Schubert D (2013) Modulation of 5-lipoxygenase in proteotoxicity and Alzheimer's disease. J Neurosci 33(25):10512–10525
386. Himeno E, Ohyagi Y, Ma L, Nakamura N, Miyoshi K, Sakae N, Motomura K, Soejima N, Yamasaki R, Hashimoto T, Tabira T, LaFerla FM, Kira J (2011) Apomorphine treatment in Alzheimer mice promoting amyloid-beta degradation. Ann Neurol 69(2):248–256
387. Marambaud P, Zhao H, Davies P (2005) Resveratrol promotes clearance of Alzheimer's disease amyloid-beta peptides. J Biol Chem 280(45):37377–37382
388. Luchsinger JA, Tang M-X, Siddiqui M, Shea S, Mayeux R (2004) Alcohol intake and risk of dementia. J Am Geriatr Soc 52(4):540–546
389. Yogev-Falach M, Amit T, Bar-Am O, Youdim MB (2003) The importance of propargylamine moiety in the anti-Parkinson drug rasagiline and its derivatives in MAPK-dependent amyloid precursor protein processing. FASEB J 17(15):2325–2327

390. Youdim MBH, Amit T, Falach-Yogev M, Am OB, Maruyama W, Naoi M (2003) The essentiality of Bcl-2, PKC and proteasome–ubiquitin complex activations in the neuroprotective-antiapoptotic action of the anti-Parkinson drug, rasagiline. Biochem Pharmacol 66(8):1635–1641
391. Alavez S, Vantipalli MC, Zucker DJ, Klang IM, Lithgow GJ (2011) Amyloid-binding compounds maintain protein homeostasis during ageing and extend lifespan. Nature 472(7342):226–229
392. Medina DX, Caccamo A, Oddo S (2011) Methylene blue reduces abeta levels and rescues early cognitive deficit by increasing proteasome activity. Brain Pathol 21(2):140–149
393. Opattova A, Filipcik P, Cente M, Novak M (2013) Intracellular degradation of misfolded tau protein induced by geldanamycin is associated with activation of proteasome. J Alzheimers Dis 33(2):339–348
394. Chakrabortee S, Liu Y, Zhang L, Matthews HR, Zhang H, Pan N, Cheng CR, Guan SH, Guo DA, Huang Z, Zheng Y, Tunnacliffe A (2012) Macromolecular and small-molecule modulation of intracellular Abeta42 aggregation and associated toxicity. Biochem J 442(3):507–515
395. Jing P, Zhang JY, Ouyang Q, Wu J, Zhang XJ (2013) Lithium treatment induces proteasomal degradation of over-expressed acetylcholinesterase (AChE-S) and inhibit GSK3beta. Chem Biol Interact 203(1):309–313
396. Bedford L, Hay D, Devoy A, Paine S, Powe DG, Seth R, Gray T, Topham I, Fone K, Rezvani N, Mee M, Soane T, Layfield R, Sheppard PW, Ebendal T, Usoskin D, Lowe J, Mayer RJ (2008) Depletion of 26S proteasomes in mouse brain neurons causes neurodegeneration and Lewy-like inclusions resembling human pale bodies. J Neurosci 28(33):8189–8198
397. Feany MB, Bender WW (2000) A Drosophila model of Parkinson's disease. Nature 404(6776):394–398
398. Lee FK, Wong AK, Lee YW, Wan OW, Chan HY, Chung KK (2009) The role of ubiquitin linkages on alpha-synuclein induced-toxicity in a Drosophila model of Parkinson's disease. J Neurochem 110(1):208–219
399. Fan GH, Zhou HY, Yang H, Chen SD (2006) Heat shock proteins reduce alpha-synuclein aggregation induced by MPP+ in SK-N-SH cells. FEBS Lett 580(13):3091–3098
400. Shang T, Kotamraju S, Zhao H, Kalivendi SV, Hillard CJ, Kalyanaraman B (2005) Sepiapterin attenuates 1-methyl-4-phenylpyridinium-induced apoptosis in neuroblastoma cells transfected with neuronal NOS: role of tetrahydrobiopterin, nitric oxide, and proteasome activation. Free Radic Biol Med 39(8):1059–1074
401. Cheng YF, Zhu GQ, Wang M, Cheng H, Zhou A, Wang N, Fang N, Wang XC, Xiao XQ, Chen ZW, Li QL (2009) Involvement of ubiquitin proteasome system in protective mechanisms of Puerarin to MPP(+)-elicited apoptosis. Neurosci Res 63(1):52–58
402. Fu RH, Harn HJ, Liu SP, Chen CS, Chang WL, Chen YM, Huang JE, Li RJ, Tsai SY, Hung HS, Shyu WC, Lin SZ, Wang YC (2014) n-Butylidenephthalide protects against dopaminergic neuron degeneration and alpha-synuclein accumulation in Caenorhabditis elegans models of Parkinson's disease. PLoS One 9(1), e85305
403. Rabey JM, Sagi I, Huberman M, Melamed E, Korczyn A, Giladi N, Inzelberg R, Djaldetti R, Klein C, Berecz G, Rasagiline Study G (2000) Rasagiline mesylate, a new MAO-B inhibitor for the treatment of Parkinson's disease: a double-blind study as adjunctive therapy to levodopa. Clin Neuropharmacol 23(6):324–330
404. Naoi M, Maruyama W, Yi H, Akao Y, Yamaoka Y, Shamoto-Nagai M (2007) Neuroprotection by propargylamines in Parkinson's disease: intracellular mechanism underlying the anti-apoptotic function and search for clinical markers. J Neural Transm Suppl 72:121–131
405. McNaught KS, Perl DP, Brownell AL, Olanow CW (2004) Systemic exposure to proteasome inhibitors causes a progressive model of Parkinson's disease. Ann Neurol 56(1):149–162
406. Zhang Z, Li X, Xie WJ, Tuo H, Hintermann S, Jankovic J, Le W (2012) Anti-parkinsonian effects of Nurr1 activator in ubiquitin-proteasome system impairment induced animal model of Parkinson's disease. CNS Neurol Disord Drug Targets 11(6):768–773
407. Li C, Guo Y, Xie W, Li X, Janokovic J, Le W (2010) Neuroprotection of pramipexole in UPS impairment induced animal model of Parkinson's disease. Neurochem Res 35(10):1546–1556
408. Li C, Biswas S, Li X, Dutta AK, Le W (2010) Novel D3 dopamine receptor-preferring agonist D-264: evidence of neuroprotective

property in Parkinson's disease animal models induced by 1-methyl-4-phenyl-1,2,3,6-tetrahydropyridine and lactacystin. J Neurosci Res 88(11):2513–2523

409. Zhu W, Xie W, Pan T, Jankovic J, Li J, Youdim MB, Le W (2008) Comparison of neuroprotective and neurorestorative capabilities of rasagiline and selegiline against lactacystin-induced nigrostriatal dopaminergic degeneration. J Neurochem 105(5):1970–1978
410. Yong-Kee CJ, Salomonczyk D, Nash JE (2011) Development and validation of a screening assay for the evaluation of putative neuroprotective agents in the treatment of Parkinson's disease. Neurotox Res 19(4):519–526
411. Yamamoto A, Lucas JJ, Hen R (2000) Reversal of neuropathology and motor dysfunction in a conditional model of Huntington's disease. Cell 101(1):57–66
412. Seo H, Sonntag KC, Kim W, Cattaneo E, Isacson O (2007) Proteasome activator enhances survival of Huntington's disease neuronal model cells. PLoS One 2(2), e238
413. Mishra A, Dikshit P, Purkayastha S, Sharma J, Nukina N, Jana NR (2008) E6-AP promotes misfolded polyglutamine proteins for proteasomal degradation and suppresses polyglutamine protein aggregation and toxicity. J Biol Chem 283(12):7648–7656
414. Miller VM, Nelson RF, Gouvion CM, Williams A, Rodriguez-Lebron E, Harper SQ, Davidson BL, Rebagliati MR, Paulson HL (2005) CHIP suppresses polyglutamine aggregation and toxicity in vitro and in vivo. J Neurosci 25(40):9152–9161
415. Yang H, Zhong X, Ballar P, Luo S, Shen Y, Rubinsztein DC, Monteiro MJ, Fang S (2007) Ubiquitin ligase Hrd1 enhances the degradation and suppresses the toxicity of polyglutamine-expanded huntingtin. Exp Cell Res 313(3):538–550
416. Hyrskyluoto A, Bruelle C, Lundh SH, Do HT, Kivinen J, Rappou E, Reijonen S, Waltimo T, Petersen A, Lindholm D, Korhonen L (2014) Ubiquitin-specific protease-14 reduces cellular aggregates and protects against mutant huntingtin-induced cell degeneration: involvement of the proteasome and ER stress-activated kinase IRE1alpha. Hum Mol Genet 23(22):5928–5939
417. Rangone H, Pardo R, Colin E, Girault JA, Saudou F, Humbert S (2005) Phosphorylation of arfaptin 2 at Ser260 by Akt Inhibits PolyQ-huntingtin-induced toxicity by rescuing proteasome impairment. J Biol Chem 280(23):22021–22028
418. Bauer PO, Nukina N (2009) Enhanced degradation of mutant huntingtin by rho kinase inhibition is mediated through activation of proteasome and macroautophagy. Autophagy 5(5):747–748
419. Kleijnen MF, Shih AH, Zhou P, Kumar S, Soccio RE, Kedersha NL, Gill G, Howley PM (2000) The hPLIC proteins may provide a link between the ubiquitination machinery and the proteasome. Mol Cell 6(2):409–419
420. Wang H, Lim PJ, Yin C, Rieckher M, Vogel BE, Monteiro MJ (2006) Suppression of polyglutamine-induced toxicity in cell and animal models of Huntington's disease by ubiquilin. Hum Mol Genet 15(6):1025–1041
421. Safren N, El Ayadi A, Chang L, Terrillion CE, Gould TD, Boehning DF, Monteiro MJ (2014) Ubiquilin-1 overexpression increases the lifespan and delays accumulation of Huntingtin aggregates in the R6/2 mouse model of Huntington's disease. PLoS One 9(1), e87513
422. Lu B, Al-Ramahi I, Valencia A, Wang Q, Berenshteyn F, Yang H, Gallego-Flores T, Ichcho S, Lacoste A, Hild M, Difiglia M, Botas J, Palacino J (2013) Identification of NUB1 as a suppressor of mutant Huntington toxicity via enhanced protein clearance. Nat Neurosci 16(5):562–570
423. Chiang MC, Chen HM, Lai HL, Chen HW, Chou SY, Chen CM, Tsai FJ, Chern Y (2009) The A2A adenosine receptor rescues the urea cycle deficiency of Huntington's disease by enhancing the activity of the ubiquitin-proteasome system. Hum Mol Genet 18(16):2929–2942
424. Wong HK, Bauer PO, Kurosawa M, Goswami A, Washizu C, Machida Y, Tosaki A, Yamada M, Knopfel T, Nakamura T, Nukina N (2008) Blocking acid-sensing ion channel 1 alleviates Huntington's disease pathology via an ubiquitin-proteasome system-dependent mechanism. Hum Mol Genet 17(20):3223–3235
425. Kim W, Seo H (2014) Baclofen, a GABAB receptor agonist, enhances ubiquitin-proteasome system functioning and neuronal survival in Huntington's disease model mice. Biochem Biophys Res Commun 443(2):706–711
426. Lai AY, Lan CP, Hasan S, Brown ME, McLaurin J (2014) Scyllo-inositol promotes robust mutant Huntingtin protein degradation. J Biol Chem 289(6):3666–3676
427. Liu Y, Hettinger CL, Zhang D, Rezvani K, Wang X, Wang H (2014) Sulforaphane enhances proteasomal and autophagic activities in mice and is a potential therapeutic

reagent for Huntington's disease. J Neurochem 129(3):539–547
428. Stark M, Behl C (2014) The Ginkgo biloba extract EGb 761 modulates proteasome activity and polyglutamine protein aggregation. Evid Based Complement Alternat Med 2014:940186
429. Niwa J, Ishigaki S, Hishikawa N, Yamamoto M, Doyu M, Murata S, Tanaka K, Taniguchi N, Sobue G (2002) Dorfin ubiquitylates mutant SOD1 and prevents mutant SOD1-mediated neurotoxicity. J Biol Chem 277(39):36793–36798
430. Sone J, Niwa J, Kawai K, Ishigaki S, Yamada S, Adachi H, Katsuno M, Tanaka F, Doyu M, Sobue G (2010) Dorfin ameliorates phenotypes in a transgenic mouse model of amyotrophic lateral sclerosis. J Neurosci Res 88(1):123–135
431. Ying Z, Wang H, Fan H, Zhu X, Zhou J, Fei E, Wang G (2009) Gp78, an ER associated E3, promotes SOD1 and ataxin-3 degradation. Hum Mol Genet 18(22):4268–4281
432. Yonashiro R, Sugiura A, Miyachi M, Fukuda T, Matsushita N, Inatome R, Ogata Y, Suzuki T, Dohmae N, Yanagi S (2009) Mitochondrial ubiquitin ligase MITOL ubiquitinates mutant SOD1 and attenuates mutant SOD1-induced reactive oxygen species generation. Mol Biol Cell 20(21):4524–4530
433. Thompson ML, Chen P, Yan X, Kim H, Borom AR, Roberts NB, Caldwell KA, Caldwell GA (2014) TorsinA rescues ER-associated stress and locomotive defects in C. elegans models of ALS. Dis Model Mech 7(2):233–243
434. Mori A, Yamashita S, Uchino K, Suga T, Ikeda T, Takamatsu K, Ishizaki M, Koide T, Kimura E, Mita S, Maeda Y, Hirano T, Uchino M (2011) Derlin-1 overexpression ameliorates mutant SOD1-induced endoplasmic reticulum stress by reducing mutant SOD1 accumulation. Neurochem Int 58(3):344–353
435. Brady OA, Meng P, Zheng Y, Mao Y, Hu F (2011) Regulation of TDP-43 aggregation by phosphorylation and p62/SQSTM1. J Neurochem 116(2):248–259
436. Trippier PC, Zhao KT, Fox SG, Schiefer IT, Benmohamed R, Moran J, Kirsch DR, Morimoto RI, Silverman RB (2014) Proteasome activation is a mechanism for pyrazolone small molecules displaying therapeutic potential in amyotrophic lateral sclerosis. ACS Chem Neurosci 5(9):823–829
437. Kim J, Kim TY, Cho KS, Kim HN, Koh JY (2013) Autophagy activation and neuroprotection by progesterone in the G93A-SOD1 transgenic mouse model of amyotrophic lateral sclerosis. Neurobiol Dis 59:80–85
438. Yang WW, Sidman RL, Taksir TV, Treleaven CM, Fidler JA, Cheng SH, Dodge JC, Shihabuddin LS (2011) Relationship between neuropathology and disease progression in the SOD1(G93A) ALS mouse. Exp Neurol 227(2):287–295
439. Luty AA, Kwok JB, Dobson-Stone C, Loy CT, Coupland KG, Karlstrom H, Sobow T, Tchorzewska J, Maruszak A, Barcikowska M, Panegyres PK, Zekanowski C, Brooks WS, Williams KL, Blair IP, Mather KA, Sachdev PS, Halliday GM, Schofield PR (2010) Sigma nonopioid intracellular receptor 1 mutations cause frontotemporal lobar degeneration-motor neuron disease. Ann Neurol 68(5):639–649
440. Tagashira H, Shinoda Y, Shioda N, Fukunaga K (2014) Methyl pyruvate rescues mitochondrial damage caused by SIGMAR1 mutation related to amyotrophic lateral sclerosis. Biochim Biophys Acta 1840(12):3320–3334
441. Sarlette A, Krampfl K, Grothe C, Neuhoff N, Dengler R, Petri S (2008) Nuclear erythroid 2-related factor 2-antioxidative response element signaling pathway in motor cortex and spinal cord in amyotrophic lateral sclerosis. J Neuropathol Exp Neurol 67(11):1055–1062
442. Pehar M, Vargas MR, Robinson KM, Cassina P, Diaz-Amarilla PJ, Hagen TM, Radi R, Barbeito L, Beckman JS (2007) Mitochondrial superoxide production and nuclear factor erythroid 2-related factor 2 activation in p75 neurotrophin receptor-induced motor neuron apoptosis. J Neurosci 27(29):7777–7785
443. Neymotin A, Calingasan NY, Wille E, Naseri N, Petri S, Damiano M, Liby KT, Risingsong R, Sporn M, Beal MF, Kiaei M (2011) Neuroprotective effect of Nrf2/ARE activators, CDDO ethylamide and CDDO trifluoroethylamide, in a mouse model of amyotrophic lateral sclerosis. Free Radic Biol Med 51(1):88–96
444. Aguzzi A, Falsig J (2012) Prion propagation, toxicity and degradation. Nat Neurosci 15(7):936–939
445. Mallucci G, Dickinson A, Linehan J, Klohn PC, Brandner S, Collinge J (2003) Depleting neuronal PrP in prion infection prevents disease and reverses spongiosis. Science 302(5646):871–874
446. Mallucci GR, White MD, Farmer M, Dickinson A, Khatun H, Powell AD, Brandner S, Jefferys JG, Collinge J (2007) Targeting cellular prion protein reverses early cognitive deficits and neurophysiological dysfunction in prion-infected mice. Neuron 53(3):325–335

447. Goold R, McKinnon C, Tabrizi SJ (2015) Prion degradation pathways: potential for therapeutic intervention. Mol Cell Neurosci 66:12–20
448. Apodaca J, Kim I, Rao H (2006) Cellular tolerance of prion protein PrP in yeast involves proteolysis and the unfolded protein response. Biochem Biophys Res Commun 347(1):319–326
449. Shao J, Choe V, Cheng H, Tsai YC, Weissman AM, Luo S, Rao H (2014) Ubiquitin ligase gp78 targets unglycosylated prion protein PrP for ubiquitylation and degradation. PLoS One 9(4), e92290
450. Webb S, Lekishvili T, Loeschner C, Sellarajah S, Prelli F, Wisniewski T, Gilbert IH, Brown DR (2007) Mechanistic insights into the cure of prion disease by novel antiprion compounds. J Virol 81(19):10729–10741
451. Goffeau A, Barrell BG, Bussey H, Davis RW, Dujon B, Feldmann H, Galibert F, Hoheisel JD, Jacq C, Johnston M, Louis EJ, Mewes HW, Murakami Y, Philippsen P, Tettelin H, Oliver SG (1996) Life with 6000 genes. Science 274(5287):546–563–547
452. Consortium CeS (1998) Genome sequence of the nematode C. elegans: a platform for investigating biology. Science 282(5396):2012–2018
453. Lander ES, Linton LM, Birren B, Nusbaum C, Zody MC, Baldwin J, Devon K, Dewar K, Doyle M, FitzHugh W, Funke R, Gage D, Harris K, Heaford A, Howland J, Kann L, Lehoczky J, LeVine R, McEwan P, McKernan K, Meldrim J, Mesirov JP, Miranda C, Morris W, Naylor J, Raymond C, Rosetti M, Santos R, Sheridan A, Sougnez C, Stange-Thomann N, Stojanovic N, Subramanian A, Wyman D, Rogers J, Sulston J, Ainscough R, Beck S, Bentley D, Burton J, Clee C, Carter N, Coulson A, Deadman R, Deloukas P, Dunham A, Dunham I, Durbin R, French L, Grafham D, Gregory S, Hubbard T, Humphray S, Hunt A, Jones M, Lloyd C, McMurray A, Matthews L, Mercer S, Milne S, Mullikin JC, Mungall A, Plumb R, Ross M, Shownkeen R, Sims S, Waterston RH, Wilson RK, Hillier LW, McPherson JD, Marra MA, Mardis ER, Fulton LA, Chinwalla AT, Pepin KH, Gish WR, Chissoe SL, Wendl MC, Delehaunty KD, Miner TL, Delehaunty A, Kramer JB, Cook LL, Fulton RS, Johnson DL, Minx PJ, Clifton SW, Hawkins T, Branscomb E, Predki P, Richardson P, Wenning S, Slezak T, Doggett N, Cheng JF, Olsen A, Lucas S, Elkin C, Uberbacher E, Frazier M, Gibbs RA, Muzny DM, Scherer SE, Bouck JB, Sodergren EJ, Worley KC, Rives CM, Gorrell JH, Metzker ML, Naylor SL, Kucherlapati RS, Nelson DL, Weinstock GM, Sakaki Y, Fujiyama A, Hattori M, Yada T, Toyoda A, Itoh T, Kawagoe C, Watanabe H, Totoki Y, Taylor T, Weissenbach J, Heilig R, Saurin W, Artiguenave F, Brottier P, Bruls T, Pelletier E, Robert C, Wincker P, Smith DR, Doucette-Stamm L, Rubenfield M, Weinstock K, Lee HM, Dubois J, Rosenthal A, Platzer M, Nyakatura G, Taudien S, Rump A, Yang H, Yu J, Wang J, Huang G, Gu J, Hood L, Rowen L, Madan A, Qin S, Davis RW, Federspiel NA, Abola AP, Proctor MJ, Myers RM, Schmutz J, Dickson M, Grimwood J, Cox DR, Olson MV, Kaul R, Raymond C, Shimizu N, Kawasaki K, Minoshima S, Evans GA, Athanasiou M, Schultz R, Roe BA, Chen F, Pan H, Ramser J, Lehrach H, Reinhardt R, McCombie WR, de la Bastide M, Dedhia N, Blocker H, Hornischer K, Nordsiek G, Agarwala R, Aravind L, Bailey JA, Bateman A, Batzoglou S, Birney E, Bork P, Brown DG, Burge CB, Cerutti L, Chen HC, Church D, Clamp M, Copley RR, Doerks T, Eddy SR, Eichler EE, Furey TS, Galagan J, Gilbert JG, Harmon C, Hayashizaki Y, Haussler D, Hermjakob H, Hokamp K, Jang W, Johnson LS, Jones TA, Kasif S, Kaspryzk A, Kennedy S, Kent WJ, Kitts P, Koonin EV, Korf I, Kulp D, Lancet D, Lowe TM, McLysaght A, Mikkelsen T, Moran JV, Mulder N, Pollara VJ, Ponting CP, Schuler G, Schultz J, Slater G, Smit AF, Stupka E, Szustakowski J, Thierry-Mieg D, Thierry-Mieg J, Wagner L, Wallis J, Wheeler R, Williams A, Wolf YI, Wolfe KH, Yang SP, Yeh RF, Collins F, Guyer MS, Peterson J, Felsenfeld A, Wetterstrand KA, Patrinos A, Morgan MJ, de Jong P, Catanese JJ, Osoegawa K, Shizuya H, Choi S, Chen YJ, International Human Genome Sequencing C (2001) Initial sequencing and analysis of the human genome. *Nature* 409(6822):860–921

Chapter 2

Review and Literature Mining on Proteostasis Factors and Cancer

Ana Sofia Carvalho, Manuel S. Rodríguez, and Rune Matthiesen

Abstract

Automatic analysis of increasingly growing literature repositories including data integration to other databases is a powerful tool to propose hypothesis that can be used to plan experiments to validate or disprove the hypothesis. Furthermore, it provides means to evaluate the redundancy of research line in comparison to the published literature. This is potentially beneficial for those developing research in a specific disease which are interested in exploring a particular pathway or set of genes/proteins. In the scope of the integrating book a case will be made addressing proteostasis factors in cancer. The maintenance of proteome homeostasis, known as proteostasis, is a process by which cells regulate protein translation, degradation, subcellular localization, and protein folding and consists of an integrated network of proteins. The ubiquitin-proteasome system plays a key role in essential biological processes such as cell cycle, DNA damage repair, membrane trafficking, and maintaining protein homeostasis. Cells maintain proteostasis by regulating protein translation, degradation, subcellular localization, and protein folding. Aberrant proteostasis leads to loss-of-function diseases (cystic fibrosis) and gain-of-toxic-function diseases (Alzheimer's, Parkinson's, and Huntington's disease). Cancer therapy on the other hand explores inhibition of proteostasis factors to trigger endoplasmic reticulum stress with subsequent apoptosis. Alternatively therapies target deubiquitinases and thereby regulate tumor promoters or suppressors. Furthermore, mutations in specific proteostasis factors are associated with higher risk for specific cancers, e.g., BRCA mutations in breast cancer. This chapter discusses proteostasis protein factors' association with cancer from a literature mining perspective.

Key words Proteostasis, Cancer, Text mining, Wordcloud

1 Introduction

The vast number of scientific publications (25 million citations in PubMed) provides an extremely valuable resource for researchers if approached by an automated analysis of the information. The background knowledge integrated with annotated content in biological databases (such as proteins in UniProt) or repositories of genes function or protein-protein interactions is fundamental for hypothesis generation. A highly comprehensive review on the latest advances on automated literature analysis for

Rune Matthiesen (ed.), *Proteostasis: Methods and Protocols*, Methods in Molecular Biology, vol. 1449,
DOI 10.1007/978-1-4939-3756-1_2, © Springer Science+Business Media New York 2016

biomedical research can be found here [1]. Our goal is to apply text mining for hypothesis generation in the case study of proteostasis and cancer.

Ubiquitin modifies proteins which target them to new cellular localization such as for example the proteasome for degradation. E1 ubiquitin activating enzymes (two in the human genome), E2 ubiquitin conjugating enzymes (~30 in the human genome), and finally E3 ubiquitin ligases (~600 in the human genome) conjugate ubiquitin through sequential actions [2–4]. Specificity is mainly provided by the E3 ubiquitin ligases which likely explain the association of specific E3 ligases with diseases. E3 ligases are both suggested as biomarkers and targets for cancer therapy [5]. Deubiquitinases (DUBs), ~100 in the human genome, cleave off ubiquitin from modified proteins [3]. DUBs like E3 ligases have also been suggested as both biomarkers and targets for cancer therapy [5]. DUBs can regulate both oncogenes and tumor suppressors. Aberrant DUBs activity, both gain and loss of function by mutation and/or altered expression, can promote cancer. DUBs associated with cancer have been described as specific for targeting proteins. Evidence suggests that DUBs specificity may depend on tissue types and stage of malignancy, thereby making it difficult to access the general role of DUBs in tumorigenesis.

The success of bortezomib (Velcade™), a proteasome inhibitor, used for the treatment of relapse or refractory patients with multiple myeloma focuses the attention of cancer biologists on potential cancer treatment strategies that target proteostasis. Examples of such strategies are listed below.

1. When the production of misfolded proteins exceeds degradation, as often occurs in damaged or aging cells, or in cells exposed to chemical agents that perturb protein folding or the endoplasmic reticulum (ER) quality control (ERQC) pathway, the ER-associated degradation (ERAD) is elicited. There are two types of molecules that affect ERQC pathway which can be used to modulate ER stress and trigger apoptosis: (a) small molecules can enhance proteostasis by binding to and stabilizing specific proteins (pharmacologic chaperones) increasing the proteostasis network capacity (proteostasis regulators) or (b) by regulating proteostasis.

   Certain cancer cells with high secretory capacities and basal levels of ER stress have been shown to be more sensitive to ER stress-induced cell death (e.g., multiple myeloma) [6, 7]. Bortezomib, a proteasome inhibitor, inhibits the chymotrypsin activity of the proteasome is approved for the treatment of mantle cell lymphoma and relapse or refractory multiple myeloma [8, 9]. The effect of bortezomib involves many pathways of which some are linked to the unfolded protein response [10–13] and others to protein factors such as p53 [14, 15] and NFκB [16].

2. Inhibition of p97 ATPase for ER membrane extraction and for subsequent transfer to the proteasome by the drug Eeyarestatin I can induce cell death in hematologic cancer cells [17]. Eeyarestatin I affects similar factors as bortezomib such as accumulation of polyubiquitinated proteins, ER stress causing downregulation of histone H2A ubiquitination with subsequent Noxa activation, and cell death [1].
3. Alternatively DUBs can also be targeted and some cancer cells are more susceptible to specific DUB inhibition than non cancer cells. This is also referred to as synthetic lethality. They may regulate the stability of key oncogenes, exemplified by USP28 stabilization of c-Myc. Alternatively DUBs can negatively regulate ubiquitin-dependent signaling cascades such as the NF-kB activation pathway [18].
4. Aberrant regulation of some E3 ligases is associated with cancer development [5]. Furthermore, cancer cells frequently overexpress E3 ligases and this correlates with increased chemoresistance and poor prognosis. E3 ligases are "drugable" and therefore potential cancer targets. Additionally, E3 ligases serve as cancer biomarkers. For example, germline mutations in the E3 ligase BRCA1 increase the predisposition for breast cancer [19]. Another example is MDM2 which targets the tumor suppressor p53 for degradation [20, 21].

In conclusion, proteostasis proteins are found to be aberrantly regulated at the expression level and mutated in cancer cells. Furthermore, proteostasis proteins are being targeted for cancer therapies. We therefore perform text mining on abstracts in PubMed to provide an overview of the most studied protein factors in connection with different cancer types.

2 Materials

The data mining was performed using the statistical programming language R. The libraries listed below were used:

1. Package "RISmed" for PubMed search.
2. Package "tm" for text mining.
3. Package "wordcloud" for graphical display.
4. Names of frequently studied proteins of the ubiquitin system were extracted from http://www.sabiosciences.com/rt_pcr_product/HTML/PAHS-3079Z.html.
5. Names of proteasome factors and DUBs were downloaded from the online database HUGO.
6. Disease-gene associations list was obtained from the DISEASES resource available at http://diseases.jensenlab.org/ [22].

3 Methods

The aim of the computer-assisted text mining approach here presented is to obtain a quick overview of the factors in the ubiquitin-proteasome system and their association with cancer (*see* **Note 1**). Furthermore, the generated wordclouds are useful to display the most important word terms related to specific ubiquitin proteasome factors. We do not provide detailed steps for the analysis since this will quickly be outdated and the manual of described software tools would always be the best source of details for computational steps.

3.1 Obtaining Text Corpus from PubMed

1. Use PubMed directly, eUtils or the R package "RISmed" to download abstracts for each of the ubiquitin and proteoasome factors of interest. We used the search term "XXX AND (leukemia OR cancer OR lymphoma)" for the analysis presented here. Where XXX is replaced with one of the ubiquitin and proteasome factors, e.g., "BRCA1 AND (leukemia OR cancer OR lymphoma)" (*see* **Notes 2–4**).
2. Use the R command grep to filter the retrieved PubMed abstracts (use the R command "?grep" to obtain the grep manual). We chose to maintain only entries that contain the ubiquitin and proteasome factors in either abstracts or title. More sensitivity can be obtained by also including abstracts having ubiquitin and proteasome factors as keyword. However, the context of ubiquitin and proteasome factors in relation to cancer becomes obscure without having access to the full paper text.

Figure 1 displays, based on the above retrieved text corpus, the association between ubiquitin ligase complexes, DUBs, and proteasome factors to different cancer types. This plot was created with the R graphical command barplot but could also be plotted in Excel or other software tool. We infer from Fig. 1 that BRCA1 and BRCA2 are the most described factors in ubiquitin ligase complexes and mainly associated with breast, ovarian, and prostate cancer. We further see that different factors in ligase complexes, DUBs, and proteasome factors are associated with different cancer types.

3.2 Wordclouds

Wordclouds are useful to obtain a visual overview of the most important terms in a text corpus. The wordclouds in Figs. 2 and 3 were created using the R packages "tm" and "wordcloud" by running the following steps.

1. myCorpus = Corpus(VectorSource(TextBRCA1)) # (*see* **Note 5**).
2. myCorpus = tm_map(myCorpus, removeWords, stopwords ("english")) # (*see* **Note 6**).
3. myCorpus = tm_map(myCorpus, removePunctuation).

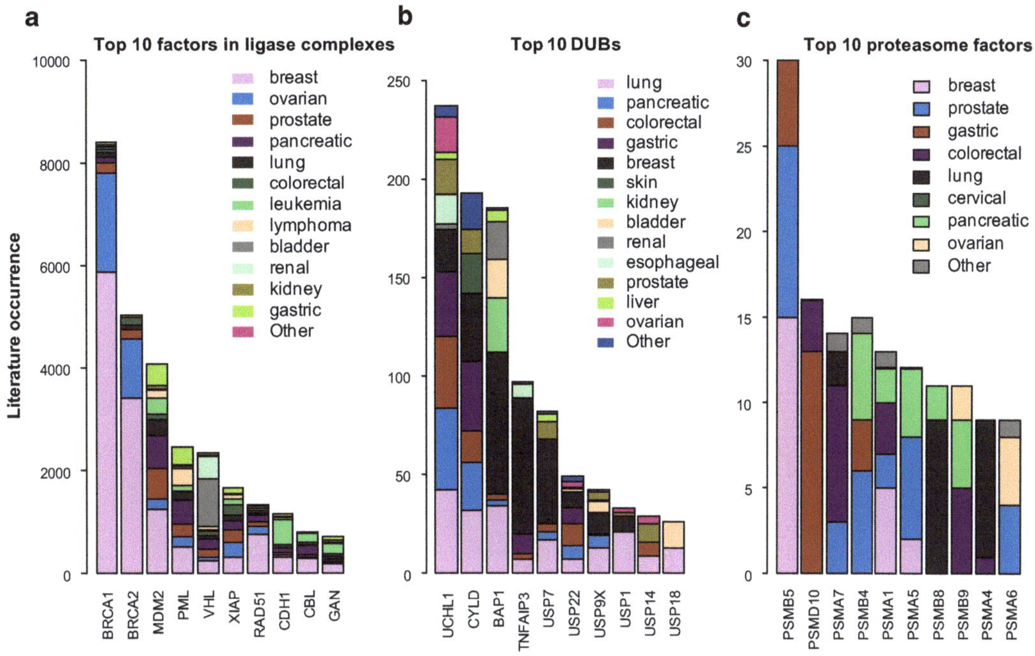

Fig. 1 Text mining inferred association between different factors in ligase complexes (**a**), Deubiquitinases (DUBs) (**b**), and proteasome factors (**c**) to different cancer types

4. myDTM = TermDocumentMatrix(myCorpus, control = list (minWordLength = 3)).

   m = as.matrix(myDTM).

5. v = sort(rowSums(m), decreasing = TRUE).
6. wordcloud(names(v),v, scale = c(5,0.5), max.words = 200, random.order = FALSE, rot.per = 0.35, use.r.layout = FALSE, colors = brewer.pal(8, "Dark2")).

The resulting wordclouds may contain duplicates such as "mutation" and "mutations". This can be resolved by using the command: "myCorpus = tm_map(myCorpus, stemDocument)". We find that the command "myCorpus = tm_map(myCorpus, removeNumbers)" has the unwanted site effect of removing all numbers resulting in BRCA1 becomes BRCA. However, for the two provided examples there was no need to remove numbers.

It is reassuring to see that the text mining and wordcloud for BRCA1 display terms like breast, ovarian, mutations, and genetic. Directly providing the valuable information that BRCA1 is associated with breast and ovarian cancers and its association to genetic predisposition. The association to other proteins such as BRCA2, p53, PARP and ATM is also informative.

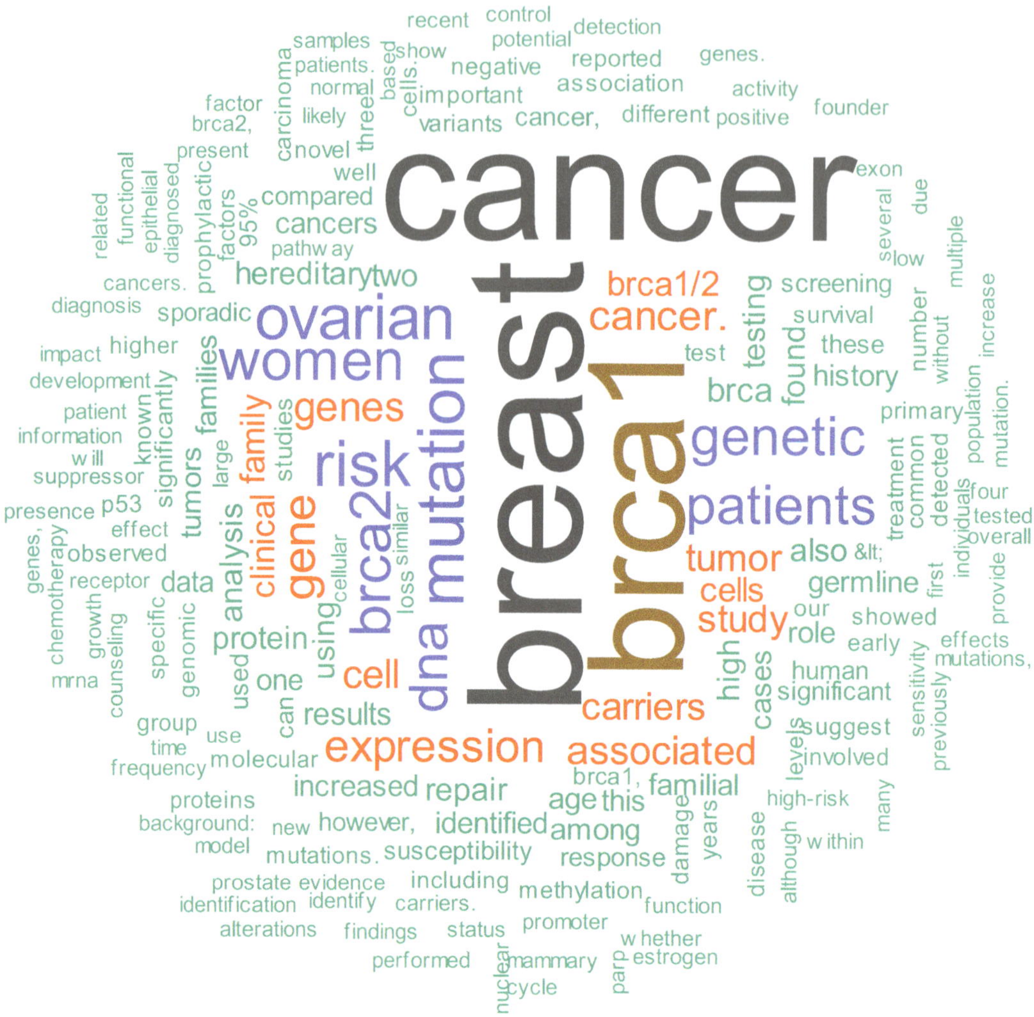

Fig. 2 Wordcloud for the search term "BRCA1"

Figure 3 shows the wordcloud for MDM2. We observe that MDM2 is associated to protein factors such as p53/TP53, bcl2, p21, nutlin3, cdk4, and HDM2. Furthermore, the more general protein families such as kinases and cyclins appear and the term apoptosis is more abundant than in the above analysis for BRCA1 (compare Figs. 2 and 3).

3.3 Gene Associations with Diseases

We next address more rigorous approaches regarding extracting text for a corpus. We reuse the DISEASES resource which lists disease-gene associations (http://diseases.jensenlab.org/). The disease-gene associations are calculated by a scoring scheme that simultaneously takes into account co-occurrences at the level of abstracts as well as individual sentences [22]. In contrast to the analysis above, we here analyzed gene association of a larger set of annotated diseases.

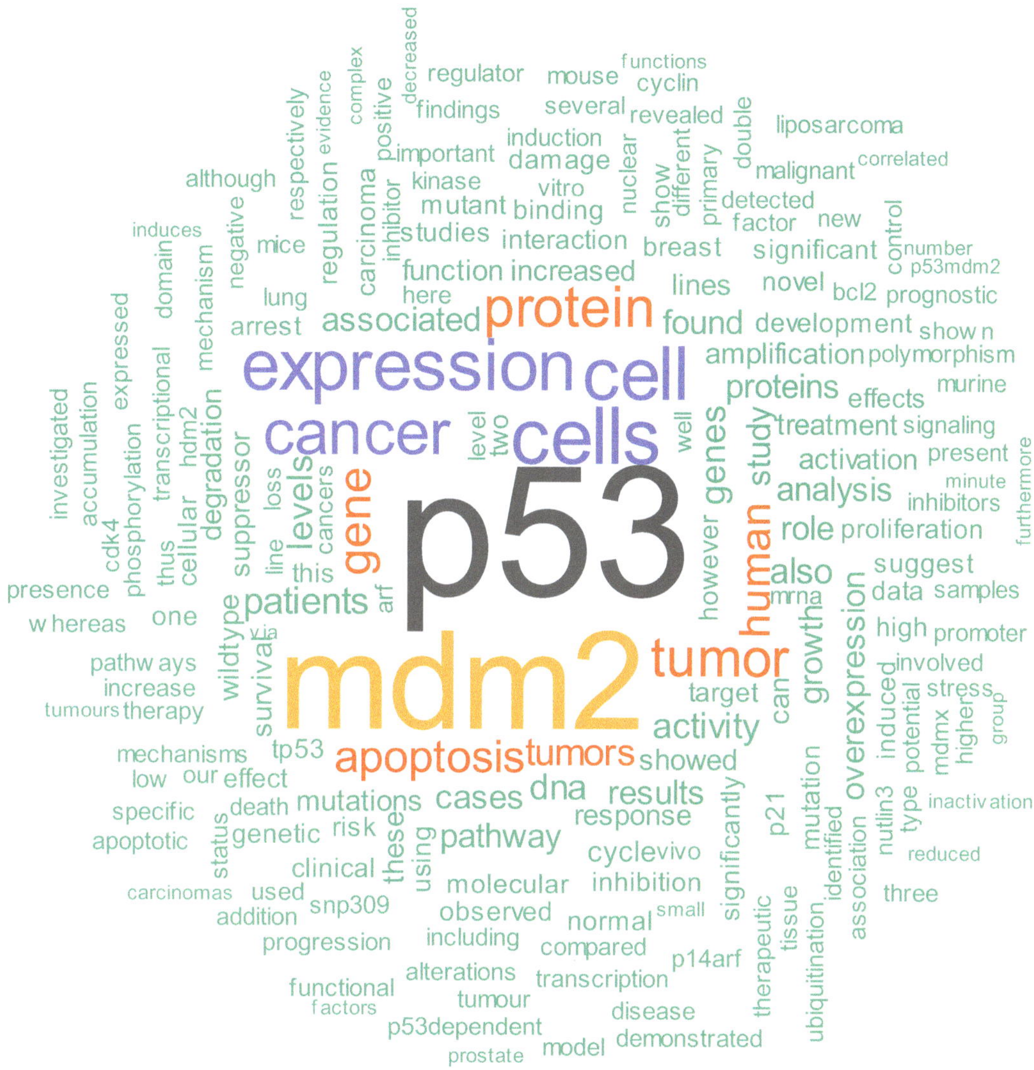

Fig. 3 Wordcloud for the search term "MDM2"

Figure 4 shows the top factors in ubiquitin ligase complexes' association with diseases by using the DISEASES resource. The height of the bars corresponds to the counts in the unfiltered text mining matrix (human_disease_textmining_full.tsv) and the subdivision is made by normalizing with each disease confidence for a specific protein factor as reported in filtered text mining matrix (human_disease_textmining_filtered.tsv). It is evident that a large number of diseases have been associated with factors in ubiquitin ligase complexes and clearly cancer ranks as the most confident associated disease to the top factors in ubiquitin ligase complexes.

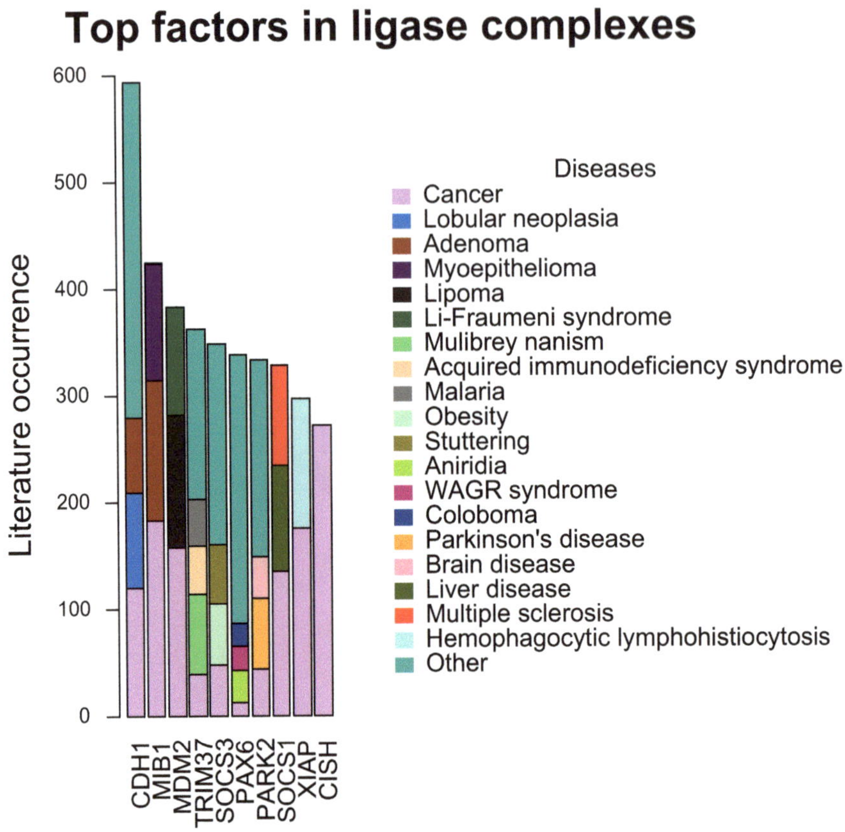

Fig. 4 Top factors in ubiquitin ligase complexes' association with diseases by using the DISEASES resource

A similar analysis for DUBs revealed that they are associated with many diseases as well where cognitive disease, inflammatory diseases, and cancer are among the most strongly associated with DUBs (Fig. 5).

The top ten proteasome factors were found to be associated with mainly infectious disease, cancer, and vascular diseases (Fig. 6).

3.4 Discussion of Results

The strongest association between the chosen genes in this analysis and a specific type of cancer was found for BRCA1's association with breast and ovarian cancers (Fig. 1). Mutations in BRCA1 increase susceptibility to breast and ovarian cancers reflecting the predominance of citations associating these two types of cancer to BRCA1 (Figs. 1 and 2). BRCA1 participates in the cellular response to DNA damage as a sensor molecule and as an effector by transcriptional regulation of genes [23]. E3 ligase activity of BRAC1 is achieved by heterodimerization through its amino-terminal (really interesting new gene) RING domain with a RING

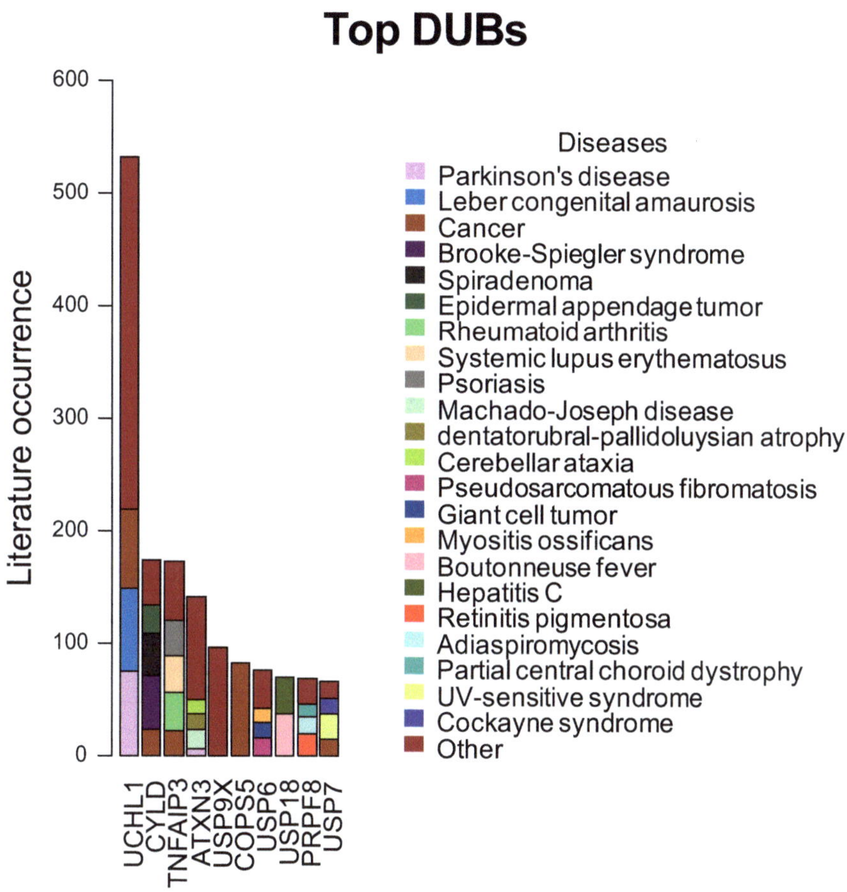

Fig. 5 Top deubiquitinases (DUBs) association with diseases by using the DISEASES resource

partner, BARD1 [24]. We do not observe BARD1 in the word-cloud presented in Fig. 2 suggesting that few studies have focused on the role of BARD1 on the E3 ubiquitin ligase activity of BRCA1. A specific mechanism of DNA damage response of BRCA1 involves ubiquitinylation of claspin, an essential activator of the CHK1 checkpoint kinase, by BRCA1 triggering homology-directed DNA repair [25]. Despite the similarities between the phenotypes induced by disruption of BRCA1 or BRCA2, they play a role in distinct functions in the biological response to DNA damage [23]. BRCA2 is a mediator of recombinase RAD51 and their role in DNA damage response is mechanistically distinct from BRCA1. MDM2 is an E3 ubiquitin-protein ligase of the RING finger class that mediates ubiquitination of the tumor suppressor p53/TP53, regulating its stability and activity [26–28]. MDM2 was the next E3 ligase, after BRCA1 and BRCA2, registering a high number of co-occurrences in the retrieved PubMed abstracts in relation with cancer, largely due to MDM2's role in

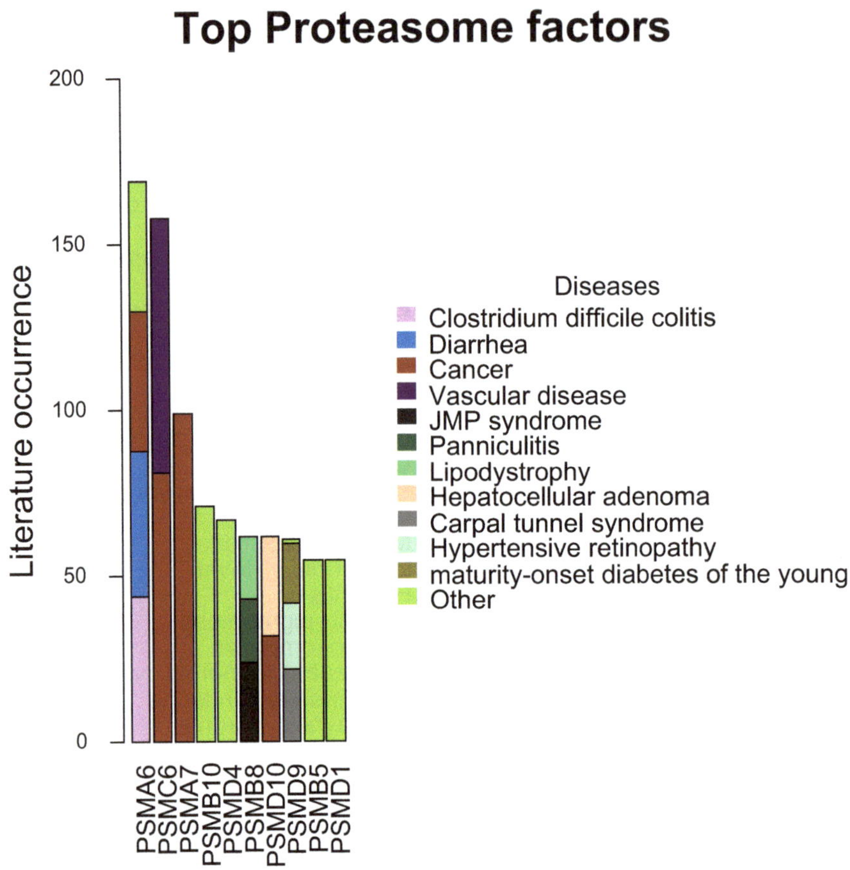

Fig. 6 Top proteasome factors' association with diseases by using the DISEASES resource

the regulation of p53 (Figs. 1 and 3). Inactivating p53 mutations occur in more than 50% of human tumors. Variations in Mdm2 due to single nucleotide polymorphism, overexpression, or amplification impact the ubiquitination levels of p53 and consequently p53 degradation. Such variations are therefore tumor-prone phenotypes. Several compounds have been designed to inhibit MDM2 E3 ubiquitin ligase such as nutlin-3. In fact an analog of nutiln-3 is in phase I trials in patients with solid tumors or leukemia [29]. MDM2-p53 interaction illustrates how targeting the ubiquitin system and its factors can potentially succeed in drug development against cancer. Another interactor of p53 is promyelocytic leukemia (PML) tumor suppressor protein, a central regulator of cell proliferation and apoptosis. PML configures as one of the top ten factors in E3 ligase complexes associated with different types of cancer (Fig. 1a). PML protects p53 from Mdm2-mediated ubiquitination and degradation, and from inhibition of apoptosis [30]. A group of other protein factors containing a RING finger domain such as the X-linked inhibitor of apoptosis (XIAP) and the Casitas

B-lineage Lymphoma (CBL) protein family are among the top factors in E3 ligase complexes associated with cancer (Fig. 1a). XIAP inhibits the activity of the cell death proteases, caspase-3, -7, and -9, and promotes the degradation of active-form caspase-3 mediated by its RING finger domain acting as an E3 ubiquitin ligase [31]. XIAP mediates an oncogenic signaling by the ubiquitination of TGF-beta-activated kinase 1 (TAK1) enabling TGF-beta to activate p65/RelA and to induce the expression of prometastatic and prosurvival genes in 4T1 breast cancer cells [32]. CBL small family of Cbl ubiquitin E3 ligases, c-Cbl, Cbl-b, and Cbl-c, regulates signaling through its N-terminal tyrosine-kinase-binding (TKB) domain composed of three different subdomains: a four-helix bundle (4H), a calcium-binding EF hand, and a divergent SH2 domain, which is followed by a RING finger and a proline-rich domain inducing a myriad of interactions [33]. Cbl proteins interact with tyrosine kinases through its TKB domain such as v-src oncogene, a preferential target of Cbl-c for degradation [34], inhibiting its oncogenic activity. Indeed Cbl-b predicts better prognosis in RANK-expressing breast cancer patients [35]. The following two examples, Von Hippel-Lindau (VHL) disease tumor suppressor gene and gigaxonin (GAN), constitute elements in ubiquitylation complexes acting upon key players in cancer-driven mechanisms. VHL is found mutated in a variety of tumors including clear cell carcinomas of the kidney, pheochromocytomas, and vascular tumors of the central nervous system and retina [36]. Under hypoxia conditions hydroxylated hypoxia-inducible factor (HIF) is recruited by the von Hippel-Lindau ubiquitination complex, leading to its ubiquitination and degradation [37]. GAN, an ubiquitin E3 ligase adaptor, and p16 protein expression contributed to senescence of cisplatin treated cells through NFkB ubiquitination. The increased nuclear p16 expression correlates with enhanced survival of head and neck cancer patients [38]. The manual validation of the text mining performed on proteostasis factors and cancer reassures the effectiveness of these approaches for assessing in an organized and nondisperse manner the vast literature on a specific subject.

4 Notes

1. We apply a text mining approach which means that the results presented are obtained semi-automatically. That is we have not extensively manually validated the extracted words for every abstract. This means that a few terms are likely to be extracted in the wrong context. Nevertheless, we are only interested in the most abundant terms so few errors are unlikely to corrupt the overall picture. However, we have performed basic validation as mentioned in the following notes.

2. We applied directly the official protein names for proteostasis factors. More sophisticated approaches could be taken to include also description of a protein and synonym protein names.
3. The keyword "AND" is very important. If not included then all abstracts with cancer, leukemia, and lymphoma will always be targeted. The key word "cancer" captures a large group of cancers such as lung cancer, stomach cancer, and breast cancer. We choose for simplicity here to only include "leukemia and lymphoma" as additional cancer types but in principle the list of cancer keywords could be longer.
4. The search terms described will also hit matches in keywords and authors fields. The hits on keywords provide more sensitivity but the hits on author fields give false matches. We therefore subsequently use the grep command to filter this first text corpus. An alternative approach could be to use the PubMed filter "[Text+Words]" to avoid matches on authors. For example, the ubiquitin E3 ligase adaptor "GAN" matches many abstracts with this author name.
5. Running the command "typeof" on the text "TextBRCA1" should give the output: "character".
6. The wordcloud still contains some noninformative words such as: use, can, and one (*see* Fig. 2). These can be filtered away by using the command: myCorpus = tm_map(myCorpus, removeWords, c("use", "can","one")).

Acknowledgements

The authors would like to acknowledge networking support by the Proteostasis COST Action (BM1307). This work is supported by Fundação para a Ciência e Tecnologia project EXPL/DTP-PIC/0616/2013. R.M. is supported by FCT investigator program 2012. A.S.C. is supported by grant SFRH/BPD/85569/2012 funded by Fundação para a Ciência e Tecnologia.

References

1. Rebholz-Schuhmann D, Oellrich A, Hoehndorf R (2012) Text-mining solutions for biomedical research: enabling integrative biology. Nat Rev Genet 13(12):829–839. doi:10.1038/nrg3337
2. Deshaies RJ, Joazeiro CA (2009) RING domain E3 ubiquitin ligases. Annu Rev Biochem 78:399–434. doi:10.1146/annurev.biochem.78.101807.093809
3. Reyes-Turcu FE, Ventii KH, Wilkinson KD (2009) Regulation and cellular roles of ubiquitin-specific deubiquitinating enzymes. Annu Rev Biochem 78:363–397. doi:10.1146/annurev.biochem.78.082307.091526
4. van Wijk SJ, Timmers HT (2010) The family of ubiquitin-conjugating enzymes (E2s): deciding between life and death of proteins. FASEB J 24(4):981–993. doi:10.1096/fj.09-136259
5. Sun Y (2006) E3 ubiquitin ligases as cancer targets and biomarkers. Neoplasia 8(8):645–654. doi:10.1593/neo.06376

6. Liu Y, Ye Y (2011) Proteostasis regulation at the endoplasmic reticulum: a new perturbation site for targeted cancer therapy. Cell Res 21(6):867–883. doi:10.1038/cr.2011.75
7. Perez-Galan P, Mora-Jensen H, Weniger MA, Shaffer AL 3rd, Rizzatti EG, Chapman CM, Mo CC, Stennett LS, Rader C, Liu P, Raghavachari N, Stetler-Stevenson M, Yuan C, Pittaluga S, Maric I, Dunleavy KM, Wilson WH, Staudt LM, Wiestner A (2011) Bortezomib resistance in mantle cell lymphoma is associated with plasmacytic differentiation. Blood 117(2):542–552. doi:10.1182/blood-2010-02-269514
8. Kane RC, Farrell AT, Sridhara R, Pazdur R (2006) United States Food and Drug Administration approval summary: bortezomib for the treatment of progressive multiple myeloma after one prior therapy. Clin Cancer Res 12(10):2955–2960. doi:10.1158/1078-0432.CCR-06-0170
9. Field-Smith A, Morgan GJ, Davies FE (2006) Bortezomib (Velcadetrade mark) in the treatment of multiple myeloma. Ther Clin Risk Manag 2(3):271–279
10. Fribley AM, Evenchik B, Zeng Q, Park BK, Guan JY, Zhang H, Hale TJ, Soengas MS, Kaufman RJ, Wang CY (2006) Proteasome inhibitor PS-341 induces apoptosis in cisplatin-resistant squamous cell carcinoma cells by induction of Noxa. J Biol Chem 281(42):31440–31447. doi:10.1074/jbc.M604356200
11. Carvalho AS, Ribeiro H, Voabil P, Penque D, Jensen ON, Molina H, Matthiesen R (2014) Global mass spectrometry and transcriptomics array based drug profiling provides novel insight into glucosamine induced endoplasmic reticulum stress. Mol Cell Proteomics 13(12):3294–3307. doi:10.1074/mcp.M113.034363
12. Fribley A, Wang CY (2006) Proteasome inhibitor induces apoptosis through induction of endoplasmic reticulum stress. Cancer Biol Ther 5(7):745–748
13. Obeng EA, Carlson LM, Gutman DM, Harrington WJ Jr, Lee KP, Boise LH (2006) Proteasome inhibitors induce a terminal unfolded protein response in multiple myeloma cells. Blood 107(12):4907–4916. doi:10.1182/blood-2005-08-3531
14. MacLaren AP, Chapman RS, Wyllie AH, Watson CJ (2001) p53-dependent apoptosis induced by proteasome inhibition in mammary epithelial cells. Cell Death Differ 8(3):210–218. doi:10.1038/sj.cdd.4400801
15. Adams J, Palombella VJ, Sausville EA, Johnson J, Destree A, Lazarus DD, Maas J, Pien CS, Prakash S, Elliott PJ (1999) Proteasome inhibitors: a novel class of potent and effective antitumor agents. Cancer Res 59(11):2615–2622
16. Hideshima T, Chauhan D, Richardson P, Mitsiades C, Mitsiades N, Hayashi T, Munshi N, Dang L, Castro A, Palombella V, Adams J, Anderson KC (2002) NF-kappa B as a therapeutic target in multiple myeloma. J Biol Chem 277(19):16639–16647. doi:10.1074/jbc.M200360200
17. Wang Q, Li L, Ye Y (2008) Inhibition of p97-dependent protein degradation by Eeyarestatin I. J Biol Chem 283(12):7445–7454. doi:10.1074/jbc.M708347200
18. Sacco JJ, Coulson JM, Clague MJ, Urbe S (2010) Emerging roles of deubiquitinases in cancer-associated pathways. IUBMB Life 62(2):140–157. doi:10.1002/iub.300
19. Petrucelli N, Daly MB, Feldman GL (2010) Hereditary breast and ovarian cancer due to mutations in BRCA1 and BRCA2. Genet Med 12(5):245–259. doi:10.1097/GIM.0b013e3181d38f2f
20. Montes de Oca Luna R, Wagner DS, Lozano G (1995) Rescue of early embryonic lethality in mdm2-deficient mice by deletion of p53. Nature 378(6553):203–206. doi:10.1038/378203a0
21. de Rozieres S, Maya R, Oren M, Lozano G (2000) The loss of mdm2 induces p53-mediated apoptosis. Oncogene 19(13):1691–1697. doi:10.1038/sj.onc.1203468
22. Pletscher-Frankild S, Palleja A, Tsafou K, Binder JX, Jensen LJ (2015) DISEASES: text mining and data integration of disease-gene associations. Methods 74:83–89. doi:10.1016/j.ymeth.2014.11.020
23. Venkitaraman AR (2002) Cancer susceptibility and the functions of BRCA1 and BRCA2. Cell 108(2):171–182
24. Hashizume R, Fukuda M, Maeda I, Nishikawa H, Oyake D, Yabuki Y, Ogata H, Ohta T (2001) The RING heterodimer BRCA1-BARD1 is a ubiquitin ligase inactivated by a breast cancer-derived mutation. J Biol Chem 276(18):14537–14540. doi:10.1074/jbc.C000881200
25. Sato K, Sundaramoorthy E, Rajendra E, Hattori H, Jeyasekharan AD, Ayoub N, Schiess R, Aebersold R, Nishikawa H, Sedukhina AS, Wada H, Ohta T, Venkitaraman AR (2012) A DNA-damage selective role for BRCA1 E3 ligase in claspin ubiquitylation, CHK1 activation, and DNA repair. Curr Biol 22(18):1659–1666. doi:10.1016/j.cub.2012.07.034
26. Haupt Y, Maya R, Kazaz A, Oren M (1997) Mdm2 promotes the rapid degradation of p53. Nature 387(6630):296–299. doi:10.1038/387296a0

27. Honda R, Tanaka H, Yasuda H (1997) Oncoprotein MDM2 is a ubiquitin ligase E3 for tumor suppressor p53. FEBS Lett 420(1):25–27
28. Kubbutat MH, Jones SN, Vousden KH (1997) Regulation of p53 stability by Mdm2. Nature 387(6630):299–303. doi:10.1038/387299a0
29. Pant V, Lozano G (2014) Dissecting the p53-Mdm2 feedback loop in vivo: uncoupling the role in p53 stability and activity. OncoTarget 5(5):1149–1156
30. Louria-Hayon I, Grossman T, Sionov RV, Alsheich O, Pandolfi PP, Haupt Y (2003) The promyelocytic leukemia protein protects p53 from Mdm2-mediated inhibition and degradation. J Biol Chem 278(35):33134–33141. doi:10.1074/jbc.M301264200
31. Suzuki Y, Nakabayashi Y, Takahashi R (2001) Ubiquitin-protein ligase activity of X-linked inhibitor of apoptosis protein promotes proteasomal degradation of caspase-3 and enhances its anti-apoptotic effect in Fas-induced cell death. Proc Natl Acad Sci U S A 98(15):8662–8667. doi:10.1073/pnas.161506698
32. Neil JR, Tian M, Schiemann WP (2009) X-linked inhibitor of apoptosis protein and its E3 ligase activity promote transforming growth factor-{beta}-mediated nuclear factor-{kappa}B activation during breast cancer progression. J Biol Chem 284(32):21209–21217. doi:10.1074/jbc.M109.018374
33. Schmidt MH, Dikic I (2005) The Cbl interactome and its functions. Nat Rev Mol Cell Biol 6(12):907–918. doi:10.1038/nrm1762
34. Kim M, Tezuka T, Tanaka K, Yamamoto T (2004) Cbl-c suppresses v-Src-induced transformation through ubiquitin-dependent protein degradation. Oncogene 23(9):1645–1655. doi:10.1038/sj.onc.1207298
35. Joazeiro CA, Wing SS, Huang H, Leverson JD, Hunter T, Liu YC (1999) The tyrosine kinase negative regulator c-Cbl as a RING-type, E2-dependent ubiquitin-protein ligase. Science 286(5438):309–312
36. Tanimoto K, Makino Y, Pereira T, Poellinger L (2000) Mechanism of regulation of the hypoxia-inducible factor-1 alpha by the von Hippel-Lindau tumor suppressor protein. EMBO J 19(16):4298–4309. doi:10.1093/emboj/19.16.4298
37. Cheng J, Kang X, Zhang S, Yeh ET (2007) SUMO-specific protease 1 is essential for stabilization of HIF1alpha during hypoxia. Cell 131(3):584–595. doi:10.1016/j.cell.2007.08.045
38. Veena MS, Wilken R, Zheng JY, Gholkar A, Venkatesan N, Vira D, Ahmed S, Basak SK, Dalgard CL, Ravichandran S, Batra RK, Kasahara N, Elashoff D, Fishbein MC, Whitelegge JP, Torres JZ, Wang MB, Srivatsan ES (2014) p16 Protein and gigaxonin are associated with the ubiquitination of NFkappaB in cisplatin-induced senescence of cancer cells. J Biol Chem 289(50):34921–34937. doi:10.1074/jbc.M114.568543

Chapter 3

Combining Zebrafish and Mouse Models to Test the Function of Deubiquitinating Enzyme (Dubs) Genes in Development: Role of USP45 in the Retina

Vasileios Toulis, Alejandro Garanto, and Gemma Marfany

Abstract

Ubiquitination is a dynamic and reversible posttranslational modification. Much effort has been devoted to characterize the function of ubiquitin pathway genes in the cell context, but much less is known on their functional role in the development and maintenance of organs and tissues in the organism. In fact, several ubiquitin ligases and deubiquitinating enzymes (DUBs) are implicated in human pathological disorders, from cancer to neurodegeneration. The aim of our work is to explore the relevance of DUBs in retinal function in health and disease, particularly since some genes related to the ubiquitin or SUMO pathways cause retinal dystrophies, a group of rare diseases that affect 1:3000 individuals worldwide. We propose zebrafish as an extremely useful and informative genetic model to characterize the function of any particular gene in the retina, and thus complement the expression data from mouse. A preliminary characterization of gene expression in mouse retinas (RT-PCR and in situ hybridization) was performed to select particularly interesting genes, and we later replicated the experiments in zebrafish. As a proof of concept, we selected *ups45* to be ***knocked down*** by morpholino injection in zebrafish embryos. Morphant phenotypic analysis showed moderate to severe eye morphological defects, with a defective formation of the retinal structures, therefore supporting the relevance of DUBs in the formation and differentiation of the vertebrate retina, and suggesting that genes encoding ubiquitin pathway enzymes are good candidates for causing hereditary retinal dystrophies.

Key words Deubiquitinating enzymes, USP45, Retina, Morpholino knockdown, Zebrafish Animal Model, Neurodegeneration

1 Introduction

More than 200 genes have been associated with different types of hereditary retinal degeneration (RetNet: https://sph.uth.edu/retnet/). Among them, mutations in genes directly involved in ubiquitin and SUMO pathways, such as *KLHL7* and *TOPORS*, have been identified. Despite the increase in the number of causative genes and mutations, the main challenge in the field is to assess their function in the retina.

Rune Matthiesen (ed.), *Proteostasis: Methods and Protocols*, Methods in Molecular Biology, vol. 1449,
DOI 10.1007/978-1-4939-3756-1_3,

Animal models have played an essential crucial role to discover gene function and test therapeutic approaches for retinal dystrophy genes [1]. The mouse (*Mus musculus*) has been the model per excellence for years because of the easy genetic manipulation, housing and handling, and the conservation of orthologous genes compared to humans [2]. However, the generation of a genetically modified mouse model is rather costly in terms of effort, time, and budget, and there is always the risk that the model does not mimic the human phenotype, as shown for several genes [3–6]. In the last decade, zebrafish (*Danio rerio*) has emerged as an extremely useful tool to rapidly assess candidate genes and mutations at the morphological level, particularly in genes that are involved in organ/tissue development [7], using large numbers of animals that enable confident statistical analyses. Besides, the cone-rich retina of zebrafish is similar to the human retina, and photoreceptor function and phototransduction pathways are highly conserved.

Here we describe some standard techniques to perform an easy and rapid screening of gene expression in the retina of mouse and zebrafish. The information gathered using both models might be extremely useful not only to identify new candidate genes and pathogenic mutations in human but also to evaluate the possibility of generating a costly modified animal model for long-term in vivo studies, ensuring that the gene is expressed in the correct tissue/cell type and shows an altered phenotype.

We have explored the possible role of two DUB genes (*Usp45* and *Usp53*) in the mouse retina, by quantification of the expression levels by real-time reverse-transcriptase PCR (qRT-PCR), determination of their spatial expression pattern by mRNA in situ hybridization in mouse as well as in zebrafish retinas, and finally, by phenotype analysis of gene knockdown in zebrafish embryos. As a proof of principle we focused on *usp45*, whose morphant (morpholino knockdown) resulted in disruption of the normal development of the retina, but also showed a severe reduction of the body size, with an anomalous development of the notochord and nervous system. Therefore, USP45 may be required for the normal development of the nervous system and, particularly, for retinal development.

Overall, this type of complementary phenotypic analysis combining several animal models can shed light on the physiological function of the ubiquitin/proteasome and other post-translational modification (sumoylation) pathways in health and disease.

2 Materials and Solutions

2.1 Dissection of Mouse Retinas and Preparation of Mouse Eye Sections

- Razor blades, scissors, and forceps.
- Stereomicroscope.
- Cryostat.
- Poly-lysine treated slides.
- Petri dishes (for acrylamide embedding).

2.2 RNA Isolation, cDNA Synthesis, qPCR, and PCR

- Polytron or similar blender for tissue samples.
- Agarose and electrophoresis system.
- LightCycler® 480 SYBR green (Roche Diagnostics, Indianapolis, IN) or similar.
- 96 or 384 well plates.
- Thermocycler.
- ImageJ/Fiji software.

2.3 In Situ Hybridization on Mouse and Zebrafish Retinal Cryosections

- Thermocycler.
- LB agar plates with 100 μg/ml ampicillin, 0.1 mM IPTG, and 40 μg/ml X-GAL.
- LB liquid medium.
- Mini-quick spin columns (Roche Diagnostics, Indianapolis, IN).
- Heat incubators at 68, 55, and 37 °C.
- Shaker.
- Microscope and camera.
- A nontransparent box to create the wet chamber for hybridizations.
- Hydrophobic pen, special for high temperatures.

2.4 Zebrafish Embryo Collection, Handling and Fixation, and Morpholino Microinjection

- Adult male and female fish.
- Fish tank with plastic separators.
- Petri dishes.
- Heat Incubator at 28 °C.
- Plastic mold.
- Microinjection equipment, needles, microscope.
- Mineral oil and micrometer.
- Morpholino antisense oligonucleotide (MO).

2.5 Solutions

- Acrylamide monomer solution (for 50 ml): 4.2 g Acrylamide, 0.007 g Bis-acrylamide, 350 μl TEMED, 5 ml 10× PBS, double distilled H2O up to 50 ml.

- *Acrylamide embedding solution*: 50 μl of 10% APS in H_2O added to 10 ml of acrylamide monomer solution.
- 50× Denhardt's solution: 1% (w/v) Ficoll 400, 1% (w/v) Polyvinylpyrrolidone, 1% (w/v) Bovine serum albumin (Fraction V). Dissolved in DEPC-H_2O.
- Prehybridization solution for mouse cryosections: 42% (v/v) Formamide, 10% (w/v) Dextran sulfate, 1× Denhardt's Solution, 0.9 M NaCl, 0.1 M Tris–HCl (pH 8.0), 5 mM EDTA (pH 8.0), 10 mM NaH_2PO_4, 1 mg/ml yeast tRNA in DEPC-treated double distilled water.
- 20× SSC: 3 M NaCl, 300 mM Sodium Citrate, 800 ml of double distilled H_2O. Adjust pH to 7.0 using HCl. Adjust volume to 1 l with double distilled H_2O. Autoclave.
- *NTE*: 0.5 M NaCl, 10 mM Tris–HCl (pH 8.0), 5 mM EDTA (pH 8.0) for this and the next buffers, (*see* **Note 1**).
- *Buffer 1*: 100 mM Tris–HCl (pH 7.5), 150 mM NaCl.
- *Buffer 2*: 100 mM Tris (pH 9.5), 100 mM NaCl.
- *Buffer 3*: 100 mM Tris (pH 9.5), 100 mM NaCl, 50 mM $MgCl_2$.
- *E3 medium*: 300 mM NaCl, 10 mM KCl, 20 mM $CaCl_2$, 20 mM $MgSO_4$ in distilled water.
- *Hybridization solution for zebrafish cryosections*: 50% formamide, 1× Denhardt's solution, 10% dextran sulfate, 0.9 M NaCl, 100 mM Tris–HCl (pH 8.0), 5 mM EDTA (pH 8.0), 10 mM NaH_2PO_4, 1 mg/ml yeast tRNA.
- *Wash solution*: 50% formamide, 1× SSC, 0.1% Tween-20.
- *MABT solution*: 100 mM $C_4H_4O_4$ (pH 7.5), 150 mM NaCl, 0.1% Tween-20.
- *Blocking Buffer*: 100 mM Tris–HCl (pH 7.5), 150 mM NaCl, 1% BSA, 0.1% Triton X-100.
- *Alkaline phosphatase staining solution*: 100 mM Tris–HCl (pH 9.5), 150 mM NaCl, 50 mM $MgCl_2$, 0.1% Tween-20.
- *Hematoxylin stock solution*: 1% w/v in alcohol 99%. Let it rest (slow oxidation) during 2 weeks before use. Before use, dilute 1:1 in distilled water and filter (paper filter) to avoid precipitates.
- *Eosin solution*: 1% w/v in distilled water. Filter (paper filter) before use.

3 Methods

3.1 Dissection of Mouse Retinas for RNA Isolation

1. Sacrifice the number of P60 adult mice required (60 postnatal days is a standard age for fully differentiated retina) (*see* **Note 2**).

2. Hold the whole eye with the forceps, make a small cut on the cornea with a razor blade and remove the lens (Fig. 1a).
3. Pull out the neural retina (pink tissue) trying to leave the retinal pigmented epithelium (RPE) out (Fig. 1a), and transfer the retina to a 1.5 ml tube (two retinas per tube, use different tubes for different animals) and freeze immediately in liquid nitrogen. Keep them at −80 °C until use.
4. To obtain the retinal RNA, disrupt and homogenize the tissue in the buffer provided in the kit for tissue RNA isolation, using a polytron or a similar electronic blender.
5. Run 2–3 μl in a 1% w/v agarose/TBE gel to assess RNA quality.
6. Perform the cDNA synthesis reaction, 1 μg of RNA per tube, using a kit that allows to mix an oligodT primer and random hexamers or decamers to ensure the complete coverage of the gene of interest (strongly suggested for large genes).
7. Depending on the protocol and the initial mRNA purity and concentration, the cDNA should be diluted between 1:2 and 1:10 times in H_2O before the qPCR.
8. Prepare the qPCR reaction following the manufacturer's instructions (e.g., Lightcycler® 480 SYBR Green Master protocol) (*see* **Note 3**).

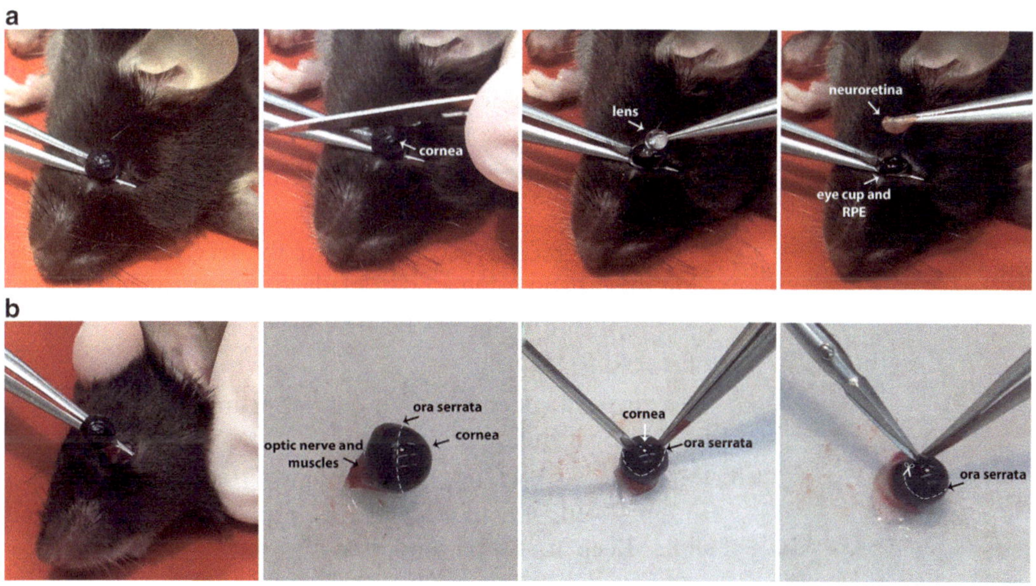

Fig. 1 Procedure for the dissection of the mouse neuroretina and eyecup. (**a**) Images illustrating four steps during the dissection of the mouse neuroretina for the purification of retinal RNA (*see* Subheading 3.1 for a complete description). (**b**) Images illustrating the dissection of the whole eyecup for the embedding and obtention of retina slides (see Subheading 3.2 for a complete description)

3.2 Dissection of Mouse Retinas for Cryosections

1. Enucleate the eye (Fig. 1b) and transfer it to a Petri dish with some drops of 4% PFA in 1× PBS.
2. Perform a small hole or cut in the cornea with a needle (Fig. 1b) to allow the PFA enter into the eye for 10 min.
3. Under the stereomicroscope and using the iridectomy scissors and forceps, cut around the iris to remove the cornea (Fig. 1b).
4. Fix the eyecups for 2 h in 4% PFA at RT.
5. Wash three times in 1× PBS for 15 min each at RT.
6. Embed the eyecups to avoid crystal formation (using either sucrose or acrylamide embedding).

3.2.1 Sucrose Embedding

- Transfer the eyecup to a tube containing a 10% w/v sucrose solution in PBS for 15 min or until the eyecup reaches the bottom at 4 °C. Repeat this step twice.
- Continue the cryoprotection by moving the eyecup to a 20% w/v sucrose solution in PBS for 15 min or until the eyecup settles down at 4 °C. Repeat this step twice.
- Finally, place the eyecup in a tube with 30% w/v sucrose solution in PBS and incubate it o/n at 4 °C.
- Proceed to **step** 7.

3.2.2 Acrylamide Embedding

- Infiltrate the eyecup in acrylamide monomer o/n at 4 °C.
- Polymerize 0.5 ml of fresh prepared acrylamide monomer in a 1.5 ml tube.
- Transfer the infiltrated eyecup on the acrylamide pad of the previous step (one eyecup per tube) and fill the tube with fresh embedding solution.
- Allow the polymerization of the acrylamide in ice (approx: 40–50 min).
- Under a stereomicroscope, remove the acrylamide surrounding the eyecup using a razor blade and iridectomy scissors on a Petri dish with double distilled H_2O.
- Proceed to **step** 7.

7. Cast the embedded tissue in a cryostat mold with OCT and freeze slowly in liquid nitrogen.
8. Using a cryostat, cut the blocks into 10–20 μm sections at -17 °C/-20 °C and place them on poly-lysine treated glass slides. Keep at -80 °C until used.

3.3 Cloning of the Riboprobe

1. Amplify with a standard Taq pol the desired region (between 400 and 800 bp in size) using gDNA or cDNA, depending on whether multiple exons are included. Check the PCR by gel electrophoresis and purify.

2. Ligate the fragment into the pGEM®-T vector following the manufacturer's protocol, and transform by heat shock in DH5α *E. coli* cells.
3. Plate onto LB supplemented with ampicillin/IPTG/XGal plates and incubate overnight at 37 °C for antibiotic and color selection
4. Pick six white colonies and grow o/n in 3 ml LB containing ampicillin (100 μg/ml) (*see* **Notes 4** and **5**).
5. Perform colony screening by plasmid minipreparation using 1.5 ml of the culture. Analyze if the plasmids are recombinant by restriction digestion.
6. Dilute the plasmid DNA down to 10 ng/μl. Use 1 μl to perform the PCR (final volume 50 μl) using the M13 primer (GTAAAACGACGGCCAGT) combined with the forward or reverse primer used in **step 1** (for each miniprep) (*see* **Note 6**).
7. Select two clones (one in the antisense and the other in the sense direction) for each gene, and sequence them for verification.

3.4 Generation of the Riboprobe

From this step onwards all the reagents must be RNase-free (*see* **Note** 7).

1. If PCRs from **step** 7 (Subheading 3.3) produced a good yield, the PCR reaction could be directly used for the generation of the riboprobes.
2. Mix 12 μl of the PCR product with 2 μl of T7 RNA polymerase, 2 μl of rNTP mix labeled with digoxigenin, 1 μl DTT (0.1 M), 1 μl RNase Inhibitor, and 2 μl of the T7 pol buffer. Incubate 2–3 h at 37 °C.
3. Add 2 μl of DNAseI and incubate 20 min at 37 °C. Separate 1 μl of the reaction for control.
4. Purify the riboprobe using mini-quick speed columns (Roche) following the manufacturer's protocol. Approximately 25–30 μl will be collected.
5. Test 1 μl of **steps 3** and **4** in a 1% w/v agarose/TBE gel.
6. If the test shows clear riboprobe production and recovery, dilute the riboprobe in 100% formamide (1:1 v/v), final concentration 50% formamide.

3.5 In Situ Hybridization on Mouse Cryosections

3.5.1 Day 1

1. Thaw the cryosections kept at −80 °C (**step 8**, Subheading 3.2) at RT for 1 h.
2. Use the hydrophobic pen (special for in situ hybridization) to surround each retina.
3. Remove OCT by washing the slides three times for 10 min in 1× PBS.
4. Incubate retinas in 2 μg/ml Proteinase K in PBS, for 20 min at 37 °C.

5. Rinse sections twice in 1× PBS for 5 min.
6. Fix retinas in 4% PFA in PBS for 20 min at RT.
7. Wash with 1× PBS.
8. Incubate 5 min in 0.1 M triethanolamine with 0.25% acetic anhydride (in PBS) at RT, followed by 5 min in 0.1 M triethanolamine with 0.5% acetic anhydride (in PBS) at RT.
9. Wash 5 min in 1× PBS at RT. Check that the hydrophobic circle drawn in **step 2** is still in a good condition. Otherwise redraw the circle.
10. Perform a prehybridization step by incubating 2–4 h in prehybridization solution at 55 °C in a nontransparent wet chamber (to avoid evaporation) (**Notes 8** and **9**).
11. Mix 150 μl of prehybridization solution with 5–10 μl of riboprobe (this is the hybridization solution).
12. Remove carefully prehybridization solution and substitute by the hybridization solution. Incubate slides o/n at 55 °C in the wet chamber.

3.5.2 Day 2

From this step onwards RNase-free conditions are not strictly required.

13. Warm 2× SSC and 2× SSC/42% formamide at 55 °C, and warm NTE, some 2× SSC and 0.2× SSC at 37 °C.
14. Wash slides 20 min in 2× SSC at 55 °C, and twice for 5 min in 2× SSC/42% formamide at 55 °C.
15. Wash three times for 5 min in NTE at 37 °C.
16. Incubate 30 min in 10 μg/ml RNaseA (in NTE) at 37 °C.
17. Rinse 15 min in NTE at 37 °C.
18. Wash twice for 15 min in 2× SSC at 37 °C, and twice for 15 min in 0.2× SSC at 37 °C.
19. Incubate 5 min in Buffer 1 at RT.
20. Block 1 h in 1% BSA+0.1% Triton X-100 in Buffer I at RT.
21. Incubate sections o/n in Buffer I containing anti-DIG-AP (1:1000) at 4 °C.

3.5.3 Day 3

22. Wash twice for 15 min in Buffer 1 at RT, 5 min in Buffer 2 at RT, and 5 min in Buffer 3 at RT.
23. Add the BMP substrate on each slide and incubate at RT in the dark (*see* **Note 10**).
24. After 30 min, check regularly the sections under the microscope (*see* **Note 11**).
25. Stop reaction by washing with PBS, and mount using Fluoprep and a coverslide (Fig. 2, positive mRNA localization is detected in blue).

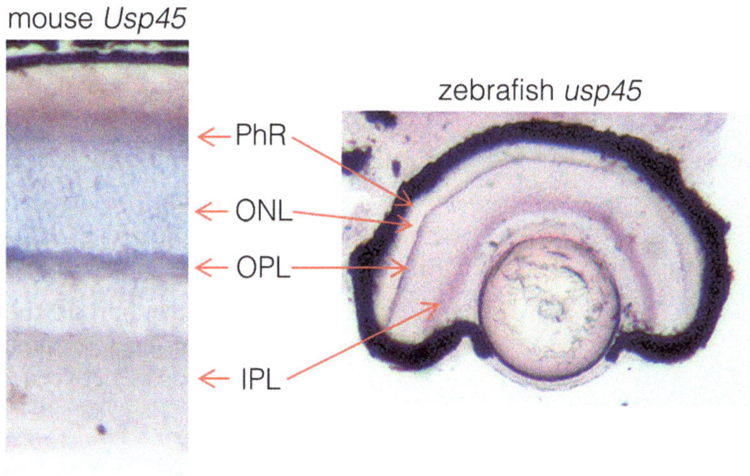

Fig. 2 Comparative *Usp45* in situ hybridizations in P60 mouse and 7 days zebrafish retinal cryosections, showing a strong correspondence of the mRNA localization in the retina of the two animal models. Of note, usp45 mRNA is found in the inner segment of the photoreceptors but also in the outer and inner plexiform layers, where most synapses occur, suggesting a role in the signal transduction pathway or in the regulation of the synaptic signal transmission rather than in photoreceptor differentiation fate. *PhR* photoreceptor cell layer, *ONL* outer nuclear layer, *OPL* outer plexiform layer, *IPL* inner plexiform layer

3.6 Zebrafish Embryo Collection, Handling and Fixation

1. Place several pairs of one male and one female adult zebrafish in a fish tank prepared for fish egg laying, but the male and female of each mating pair should be set close but separated with a plastic separator. Set a cycle of 14-h light/10-h dark cycle room for 24 h (*see* **Note 12**).
2. Next day remove the separator and allow the pair to mate. After fertilization, collect the eggs with a plastic pipette and allow them to develop in a Petri dish with 1× E3 medium.
3. Breed the eggs in an incubator at 28 °C until they reach the desired developmental stages (we obtained embryos at 12 h, 24 h, 36 h, 48 h and 72 hpf.) (*see* **Note 13**) [8].
4. Transfer the selected embryos in an Eppendorf tube. Fixation is performed with a 4% Paraformaldehyde (PFA) solution (w/v) in PBS for 2 h at room temperature.
5. Wash three times with PBS 1× for 10 min each.
6. Immerse successively the fixated embryos into 20 and 30% sucrose in PBS (w/v) solutions for 30 min each at room temperature. Finally, immerse the embryos o/n in 40% sucrose w/v at 4 °C.
7. Embed the embryos in OCT for 1 h, freeze them in liquid nitrogen, and store them at −80 °C.

3.7 Semi-Quantitative PCR

1. Add the components of the PCR to a 50 μl final volume reaction (standard reactions contain: 0.2 μM of each primer, 1.5 mM $MgCl_2$, 0.2 mM dNTPs, Taq Buffer 1×, and 1 U Taq DNA Polymerase). The sequences of the primer pairs used, including those of *β-actin*, a normalization control are listed in Table 1.
2. Mix gently the reaction components and set the tubes into a thermocycler. The standard PCR conditions used are shown in Table 2.
3. After agarose gel electrophoresis, the amplified bands are visualized and quantified using appropriate software (e.g., ImageJ/Fiji) to allow comparison between genes and developmental stages (Fig. 3).

3.8 In Situ Hybridization on Zebrafish Retinal Sections

1. Thaw the retinal sections stored at −80 °C, and let them air dry for 1 h at room temperature (RT).
2. Rinse them three times for 10 min with 1× PBS.

Table 1
Sequences and characteristics of the primer pairs used in semi-quantitative PCR and *in-situ* hybridizations

Real-Time qPCR (mouse retinas)			
Gene	**Orientation**	**Sequence (5′-3′)**	**Tm (°C)**
Usp45	Forward	AGCCTCACTGACGGCAGCG	71.5
	Reverse	AGGCTGCTTGGAAGCGATC	66.8
Usp53	Forward	GGAGTCCATGCATGACCCAGG	71.1
	Reverse	TGAACAACTGGACGGGTAGCTG	68.3
Gapdh	Forward	TGACAATGAATACGGCTACAGCAA	67.2
	Reverse	TACTCCTTGGAGGCCATGTAGG	66.1
Rho	Forward	GCCCTTCTCCAACGTCACAG	67.1
	Reverse	GCAGCTTCTTGTGCTGTACGG	67.1
In-situ Hybridization (mouse retinal cryosections)			
Gene	**Orientation**	**Sequence (5′-3′)**	**Tm (°C)**
Usp45	Forward	AGCCTCACTGACGGCAGCG	71.5
	Reverse	GACAGGACTGGACTGAGCAT	62
Usp53	Forward	CATCTGTGAGAACTGCTGGGCT	67.9
	Reverse	TGAACAACTGGACGGGTAGCTG	68.3
Rho	Forward	GCCCTTCTCCAACGTCACAG	67.1
	Reverse	GCAGCTTCTTGTGCTGTACGG	67.1

(continued)

Table 1
(continued)

Semi-quantitative PCR (zebrafish embryos)			
Gene	**Orientation**	**Sequence (5′-3′)**	**Tm (°C)**
usp45	Forward	CAGTCAGGAATTGCTGCATTACC	66.6
	Reverse	TGGGCAGCTAATGAGTCATCATG	68.1
usp53a	Forward	CTGACGCCTGCACGTCCAAG	71.6
	Reverse	AGTGAGGTCGGACTGCTCCGA	70.8
usp53b	Forward	GTCTCATGGATGATGCAGCGGA	71.7
	Reverse	TTGATACTCTGCCCACAGTTAC	60.6
β-actin	Forward	CTACAACGAGCTGCGTGTTGC	68.1
	Reverse	CGGTCAGGATCTTCATGAGGT	65.3

In-situ Hybridization (zebrafish retinal cryosections)			
Gene	**Orientation**	**Sequence (5′-3′)**	**Tm (°C)**
usp45	Forward	TCTCAGACCCACATGCTGAATG	67.4
	Reverse	GTCCACTGAGCCTCCTGCTGT	68.2
crx	Forward	CCTTCCCGAGTCCAGAGTTC	65.1
	Reverse	AAGAGCCATAGCCCTGGCTG	67.6

Control of knockdown			
Gene	**Orientation**	**Sequence (5′-3′)**	**Tm (°C)**
usp45	Forward	TCTCAGACCCACATGCTGAATG	67.4
	Reverse	CCTCCACTCTCATAGAGTCCAG	61.8

!*β-actin* was used as a normalization control

Table 2
Semi-quantitative PCR conditions

Step	Temperature (°C)	Time	
Hot start	94	3 min	
Denaturation	94	10 s	×35
Annealing	58	30 s	
Elongation	72	25 s	
Stop	12	∞	

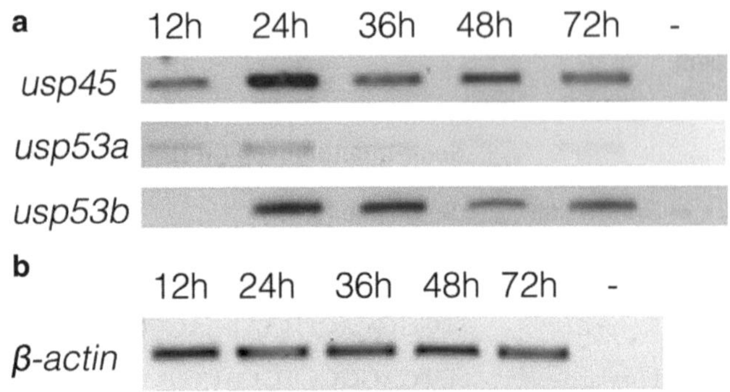

Fig. 3 Semi-quantitative expression analysis during embryonic development in zebrafish. Expression levels of (**a**) three studied genes, *usp45*, *usp53a*, and *usp53b*. There is a single *usp45* gene in zebrafish, which is expressed in the five studied embryonic developmental stages at a high level of expression. *usp53* has two highly similar paralogues in zebrafish, *usp53a* and *usp53b*. Note that *usp53a* is expressed at lower levels than *usp53b*, at the analyzed stages; (**b**) *β-actin*, used for normalization

3. Denature the riboprobes (antisense and sense) for 5 min at 68 °C and add 0.1–1 μg/ml of each one to their corresponding in situ hybridization solution. (Riboprobes are prepared as in Subheadings 3.3 and 3.4). Incubate overnight (at least 16 h) at 68 °C in a wet chamber protected from light.
4. After hybridization, wash the slides thrice for 30 min each at 68 °C in wash solution, and thrice for 30 min at RT in MABT.
5. Block them in blocking buffer for 4 h at RT.
6. Incubate o/n at 4 °C with an anti-digoxigenin-AP conjugate antibody (dilution 1:1000) in Blocking Buffer.
7. Wash the sections once in MABT for 30 min at RT, and twice for 10 min each in staining solution of alkaline phosphatase, in a shaker.
8. Incubate with freshly filtered BMP and allow the reaction to develop until a clear expression signal is obtained in the antisense hybridized sections or if staining appears in the sense sections. The reaction is stopped by washing in 1× PBS.
9. Mount the sections in fluoprep before making photographs with a camera attached to a light microscope (Fig. 2, positive mRNA localization is detected in blue).

3.9 Morpholino Microinjection in Zebrafish Embryos

1. Collect the embryos as described above, and place the embryos in chambers and align them in the same direction. To prepare the chambers set a plastic mold into a Petri dish containing 1.5 % (w/v) liquid agarose with 1× E3 medium. Once the agarose is gelified, remove the plastic mold and keep it at 4 °C.
2. Turn on the air source and the microinjector and insert the needle. Pinch off the needle at the point of interest using a

microscope and a pair of sharp forceps. To calculate the volume of each microinjection, use a drop of mineral oil on a micrometer (*see* **Note 14**).

3. Mix the morpholino antisense oligonucleotide (MO) of interest (designed and synthesized by Gene Tools, *see* **Note 15**) with 0.5 % phenol red, which serves as a visible marker for the injection of the solution into the embryo.
4. Microinject the MO of interest and the standard scrambled MO (negative control) into the yolk of the aligned 1- to 4-cell stage

Table 3
Injected volumes and final concentrations of MO-USP45

MO	MO injected volumes (pl)	MO final concentration (μM)
MO-USP45 (1)	65	0.036
MO-USP45 (2)	100	0.29
MO-USP45 (3)	150	1

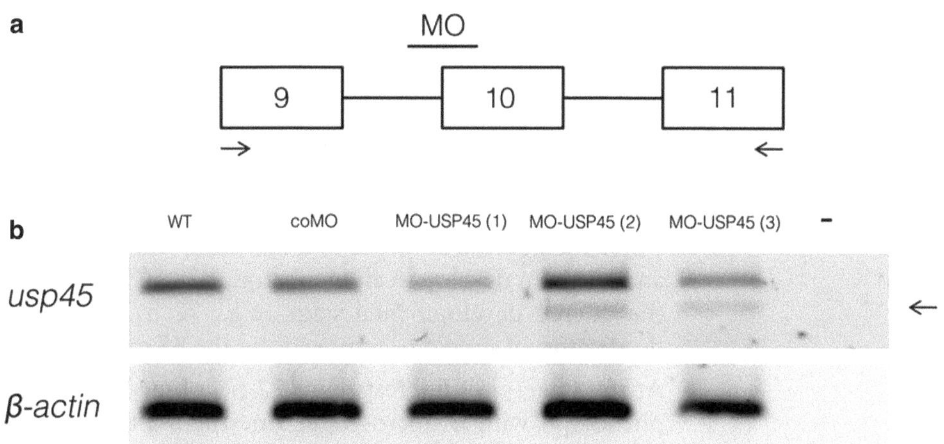

Fig. 4 Knockdown of *usp45* in zebrafish by morpholino microinjection. (**a**) The position of the morpholino in the unprocessed RNA and the primers used for PCR test are indicated. The MO targeted the acceptor splice site of intron 9 and the beginning of exon 10 of *usp45* (MO-USP45: 5′-AATGCGCTGTCAGTGAAAACACAAT-3′). A scrambled MO was used as negative control (coMO: 5′-CCTCTTACCTCAGTTACAATTTATA-3′); (**b**) Effects of the morpholino knockdown on the transcription of *usp45* detected by semi-quantitative RT-PCR, showing inhibition of intron 9 splicing (causing the introduction of a STOP codon, which in turn would result in a premature protein truncation and probably, non-sense-mediated decay of the misprocessed mRNA). Several morpholino concentrations were tested. The lowest tested concentration (0.036 μM) was the most efficient, as it knocked down *usp45* expression to 51.5 % while still being compatible with viability. *β-Actin* was used for normalization. coMO: control standard scramble MO; MO-USP45 (1): 0.036 μM; MO-USP45 (2): 0.29 μM; MO-USP45 (3): 1 μM. The *arrow* indicates the band produced when exon 10 is skipped by the morpholino action

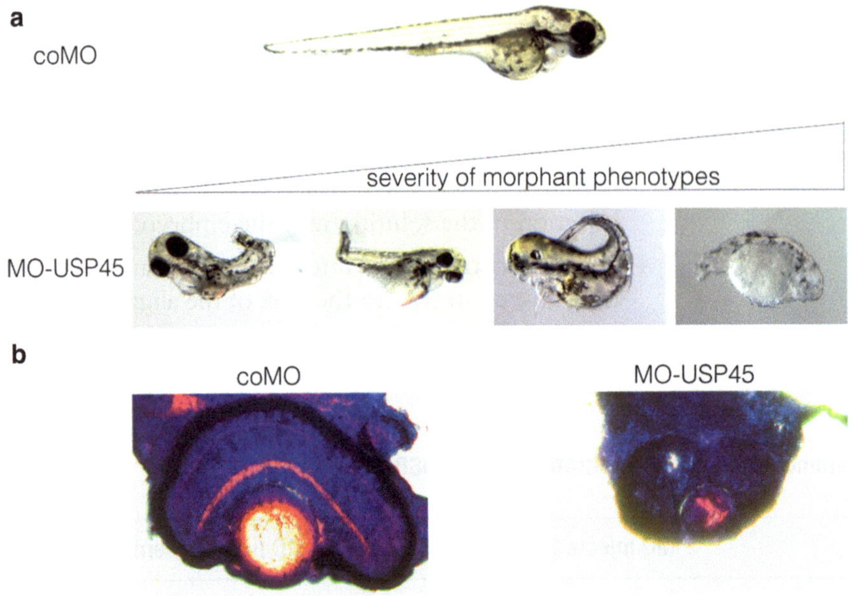

Fig. 5 (**a**) Morphant phenotypes observed in 72 hpf embryos after MO microinjection in eggs (0.036 μM). The main traits are: eye size reduction, small body size with small or no tail, and disruption in the formation of notochord (mild phenotype). 6 % of the embryos show a very severe phenotype with no eyes; (**b**) Hematoxylin and eosin stained eye sections in coMO (control embryos, injected with a standard scramble morpholino) and MO-USP45 injected embryos (72 hpf). MO-USP45 injected morphants show defects in eye formation and the lamination of the retina, with no distinguishable photoreceptors or plexiform layers (IPL and OPL) and with smaller retinas (low number of neurons), compared with the coMO retinas

embryos [9] (*see* **Note 16**). The volumes and inferred final concentrations of injected MO-USP45 are shown in Table 3.

5. Move the injected embryos to a Petri dish with 1× E3 medium and let them develop in an incubator at 28 °C until they reach the desired developmental stage (e.g., 72 h). Every day, the dead embryos should be removed and the E3 medium changed.
6. At the stage(s) of interest (e.g., 72 h), observe the morphant phenotype with a microscope. Anesthetize the embryos by immersion in a tricaine solution (4.2 % v/v in 1× E3 medium) and photograph them with a camera attached to a light microscope (Fig. 5a). After setting them back to 1× E3 medium, the embryos recover from the anesthesia.
7. Select half of the animals to perform a semi-quantitative RT-PCR assay to evaluate the knockdown effect of the *usp45* MO (Fig. 4).
8. Fix the other half of the embryos for histological morphological analysis, as described in the **step 4** (Subheading 3.6), and obtain retinal cryosections (14–16 μm width).

3.10 Hematoxylin/Eosin Staining of Zebrafish Retinal Sections

1. Thaw the retinal sections (14–16 μm width) stored at −80 °C, and let them air dry for 1 h at room temperature (RT).
2. Rinse retinal sections with 1× PBS for 10 min.
3. Stain with freshly diluted and filtered hematoxylin solution for 90 s, and wash with distilled H_2O for 10 min.
4. Stain with freshly filtered eosin solution for 4 min and 30 s, and wash quickly with distilled H_2O.
5. Mount the sections with fluoprep and take photographs with a camera attached to a light microscope.
6. Compare the phenotype qualitatively between scramble and morpholino injected animals (Fig. 5b).
7. Count and compare the number of nuclei rows in the OPL with the help of ImageJ, and compare the relative width of the retina, the outer photoreceptor segment, and the outer plexiform layers (*see* **Note 17**).

4 Notes

1. The pH of Tris buffers of these protocols must be accurate to obtain good results.
2. Due to the inter-individual differences in transcription, we suggest 3–6 animals, depending on the amount of genes to be screened and the number of replicates required for statistical significance.
3. For the qPCR, oligonucleotides to amplify fragments of around 100 bp (preferably, primers should map at different exons to prevent amplification due to genomic contamination) should be used. The annealing (melting) temperature should be close to or higher than 60 °C. Primers of control genes to normalize expression values should be also designed. All primers should be checked first to assess that they amplify a single amplicon.
4. For each gene two riboprobes, sense (negative control) and antisense (assay), are required.
5. Fragments shorter or equal to 300 bp may result in light blue or blue colonies if they are in frame.
6. This will allow the identification of the direction of the probe. If a band with the correct size is observed using the M13 and forward primers the probe is cloned in the antisense direction, while amplification using the M13 and reverse primers indicates sense probes.
7. Before starting, all tubes and stable solutions must be autoclaved twice. Bench must be clean and RNase/DNase-free filter tips are recommended. Non-autoclavable solutions must be freshly prepared and filtered (Ø 22 μm), and only used for RNA-related purposes.

8. As a wet chamber, use a nontransparent flat box, with the bottom covered by a wet paper filter. Place the slides on top, retinas facing up.
9. Never apply any solution directly on the tissue section, morphology might be damaged or the retina section might detach.
10. Use a freshly clarified BM Purple AP (BMP) substrate, either by filtering through a 0.45 μm filter or by centrifuging at maximum speed for 5 min.
11. This step might take from minutes to hours. Replace every 2 h the BMP solution to avoid precipitation. The reaction is faster at RT, but it can also proceed more slowly by incubating at 4 °C using longer incubation times.
12. Three days before mating and egg laying, zebrafish animals should be fed with dry *Artemia salina* pellets to increase the metabolism and favor egg production. Animals should be young and well fed to lay eggs. If the mating pairs do not lay eggs, or the eggs are too fragile and do not survive microinjection, buy fresh younger animals.
13. Several embryonic developmental stages should be analyzed to assess the expression of our genes of interest. Although adulthood and sexual maturity is reached at 90 days, most tissues and organs are developed during the first 72 h of the larval development.
14. For instance, an oil drop of Ø 100 μm contains 520 pl of injection material.
15. Many efficient morpholinos are directed against splice acceptor or donor sites, so that they inhibit proper mRNA splicing.
16. Several concentrations of MO-USP45 (different volumes) were microinjected in order to find the more efficient in knocking down *usp45* (the concentration is calculated as the final DNA amount per embryo).
17. Additionally, confocal microscopy for immunodetection of specific proteins and retinal markers (e.g., rhodopsin for rod photoreceptors) could also be performed for detailed analysis.

Acknowledgments

We are indebted to S. García-Monclús and D. Bruguera for technical advice. Funding for the work was provided by grants BFU2010-15656 (MICINN), SAF2013-49069-C2-1-R (MINECO) to G.M. and 2014SGR-932. The group is a member of the Proteostasis COST Action (BM1307).

References

1. Zeiss CJ (2013) Translational models of ocular disease. Vet Ophthalmol 16(Suppl 1):15–33
2. Song BJ, Tsang SH, Lin C-S (2007) Genetic models of retinal degeneration and targets for gene therapy. Gene Ther Mol Biol 11:229–262
3. Weskamp G, Cai H, Brodie TA, Higashyama S, Manova K, Ludwig T, Blobel CP (2002) Mice lacking the metalloprotease-disintegrin MDC9 (ADAM9) have no evident major abnormalities during development or adult life. Mol Cell Biol 22(5):1537–1544
4. Kurth I, Thompson DA, Ruther K, Feathers KL, Chrispell JD, Schroth J, McHenry CL, Schweizer M, Skosyrski S, Gal A, Hubner CA (2007) Targeted disruption of the murine retinal dehydrogenase gene Rdh12 does not limit visual cycle function. Mol Cell Biol 27(4):1370–1379
5. Liu X, Bulgakov OV, Darrow KN, Pawlyk B, Adamian M, Liberman MC, Li T (2007) Usherin is required for maintenance of retinal photoreceptors and normal development of cochlear hair cells. Proc Natl Acad Sci U S A 104(11):4413–4418
6. Garanto A, Vicente-Tejedor J, Riera M, de la Villa P, Gonzalez-Duarte R, Blanco R, Marfany G (2012) Targeted knockdown of Cerkl, a retinal dystrophy gene, causes mild affectation of the retinal ganglion cell layer. Biochim Biophys Acta 1822(8):1258–1269
7. Raghupathy RK, McCulloch DL, Akhtar S, Al-mubrad TM, Shu X (2013) Zebrafish model for the genetic basis of X-linked retinitis pigmentosa. Zebrafish 10(1):62–69
8. Kimmel CB, Ballard WW, Kimmel SR, Ullmann B, Schilling TF (1995) Stages of embryonic development of the zebrafish. Dev Dynam 203:253–310
9. Rosen JN, Sweeney MF, Mably JD (2009) Microinjection of zebrafish embryos to analyze gene function. J Vis Exp 25:115, http://www.jove.com/index/Details.stp?ID=1115, doi: 10.3791/1115

Chapter 4

Immunodepletion and Immunopurification as Approaches for CSN Research

Amnon Golan, Ning Wei, and Elah Pick

Abstract

The COP9 signalosome (CSN) is an evolutionary conserved complex that is found in all eukaryotes, and implicated in regulating the activity of Cullin-RING ubiquitin Ligases (CRLs). Activity of CRLs is highly regulated; complexes are active when the cullin subunit is covalently attached to the ubiquitin like modifier, Nedd8. Neddylation/deneddylation cycles are required for proper CRLs activity, and deneddylation is performed by the CSN complex.

We describe here a method utilizing resin-coupled antibodies to deplete the CSN from human cell extracts, and to obtain endogenous CSN complexes by immunopurification. In the first step, the cross-linked primary antibodies recognize endogenous CSN complexes, and deplete them from cell extract as the extract passes through the immunoaffinity column. The resulting "CSN-depleted extract" (CDP) is rich in neddylated cullins that can be used as a substrate for cullin-deneddylation assay for CSN complexes purified from various eukaryotes. Consequently, regeneration of the column results in dissociation of a highly purified CSN complex, together with its associated proteins. Immunopurification of the CSN from various human tissues or experimental conditions is advantageous for the generation of numerous CSN-interaction maps.

Key words Nedd8, COP9 signalosome, Cullin-RING ubiquitin ligase, Immunodepletion, Immunopurification

1 Introduction

The COP9 signalosome (CSN) is an evolutionary conserved 8-subunit (Csn1-8) complex [1]. Classic CSNs (with eight subunits) are found in most eukaryotes and have been shown to regulate the activity of cullin-RING ubiquitin Ligases (CRLs). CRLs are the largest family of ubiquitin E3 ligases, responsible for fifth of all ubiquitinated substrates within a cell in human [2]. CRLs are modular complexes represented by the archetypical Skp1-Cullin1-F-box (SCF) complex [3, 4]. The SCF consists of a Cullin-1 scaffold subunit that interacts with the RING domain protein Rbx1 via its C-terminus, and with the cullin-specific adaptor protein (Skp1) via its N-terminus. Skp1 binds to an F-box protein (FBP) that serves as a substrate receptor (SR), which in turn recruits

Rune Matthiesen (ed.), *Proteostasis: Methods and Protocols*, Methods in Molecular Biology, vol. 1449,
DOI 10.1007/978-1-4939-3756-1_4, © Springer Science+Business Media New York 2016

substrates, for ubiquitination (Fig. 1) [3, 4]. Activity of CRLs is highly regulated; complexes are active when the cullin subunit is covalently attached by the ubiquitin-related modifier, Nedd8 (a.k.a. neddylation) [5–7]. Removal of Nedd8 is carried out by the CSN (a.k.a. deneddylation). Neddylation/deneddylation cycles are required for the dynamic regulation of CRLs function in vivo (Fig. 1) [2, 8, 9].

Deneddylation of cullins by the CSN depends on the metal-binding MPN^{+}/JAMM metalloprotease motif harbored in the fifth subunit Csn5, which is active only when integrated in the CSN holoenzyme [6, 10, 11]. CRL-free CSN complex is kept inactive, while the CSN-CRL interactions result in a substantial rearrangement that triggers a cascade of conformational changes, leading to the activation of Csn5 [12]. The CSN also controls CRL activity in a nonenzymatic manner by steric effects, possibly by preventing interactions between the substrate and SR, and between Rbx1 and the E2 [13–15]. In all studied multicellular organisms, CSN subunits are required for viability; and loss-of-function mutants display critical pleiotropic defects such as abnormal response to DNA damage, defects in cell cycle progression or in development [1, 16–19]. CSN deficiency is also associated with altered half-life of many transcription factors (TFs), suggesting a role in regulating gene expression [20, 21]. In addition, increased expression of several CRLs and CSN subunits is correlated with various tumors [9, 22–32]. As a result, CSN, CRLs, and the Nedd8 conjugation pathway have recently emerged as drug targets for cancer chemotherapy [31, 33, 34].

We describe here a powerful set of resin-coupled antibodies, which are suitable for immunodepletion and immunoaffinity

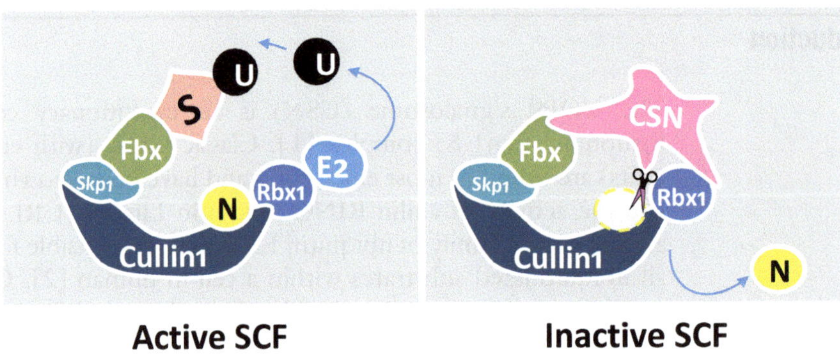

Fig. 1 Inactivation of the SCF by the CSN complex. CRLs are represented by the archetypical Skp1-Cullin1-F-box (SCF) complex. The Fbox protein (Fbx) is a substrate receptor that recognizes a specific substrate (S). The E2 enzyme is attached to the E3 subunit Rbx1 and donates ubiquitin (U) to the substrate (*left*). CRLs are active when the cullin scaffold subunit is covalently attached to Nedd8 (N). The eight-subunit CSN complex inactivates the SCF through enzymatic hydrolysis of Nedd8 and through steric clashes between Fbx and substrate and between Rbx1 and the E2 (*right*)

purification of the CSN complex. The method involves cross-linking of primary antibodies against CSN subunits to sepharose resin, producing CSN immunoaffinity columns. These antibody columns can be used to generate “CSN-depleted extract” (CDP). The antibodies recognize endogenous CSN assemblages and deplete them.

The following step regenerates the column, which includes thorough washing of the column, together with its associated proteins.

The CDP produced by this method is enriched in endogenous Nedd8-cullin conjugates in a most natural but cell-free state, and is suitable for deneddylation activity assay (Fig. 2, I). The method also enables elution of the immunoaffinity purified CSN complex that can be used to study the interactome of endogenous CSN, using various conditions or treatments (Fig. 2, II). Notably, using this method, we have been able to deplete CSN, and approach cullins for deneddylation activity [35–37]. This has proven that the method is valuable for CSN functional studies.

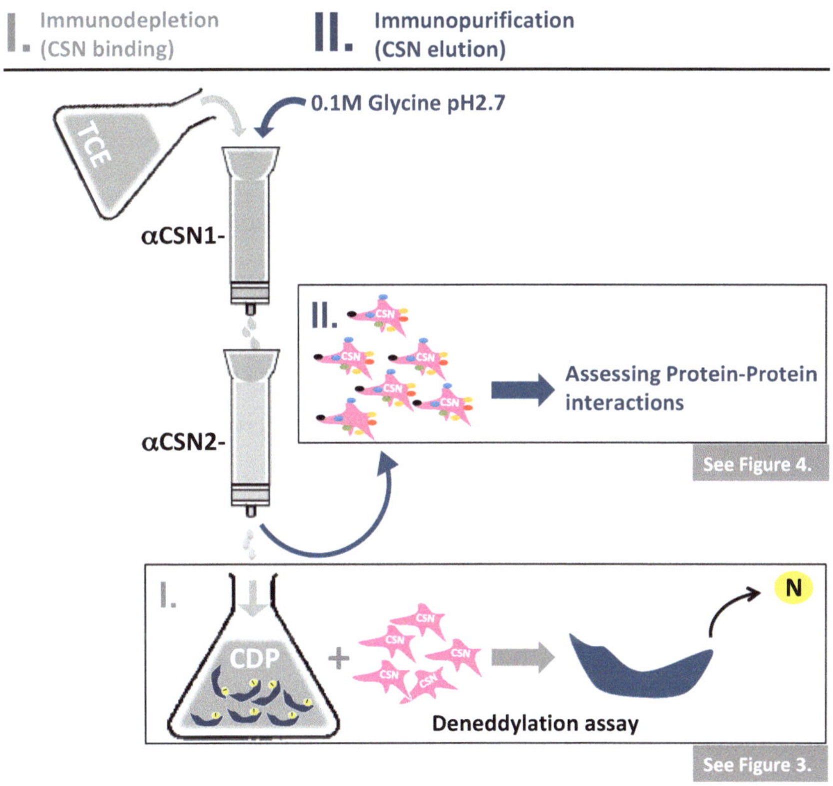

Fig. 2 Illustration of CSN immunodepletion and immunoaffinity purification. (*I*) A selective removal of the CSN from total cells’ extract by immunodepletion. CSN-depleted extract can be approached to determine deneddylation activity of CSN complexes purified from various eukaryotes. (*II*) Immunoaffinity purification of the CSN and associated proteins is accessible for CSN studies using various conditions, treatments, or human cell types

2 Materials

2.1 Production of Immunoaffinity Columns

1. HiTrap NHS-Activated HP, # 17-0716-01, GE Healthcare.
2. ÄKTA Protein Purification Systems, GE Healthcare.
3. Anti-Csn1 and anti-Csn2 antibodies, as described in Wei et al. [38].

2.2 Cells and TCE Preparation

1. Hek293 cells—293T (ATCC® CRL-3216™).
2. CO_2 incubator, temperature setting 37 °C incubator with 5% CO_2.
3. DMEM media.
4. Heat inactivated Fetal Bovine serum (HI FBS).
5. Sterile PBS (137 mM NaCl, 2.7 mM KCl, 10 mM Na_2HPO_4, 2 mM KH_2PO_4, pH adjusted to 7.4).
6. 150 cm^2 cell culture plastic dishes.
7. Single-Edge Blades.
8. Lysis buffer (0.5% NP-40, 50 mM Tris (pH 7.4), 150 Mm NaCl). Add to the buffer immediately before use: Protease Inhibitor Cocktail (EDTA-Free, B14001, Sigma), PhosSTOP (#04906845001, Roche).
9. Dounce homogenizer, 10 ml.
10. 15, 50 ml Conical plastic tubes.
11. 15 ml Centrifuge tubes.
12. 1.5 ml microcentrifuge tubes.
13. 10 ml Syringe.
14. Nonsterile Syringe Filters; 0.22 μM filter.
15. 5× Laemmli sample buffer (375 mM Tris (pH 6.8), 50% glycerol, 350 mM SDS, 250 mM DTT, 0.1% Bromophenol Blue).
16. 95 °C dry block heater for microcentrifuge tubes.

2.3 Preparation of CDP

1. EtOH 20% (HPLC grade)—minimum 1 l.
2. MQ (ultrapure water)—minimum 1 l.
3. ÄKTA Protein Purification Systems, GE Healthcare.
4. Buffer A (0.05% NP-40, 50 mM Tris (pH 7.4), 150 mM NaCl).
5. Buffer B (50 mM Tris (pH 7.4), 150 mM NaCl).
6. 10 ml loop, or two loops of 5 ml connected to each other.
7. Microcentrifuge tubes.
8. 15, 50 ml Conical plastic tubes.
9. Liquid nitrogen tank.

2.4 Eluting the Immunoaffinity Purified CSN Complex

1. Tris–HCl buffer 1 M pH 8.5.
2. Elution buffer (100 mM Glycine (pH 2.7)).
3. Buffer C (50 mM Tris (pH 11), 150 mM NaCl).

4. Buffer D (50 mM Tris (pH 7.4), 150 mM NaCl, 0.02 % NaN_3).
5. 5× Laemmli sample buffer (375 mM Tris (pH 6.8), 50 % glycerol, 350 mM SDS, 250 mM DTT, 0.1 % Bromophenol Blue).
6. 95 °C dry block heater for microcentrifuge tubes.

2.5 Clean-Up of FPLC

1. 20 % EtOH (HPLC grade)—minimum 1 l.
2. MQ (ultrapure water)—minimum 1 l.

2.6 Evaluation of immunodepletion

1. Csn1, Csn2 antibody, as described by Wei et al. [38].
2. Csn3 antibody (Santa Cruz Biotechnology (RR12): sc-100693).
3. Cullin 4a and Ddb1 antibodies, as described by Pick et al. [39].
4. Cullin 1 antibody (Santa Cruz Biotechnology (H-213): sc-11384).
5. Cullin 2 antibody (Santa Cruz Biotechnology (N-19): sc-8554).
6. Cullin 3 antibody(Santa Cruz Biotechnology, UL-3 (C-18): sc-8556).

2.7 Deneddylation Activity Assay

1. Liquid nitrogen.
2. (−80 °C) Freezer.
3. BSA (Bovine Serum Albumin) 0.4 μg/μl.
4. Purified CSN isolated from human erythrocytes (Enzo Life Sciences Cat.# BML-PW9425-0020; LifeSensors Cat.#: CP009).
5. 5× Laemmli sample buffer (375 mM Tris (pH 6.8), 50 % glycerol, 350 mM SDS, 250 mM DTT, 0.1 % Bromophenol Blue).
6. 30, 95 °C dry block heater for microcentrifuge tubes.

3 Methods

Herein, we describe a protocol for CSN immunodepletion (Subheading 3.2) followed by its immunoaffinity purification (Subheading 3.3). Next, we will illustrate the use of these techniques for CSN research by describing deneddylation activity assay (Subheading 3.7) and the analysis of CSN interactions with associated proteins (Subheading 3.8).

3.1 The Production of Affinity Columns

The immunoaffinity columns containing immobilized antibody against CSN subunits are required in this method. Here, you will prepare two columns, by covalent coupling of 3 mg of affinity purified anti-Csn1 or anti-Csn2 antibodies, to a prepacked ready-to-use column of activated sepharose resin (HiTrap NHS-Activated HP, # 17-0716-01, GE Healthcare).

1. Perform antibody coupling precisely as detailed by the manufacturer (https://www.gelifesciences.com/gehcls_images/GELS/Related%20Content/Files/1335359522418/litdoc18113480_20120425152132.pdf).

3.2 Preparation of TCE

This method enables preparation of TCE out of any mammalian cells or tissues, but we frequently used the Hek293 cells—293T (ATCC® CRL-3216™), which were often approached for CRLs-CSN studies [5, 39–42].

1. Split 90% confluent cells culture in 1:10 ratio, and seed five dishes of 150 cm^2 with Hek293 cells in Dulbecco's Modified Eagle's Medium (DMEM), supplemented with 10% heat inactivated Fetal Bovine serum, 2 mM L-Glutamine, 50 μg/ml of Penicillin Streptomycin mixture.
2. Culture cells at 37 °C with 5% CO_2 until 80% confluence (approximately 3 days).
3. Wash gently the culture dishes with 10 ml of ice-cold Phosphate Buffer Saline (PBS), by pipetting the buffer slowly on the dish walls, and not directly on cells.
4. Tilt the plate gently to wash the cells.
5. Pump off used PBS and add 10 ml of fresh buffer.
6. Harvest cells by releasing them with Single edge safety razor blades.
7. Collect cells and buffer with 5 ml sterile pipettes into prechilled 15 ml falcon tubes.
8. From this point, carry out all procedures at 4 °C, unless otherwise stated.
9. Centrifuge 5 min, at 4000 × *g* at 4 °C.
10. Draw out the buffer, and keep the pellet on ice. It is possible to stop here, and freeze down the pellet in liquid nitrogen; keep at −80 °C.
11. By gentle pipetting, resuspend cell pellet into 9 ml of lysis buffer complemented with PhosSTOP, EDTA-free protease inhibitors.
12. Transfer the suspension into a prechilled 10 ml Dounce homogenizer.
13. Homogenize in Dounce homogenizer 25–30 times, avoiding the generation of foam, which can cause denaturation of proteins (the chromatin should be visualized).
14. Take the pistil off the homogenizer and keep the extract on ice for 10 min.
15. Repeat **steps 13–14** one more time.
16. Transfer the cells extract into prechilled 15 ml centrifuge tubes.
17. Centrifuge at 10,000 × *g* for 20 min at 4 °C.
18. Collect the clear lysate and transfer to new (or recycled) conical tubes.
19. Keep the pellet for Subheading 3.2, **step 26**.

20. Filter the clear lysate by a syringe filter, pore size 0.22 μM into a 50 ml conical plastic tube. This step is to prevent obstruction of a high pressure FPLC system (Subheading 3.3).
21. Keep the lysate tube on ice. This is your untreated (UT) sample.
22. Transfer 100 μl of the UT sample to a prechilled microcentrifuge tube.
23. Add 25 μl of 5× Laemmli sample buffer.
24. Heat to 95 °C for 5 min. Mark the tube with UT. This will be used later on to compare with the CDP by immunoblotting analysis.
25. Make ten additional aliquots of 50 μl UT of **step 21** of Subheading 3.2 in prechilled microcentrifuge tubes, mark all of them as "UT," freeze down in liquid nitrogen, and keep at −80 °C freezer. You will use them later as controls for deneddylation assay in Subheading 3.4.
26. Keep the rest of UT on ice; you will use it in Subheading 3.3.
27. To prepare a sample of the pellet for immunoblotting, resuspend the pellet of **step 19** of Subheading 3.2 in the initial volume of 10 ml with the hypotonic buffer.
28. Take a sample of 100 μl and transfer to microcentrifuge tubes.
29. Add 25 μl of 5× Laemmli sample buffer.
30. Mark the tube with "P" for pellet, and heat to 95 °C for 5 min in a test tube dry heater.

3.3 Preparation of CDP

A typical laboratory Fast Performance Liquid Chromatography (FPLC) used for protein purification is commonly refrigerated. Otherwise, keep the column area and the buffers chilled. Always degas and filter all FPLC buffers.

To start the procedure you first need to wash the system from previous users, and to make sure that the flow pressure is lower than 0.3 MPa.

1. Start a manual run to wash the system from 20% Ethanol with MQ water in a flow rate of 0.5 ml/min. Make sure to turn on both pump "A" and pump "B."
2. Transfer the pump filters from MQ to appropriate buffer "A" bottle and buffer "B" bottle.
3. Wash pump "A" with buffer "A."
4. Wash pump "B" with buffer "B."
5. Pre-equilibrate the system with buffer "A" using pump-wash basic program and run for 10 min.
6. Connect a 10 ml loop via ports 2 and 6 on injection valve of your FPLC system.
7. Set injection valve to the "LOAD" position.

8. Connect round-tip needle to the syringe and wash the loop with buffer A. This also frees the trapped bubbles in the injection port, and releases them directly to the waste.
9. Connect the columns:
   (a) Connect first the bottom of the anti-Csn2 column and keep the drip of buffer on the top of the column in a flow rate of 0.5 ml/min.
   (b) Connect the bottom of the anti-Csn1 column to the top of the anti-Csn2 column while dripping on the top of anti-Csn1 column to avoid bubbles.
   (c) Connect the top of the anti-Csn1 column to the tube.
   (d) If no bubbles appeared, and the columns are well connected, change the flow rate to 1 ml/min.
10. Wash column with 10 column volumes (CV) of buffer "A" (or continue to wash the column until the UV is <5 mAu and constant).
11. Wash manually the syringe and the injection loop with buffer "A" too.
12. Change the Flow Path of your FPLC to "inject," using injection rate of 0.5 ml/min with a 0.3 MPa of pressure alarm.
13. Load UT supernatant from **step 21** of Subheading 3.2 onto the loop.
14. During injection of UT, collect the buffer from the loop, which goes to the waste tube (flow through is kept only to avoid the loss of UT sample because of an unexpected working mistake).
15. To load the sample on the column change the Flow Path from "LOAD" to "INJECT," using injection rate of 0.5 ml/min with a 0.3 MPa of pressure alarm.
16. Hold a clean 50 ml conical tube at the waste to collect 10 ml of the unbound (UB) lysate. To prevent dilution of UB, it is possible to follow after the UV curve, and stop collecting the sample as soon as UV is <5 mAu and constant.
17. Seal your UB tube and keep on ice.
18. Keep washing the column with additional 20 CV of buffer "A" (40 ml).
19. Repeat **steps 13–18** of Subheading 3.3 three more times, by reloading the UB fraction again and again in order to deplete most of the CSN (see Fig. 4a).
20. Keep the latest UB fraction. This is your CDP.
21. Aliquot your CDP into 500 μl fractions, and keep in prechilled microcentrifuge tubes.

22. Mark the tubes as "CDP1" with date and volume notification.
23. Freeze down CDP1 aliquots with liquid nitrogen, and keep frozen in −80 °C until use.

3.4 Eluting the Immunoaffinity Purified CSN Complex

At this point, immunodepletion is completed (Fig. 2, I), and you are starting the immunopurification part (Fig. 2, II). The set of columns is still connected to the FPLC, and is unwashed. In general, antibody-antigen binding is most effective in aqueous buffers at physiological pH. Accordingly, elution often occurs by raising/lowering the pH to disturb this interaction. The most widely utilized elution buffer for immunoaffinity purification is 0.1 M Glycine·HCl, pH 2.5–3.0. This buffer effectively dissociates most antibody-antigen interactions.

Wash the column with 20 CV of buffer "A," or until UV is <5 mAu and constant. Use flow rate of 1 ml/min and a "LOAD" flow path.

1. Switch to pump "B" and wash the column with 10 CV of buffer "B" with a flow rate of 1 ml/min. This step is necessary if you plan to analyze your CSN sample by mass spectrometry.
2. Wash the pump filters "A" and "B" from the buffers, and transfer them into the Elution buffer (Pump "A"), and buffer "C" (Pump "B").
3. Wash the pumps with the corresponding buffers.
4. Before eluting the CSN, set 20 microcentrifuge tubes on the fraction collection system, and each of them includes 100 μl of 1 M Tris–HCl buffer (pH 8.5) (to prevent acidic hydrolysis of the eluted proteins).
5. Set the elution size to 0.5 CV (1 ml).
6. Elute the CSN into the tubes.
7. To avoid damage to the cross-linked antibodies, you need to wash away the acidic buffer with 10 CV (20 ml) of alkaline Tris–HCl buffer (Buffer "C") immediately after elution.
8. Wash pump "A" with buffer "D." Buffer "D" is similar to Buffer "A," but includes also azide.
9. Prepare 100 μl of samples for immunoblotting, by taking a sample of 100 μl from each of the elution fractions, and add 25 μl of 5× Laemmli sample buffer.
10. Heat to 95 °C for 10 min in a dry block heater.
11. The rest of your elution sample will be restored at −80 °C until you identify peak of CSN (Fig. 4).

3.5 Clean-Up of FPLC

1. Disconnect and recap the columns.
2. Label column's last day of use and store in 4 °C for the next use.

3. Replace line to port using "drop-to-drop" connection.
4. Pump-wash the system with DDW in a flow rate of 1 ml/min.
5. Pump-wash the system with 20% Ethanol in a flow rate of 1 ml/min. Make sure that Ethanol is not flowing through the columns, since it could destroy antibody function.

3.6 Evaluating CSN Immunodepletion

Approach the UT and P samples from **steps 21** and **19** of Subheading 3.2 for immunoblotting with CSN antibodies to evaluate depletion levels, and with antibodies recognizing cullins, to evaluate the high accumulation of neddylated cullins in CDP (Fig. 3).

1. Accordingly, you can move to Subheading 3.7.

3.7 Deneddylation Activity Assay

The CSN/cullin enzyme-substrate interplay is highly conserved between yeast and mammals [35], and deneddylation of mammalian cullins can be carried out by the most diverged CSN complex, such

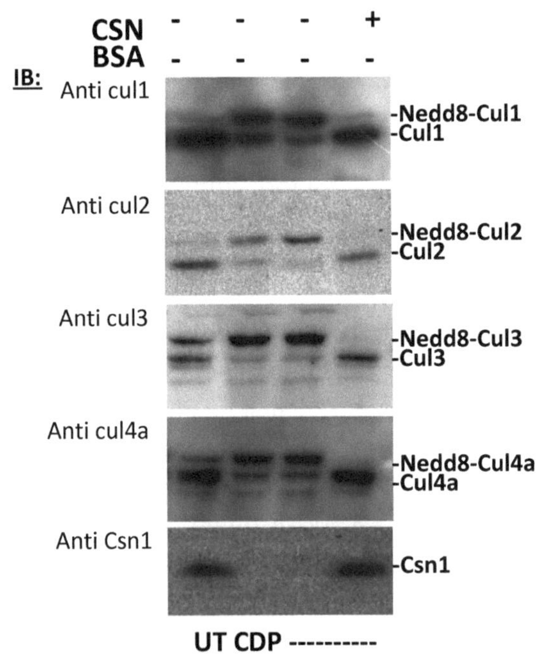

Fig. 3 CSN deneddylation assay. CSN activity assays were performed using CSN-depleted cell extracts as a source of neddylated cullin substrates (*lanes 2–4*). CSN isolated from human erythrocytes was tested for deneddylation activity (*lane 4*). Both untreated extract (UT) that was taken before CSN depletion (*lane 1*), and BSA (*lane 3*) served as negative controls. Neddylation levels of Cul4a and Cul1-3 were examined by immunoblotting with respective antibodies. Depletion had confirmed by immunoblotting with anti-Csn1

as the CSN of *S. cerevisiae*. This suggests that human Nedd8-CRL conjugates may be used as an efficient substrate to study mechanistic aspects of CSNs purified from any eukaryotic organism.

1. Purified CSN isolated from human erythrocytes (Enzo Life Sciences Cat.# BML-PW9425-0020; LifeSensors Cat.#: CP009).
2. Thaw a tube of 500 μl of CDP from **step 19** of Subheading 3.3, and aliquot it into ten microcentrifuge tubes, 50 μl in each.
3. Use three aliquots from the previous step for the experiment, mark the remaining tubes, and freeze back with liquid nitrogen and keep in −80 °C.
4. Add 1 μl of 0.4 μg/μl purified CSN complex to one of the 50 μl CDP tubes.
5. In parallel, 0.4 μg of a control protein in the same concentration (BSA or any recombinant protein you may have available in your lab).
6. To the third tube add 1 μl of buffer A.
7. Transfer the three tubes to 30 °C for 20 min.
8. After 10 min add 15 μl of 5× Laemmli sample buffer and heat to 95 °C for 10 min in a dry block heater.
9. Use the samples for immunoblotting with CSN cullins antibodies to confirm CSN existence in the deneddylation activity assay (Fig. 3).
10. As for a control, load on the gel also a sample of UT (Fig. 3).

3.8 Assessment of CSN Interactions

Immunopurification of endogenous CSNs could be also a useful approach allowing the isolation of CSN complex out of variety of experimental conditions such as stress, treatments with drugs, different cell types, etc.). Combining immunopurification with advanced mass spectrometry based proteomics could assess the dynamics of CSN interaction map and regulation of CRLs activity/integrity, and help pinpoint the involved mechanism.

1. To find fractions that include the CSN complex and interacting proteins, assess each of the elution fractions (Subheading 3.4, **step 7**) by immunoblotting with CSN antibodies, as well as by silver staining (Fig. 4).
2. Fractions that include the peak of CSN could be further analyzed by mass spectrometry.

4 Notes

1. Always set your buffer pH at 4 °C, since the pH value of Tris–HCl buffer changes.

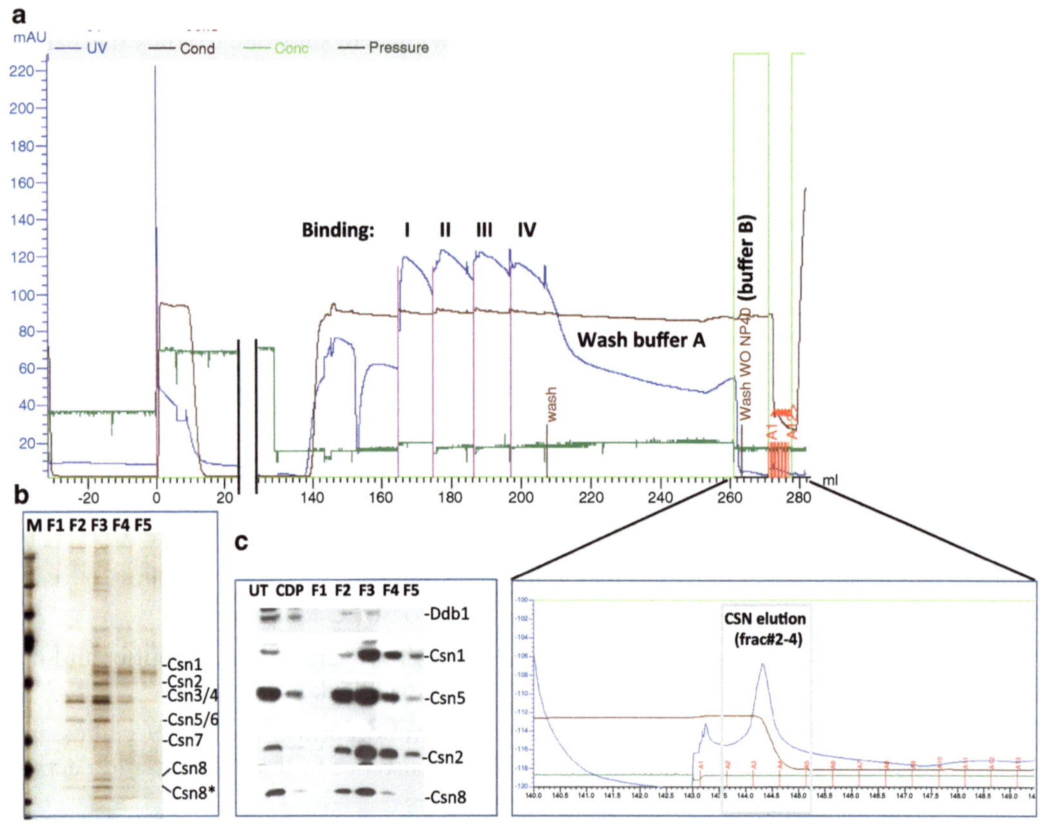

Fig. 4 Immunoaffinity purification of the CSN complex. (**a**) FPLC chromatogram displaying the immunoaffinity purification of CSN from Hek293 cells. CSN-containing fractions are enlarged below. SDS-PAGE followed by either silver staining (**b**), or western blot (**c**) of CSN-containing fractions is also shown

2. If the capacity of the column is not enough for immunodepletion, it is possible to repeat immunodepletion (Subheading 3.3) and column regeneration (Subheading 3.4), to deplete the remaining CSN complex from the first CDP. It is required to repeat this procedure up to three times in cell lines with high CSN expression. For that reason it is better to keep CDP refrigerated until you confirmed depletion and aliquot it only after completing the western blot.
3. Avoid using CDP lysate that froze/thawed more than twice.
4. It is important to notice that the eluted CSN complex is inactive.
5. Reducing agents (such as DTT or 2-mercaptoethanol) will destroy the columns and must be avoided.
6. Extremes in pH and excessive detergent concentrations can interfere with the antibody-antigen interaction.
7. At 3–4.4, before elution with 0.1 M Glycine (pH 2.7), it is suggested to re-estimate how much Tris (pH 8.5) will be required to neutralize your fractions.

Acknowledgements

We would like to thank the Technion-Haifa Joint Research Grant Program for funding our studies; the laboratory of Prof. Yuval Shoham, at the Biotechnology and Food Engineering Faculty at the Technion, Israel, for professional support. The authors would like to acknowledge networking support by the Proteostasis COST Action (BM1307).

References

1. Wei N, Serino G, Deng XW (2008) The COP9 signalosome: more than a protease. Trends Biochem Sci 33(12):592–600
2. Lydeard JR, Schulman BA, Harper JW (2013) Building and remodelling Cullin-RING E3 ubiquitin ligases. EMBO Rep 14(12):1050–1061. doi:10.1038/embor.2013.173
3. Sarikas A, Hartmann T, Pan ZQ (2011) The cullin protein family. Genome Biol 12(4):220. doi:10.1186/gb-2011-12-4-220
4. Deshaies RJ, Joazeiro CA (2009) RING domain E3 ubiquitin ligases. Annu Rev Biochem 78:399–434. doi:10.1146/annurev.biochem.78.101807.093809
5. Xirodimas DP (2008) Novel substrates and functions for the ubiquitin-like molecule NEDD8. Biochem Soc Trans 36(Pt 5):802–806
6. Cope GA, Suh GS, Aravind L, Schwarz SE, Zipursky SL, Koonin EV, Deshaies RJ (2002) Role of predicted metalloprotease motif of Jab1/Csn5 in cleavage of NEDD8 from CUL1. Science 298(5593):608–611
7. Lyapina S, Cope G, Shevchenko A, Serino G, Tsuge T, Zhou C, Wolf DA, Wei N, Shevchenko A, Deshaies RJ (2001) Promotion of NEDD-CUL1 conjugate cleavage by COP9 signalosome. Science 292(5520):1382–1385
8. Wee S, Geyer RK, Toda T, Wolf DA (2005) CSN facilitates Cullin-RING ubiquitin ligase function by counteracting autocatalytic adapter instability. Nat Cell Biol 7(4):387–391
9. Gummlich L, Rabien A, Jung K, Dubiel W (2013) Deregulation of the COP9 signalosome-cullin-RING ubiquitin-ligase pathway: mechanisms and roles in urological cancers. Int J Biochem Cell Biol 45(7):1327–1337. doi:10.1016/j.biocel.2013.03.023
10. Ambroggio XI, Rees DC, Deshaies RJ (2004) JAMM: a metalloprotease-like zinc site in the proteasome and signalosome. PLoS Biol 2(1), e2
11. Schmaler T, Dubiel W (2010) Control of deneddylation by the COP9 signalosome. Subcell Biochem 54:57–68. doi:10.1007/978-1-4419-6676-6_5
12. Lingaraju GM, Bunker RD, Cavadini S, Hess D, Hassiepen U, Renatus M, Fischer ES, Thoma NH (2014) Crystal structure of the human COP9 signalosome. Nature 512(7513):161–165. doi:10.1038/nature13566
13. Enchev RI, Scott DC, da Fonseca PC, Schreiber A, Monda JK, Schulman BA, Peter M, Morris EP (2012) Structural basis for a reciprocal regulation between SCF and CSN. Cell Rep 2(3):616–627. doi:10.1016/j.celrep.2012.08.019
14. Bornstein G, Ganoth D, Hershko A (2006) Regulation of neddylation and deneddylation of cullin1 in SCFSkp2 ubiquitin ligase by F-box protein and substrate. Proc Natl Acad Sci U S A 103(31):11515–11520. doi:10.1073/pnas.0603921103
15. Fischer ES, Scrima A, Bohm K, Matsumoto S, Lingaraju GM, Faty M, Yasuda T, Cavadini S, Wakasugi M, Hanaoka F, Iwai S, Gut H, Sugasawa K, Thoma NH (2011) The molecular basis of CRL4DDB2/CSA ubiquitin ligase architecture, targeting, and activation. Cell 147(5):1024–1039. doi:10.1016/j.cell.2011.10.035
16. Menon S, Chi H, Zhang H, Deng XW, Flavell RA, Wei N (2007) COP9 signalosome subunit 8 is essential for peripheral T cell homeostasis and antigen receptor-induced entry into the cell cycle from quiescence. Nat Immunol 8(11):1236–1245
17. Nezames CD, Deng XW (2012) The COP9 signalosome: its regulation of cullin-based E3 ubiquitin ligases and role in photomorphogenesis. Plant Physiol. doi:10.1104/pp.112.198879
18. Hannss R, Dubiel W (2011) COP9 signalosome function in the DDR. FEBS Lett 585(18):2845–2852, doi:S0014-5793(11)00274-2 [pii] 10.1016/j.febslet.2011.04.027
19. Kato JY, Yoneda-Kato N (2009) Mammalian COP9 signalosome. Genes Cells 14(11):1209–1225
20. Chamovitz DA (2009) Revisiting the COP9 signalosome as a transcriptional regulator. EMBO Rep 10(4):352–358
21. Singer R, Atar S, Atias O, Oron E, Segal D, Hirsch JA, Tuller T, Orian A, Chamovitz DA (2014) Drosophila COP9 signalosome subunit

7 interacts with multiple genomic loci to regulate development. Nucleic Acids Res. doi:10.1093/nar/gku723

22. Richardson KS, Zundel W (2005) The emerging role of the COP9 signalosome in cancer. Mol Cancer Res 3(12):645–653
23. Genschik P, Sumara I, Lechner E (2013) The emerging family of CULLIN3-RING ubiquitin ligases (CRL3s): cellular functions and disease implications. EMBO J 32(17):2307–2320. doi:10.1038/emboj.2013.173
24. Singhal S, Amin KM, Kruklitis R, DeLong P, Friscia ME, Litzky LA, Putt ME, Kaiser LR, Albelda SM (2003) Alterations in cell cycle genes in early stage lung adenocarcinoma identified by expression profiling. Cancer Biol Ther 2(3):291–298
25. Salon C, Brambilla E, Brambilla C, Lantuejoul S, Gazzeri S, Eymin B (2007) Altered pattern of Cul-1 protein expression and neddylation in human lung tumours: relationships with CAND1 and cyclin E protein levels. J Pathol 213(3):303–310. doi:10.1002/path.2223
26. Schindl M, Gnant M, Schoppmann SF, Horvat R, Birner P (2007) Overexpression of the human homologue for Caenorhabditis elegans cul-4 gene is associated with poor outcome in node-negative breast cancer. Anticancer Res 27(2):949–952
27. Pan Y, Yang H, Claret FX (2014) Emerging roles of Jab1/CSN5 in DNA damage response, DNA repair, and cancer. Cancer Biol Ther 15(3):256–262. doi:10.4161/cbt.27823
28. Adler AS, Lin M, Horlings H, Nuyten DS, van de Vijver MJ, Chang HY (2006) Genetic regulators of large-scale transcriptional signatures in cancer. Nat Genet 38(4):421–430. doi:10.1038/ng1752
29. Lee MH, Zhao R, Phan L, Yeung SC (2011) Roles of COP9 signalosome in cancer. Cell Cycle 10(18):3057–3066
30. Yoshida A, Yoneda-Kato N, Kato JY (2013) CSN5 specifically interacts with CDK2 and controls senescence in a cytoplasmic cyclin E-mediated manner. Sci Rep 3:1054. doi:10.1038/srep01054
31. Xue Y, Chen J, Choi HH, Phan L, Chou PC, Zhao R, Yang H, Santiago J, Liu M, Yeung GE, Yeung SC, Lee MH (2012) HER2-Akt signaling in regulating COP9 signalosome subunit 6 and p53. Cell Cycle 11(22):4181–4190. doi:10.4161/cc.22413
32. Bech-Otschir D, Kraft R, Huang X, Henklein P, Kapelari B, Pollmann C, Dubiel W (2001) COP9 signalosome-specific phosphorylation targets p53 to degradation by the ubiquitin system. EMBO J 20(7):1630–1639
33. Brownell JE, Sintchak MD, Gavin JM, Liao H, Bruzzese FJ, Bump NJ, Soucy TA, Milhollen MA, Yang X, Burkhardt AL, Ma J, Loke HK, Lingaraj T, Wu D, Hamman KB, Spelman JJ, Cullis CA, Langston SP, Vyskocil S, Sells TB, Mallender WD, Visiers I, Li P, Claiborne CF, Rolfe M, Bolen JB, Dick LR (2010) Substrate-assisted inhibition of ubiquitin-like protein-activating enzymes: the NEDD8 E1 inhibitor MLN4924 forms a NEDD8-AMP mimetic in situ. Mol Cell 37(1):102–111, doi:S1097-2765(09)00955-1 [pii] 10.1016/j.molcel.2009.12.024
34. Zhao Y, Sun Y (2013) Cullin-RING ligases as attractive anti-cancer targets. Curr Pharm Des 19(18):3215–3225
35. Pick E, Golan A, Zimbler JZ, Guo L, Sharaby Y, Tsuge T, Hofmann K, Wei N (2012) The minimal deneddylase core of the COP9 signalosome excludes the Csn6 MPN(−) domain. PLoS One 7(8), e43980. doi:10.1371/journal.pone.0043980
36. Menon S, Rubio V, Wang X, Deng XW, Wei N (2005) Purification of the COP9 signalosome from porcine spleen, human cell lines, and Arabidopsis thaliana plants. Methods Enzymol 398:468–481
37. Yang X, Menon S, Lykke-Andersen K, Tsuge T, Di X, Wang X, Rodriguez-Suarez RJ, Zhang H, Wei N (2002) The COP9 signalosome inhibits p27(kip1) degradation and impedes G1-S phase progression via deneddylation of SCF Cul1. Curr Biol 12(8):667–672
38. Wei N, Tsuge T, Serino G, Dohmae N, Takio K, Matsui M, Deng XW (1998) The COP9 complex is conserved between plants and mammals and is related to the 26S proteasome regulatory complex. Curr Biol 8:919–922
39. Pick E, Lau O, Tsuge T, Menon S, Tong Y, Dohmae N, Plafker S, Deng X, Wei N (2007) Mammalian DET1 regulates Cul4A activity and forms stable complexes with E2 ubiquitin-conjugating enzymes. Mol Cell Biol 27(13):4708–4719
40. Cope GA, Deshaies RJ (2006) Targeted silencing of Jab1/Csn5 in human cells downregulates SCF activity through reduction of F-box protein levels. BMC Biochem 7:1
41. Liu Y, Shah SV, Xiang X, Wang J, Deng ZB, Liu C, Zhang L, Wu J, Edmonds T, Jambor C, Kappes JC, Zhang HG (2009) COP9-associated CSN5 regulates exosomal protein deubiquitination and sorting. Am J Pathol 174(4):1415–1425
42. Bhattacharya S, Garriga J, Calbo J, Yong T, Haines DS, Grana X (2003) SKP2 associates with p130 and accelerates p130 ubiquitylation and degradation in human cells. Oncogene 22(16):2443–2451

Chapter 5

Studying Protein Ubiquitylation in Yeast

Junie Hovsepian, Michel Becuwe, Oded Kleifeld, Michael H. Glickman, and Sébastien Léon

Abstract

Ubiquitylation is a reversible posttranslational modification that is critical for most, if not all, cellular processes and essential for viability. Ubiquitin conjugates to substrate proteins either as a single moiety (monoubiquitylation) or as polymers composed of ubiquitin molecules linked to each other with various topologies and structures (polyubiquitylation). This contributes to an elaborate ubiquitin code that is decrypted by specific ubiquitin-binding proteins. Indeed, these different types of ubiquitylation have different functional outcomes, notably affecting the stability of the substrate, its interactions, its activity, or its subcellular localization. In this chapter, we describe protocols to determine whether a protein is ubiquitylated, to identify the site that is ubiquitylated, and provide direction to study the topology of the ubiquitin modification, in the yeast *Saccharomyces cerevisiae*.

Key words Yeast, Ubiquitin, Histidine-tagged ubiquitin purification, Immunoprecipitation in denaturing conditions, Ubiquitylation site mapping, Ubiquitin chain topology

1 Introduction

The conjugation of ubiquitin (hereafter referred to as "ubiquitylation") is a complex modification that can alter a protein's ability to interact with other proteins, and as such impacts on its stability, localization, or function [1]. Ubiquitin is generally conjugated on the ε-amino group of lysine residues of the target substrates, also called monoubiquitylation. However, ubiquitin possesses itself 8 amino groups (N-terminal + 7 lysines) that can be used for ubiquitin conjugation, thereby generating a polyubiquitin chain (polyubiquitylation). Hence, several factors will contribute to the signal generated by ubiquitylation, such as the identity of the residue targeted by ubiquitin on the substrate, the type of modification: mono- vs. polyubiquitylation, and in the latter case, the topology of the chain (i.e., which lysine residue of ubiquitin is used for chain elongation) and its length [1]. Various ubiquitin-binding domains contribute to the decoding of this modification [2].

Rune Matthiesen (ed.), *Proteostasis: Methods and Protocols*, Methods in Molecular Biology, vol. 1449,
DOI 10.1007/978-1-4939-3756-1_5,

Consequently, different types of ubiquitylation create a wide range of molecular signals that can lead to various functional outcomes, and altogether ubiquitylation contributes to many, if not all, cellular pathways.

Ubiquitin is best known for its role as a signal for the recognition and subsequent degradation of target proteins by the proteasome, which is mainly mediated by K48-linked polyubiquitin chains. However, chains of all ubiquitin linkages have been identified and the study of their structure and function is still undergoing [3]. K63-linked polyubiquitylation is notably well known for its role in DNA repair [4–6], the regulation of signaling pathways [7], or membrane trafficking [8, 9]. Although K48-linked polyubiquitylation represents only a fraction of the possible combinations of the ubiquitin code, it is the most abundant in all cells studied to date [10–14]. This type of ubiquitylation is often a transient event in the cell, occurring before the target protein is degraded, and can therefore be difficult to detect experimentally.

Furthermore, ubiquitylation is a reversible modification. Several family of proteases act as ubiquitin isopeptidases that cleave ubiquitin off substrates and process ubiquitin chains, allowing a fine regulation of the ubiquitylation signal [15]. Therefore, another technical problem when studying protein ubiquitylation is due to the activity of these enzymes when preparing a protein extract in native conditions (e.g., during an immunoprecipitation).

Here, we describe a protocol that allows the identification of ubiquitylated proteins in denaturing conditions in yeast. This is based on a construct allowing the expression of polyhistidine-tagged ubiquitin, which allows the purification of ubiquitylated proteins in denaturing conditions by immobilized metal-ion affinity chromatography (IMAC).

Evidence that a protein is ubiquitylated can also be obtained by performing the reverse experiment, in which a protein of interest is immunoprecipitated and the resulting sample is blotted with anti-ubiquitin antibodies. In this type of experiment, the immunoprecipitation should be performed on denatured samples, to ensure that the ubiquitylation signal detected originates from the protein considered, and not from its potential interactants. Therefore, we describe another protocol to achieve immunoprecipitation in denaturing conditions.

The identification of ubiquitylated sites on a protein of interest is often instrumental to understand the functional contribution of this modification. We describe two possible approaches that can lead to the identification of the ubiquitylation site on a protein. The first approach is a tandem purification procedure to purify the protein of interest in denaturing conditions. This may be helpful to identify ubiquitylation sites by mass spectrometry. We also describe a general, truncation-based genetic method to identify the ubiquitylated sites on a protein.

Finally, we will discuss on the possibilities offered to define the topology of ubiquitin modification (linkage used within an ubiquitin chain on a substrate) and will provide an example using a set of yeast strains carrying ubiquitin mutations, developed in the Finley lab [5].

2 Materials

Prepare all solutions using ultrapure water (prepared by purifying deionized water to attain a sensitivity of 18 MΩ cm at 25 °C) and analytical grade reagents. Prepare and store all reagents at room temperature (unless indicated otherwise).

2.1 Components Required for the Visualization of Ubiquitylation on Crude Extracts (Subheading 3.1)

1. Yeast culture medium (synthetic complete medium, SC): 1.7 g/L yeast nitrogen base, 5 g/L ammonium sulfate, 20 g/L glucose. Autoclave. Add sterile-filtered amino acid solution as required (e.g., dropout bases from Elis Solutions, Erpent, Belgium).
2. Plasmid encoding His-tagged ubiquitin (*see* Subheading 3.1).
3. 100% TCA (Trichloroacetic acid) solution (w/v, i.e., 6.1 M). Store at 4 °C.
4. 10% TCA (Trichloroacetic acid) solution (w/v, i.e., 0.61 M). Store at 4 °C.
5. Vortex equipped with a microtube foam insert (placed in a cold room).
6. Glass beads: 0.4–0.6 mm (BBI-8541701, Sartorius Mechatronics).
7. Needles: 23 G×1″—0.6 ×25 mm (Terumo Medical Corporation).
8. 5× Sample buffer (250 mM Tris–HCl pH 6.8, 500 mM dithiothreitol, 10% SDS, 0.01% bromophenol blue, 50% glycerol). Store at −20 °C.
9. 1 M Tris-base: 1 M Tris, non-buffered, in water.
10. 1× Sample buffer for TCA precipitates: 1 vol. of 5× Sample buffer, 1 vol. of Tris-base, and 3 vol. water. Store at −20 °C.
11. General equipment for SDS-PAGE and western blotting (gel casting, solutions, tanks for migration/transfer, nitrocellulose membrane etc.).
12. Antibody directed against the protein of interest (or its tag if applicable).

2.2 Components Required for the Purification of Ubiquitin Conjugates Using PolyHis-Tagged Ubiquitin (Subheading 3.2)

1. Yeast culture medium (synthetic complete medium, SC): 1.7 g/L yeast nitrogen base, 5 g/L ammonium sulfate, 20 g/L glucose. Autoclave. Add sterile-filtered amino acid solution as required (e.g., dropout bases from Elis Solutions, Erpent, Belgium).
2. Plasmid encoding His-tagged ubiquitin (*see* Subheading 3.1).
3. 100% TCA (Trichloroacetic acid) solution (w/v, i.e., 6.1 M). Store at 4 °C.
4. 10% TCA (Trichloroacetic acid) solution (w/v, i.e., 0.61 M). Keep at 4 °C.
5. Glass beads: 0.4–0.6 mm (BBI-8541701, Sartorius Mechatronics).
6. Vortex equipped with a microtube foam insert (placed in a cold room).
7. Needles: 23 G × 1″—0.6 × 25 mm (Terumo Medical Corporation).
8. 1 M Tris-base: 1 M Tris, non-buffered, in water.
9. Buffer A: 6 M guanidinium-HCl, 20 mM Tris pH 8.0, 100 mM K_2HPO_4, 10 mM imidazole, 100 mM NaCl, 0.1% Triton X-100 (v/v).
10. Ni-NTA superflow (Qiagen).
11. Micro Bio-Spin™ Chromatography Column (Bio-Rad).
12. Wash1 buffer: 20 mM Tris pH 8.0, 100 mM K_2HPO_4, 20 mM imidazole, 100 mM NaCl, 0.1% Triton X-100 (v/v).
13. Wash2 buffer: 20 mM Tris pH 8.0, 100 mM K_2HPO_4, 10 mM imidazole, 1 M NaCl, 0.1% Triton X-100 (v/v).
14. Elution buffer: 20 mM Tris pH 8.0, 100 mM K_2HPO_4, 500 mM imidazole, 100 mM NaCl. Store at 4 °C for no longer than a month.
15. 5× Sample buffer: Tris–HCl 250 mM pH 6.8, dithiothreitol 500 mM, SDS 10%, bromophenol blue 0.01%, glycerol 50%. Store at −20 °C.
16. 100% acetone, ice-cold.
17. Ponceau S solution: 0.2% ponceau S (w/v), 3% TCA (w/v). Store at 4 °C.
18. General equipment for SDS-PAGE and western blotting (gel casting, solutions, tanks for migration/transfer, nitrocellulose membrane etc.).
19. Antibody directed against ubiquitin (e.g., mouse monoclonal P4D1 antibody: sc-8017, Santa Cruz Biotech; use at 1/10,000) and the protein of interest (or its tag if applicable). Store at 4 or −20 °C.

2.3 Components Required for Immunoprecipitation in Denaturing Conditions (Subheading 3.3)

1. Yeast culture medium (synthetic complete medium, SC): yeast nitrogen base 1.7 g/L, ammonium sulfate 5 g/L, glucose 2 % w/v. Autoclave. Add sterile-filtered amino acid solution as required (e.g., dropout bases from Elis Solutions, Erpent, Belgium).
2. 10 % TCA (Trichloroacetic acid) solution (w/v, i.e., 0.61 M). Keep at 4 °C.
3. Vortex equipped with a microtube foam insert (placed in a cold room).
4. Glass beads: 0.4–0.6 mm (BBI-8541701, Sartorius Mechatronics).
5. Needles: 23 G × 1″—0.6 × 25 mm (Terumo Medical Corporation).
6. Protein-G-coupled Sepharose (e.g., GammaBind™ G sepharose™, GE Healthcare), or if using an HA-tag, anti-HA affinity matrix (Roche).
7. IP dilution buffer: 50 mM Tris pH 7.5, 2 mM EDTA, 100 mM NaCl, 1.2 % Triton X-100, 0.5 % Bovine Serum Albumin, yeast protease inhibitor cocktail (e.g., Sigma P821520, diluted 1:100), 20 mM N-ethylmaleimide. Store at 4 °C for no more than a week.
8. Buffer B0: SDS 2 % (w/v), bromophenol blue 0.05 % (w/v).
9. Buffer B1: 50 mM Tris pH 7.5, 2 mM EDTA, 100 mM NaCl, 1 % Triton X-100, 0.2 % SDS. Store at 4 °C.
10. Buffer B2: 50 mM Tris pH 7.5, 2 mM EDTA, 100 mM NaCl, 0.1 % Triton X-100, 0.5 % SDS, 0.5 % Na-deoxycholate. Store at 4 °C.
11. Buffer B3: 50 mM Tris pH 7.5, 2 mM EDTA, 500 mM NaCl, 0.1 % Triton X-100. Store at 4 °C.
12. Buffer B4: 50 mM Tris pH 7.5, 2 mM EDTA, 100 mM NaCl. Store at 4 °C.
13. 5x Sample buffer: Tris–HCl 250 mM pH 6.8, dithiothreitol 500 mM, SDS 10 %, bromophenol blue 0.01 %, glycerol 50 %. Store at −20 °C.
14. 100 % acetone, ice-cold.
15. 1x Sample buffer: 1 vol. 5× Sample buffer, 4 vol. water. Store at −20 °C.
16. General equipment for SDS-PAGE and western blotting (gel casting, solutions, tanks for migration/transfer, nitrocellulose membrane etc.).
17. Antibody directed against the protein of interest (or its tag if applicable) and against ubiquitin (e.g., mouse monoclonal P4D1 antibody: sc-8017, Santa Cruz Biotech; use at 1/10,000; or sc-8017 HRP, Santa Cruz Biotech; use at 1/2000). Store at 4 or −20 °C.

2.4 Components Required for the Tandem Purification of a Protein in Denaturing Conditions for the Identification of Ubiquitin-Conjugated Sites (Subheading 3.4)

1. Yeast culture medium: synthetic complete medium (SC): yeast nitrogen base 1.7 g/L, ammonium sulfate 5 g/L, glucose 2% w/v. Autoclave. Add required amino acid as sterile-filtered solutions (e.g., CSM-Ura dropout, Elis Solutions, Erpent, Belgium).
2. Yeast culture medium: synthetic complete raffinose medium: yeast nitrogen base 1.7 g/L, ammonium sulfate 5 g/L, raffinose (2% w/v), 0.02% glucose (w/v) (*see* **Note 1**).
3. Galactose solution, 20% (w/v), sterile-filtered.
4. pBG1805-based plasmid containing the protein of interest (commercially available at OpenBiosystems/GE Healthcare Dharmacon).
5. Protein-G-coupled Sepharose (e.g., GammaBind™ G sepharose™, GE Healthcare).
6. IP lysis buffer: 50 mM HEPES-KOH pH 7.5, 0.25 M NaCl, 10% glycerol, 1 mM EDTA. Filter sterilize and store at 4 °C.
7. Protease inhibitors: yeast protease inhibitor cocktail (Sigma P821520; use at 1:100 dilution, store at −20 °C), phenylmethanesulfonyl fluoride (PMSF) (100 mM stock in EtOH; store at −80 °C), N-ethylmaleimide (NEM) (1 M stock in EtOH; store at −80 °C), and MG-132 (Enzo Life Sciences; 100 mM stock in DMSO; store at −80 °C).
8. Antibodies directed against the HA tag (e.g., mouse monoclonal F7 antibody: sc-7392, Santa Cruz Biotech). Store at 4 or −20 °C.
9. Glass beads: 0.4–0.6 mm (BBI-8541701, Sartorius Mechatronics).
10. Vortex equipped with a microtube foam insert (placed in a cold room).
11. Needles: 23 G×1″—0.6×25 mm (Terumo Medical Corporation).
12. Triton X-100 solution: 25% (v/v) in water.
13. Urea buffer: urea 8 M, 10 mM Tris pH 6.3; 5 mM imidazole, 10 mM beta-mercaptoethanol, 10 mM N-ethylmaleimide, 100 μM MG-132.
14. Ni-NTA superflow (Qiagen).
15. 5x Sample buffer: 250 mM Tris–HCl pH 6.8, 500 mM dithiothreitol, 10% SDS, 0.01% bromophenol blue, 50% glycerol. Store at −20 °C.
16. Urea elution buffer: 8 M urea, 10 mM Tris pH 6.3, 600 mM imidazole. Store at 4 °C.
17. General equipment for SDS-PAGE and western blotting (gel casting, solutions, tanks for migration/transfer, nitrocellulose membrane etc.).

18. Antibodies directed against ubiquitin (e.g., mouse monoclonal P4D1 antibody: sc-8017, Santa Cruz Biotech; use at 1/10,000; or sc-8017 HRP, Santa Cruz Biotech; use at 1/2000). Store at 4 or −20 °C.

3 Methods

3.1 Obtaining Evidence of Ubiquitylation on Crude Extracts

When detecting the protein of interest by western blot, higher molecular weight species are sometimes observed, which may be caused by posttranslational modifications such as ubiquitylation. The expression of a tagged (i.e., heavier) ubiquitin causes a shift in all ubiquitin-conjugated species, which can sometimes be sufficient to obtain a first indication that a given protein is ubiquitylated. Several plasmids available for this application have been described in the literature, driving the expression of His-tagged ubiquitin under the control of either a copper inducible promoter (pJD421, pYEp96–6His-Ub) [16, 17] or the strong *ADH1* promoter (m886) [11]. Note that His-tagged ubiquitin was previously shown to be functionally comparable to non-tagged ubiquitin [18]. For expression of His-tagged ubiquitin driven by the *CUP1* promoter, we noticed that the traces of copper in the synthetic medium (DO) are sufficient to drive the expression of His-Ub to a level comparable as endogenous Ub and therefore we recommend not to add extra copper in the medium to avoid potential artifacts that may be due to a massive ubiquitin overexpression.

Therefore, this protocol describes a classical method to prepare crude extracts from cells expressing tagged ubiquitin or not that may allow to determine whether a shift in molecular weight is due to ubiquitylation. The example of a result is illustrated in Fig. 1.

1. *DAY 1*. In the morning, inoculate freshly streaked yeast cells in synthetic medium (SC) medium supplemented with the required amino acids (preculture). Use a strain expressing your protein of interest together with His-tagged ubiquitin, and as a control, use the same strain but that does not contain the His-tagged ubiquitin plasmid.
2. In the evening, measure the OD_{600} of the cultures and inoculate yeast cells in 5 mL SC medium, at $OD_{600} = 0.001$. Grow overnight under agitation (200 rpm) at 30 °C.
3. *DAY 2*. Measure the OD_{600} of the cultures. The cultures should have reached an OD_{600} of ~0.3–0.5 (*see* **Notes 2** and **3**).
4. Take 1 mL of culture and place in a 1.5-mL tube.
5. Add 100 μL of a 100% (w/v) TCA solution.
6. Incubate on ice for 10 min.
7. Spin down the cells by centrifugation for 1 min at 13,000 × *g* at room temperature.

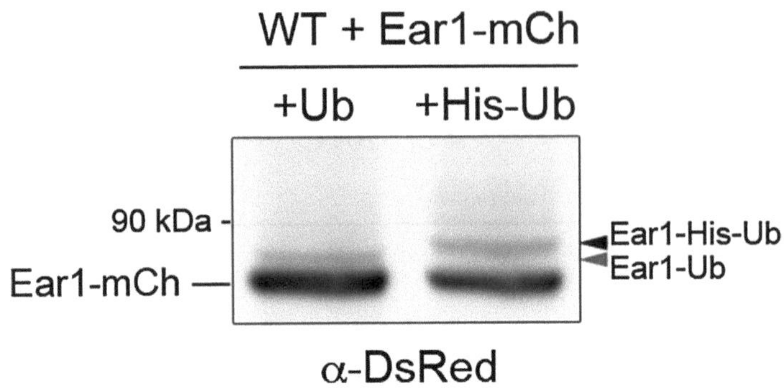

Fig. 1 Obtaining evidence of ubiquitylation on crude extracts. Protein extracts from wild-type (BY4741) cells expressing the protein Ear1 tagged with mCherry (pSL22) [39] together with either pCUP1:Ub (pRS426-based plasmid: pRHT79; *left*) or pCUP1-His-Ub (pJD421; *right*) [16] were prepared as described in the text, electrophoresed on SDS-PAGE and blotted with monoclonal anti-DsRed antibodies (Clontech). The *upper band* observed on *top* of Ear1-mCh (*gray arrowhead*) is shifted up in apparent size when expressing His-tagged ubiquitin (*black arrowhead*)

8. Discard the supernatant.
9. Resuspend yeast cells in 100 μL of a 10% TCA solution.
10. Add glass beads up to 1–2 mm below the meniscus.
11. Lyse cells on a vortex equipped with a microtube foam insert for 10 min at 4 °C.
12. After lysis, open the tube, poke a small hole at the bottom of the tube using a needle (Terumo 23 G×1″—0.6×25 mm), place into a clean 1.5-mL tube to collect the lysate, and close the cap.
13. Centrifuge for a few seconds on a mini-centrifuge to collect the lysate and leave the beads in the upper tube. Discard the tube with the beads (they should be dry).
14. Centrifuge the lysate for 1 min at 13,000×*g* at RT.
15. Remove as much supernatant as possible.
16. Resuspend the pellet in 1× Laemmli buffer for TCA precipitates, considering that 1 OD_{600} unit should be resuspended in 50 μL.
17. Denature at the desired temperature (37–95 °C) for 5–10 min.
18. Use 7 μL for SDS-PAGE and transfer onto nitrocellulose membrane.
19. Blot with antibodies directed against the protein of interest.

3.2 Purification of Ubiquitin Conjugates Using PolyHis-Tagged Ubiquitin

In this protocol, total ubiquitin conjugates from cells will be purified by immobilized metal-ion affinity chromatography. The presence of the protein of interest in this fraction is addressed using a specific antibody (against the protein of interest or a tag). The advantage of this purification is that it does not involve antibodies during the purification procedure, which may interfere when revealing the presence of the protein of interest by western blotting. It is also amenable to proteome-wide, mass spectrometry-based studies either alone [11, 19] or in combination with additional purification procedures [20–22].

This protocol (based on [11], and slightly modified in [23]) requires the prior transformation of the yeast strain with a plasmid expressing His-tagged ubiquitin, described in the previous section. Other protocols using slightly different procedures have been described in the past [24, 25]. Please note that this technique is not suitable to study the ubiquitylation of proteins that contains an endogenous His-stretch, as their purification on the Ni-NTA column will not depend on their ubiquitylation. A list of the *S. cerevisiae* proteins containing endogenous polyhistidine (≥5) stretches is displayed in Table 1. Of note, Fig. 2 shows the purification of one of these proteins, Snf1, on Ni-NTA beads even in the absence of His-tagged ubiquitin.

1. *DAY 1.* Preculture: inoculate freshly streaked yeast cells in synthetic medium (DO) supplemented with the required amino acids in the morning. Use a strain expressing your protein of interest together with His-tagged ubiquitin, and as a control, use the same strain but that does not contain the His-tagged ubiquitin plasmid.
2. In the evening, measure the OD_{600} of the cultures and inoculate yeast cells in 100 mL DO medium, at $OD_{600} = 0.001$. Grow overnight under agitation (200 rpm) at 30 °C.
3. *DAY 2.* Measure the OD_{600} of the cultures. The cultures should have reached an OD_{600} of ~0.3–0.5 (*see* **Note 2**). Spin down the cells in 2 × 50-mL tubes (3000 × *g*, 3 min), remove the supernatant, place on ice.
4. Resuspend each pellet in 500 μL 10% TCA (w/v; i.e., 0.61 N), pool into a 1.5-mL tube.
5. Leave on ice for at least 10 min (up to a few days).
6. Pellet cells: 1 min, 13,000 × *g* at room temperature.
7. Resuspend the precipitated cells with 100 μL of 10% TCA by pipetting.
8. Add glass beads up to 1–2 mm below the meniscus.
9. Lyse cells on a vortex equipped with a microtube foam insert for 20 min at 4 °C.
10. After lysis, open the cap, poke a small hole at the bottom of the tube using a needle, place into a clean 1.5-mL tube to collect the lysate, and close the cap.
11. Centrifuge for a few seconds on a mini-centrifuge to collect the lysate and leave the beads in the upper tube. Discard the

Table 1
List of *S. cerevisiae* proteins containing endogenous polyhistidine (≥5) stretches

ORF	Protein	Sequence	Locus Information
YBR086C	IST2	892-ATQP**HHHHHHHH**RHRD-906	Involved in ER-plasma membrane tethering
YBR129C	OPY1	48-PQGY**HHHHHHHH**RHLW-62	Protein of unknown function
YDL025C	RTK1	251-YHDN**HHHHHHHH**NRGS-263	Putative protein kinase
YDR475C	JIP4	759-QQQQ**HHHHHH**RDTD-771	Protein of unknown function
YDR477W	SNF1	18-ANSS**HHHHHHHHHHHHH**GHGG-34	AMP-activated serine/threonine protein kinase
YER132C	PMD1	319-KLPK**HHHHHHH**GDLK-327	Protein with an N-terminal kelch-like domain
YGR237C	YGR237C	724-QRKA**HHHHHHHH**NHVS-738	Putative protein of unknown function
YJL042W	MHP1	124-ADSG**HHHHHRHHHH**TEDA-141	Protein involved in microtubule organization
YJL083W	TAX4	393-LPFP**HHHHHHH**QLHN-407	EH domain-containing protein
YKR075C	YKR075C	251-DIQHSRHHRRHHRR**HHHHHHH**QNSS-273	Protein of unknown function
YKR098C	UBP11	553-SKSP**HHHHHHHH**SSDD-568	Ubiquitin-specific protease
YLR328W	NMA1	57-KHPK**HHHHHHHH**SRKE-71	Nicotinic acid mononucleotide adenylyltransferase
YLR371W	ROM2	325-FSDP**HHHHHHHHH**SSNS-340	GDP/GTP exchange factor (GEF) for Rho1p and Rho2p
YMR070W	MOT3	236-PGPP**HHHHHHH**SNTH-249	Transcriptional regulator with two C2H2 zinc fingers
YOL087C	DUF1	360-FKPD**HHHHHHHHHH**EHEE-376	Ubiquitin-binding protein of unknown function
YOR134W	BAG7	315-NFTI**HHHHHHHHH**ALFP-330	Rho GTPase activating protein (RhoGAP)

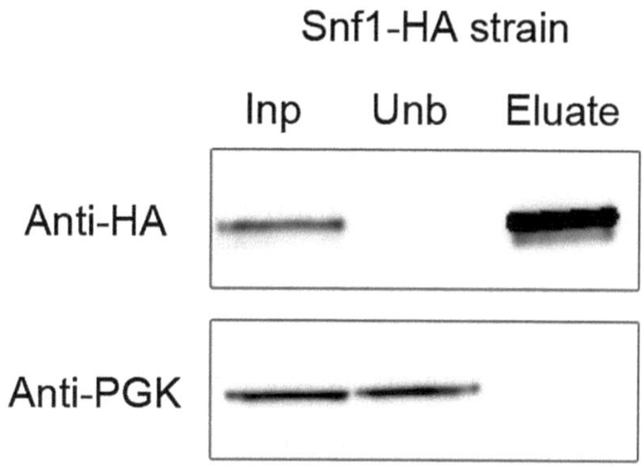

Fig. 2 Purification of the endogenously, polyhistidine-tagged protein Snf1 on Ni-NTA column in the absence of His-tagged ubiquitin. Proteins from a WT strain (BY4741) expressing Snf1-HA from a plasmid (pSL204) were prepared as described and purified on Ni-NTA resin in the absence of His-tagged ubiquitin. Equal amounts of extract (input, *first lane*) and unbound fraction (*second lane*) were loaded, showing the total depletion of Snf1, which is recovered in the eluate (*third lane*). Phosphoglycerate kinase (PGK) is used as a control and does not bind to the beads. Other proteins that may show a similar behavior are displayed in Table 1

tube with the beads (they should be dry). For complete recovery, the glass beads can be washed with 200 μL of cold 10% TCA if necessary, in this case, repeat the spin and pool with the first lysate.

12. Centrifuge the lysate for 10 min at 13,000 ×*g* at 4 °C.
13. Remove as much supernatant as possible.
14. Wash the pellet with 100% ice-cold acetone, and centrifuge again for 10 min at 13,000 ×*g* at 4 °C.
15. Neutralize the residual TCA present in the pellet by adding 30 μL of 1 M Tris (non-buffered) to the pellet (*see* **Note 4**).
16. Add 200 μl of buffer A.
17. Resuspend the pellet. Add a few glass beads that will help to break the pellet, and vortex (*see* **Note 5**).
18. Add 800 μL buffer A to the resuspended pellet.
19. Let solubilize further by rotating for 1 h at room temperature. Resuspension should be complete.
20. Transfer the lysate to a new tube and centrifuge for 10 min at 13,000 ×*g*, at room temperature (*see* **Note 6**).
21. During the centrifugation, prepare Ni-NTA beads (Qiagen): transfer 200 μL slurry to 2-mL tubes, wash the beads twice with 1.8 mL buffer A. Centrifuge for 1 min at 1000 ×*g* between each wash (*see* **Note 7**).
22. From the centrifugation (**step 20**), keep the supernatant, which corresponds to the solubilized lysate, and transfer to a

new tube. Keep 25 μL aside, at room temperature, for TCA precipitation ("input fraction") (*see* **step 35**).

23. Add the remaining sample to the tube containing the washed beads (*see* **step 21**), and rotate for 2 h at RT.
24. Centrifuge for 30 s at 500 × *g* to spin down the beads. Collect 25 μL of unbound fraction and keep at room temperature for TCA precipitation ("Unbound fraction") (*see* **step 35**). Discard remaining supernatant.
25. Resuspend the beads in 1.8 mL buffer A.
26. Transfer to a micro-chromatography column placed in a tube holder (e.g., floating foam rack placed on top of a beaker), and let the wash solution drop by gravity.
27. Wash again with 1.8 mL buffer A, twice.
28. Wash with 1.8 mL Wash1 buffer, three times.
29. Wash with 1.8 mL Wash2 buffer, three times.
30. Place the micro-chromatography column in a 1.5-mL tube, centrifuge for 10 s at 1000 × *g*.
31. Place the micro-chromatography column in a clean 1.5-mL tube, and add 100 μL elution buffer to the beads.
32. Incubate for 5 min at RT.
33. Spin down eluate for 1 min at 1000 × *g*, at room temperature.
34. Add 25 μL of 5× SDS sample buffer and denature at the desired temperature (37–95 °C) for 5–10 min.
35. To prepare the "input" (*see* **step 22**) and "unbound" (*see* **step 24**) fraction, proceed as follows. Dilute the 25 μL samples with 1.8 mL water, add 200 μL of 100% TCA, and keep on ice for 10 min.
36. Spin down for 10 min at 13,000 × *g*, at 4 °C.
37. Discard supernatant. Wash the pellet with 1 mL ice-cold acetone (100%).
38. Spin down for 1 min at 13,000 × *g*, at 4 °C.
39. Discard supernatant. Let at room temperature with cap open until acetone has evaporated completely.
40. Resuspend in 25 μL of sample buffer for TCA precipitates and denature at the desired temperature (37–95 °C) for 5–10 min.
41. For SDS-PAGE, use 4 μL of input/unbound fractions as prepared above, and 3–7 μL of the eluate (depending on the expression level of the protein and its rate of ubiquitylation) (*see* **Notes 8** and **9**).
42. Transfer on nitrocellulose membrane. After transfer, performing a staining of the transferred proteins with a Ponceau S solution should allow to see His-Ub and a ladder of ubiquitylated proteins (Fig. 3a).

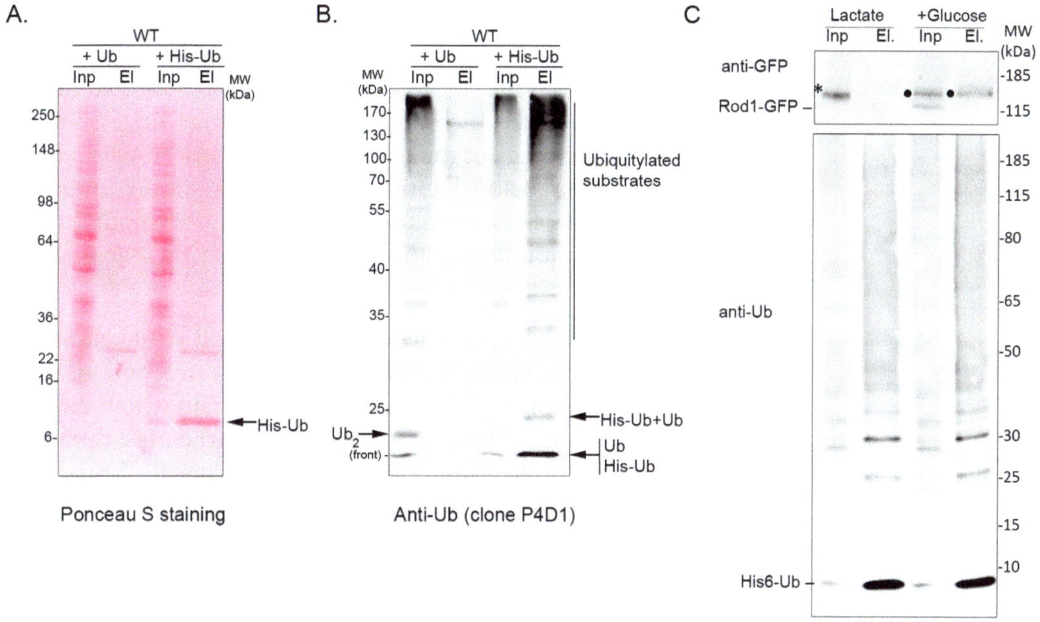

Fig. 3 Purification of ubiquitin conjugates using polyHis-tagged ubiquitin. Protein extracts from WT cells (BY4741) expressing either Ub or His-tagged Ub (*see* Fig. 1) were prepared and treated as described for the purification of His-tagged ubiquitin. (**a**) Ponceau S staining of the nitrocellulose membrane after transfer of the indicated sample, showing the purification of His-Ub in the eluate fraction (El) (*arrow*). (**b**) Western blot using anti-ubiquitin antibodies on a similar experiment. The gel shows ubiquitin dimers (Ub_2) in both cell types, and His-Ub + Ub dimers in cells expressing His-Ub. **C.** Example of a purification of His-tagged ubiquitin from yeast cells grown either in lactate or glucose medium and expressing the arrestin-related protein Rod1 tagged with GFP. Displayed are the "input" (Inp) and eluate fractions (El.). Rod1 is phosphorylated in lactate medium (*asterisk*) and becomes ubiquitylated rapidly after glucose addition (*black circle*) (*see* also [23])

43. Blot with the antibody used to detect the protein of interest and with anti-Ubiquitin antibodies (e.g., mouse monoclonal P4D1 antibody: sc-8017, Santa Cruz Biotech; 1/10,000) to reveal Ub conjugates.

3.3 Assessment of Protein Ubiquitylation After Immuno-precipitation in Denaturing Conditions

The previous method involves the expression of tagged ubiquitin and the affinity purification of ubiquitin conjugates, prior to detecting the presence of the protein of interest within this fraction. Performing the opposite experiment, i.e., by immunoprecipitating the protein of interest and studying the presence of ubiquitin in the immunoprecipitate, can also provide evidence of the ubiquitylation of a given protein. A main caveat of this approach is the fact that the ubiquitin signal detected in the immunoprecipitated sample may either be due to the protein that was immunoprecipitated or to potential interactants that were co-immunoprecipitated. Therefore, we provide a protocol that allows the immunoprecipitation of proteins in denaturing conditions, which was successfully used in the past [26, 27]. These immunoprecipitates can be used to reveal the ubiquitylation of the protein of interest.

An advantage of this technique is that is does not involve the use of overexpressed tagged ubiquitin, and as such may be more physiologically relevant. However, it depends on the use of anti-ubiquitin antibodies, which also have their limits since some are specific of the topology of the ubiquitin modification. We routinely use the P4D1 monoclonal antibody, which is a good compromise for the detection of both mono- and polyubiquitin modifications, although the monoubiquitylation signal is sometimes weak (*see* for instance [28]).

1. *DAY 1.* Preculture: inoculate freshly streaked yeast cells in synthetic medium (DO) supplemented with the required amino acids in the morning. As a control, use a strain which does not express the tagged protein, to confirm that the ubiquitin signal observed after the immunoprecipitation is specific for the presence of the tagged protein.
2. In the evening, measure the OD_{600} of the cultures and inoculate the yeast cells in 100 mL DO medium, at $OD_{600} = 0.001$. Grow overnight under agitation (200 rpm) at 30 °C.
3. *DAY 2.* Measure the OD_{600} of the cultures. The cultures should have reached an OD_{600} of ~0.3–0.5 (*see* **Note 2**). Spin down the cells in 2 × 50-mL tubes (3000 × *g*, 3 min), remove the supernatant, place on ice.
4. Resuspend each pellet in 500 μL 10% TCA (w/v; i.e., 0.61 N), pool into a 1.5-mL tube.
5. Leave on ice for at least 10 min (up to a few days).
6. Prepare antibody-coupled beads for the immunoprecipitation. Take 50 μL of Protein-G-coupled Sepharose slurry (*see* **Note 10**), wash in 1 mL IP dilution buffer, resuspend in 1 mL IP dilution buffer, and use the appropriate amount of antibodies required for immunoprecipitation. Incubate for ≥1 h at 4 °C on a rotating wheel.
7. Pool the precipitated cells into one single 1.5-mL tube.
8. Pellet the precipitated cells: 1 min, 13,000 × *g* at room temperature.
9. Resuspend the precipitated cells with 100 μL of 10% TCA by pipetting up and down.
10. Add glass beads up to 1–2 mm below the meniscus.
11. Lyse cells on a vortex equipped with a microtube foam insert for 20 min at 4 °C.
12. After lysis, open the cap, poke a small hole at the bottom of the tube using a needle, place into a clean 1.5-mL tube to collect the lysate, and close the cap.
13. Centrifuge for a few seconds on a mini-centrifuge to collect the lysate and leave the beads in the upper tube. Discard the

tube with the beads (they should be dry). The glass beads can be washed with 200 μL of cold 10% TCA if necessary, in this case, repeat the spin and pool with the first lysate.

14. Centrifuge the lysate for 1 min at 13,000 × *g* at RT.
15. Remove as much supernatant as possible.
16. Wash the pellet with 1 mL ice-cold acetone (100%) (*see* **Note 11**).
17. Spin down for 1 min at 13,000 × *g*, at 4 °C.
18. Discard supernatant. Let at room temperature with cap open until acetone has evaporated completely.
19. Resuspend in 100 μL buffer B0. Make sure that the sample is blue (i.e., sample is not acidic). If the sample is yellow, add increments of 1 μL Tris–HCl 1 M pH 7.5 until the pH is neutralized.
20. Denature the sample at the desired temperature (37–95 °C) for 5–10 min in a thermomixer.
21. Spin down unsolubilized material by centrifuging for 5 min at 16,000 × *g*, at room temperature.
22. Transfer the supernatant to a clean tube.
23. Add 400 μL of IP dilution buffer.
24. Keep 50 μL aside for the "input" fraction; keep on ice.
25. Spin down the antibody-coated beads (*see* **step 6**) for 1 min at 300 × *g* (4 °C) and remove the supernatant.
26. Add the protein sample prepared in **step 23** to the antibody-coated beads.
27. Incubate on a rotating wheel at 4 °C for 2 h.
28. Spin down the beads for 1 min at 300 × *g* (4 °C).
29. Take 50 μL of the supernatant and keep 50 μL aside for the "unbound" fraction; keep on ice.
30. Carefully remove and discard remaining supernatant.
31. First wash: Resuspend the beads in 1 mL buffer B1, and spin down the beads for 1 min at 300 × *g* (4 °C).
32. Repeat the first wash.
33. Second wash: Resuspend the beads in 1 mL buffer B2 and spin down the beads for 1 min at 300 × *g* (4 °C).
34. Repeat the second wash.
35. Third wash: Resuspend the beads in 1 mL buffer B3 and spin down the beads for 1 min at 300 × *g* (4 °C).
36. Repeat the third wash.
37. Last wash: Resuspend the beads in 1 mL buffer B4, and spin down the beads for 1 min at 300 × *g* (4 °C).
38. Repeat the last wash, and remove as much supernatant as possible.

39. Add 50 μL of 1× sample buffer to the beads.
40. To the "input" and "unbound" fractions, add 12.5 μL of 5× Laemmli sample buffer.
41. Denature at the desired temperature (37–95 °C) for 5–10 min.
42. Load 7 μL of input/unbound fractions as prepared above, and 3–7 μL of the eluate (depending on the expression level of the protein and its rate of ubiquitylation).
43. Blot with the antibody against the protein of interest or its epitope, to evaluate the efficiency of the immunoprecipitation, and with anti-Ubiquitin antibodies (1/10,000 mouse monoclonal P4D1 antibody: sc-8017, Santa Cruz Biotech) to reveal Ub conjugates (*see* **Note 12**).

3.4 Tandem Purification of a Protein in Denaturing Conditions for the Identification of Ubiquitin-Conjugated Sites

Once ubiquitylation of a protein has been documented, a traditional approach to understand the consequences of this modification resides in identifying the sites targeted for ubiquitylation, followed by their mutation to evaluate the outcome of the lack of ubiquitylation on the protein's stability or function.

This protocol is based on a two-step purification. It involves the use of pBG1805-derived constructs (*URA3*-based) [29], in which all yeast genes have been cloned and that are commercially available. This allows the galactose-driven expression of the protein of interest tagged at its C-terminus with multiple epitopes: 6xHis, HA, and a ProtA (ZZ domains) preceded by a Protease 3C cleavage site (the size of the tag is 19 kDa). The first step is a regular immunoprecipitation in native conditions using anti-HA antibodies, followed by a Ni-NTA-based purification of the His-tag in denaturing conditions. It allows to purify the protein (both ubiquitylated and non-ubiquitylated, *see* Fig. 4) for further, mass spectrometry-based, identification of the ubiquitylation sites (*see* **Note 13**). Note that Protein A itself can be ubiquitylated in yeast cells (S. Léon, unpublished results). It is therefore necessary to identify the ubiquitylated sites by mass spectrometry to make sure that ubiquitin is carried on the substrate, and not the tag. Consequently, this protocol should merely be used to identify the ubiquitylation sites, and not to prove that a protein is ubiquitylated.

DAY 1. Preculture: in the morning, inoculate freshly streaked yeast cells in 10 mL synthetic glucose medium lacking uracil.

1. In the evening, measure the OD_{600} of the cultures and inoculate the yeast cells in 180 mL synthetic raffinose medium, at OD_{600} = 0.004. Grow overnight under agitation (200 rpm) at 30 °C.
2. *DAY 2.* Measure the OD_{600} of the cultures. The cultures should have reached an OD_{600} of ~0.3–0.5.
3. Add galactose (2 % final) in the medium for the desired amount of time (*see* **Note 14**).

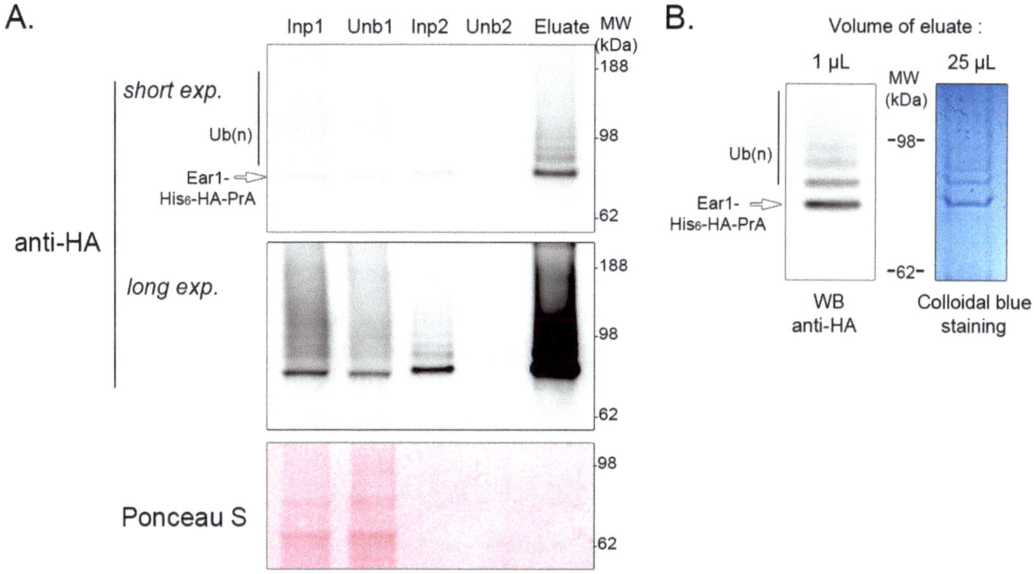

Fig. 4 Tandem purification of a protein in denaturing conditions for the identification of ubiquitin-conjugated sites. (**a**) WT cells (BY4741) expressing the galactose-inducible Ear1-His$_6$-HA-ProtA construct were subjected to a double purification as described. First, an immunoprecipitation using anti-HA antibodies was performed on a crude extract in native conditions (input 1 and unbound 1). The eluate of this immunoprecipitation was used as the input fraction (input 2) for the second, Ni-NTA based purification of the protein in denaturing conditions (unbound 2 and final eluate). A western blot using anti-HA antibodies allows to evaluate the purification procedure. (**b**) From the same fraction, 1 or 25 μL of eluate was loaded on the same gel. The gel was cut, and the left part was subjected to a western blot using anti-HA antibodies, whereas the right part was used for colloidal blue staining, showing the purity of the protein obtained

4. Measure the OD of the culture (*see* **Note 2**).
5. Spin down the equivalent of 150 OD units of cells at 3000 × *g*, 3 min at room temperature.
6. Remove the supernatant, resuspend the pellet in 10 mL cold water.
7. Spin down for 30 s at 13,000 × *g*, 4 °C.
8. Discard supernatant and keep the pellet on ice.
9. Prepare antibody-coupled beads for the immunoprecipitation (*see* **step 23**). Take 200 μL of Protein-G-coupled Sepharose slurry, wash in 1 mL IP lysis buffer, resuspend in 1 mL IP lysis buffer, and use the appropriate amount of antibodies required for immunoprecipitation (e.g., 10 μL of mouse monoclonal F7 antibody: sc-7392 from Santa Cruz Biotech). Incubate for ≥1 h at 4 °C on a rotating wheel.
10. Resuspend the pellet in 2.5 mL IP lysis buffer containing protease inhibitors (yeast protease inhibitor cocktail, 1 mM PMSF, 10 mM N-ethylmaleimide, 100 μM MG-132).
11. Dispatch the resuspended pellet in 5 × 600 μL fractions, in 1.5-mL tubes.

12. Add glass beads (0.4–0.6 mm) up to 1–2 mm below the meniscus.
13. Lyse cells on a vortex equipped with a microtube foam insert for 4 × 30 s at 4 °C (cold room), with 1 min incubation on ice between each pulse.
14. After lysis, in the cold room, open the tubes, poke a small hole at the bottom using a needle, place into a clean 1.5-mL tube to collect the lysate, and close the cap.
15. In the cold room, centrifuge for a few seconds on a mini-centrifuge to collect the lysate and leave the beads in the upper tube. Discard the tube with the beads (they should be dry).
16. Pool lysates in a clean, cold centrifugation tube.
17. To the lysate, add 100 μL of a 25 % solution of Triton X-100, and invert to mix (*see* **Note 15**).
18. Let solubilize on ice for 10 min.
19. Centrifuge the lysate for 5 min at 3000 × *g* at 4 °C.
20. Transfer supernatant to a clean, cold 15-mL tube and keep on ice.
21. Keep 100 μL aside for the "input 1" fraction, keep on ice.
22. Spin down the beads coupled to antibodies (**step 10**) for 30 s at 1000 × *g*, 4 °C.
23. Discard the beads' supernatant.
24. Add the beads to the lysate.
25. Incubate on a rotating wheel at 4 °C for 3 h.
26. Spin down the beads for 30 s at 1000 × *g*, 4 °C.
27. Keep 100 μL of the supernatant aside for the "unbound 1" fraction, keep on ice.
28. Discard remaining supernatant.
29. Add 1 mL of IP lysis buffer to the beads
30. Transfer to a cold, 1.5-mL tube
31. Incubate on a rotating wheel at 4 °C for 10 min.
32. Spin down the beads for 30 s at 1000 × *g*, 4 °C. Discard supernatant.
33. Repeat the wash twice and carefully remove supernatant.
34. Add 1 mL of urea buffer.
35. Incubate for 5 min at room temperature under agitation (e.g., in a thermomixer, 1000 rpm).
36. Open the tube, poke a small hole at the bottom of the tube using a needle, place into a clean 1.5-mL tube to collect the sample, and close the cap.
37. Centrifuge for a few seconds on a mini-centrifuge to collect the eluate and leave the beads in the upper tube. Discard the tube with the beads (they should be dry).

38. Keep 30 μL of the eluate aside for the "input 2" fraction, keep on ice.
39. To the remaining sample, add 500 μL Ni-NTA beads pre-washed in urea buffer.
40. Incubate overnight on a rotating wheel at 4 °C.
41. To the "Input 1," "Unbound 1" and "Input 2" fractions (**steps 21**, **27**, and **38**), add 1/5 volume of 5× sample buffer, denature all samples at 55 °C for 10 min, and store at −20 °C.
42. *DAY 3*. Spin down the beads (*see* **step 40**) for 30 s at 1000 × *g*, 4 °C.
43. Keep 30 μL aside for the "unbound 2" fraction, keep on ice.
44. Discard remaining supernatant.
45. Add 2.5 mL of Urea buffer to the beads, incubate on a rotating wheel at 4 °C for 10 min.
46. Spin down the beads for 30 s at 1000 × *g*, 4 °C. Discard supernatant.
47. Repeat **steps 45–46** and carefully remove supernatant.
48. Add 1 mL of Urea buffer to the beads, and transfer to a 1.5-mL tube.
49. Incubate on a rotating wheel at 4 °C for 10 min.
50. Spin down the beads for 30 s at 1000 × *g*, 4 °C. Discard as much supernatant as possible.
51. Add 100 μL of Urea elution buffer.
52. Incubate for 5 min at room temperature under agitation (e.g., in a thermomixer, 1000 rpm).
53. Open the tube, poke a small hole at the bottom of the tube using a needle, place into a clean 1.5-mL tube to collect the eluate, and close the cap.
54. Centrifuge for a few seconds on a mini-centrifuge to collect the sample and leave the beads in the upper tube. Discard the tube with the beads (they should be dry).
55. The total volume of the eluate (incl. the bed volume of the beads) should be approximately 175 μL.
56. To the "unbound 2" and the "eluate" samples (**steps 43** and **55**), add 1/5 volume of 5× sample buffer.
57. Denature all samples (input 1, output 1, input 2, output 2, eluate) at 55 °C for 10 min.
58. Load 5 μL of input 1, output 1, input 2, output 2 and 10 μL of the final elution onto an SDS-PAGE gel.
59. Blot with the antibody against your protein of interest or its epitope, to evaluate the purification procedure, and with anti-Ubiquitin antibodies (1/10,000 mouse monoclonal P4D1 antibody: sc-8017, Santa Cruz Biotech) to reveal Ub conjugates.

3.5 Identification of Ubiquitylated Sites Within a Protein of Interest by a Genetic Approach

Ubiquitin conjugation can occur at defined sites within a protein, and the identification of those sites (e.g., using the approach described in the above section) can help to generate non-ubiquitylatable mutants after its/their mutation. However, the mutation of the target lysine is sometimes compensated by the ubiquitylation of yet another site, because ubiquitylation can show little specificity towards the sites targeted [30]. Therefore, we describe a mutant-based approach that should lead to a construct that is not ubiquitylatable, regardless of the prior identification of the ubiquitylation sites, that we successfully used in the past [23]. This is based on a synthetic version of the ORFs encoding your protein of interest, in which all lysine residues have been substituted to arginine (K0 construct) (Fig. 5a). Then, endogenous restriction sites can be used to construct chimeras. If no restriction sites are available, silent restriction sites should be introduced at various places within the sequence when designing the construct for gene synthesis, in which case a synthetic WT ORF must also be ordered, with the same restriction sites. The ubiquitylation of the resulting chimeras can be evaluated individually, to narrow down to the region and/or sites that are targeted by ubiquitylation (Fig. 5b).

Although it involves several cloning steps, this approach is still fairly efficient when dealing with proteins carrying many lysines. Note that this approach requires that the ubiquitylated site(s) lie(s) within a defined region of the primary structure of a protein. Below, we provide the general directions to properly map the ubiquitylated site(s) using this approach.

1. Order synthetic gene versions encoding the protein of interest: WT and K0. Consider adding silent restriction sites within the sequence that are not present on the plasmid in which these synthetic genes will be cloned.
2. Clone the synthetic genes in the plasmid of interest. It is advisable to use a plasmid allowing the expression of epitope-tagged protein, even if antibodies directed against the protein are available, because the KR mutations may affect the antigenicity of the protein.
3. Assay for the ubiquitylation of the WT construct and the K0 mutant. The K0 should have lost all traces of ubiquitylation, if ubiquitin is conjugated on a lysine.
4. Construct chimeras (Nt-WT/Ct-K0 and Nt-K0/Ct-WT) and assay again the ubiquitylation of the chimeric proteins. The disappearance of ubiquitylated species for one of these constructs indicates that the target lysine(s) are contained within those that have been mutated (*see* **Note 16**).
5. Prepare additional chimeras to narrow down to the minimal region.
6. If the density of lysines within the initial sequence is such that the final construct still bears several lysine residues, then site-directed mutagenesis can be used to further discriminate between each of those.

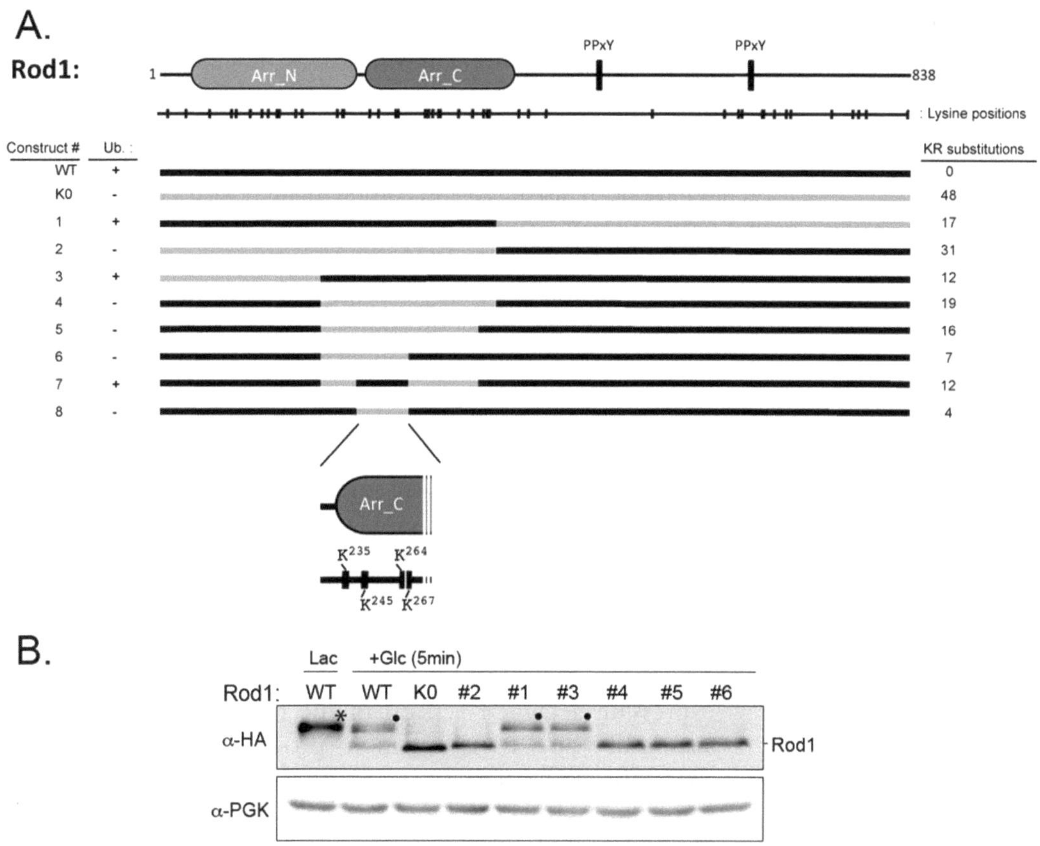

Fig. 5 Identification of ubiquitylated sites within a protein of interest by a genetic approach. (**a**) Example of the method: identification of the ubiquitylated region within the arrestin-related protein, Rod1, using a synthetic construct in which all lysines have been mutated (K0) and its derived chimeric constructs [23]. (**b**) Western blot showing the ubiquitylation status of several of the chimeric constructs depicted in (**a**). The *asterisk* indicates phosphorylated Rod1, which is only observed when yeast cells are grown in the absence of glucose, such as in lactate medium (Lac) [23]. The *black circles* indicate Rod1 ubiquitylation, which occurs only after glucose exposure and for constructs in which the ubiquitylation site is still intact. PGK is used as a loading control

3.6 Studying the Topology of Ubiquitin Modification

When dealing with a polyubiquitylated substrate, it may be interesting to determine the topology of the ubiquitin chain, because the various structures obtained through various ubiquitin linkages will lead to various functional outcomes.

Various tools have been developed to do so, such as antibodies directed against the most abundant polyubiquitin chain linkages, namely K48- and K63-linked polyubiquitin, whose use has already been described [31]. The development of tandem ubiquitin-binding entities (TUBEs) [32, 33] and other sensor proteins [34], consisting of multiple ubiquitin-binding domains, also allows the purification of substrates that are polyubiquitylated with various topologies and are commercially available.

An alternative approach consists in using ubiquitin mutants in which either of the ubiquitin lysine residues has been mutated into an arginine, to hamper the ability to make a chain. These chain-terminating mutants can be individually overexpressed in the context of a wild-type strain in which these mutants will act as dominant negatives, as previously described [24]. Instead, here, we provide an example of the use of a set of yeast strains ("SUB" strains) engineered by the Finley lab, in which all endogenous ubiquitin genes have been disrupted and complemented with a plasmid encoding either WT or lysine point mutants of ubiquitin [5]. Below are guidelines on how to use these strains to determine ubiquitin linkage. Note that this approach is limited to the study of non-K48-based polyubiquitin chain substrates, because the expression of a K48R mutant in yeast as a sole source of ubiquitin is not viable [35]. For this specific case, the overexpression of Ub-K48R mutant in the context of a wild-type strain is therefore preferable [24]. Also, because the ubiquitin is not tagged in these strains, it cannot be purified by Ni-NTA based purification. Therefore, the analysis relies on the ability to see protein ubiquitylation on a crude extract (Subheading 3.1) (Fig. 6). Alternatively, the protein may be purified from each of these strains by denaturing IP, followed by a western blot using anti-ubiquitin antibodies as described in Subheading 3.3.

1. Transform the SUB strains with an *URA3*-based plasmid expressing a tagged version of the protein of interest using standard yeast transformation procedure. These includes SUB280 (WT ubiquitin), SUB515 (K6R Ub), SUB516 (K11R), SUB517 (K27R), SUB518 (K29R), SUB519 (K33R), and SUB413 (K63 Ub) [5] (*see* **Note 17**).
2. Grow cells as described in Subheading 3.1 (for crude extracts) or in Subheading 3.3 (for denaturing IPs), in synthetic medium lacking uracil, starting at OD = 0.010 for SUB280 and OD = 0.020 for the mutants.

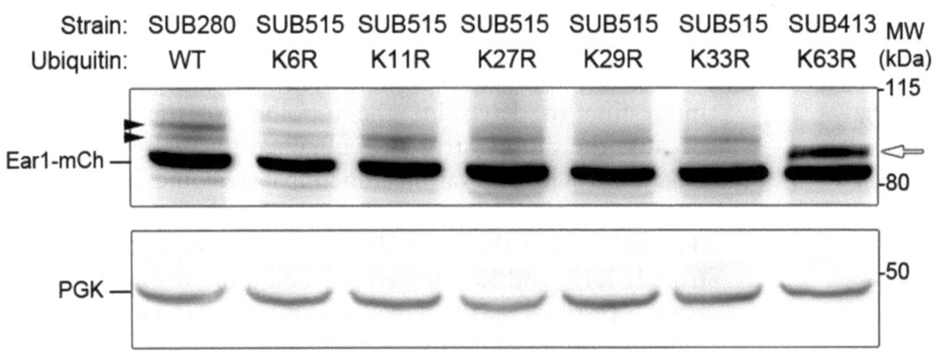

Fig. 6 Studying the topology of ubiquitin modification. An example of the use of the SUB strains (which express K-to-R ubiquitin mutants) to gain information on the topology of the ubiquitin modification on a substrate (here, the endosomal Rsp5 adaptor protein, Ear1). *Black arrowheads*: Ear1 ubiquitylated adducts. *White arrow*: the ubiquitin modification appears as a single band in the SUB413 strain, suggesting that Ear1 is polyubiquitylated with a short K63 chain (for more details, *see* also [40]). PGK is used as a loading control

3. For each strain, perform either a crude extract as described in Subheading 3.1, or a denaturing IP as described in Subheading 3.3.
4. Load samples on SDS-PAGE.
5. Blot with the antibodies used to detect the protein of interest.

4 Notes

1. The presence of 0.02 % glucose allows a faster initiation of growth in raffinose medium.
2. Cells will not lyse efficiently if they are not in exponential phase. We advise never to reach an OD > 1.0.
3. For proteins suspected to be modified by a K48-linked polyubiquitin chain, and/or to be proteasomal substrates, the treatment of cells with a proteasome inhibitor (e.g., MG132, 100 μM final) prior harvest can lead to their stabilization, sometimes in its ubiquitylated form (although this is usually counteracted by the presence of endogenous ubiquitin isopeptidases). However, the use of MG132 in yeast cultures at this concentration is only efficient in strains in which either *PDR5* or *ERG6* is disrupted, leading to an increased permeability to various drugs [36, 37]. An alternative method to promote the efficiency of MG-132 in WT yeast strains has been described and is based on the use of a different culture medium [38].
4. This step is essential, because His-tagged proteins will not bind to Ni-NTA beads in acidic conditions.
5. A pipette tip can be used too, but we advise not to pipette up and down because the pellet is sticky.
6. The guanidine-HCl in buffer A may precipitate upon sample cooling.
7. Do not incubate the beads with buffer A for too long, because they will titrate the imidazole in buffer A, which may hinder their ability to bind to His-tagged proteins later on.
8. Preferentially use precast gradient gels (e.g., NuPAGE® 4–12 %, Bis-Tris, Life Technologies).
9. Make sure that samples do not run out of the gel, because His-tagged ubiquitin is about 10 kDa and will be used as a purification control.
10. It is advised to use Protein G- rather than Protein A-coupled beads except when the source of the antibody is guinea pig. Neither Protein A nor Protein G binds chicken IgY. Alternatively, when using HA-tagged proteins, use the ready-to-use anti-HA affinity matrix (Roche): use 30 μL, resuspend in 1 mL IP dilution buffer, and keep on ice.

11. The use of a water bath sonicator to resuspend the pellet helps for the final solubilization of the sample.
12. Because of the presence of antibodies in the immunoprecipitate, we advise to use antibodies to ubiquitin that are directly coupled to HRP to avoid the use of a secondary antibody (e.g., mouse monoclonal P4D1 antibody: sc-8017 HRP, Santa Cruz Biotech; use at 1/2000).
13. Identifying posttranslational modification sites by mass spectrometry can be hindered by the spacing of Arg/Lys residues, at which the cleavage with trypsin occurs within the protein of interest. Tryptic fragments that are either too long or too short will not be analyzed, leading to a poor coverage and eventually, the inability to detect the posttranslational modification. The coverage of a given protein by mass spectrometry can be visualized in repository databases, such as the Global Proteome Machine Database (http://gpmdb.thegpm.org). In this case, other proteases than trypsin should be considered prior to MS analysis.
14. The plasmid used is a 2μ-based (high-copy) plasmid, and the expression of the protein is driven by the strong galactose-inducible promoter. This can lead to a massive overexpression of the protein, and the ubiquitylation detected may not be physiologically relevant. To circumvent this problem, the expression level of the protein at various times after galactose induction can be checked and compared with endogenous protein level. Protein expression can usually be observed within 15 min of galactose induction.
15. **Steps 17** and **18** are dispensable if dealing with a soluble protein. Adding TX-100 may lead to smears when visualizing the protein of interest by western blot. Note that 1 % TX-100 solubilizes many membrane proteins, but not all, so this should be checked prior to the experiment.
16. If both chimeric proteins appear ubiquitylated, then it is likely that the protein is ubiquitylated on multiple sites throughout the primary structure of the protein, or that ubiquitylation can occur on any of those sites. In both cases, this approach should be discontinued, and we advise to initiate a mapping of the ubiquitylated lysines using biochemical purification and mass spectrometry (*see* Subheading 3.4).
17. Note that only the *URA3* gene can be used for selection of transformants in the SUB strains genetic backgrounds.

Acknowledgments

We thank Dan Finley (Harvard Medical School, Boston, MA, USA) for providing us with the SUB strains. This work was funded by the CNRS, the Fondation ARC pour la Recherche sur le Cancer (SFI20101201844 and SFI20121205762) and the Ligue contre le Cancer – Comité de Paris (RS13/75-45 and RS14/75-120) to SL. The authors would like to acknowledge networking support by the Proteostasis COST Action (BM1307).

References

1. Komander D, Rape M (2012) The ubiquitin code. Annu Rev Biochem 81:203–229. doi:10.1146/annurev-biochem-060310-170328
2. Husnjak K, Dikic I (2012) Ubiquitin-binding proteins: decoders of ubiquitin-mediated cellular functions. Annu Rev Biochem 81:291–322. doi:10.1146/annurev-biochem-051810-094654
3. Kulathu Y, Komander D (2012) Atypical ubiquitylation – the unexplored world of polyubiquitin beyond Lys48 and Lys63 linkages. Nat Rev Mol Cell Biol 13:508. doi:10.1038/nrm3394
4. Ulrich HD, Jentsch S (2000) Two RING finger proteins mediate cooperation between ubiquitin-conjugating enzymes in DNA repair. EMBO J 19:3388–3397. doi:10.1093/emboj/19.13.3388
5. Spence J, Sadis S, Haas AL, Finley D (1995) A ubiquitin mutant with specific defects in DNA repair and multiubiquitination. Mol Cell Biol 15:1265–1273
6. Hofmann RM, Pickart CM (1999) Noncanonical MMS2-encoded ubiquitin-conjugating enzyme functions in assembly of novel polyubiquitin chains for DNA repair. Cell 96:645–653
7. Signaling to NF-kappaB: regulation by ubiquitination (2010) 2:a003350. http://eutils.ncbi.nlm.nih.gov/entrez/eutils/elink.fcgi?dbfrom=pubmed&id=20300215&retmode=ref&cmd=prlinks
8. Versatile roles of k63-linked ubiquitin chains in trafficking (2014) 3:1027–1088. http://eutils.ncbi.nlm.nih.gov/entrez/eutils/elink.fcgi?dbfrom=pubmed&id=25396681&retmode=ref&cmd=prlinks.
9. Galan JM, Haguenauer-Tsapis R (1997) Ubiquitin lys63 is involved in ubiquitination of a yeast plasma membrane protein. EMBO J 16:5847–5854. doi:10.1093/emboj/16.19.5847
10. Xu P, Duong DM, Seyfried NT, Cheng D, Xie Y et al (2009) Quantitative proteomics reveals the function of unconventional ubiquitin chains in proteasomal degradation. Cell 137:133–145. doi:10.1016/j.cell.2009.01.041
11. Ziv I, Matiuhin Y, Kirkpatrick DS, Erpapazoglou Z, Léon S et al (2011) A perturbed ubiquitin landscape distinguishes between ubiquitin in trafficking and in proteolysis. Mol Cell Proteomics 10:M111.009753. doi:10.1074/mcp.M111.009753
12. Wagner SA, Beli P, Weinert BT, Nielsen ML, Cox J et al (2011) A proteome-wide, quantitative survey of in vivo ubiquitylation sites reveals widespread regulatory roles. Mol Cell Proteomics 10:M111.013284. doi:10.1074/mcp.M111.013284
13. Kim W, Bennett EJ, Huttlin EL, Guo A, Li J et al (2011) Systematic and quantitative assessment of the ubiquitin-modified proteome. Mol Cell 44:325–340. doi:10.1016/j.molcel.2011.08.025
14. Polyubiquitin linkage profiles in three models of proteolytic stress suggest the etiology of Alzheimer disease (2011) 286:10457–10465. http://eutils.ncbi.nlm.nih.gov/entrez/eutils/elink.fcgi?dbfrom=pubmed&id=21278249&retmode=ref&cmd=prlinks
15. Clague MJ, Coulson JM, Urbé S (2012) Cellular functions of the DUBs. J Cell Sci 125:277–286. doi:10.1242/jcs.090985
16. Dohmen RJ, Stappen R, McGrath JP, Forrová H, Kolarov J et al (1995) An essential yeast gene encoding a homolog of ubiquitin-activating enzyme. J Biol Chem 270:18099–18109
17. Gwizdek C, Iglesias N, Rodriguez MS, Ossareh-Nazari B, Hobeika M et al (2006) Ubiquitin-associated domain of Mex67 synchronizes recruitment of the mRNA export machinery with transcription. Proc Natl Acad Sci U S A 103:16376–16381. doi:10.1073/pnas.0607941103
18. Histidine-tagged ubiquitin substitutes for wild-type ubiquitin in Saccharomyces cerevisiae and facilitates isolation and identification of in vivo substrates of the ubiquitin pathway (2000)

282:54–64. http://eutils.ncbi.nlm.nih.gov/entrez/eutils/elink.fcgi?dbfrom=pubmed&id=10860499&retmode=ref&cmd=prlinks.

19. Peng J, Schwartz D, Elias JE, Thoreen CC, Cheng D et al (2003) A proteomics approach to understanding protein ubiquitination. Nat Biotechnol 21:921–926. doi:10.1038/nbt849
20. Mayor T, Deshaies RJ (2005) Two-step affinity purification of multiubiquitylated proteins from Saccharomyces cerevisiae. Meth Enzymol 399:385–392. doi:10.1016/S0076-6879(05)99026-5
21. Guerrero C, Tagwerker C, Kaiser P, Huang L (2006) An integrated mass spectrometry-based proteomic approach: quantitative analysis of tandem affinity-purified in vivo cross-linked protein complexes (QTAX) to decipher the 26S proteasome-interacting network. Mol Cell Proteomics 5:366–378. doi:10.1074/mcp.M500303-MCP200
22. Tagwerker C, Flick K, Cui M, Guerrero C, Dou Y et al (2006) A tandem affinity tag for two-step purification under fully denaturing conditions: application in ubiquitin profiling and protein complex identification combined with in vivo cross-linking. Mol Cell Proteomics 5:737–748. doi:10.1074/mcp.M500368-MCP200
23. Becuwe M, Vieira N, Lara D, Gomes-Rezende J, Soares-Cunha C et al (2012) A molecular switch on an arrestin-like protein relays glucose signaling to transporter endocytosis. J Cell Biol 196:247–259. doi:10.1083/jcb.201109113
24. Laney JD, Hochstrasser M (2002) Assaying protein ubiquitination in Saccharomyces cerevisiae. Meth Enzymol 351:248–257
25. Kaiser P, Tagwerker C (2005) Is this protein ubiquitinated? Meth Enzymol 399:243–248. doi:10.1016/S0076-6879(05)99016-2
26. Kragt A, Voorn-Brouwer T, van den Berg M, Distel B (2005) The Saccharomyces cerevisiae peroxisomal import receptor Pex5p is monoubiquitinated in wild type cells. J Biol Chem 280:7867–7874. doi:10.1074/jbc.M413553200
27. Léon S, Zhang L, McDonald WH, Yates J, Cregg JM et al (2006) Dynamics of the peroxisomal import cycle of PpPex20p: ubiquitin-dependent localization and regulation. J Cell Biol 172:67–78. doi:10.1083/jcb.200508096
28. Stawiecka-Mirota M, Pokrzywa W, Morvan J, Zoladek T, Haguenauer-Tsapis R et al (2007) Targeting of Sna3p to the endosomal pathway depends on its interaction with Rsp5p and multivesicular body sorting on its ubiquitylation. Traffic 8:1280–1296. doi:10.1111/j.1600-0854.2007.00610.x
29. Gelperin DM, White MA, Wilkinson ML, Kon Y, Kung LA et al (2005) Biochemical and genetic analysis of the yeast proteome with a movable ORF collection. Genes Dev 19:2816–2826. doi:10.1101/gad.1362105
30. Danielsen JMR, Sylvestersen KB, Bekker-Jensen S, Szklarczyk D, Poulsen JW et al (2011) Mass spectrometric analysis of lysine ubiquitylation reveals promiscuity at site level. Mol Cell Proteomics 10:M110.003590. doi:10.1074/mcp.M110.003590
31. Newton K, Matsumoto ML, Ferrando RE, Wickliffe KE, Rape M et al (2012) Using linkage-specific monoclonal antibodies to analyze cellular ubiquitylation. Methods Mol Biol 832:185–196. doi:10.1007/978-1-61779-474-2_13
32. Aillet F, Lopitz-Otsoa F, Hjerpe R, Torres-Ramos M, Lang V et al (2012) Isolation of ubiquitylated proteins using tandem ubiquitin-binding entities. Methods Mol Biol 832:173–183. doi:10.1007/978-1-61779-474-2_12
33. Hjerpe R, Aillet F, Lopitz-Otsoa F, Lang V, England P et al (2009) Efficient protection and isolation of ubiquitylated proteins using tandem ubiquitin-binding entities. EMBO Rep 10:1250–1258. doi:10.1038/embor.2009.192
34. Sims JJ, Scavone F, Cooper EM, Kane LA, Youle RJ et al (2012) Polyubiquitin-sensor proteins reveal localization and linkage-type dependence of cellular ubiquitin signaling. Nat Meth 9:303–309. doi:10.1038/nmeth.1888
35. Inhibition of proteolysis and cell cycle progression in a multiubiquitination-deficient yeast mutant (1994) 14:5501–5509. http://eutils.ncbi.nlm.nih.gov/entrez/eutils/elink.fcgi?dbfrom=pubmed&id=8035826&retmode=ref&cmd=prlinks
36. Selective inhibitors of the proteasome-dependent and vacuolar pathways of protein degradation in Saccharomyces cerevisiae (1996) 271:27280–27284. http://eutils.ncbi.nlm.nih.gov/entrez/eutils/elink.fcgi?dbfrom=pubmed&id=8910302&retmode=ref&cmd=prlinks
37. Combined chemical and genetic approach to inhibit proteolysis by the proteasome (2010) 27:965–974. http://eutils.ncbi.nlm.nih.gov/entrez/eutils/elink.fcgi?dbfrom=pubmed&id=20625982&retmode=ref&cmd=prlinks
38. Liu C, Apodaca J, Davis LE, Rao H (2007) Proteasome inhibition in wild-type yeast Saccharomyces cerevisiae cells. Biotechniques 42:158–160
39. Léon S, Erpapazoglou Z, Haguenauer-Tsapis R (2008) Ear1p and Ssh4p are new adaptors of the ubiquitin ligase rsp5p for cargo ubiquitylation and sorting at multivesicular bodies. Mol Biol Cell 19:2379–2388. doi:10.1091/mbc.E08-01-0068
40. Erpapazoglou Z, Dhaoui M, Pantazopoulou M, Giordano F, Mari M et al (2012) A dual role for K63-linked ubiquitin chains in multivesicular body biogenesis and cargo sorting. Mol Biol Cell 23:2170–2183. doi:10.1091/mbc.E11-10-0891

Chapter 6

Strategies to Detect Endogenous Ubiquitination of a Target Mammalian Protein

Sara Sigismund and Simona Polo

Abstract

Different biochemical techniques are well established to investigate target's ubiquitination in mammals without overexpressing a tagged version of ubiquitin (Ub). The simplest and more direct approach is to immunoprecipitate (IP) your target protein from cell lysate (stimulated and/or properly treated), followed by western blot analysis utilizing specific antibodies against Ub (*see* Subheading 3.1). This approach requires a good antibody against the target working in IP; alternatively, one could express a tagged version of the protein, possibly at the endogenous level. Another approach consists in IP ubiquitinated proteins from total cell lysate followed by detection with the antibody against the protein of interest. This second method relies on the availability of specific and very efficient antibodies against Ub (*see* Subheading 3.2). A more quantitative approach is the DELFIA assay (Perkin Elmer), an ELISA-based assay, which allows comparing more samples and conditions (*see* Subheading 3.3). Cross-validation with more than one approach is usually recommended in order to prove that your protein is modified by ubiquitin.

Here we will use the EGFR as model system but protocols can be easily modified according to the protein of interest.

Key words Endogenous ubiquitination, Immunoprecipitation, Western blot, ELISA, EGFR, Endocytosis

1 Introduction

Most proteins—if not all—are regulated by the ubiquitin pathway. The abundance of ubiquitinated proteins is often low in cells; thus an essential step for their analysis is represented by a pre-enrichment of the ubiquitinated species. Many affinity approaches have been tested to isolate Ub conjugates under native and denaturing conditions, including ubiquitin antibodies, ubiquitin-binding proteins, and epitope-tagged ubiquitin. Ubiquitin overexpression often renders ubiquitination a constitutive process that, at least in some cases, could be a disadvantage for the analysis (*see* **Note 1**). In addition, antibodies recognizing the Ub-modified peptides (anti-GlyGly antibodies) can be employed to facilitate the identification of Ub acceptor sites by mass spectrometry analysis. These repre-

Rune Matthiesen (ed.), *Proteostasis: Methods and Protocols*, Methods in Molecular Biology, vol. 1449, DOI 10.1007/978-1-4939-3756-1_6, © Springer Science+Business Media New York 2016

sent complementary approaches in addition to the ones described in this chapter, to confirm that your protein of interest is indeed ubiquitinated in vivo.

Ligand-induced trafficking of the EGFR is one of the best characterized examples of how the regulation of receptor turnover is modulated by the Ub signal. EGFR-Ub occurs at the plasma membrane (PM) and is catalyzed by the E3 ligase Cbl in complex with the adaptor molecule Grb2, recruited to the phosphorylated/active receptor [1, 2]. It is regulated by ligand concentration, being sharply activated at high dose of EGF [3]. Importantly, EGFR-Ub proceeds all along the endocytic pathway since Cbl remains bound to the EGFR after internalization [4]. Mass-spectrometry analysis has revealed that EGFR is both mono- and polyubiquitinated through Lys63-linked chains [5]. Whether these two types of Ub modifications may act at different steps and/or have different impact on EGFR fate is currently still unknown. EGFR ubiquitination is a reversible modification, being regulated by several deubiquitinating enzymes (DUBs). Up to now, several DUBs acting on the EGFR pathway have been identified. They act downstream the internalization step, at the level of endosomes and MVBs sorting stations [6].

Ub plays a pivotal role in determining the fate and the signaling ability of the EGFR thanks to the accurate recognition exerted by UBD-containing endocytic "route controllers" that inexorably ferry the internalized receptor towards a degradative fate in lysosomes and away from a recycling pathway. As such, the Ub signal is relevant at multiple endocytic stations and on different targets. At the PM, receptor ubiquitination is essential for internalization of EGFR via non-clathrin endocytosis (NCE). NCE is activated at high dose of ligand and targets the majority of EGFRs to degradation [7, 8]. In clathrin-mediated endocytosis (CME) EGFR-Ub is not essential [8, 9]; however, it may participate with other internalization signals to render the system more robust [10]. At the endosomal level, EGFR ubiquitination is critical to target receptor to the ESCRT complex, destining it to intraluminal vesicles of MVBs and, finally, to lysosomal degradation. This last step leads to signal extinction that is the final goal of EGFR ubiquitination [11, 12].

2 Materials

2.1 Buffers and Solutions

1. RIPA lysis buffer: 50 mM Tris–HCl, 150 mM NaCl, 1 mM EDTA, 1 % triton, 1 % Na deoxycholate, 0.1 % SDS (or 1 % *see* **Note 2**), supplemented with a cocktail of proteases, phosphatases, and DUB inhibitors [phosphatases, proteases, and DUB inhibitors were freshly added to the buffer prior to lysis: 20 mM Na pyrophosphate pH 7.5, protease cocktail CALBIOCHEM

(200×), 50 mM NaF, 2 mM PMSF, 10 mM Na vanadate in HEPES pH 7.5 (*see* **Note 3**), 5 mM NEM or 25 μM PR619].

2. Laemmli buffer: before loading on SDS-PAGE gel, samples are resuspended in 2× Laemmli buffer [4 % SDS, 125 mM Tris pH 6.8, 20 % glycerol, 0.002 % saturated bromophenol blue, 10 % (v/v) β-mercaptoethanol (14 M)].
3. Denaturing solution: 6 M Guanidinium chloride, 20 mM Tris pH 7.5, and, freshly added, 1 mM PMSF and 5 mM β-mercaptoethanol.

2.2 Reagents and Antibodies

Mouse monoclonal anti-Ub antibodies: P4D1 (Santa Cruz) and FK2 (EnzoLifescience or MBL); anti-EGFR antibodies: rabbit polyclonal anti-EGFR, against a 1172–1186 of human EGFR (EGFR inTra, Eurogentech), and mouse monoclonal anti-EGFR (m108 hybridoma, directed against the extracellular domain of human EGFR, EGFR exTra, ATCC, [13]); mouse monoclonal anti-β-catenin (BD); anti-eps15 (homemade monoclonal antibody directed against EH domain); protein G-conjugated sepharose beads (Zymed).

2.3 DELFIA (Perkin Elmer)

DELFIA kit includes microwell plates, wash buffer, assay buffer, Europium-labeled secondary antibodies, and enhancement solution.

Plate coating buffer: 3.03 g $NaCO_3$, 6.0 g $NaHCO_3$ pH 9.6 dissolved in 1 l of H_2O.

2.4 EnVision Instrument (Perkin Elmer)

3 Methods

3.1 Analysis by IP of a Specific Substrate (i.e., EGFR) and Western Blot Anti-ubiquitin

For this experiment, HeLa cells (10 cm plate for each condition, 70 % of confluence) are serum starved for 16 h and then stimulated with 100 ng/ml of EGF for 0, 2, 10, 30, and 120 min. Ligand stimulation induces the rapid ubiquitination of the EGFR (peak at 2 min) followed by its endocytosis and lysosomal degradation. A trick to block receptor degradation and to accumulate ubiquinated EGFR species is to treat cells with chloroquine (100 μM, pretreatment of 1 h and kept during all experiment) or NH_4Cl (20 mM, pretreatment of 1 h and kept during all experiment), both affecting lysosomal acidification and function. Cell lysates are subjected to IP anti-EGFR and western blot anti-Ub and anti-EGFR, to check for IP efficiency. In case of target proteins degraded by the proteasome, treatment with MG132 inhibitor can be used to improve the signal. Dose and time depend on the protein of interest and on cell type (you can start testing 3–6 h at 5 μM).

To control for the specificity of the anti-Ub signal, a parallel IP with lysates of cell knockdown for the protein of interest (e.g., HeLa cell knockdown for the EGFR) should be performed. In addition, an IP with an unrelated antibody of the same specie and isotype of the antibody against your protein of interest is also recommended.

3.1.1 Cell Lysis and IP Reaction

1. Cells are washed twice in cold PBS. Remove accurately PBS from plates by letting them tilted for 2/3 min.
2. Cell lysis is performed directly on the plate by scraping in RIPA buffer, plus inhibitors, 150–200 μl/plate (Subheading 2.1). In case of RIPA w/1 % SDS (*see* **Note 2**) use the minimal amount of buffer required to cover the plate (≤80–100 μl/plate).
3. Lysates are let on ice for 10 min and, then, spin at 16,000 × *g* for 15 min at 4 °C (standard RIPA) or subjected to ultracentrifugation (RIPA 1 % SDS) at 45,000 × *g* for 45 min at 4 °C.
4. Supernatants are collected and protein content was determined by Bradford protein assay (not ideal due to the presence of SDS) or BCA (*see* **Note 4**).
5. Incubate 250 μg of lysates for each condition (IP volume of around 250 μl, 1 μg/ml) in the presence of anti-EGFR intra-polyclonal antibody.
6. Incubate the samples with gentle agitation at 4 °C for 2 h.
7. Add 30 μl of 50 % slurry Protein G-conjugated sepharose beads and incubate at 4 °C for additional 2 h.
8. Wash the beads four times with 0.5 ml of RIPA buffer/each, through centrifugation for 1 min at 400 × *g*.
9. Add 20 μl 2× Laemmli buffer to the beads (previously dried), boil for 5 min at 95 °C, and spin for 1 min at 13,000 rpm. Samples are ready to be loaded on SDS-PAGE gel.

3.1.2 Western Blot Anti-Ub

1. Keep 1/10 of the immunopurified samples to perform direct western blot anti-EGFR (MW of ~180 kDa) by loading on 7 % acrylamide gels, transferring on nitrocellulose filter, and performing anti-EGFR western blot by standard procedure ([14] and **Note 5**).
2. The remaining 9/10 of the samples are loaded on 7% acrylamide gels in order to probe for the presence of ubiquitinated EGFR by western blot anti-Ub.
3. Transfer proteins to polyvinylidene fluoride (PVDF) membrane previously activated by incubation in 100% MeOH for 5 min at RT (*see* **Note 6**), followed by extensive washing in TBS-T buffer (TBS, 0.1 % tween).
4. After transfer (*see* **Note 7**), treat the filters in denaturing solution (Subheading 2.1) for 30 min at 4 °C.

5. After extensive washing in TBS-T, incubate the filters overnight at 4 °C in 5 % BSA (in TBS-T).
6. Incubate the filters with anti-Ub antibody: P4D1 antibody diluted 1:1000 in TBS-T 5 % BSA and incubated for 1 h at room temperature.
7. After three washes of 10 min each in TBS-T, incubate the filters with the anti-mouse horseradish peroxidase-conjugated secondary antibody diluted in TBS-T 3 % BSA for 30 min at RT.
8. Wash the filters three times in TBS-T (5 min each). The bound secondary antibody is revealed using the enhanced chemiluminescence (ECL) method (Amersham).

3.2 Immunoprecipitation of Ub-Modified Proteins from Cellular Lysate

For this experiment, growing Hela cells (10 cm plate for each condition, 70 % of confluence) are treated as in Subheading 3.1. Cells are then lysed in RIPA buffer supplemented with a cocktail of protease, phosphatase, and DUB inhibitors as previously described (Subheading 2.1). Lysates are subjected to IP with FK2 monoclonal anti-Ub antibody. Western blot anti-EGFR and eps15 are performed on the IP as controls for EGF-induced Ub. For MG132-treated samples, anti-βcatenin can be used as internal control.

As a control for unspecific binding of your protein to FK2/Protein G-sepharose beads, a parallel IP with an unrelated antibody of the same specie and isotype of FK2 antibody (mouse IgG1) is recommended.

3.2.1 IP Anti-Ub

1. Incubate 500 μg of lysates for each condition (EGF, MG132, or other) in the presence of 10 μg of anti-FK2 monoclonal antibody for 1 h at 4 °C.
2. Add 30 μl of 50 % slurry protein G-conjugated sepharose beads and incubate for an additional hour at 4 °C.
3. Wash the beads four times with 0.5 ml of RIPA buffer (Subheading 2.1), through centrifugation for 1 min at 400 × *g*.
4. Add 20 μl 2× Laemmli buffer to the dried beads (Subheading 2.1), boil for 5 min at 95 °C, and spin for 1 min at 16,000 × *g*.
5. Samples are ready to be loaded on SDS-PAGE gel.
6. IP are transferred to nitrocellulose, for western blot anti-eps15 (150 kDa), anti-EGFR intra (180 kDa), or anti-β-catenin (90 kDa).

3.3 ELISA Assays for EGFR Ubiquitination

For this assay, the dissociation-enhanced lanthanide fluoroimmunoassay (DELFIA) technology from Perkin Elmer is employed [3]. It is based on sandwich recognition of a target protein by a capture antibody and a detection antibody. The capture antibody is immobilized on a solid surface (microwells) directly through non-covalent bonds. After the addition of the analyte (cell lysate), the detection of signals relies on a lanthanide (Europium)-conjugated antibody that is able to produce a fluorescent signal upon enhancement with acidic enhancement buffer. Lanthanide ions are released in solution at low pH and they rapidly form new, highly stable fluorescent chelates. The fluorescence of the lanthanide chelate is amplified one to ten million times by this enhancement step and it develops a signal in 5 min that is stable for up to 8 h.

As specificity control, incubation of the antibody-coated plate with lysates of cell knockdown for the protein of interest (e.g., HeLa cell knockdown for the EGFR), followed by the detection step, is recommended.

3.3.1 Plate Coating, Incubation, and Detection

1. Microwell plates are coated with the capturing antibody diluted in coating buffer (Subheading 2.3). *See* Table 1 for antibody concentration.
2. Blocking is performed for 2 h with BSA 2% in PBS.
3. 25–50 μg of lysates from HeLa cells, stimulated with the indicated concentration of EGF, are incubated overnight at 4 °C. Lysates are prepared in RIPA w/1% SDS buffer and diluted to 0.2% SDS before incubation step (Subheading 2.1 and **Notes 2** and **4**).
4. After three washes, wells are incubated with primary antibodies, diluted at 1 μg/ml in assay buffer (provided in the DELFIA kit), for 1 h at RT.
5. After three washes, anti-mouse or rabbit Europium-labeled secondary antibodies (1 μg/ml in assay buffer) are added for an additional hour.
6. After three washes and treatment with enhancement solution for 15 min at RT, fluorescence is measured with EnVision instrument (excitation at 340 nm and emission at 615 nm).

Capturing and detecting antibodies differ depending on whether a forward or reverse approach is performed (*see* also Fig. 1 for a scheme of the two procedures). Note that in the case of the EGFR substrate the reverse approach is more sensitive.

Table 1
Antibodies' scheme for the ELISA

	Capturing antibodies	Detecting antibodies
Forward ELISA	Rabbit anti-EGFR intra (5 μg/ml)	Mouse monoclonal antibodies against Ub (FK2) or EGFR extra (both diluted at 1 μg/ml)
Reverse ELISA	Monoclonal antibodies against Ub (FK2, 5 μg/ml) or EGFR extra (1 μg/ml)	Rabbit anti-EGFR intra (1 μg/ml)

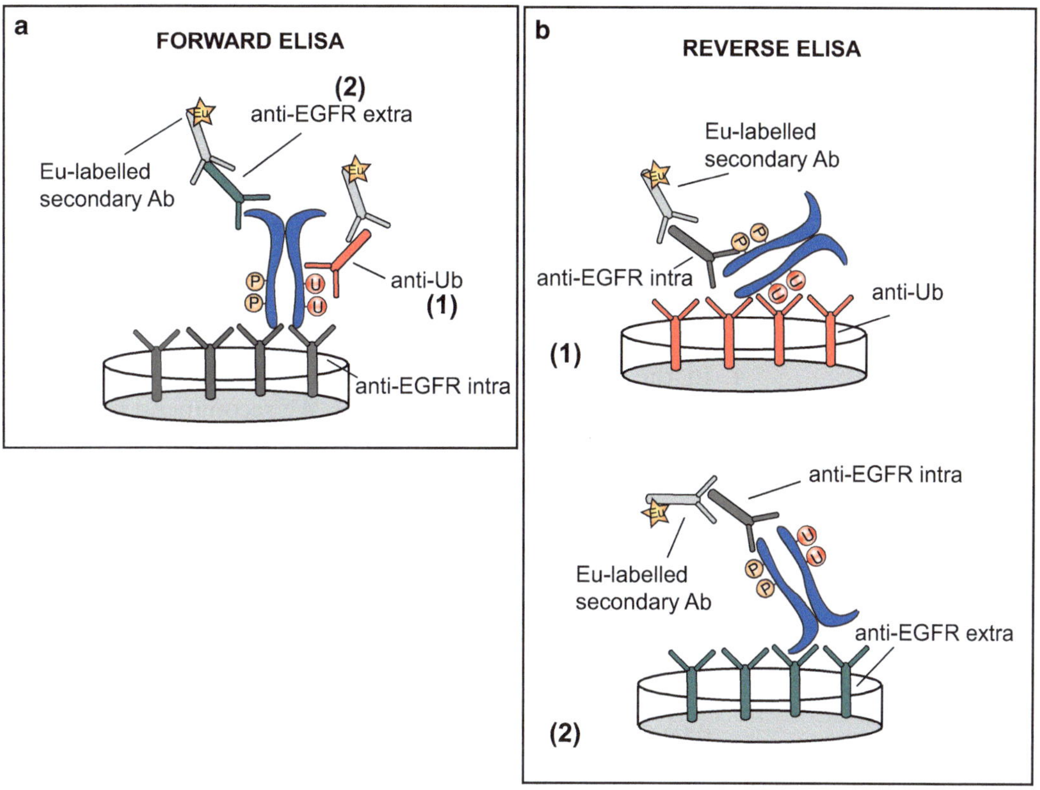

Fig. 1 Scheme of the ELISA assay. (**a**) In the "forward" approach, microwell plates are coated with a polyclonal anti-EGFR intra, which captures the receptor from the lysate (1). Detection of Ub-EGFR or total EGFR (to normalize) is performed with primary monoclonal antibodies directed against Ub (FK2, 1) or EGFR extra (2), respectively, followed by Europium (Eu)-labeled secondary antibodies. (**b**) In the reverse approach, microwell plates are coated with monoclonal antibodies directed against Ub (FK2, 1) or EGFR extra (2) that capture Ub-EGFR (1) or total EGFR (2), respectively. Detection is performed with anti-EGFR intra followed by a Eu-labeled secondary antibody

4 Notes

1. Ectopically expressed epitope-tagged forms of Ub (HA, FLAG, Myc-tagged; single Ub or in tandem) can be used. Although it is not easy to overexpress Ub compared to the endogenous level (cellular Ub level are homeostatically regulated), the tagged Ub are well incorporated into proteins and may be useful to get an initial idea if Ub modifies a specific protein. A major caveat is that Ub overexpression often renders ubiquitination a constitutive process. This could be an advantage if there are no clues on the signal regulating the process but it is not recommended to study the physiological role of the Ub modification of a given target. In principle, target's ubiquitination should always be validated with endogenous ubiquitin.

2. In order to eliminate interacting proteins, lysates prepared in modified RIPA buffer containing 1 % of SDS are recommended. Lysates are subjected to ultracentrifugation (1 % SDS) at 45,000 × *g* for 45 min at 4 °C and supernatants are collected. Dilution to reach 0.2 % SDS (using RIPA buffer w/o SDS) is performed prior to quantification (*see* also **Note 4**) and IP reaction. Performance of target's antibody in this condition should be set before proceeding.
3. To prepare 0.5 M Na vanadate stock, dissolve the powder in 1 M Hepes pH 7.5 and boil for 2 min at 95 °C.
4. In the case of RIPA with 1 % SDS, Bradford protein assay quantitation method is not ideal. We recommend to use BCA method and to dilute the lysate to a final concentration of 0.2 % SDS prior to quantification.
5. For western blot anti-EGFR, we know that our antibody perform better with nitrocellulose than with PVDF, but this should be verified case by case. Instead for anti-Ub, PVDF membrane is required.
6. Staining the filter with Ponceau solution is not recommended, as this might interfere with antibody recognition. Pre-stained molecular weight markers are used to check the transferring efficiency.
7. It is particularly important to avoid drying the membrane during these treatments. In case of drying it is possible to hydrate PVDF membrane again using MeOH for 2 min followed by extensive washing in TBS-T buffer.

Acknowledgements

The authors would like to acknowledge networking support by the Proteostasis COST Action (BM1307).

References

1. Huang F, Sorkin A (2005) Growth factor receptor binding protein 2-mediated recruitment of the RING domain of Cbl to the epidermal growth factor receptor is essential and sufficient to support receptor endocytosis. Mol Biol Cell 16:1268–1281
2. Levkowitz G, Waterman H, Ettenberg SA, Katz M, Tsygankov AY, Alroy I, Lavi S, Iwai K, Reiss Y, Ciechanover A, Lipkowitz S, Yarden Y (1999) Ubiquitin ligase activity and tyrosine phosphorylation underlie suppression of growth factor signaling by c-Cbl/Sli-1. Mol Cell 4: 1029–1040
3. Sigismund S, Algisi V, Nappo G, Conte A, Pascolutti R, Cuomo A, Bonaldi T, Argenzio E, Verhoef LG, Maspero E, Bianchi F, Capuani F, Ciliberto A, Polo S, Di Fiore PP (2013) Threshold-controlled ubiquitination of the EGFR directs receptor fate. EMBO J 32: 2140–2157
4. Umebayashi K, Stenmark H, Yoshimori T (2008) Ubc4/5 and c-Cbl continue to ubiquitinate EGF receptor after internalization to facilitate polyubiquitination and degradation. Mol Biol Cell 19:3454–3462. doi:10.1091/mbc.E07-10-0988

5. Huang F, Kirkpatrick D, Jiang X, Gygi S, Sorkin A (2006) Differential regulation of EGF receptor internalization and degradation by multiubiquitination within the kinase domain. Mol Cell 21:737–748
6. Clague MJ, Liu H, Urbe S (2012) Governance of endocytic trafficking and signaling by reversible ubiquitylation. Dev Cell 23: 457–467
7. Sigismund S, Argenzio E, Tosoni D, Cavallaro E, Polo S, Di Fiore PP (2008) Clathrin-mediated internalization is essential for sustained EGFR signaling but dispensable for degradation. Dev Cell 15:209–219
8. Sigismund S, Woelk T, Puri C, Maspero E, Tacchetti C, Transidico P, Di Fiore PP, Polo S (2005) Clathrin-independent endocytosis of ubiquitinated cargos. Proc Natl Acad Sci U S A 102:2760–2765
9. Huang F, Goh LK, Sorkin A (2007) EGF receptor ubiquitination is not necessary for its internalization. Proc Natl Acad Sci U S A 104:16904–16909
10. Goh LK, Huang F, Kim W, Gygi S, Sorkin A (2010) Multiple mechanisms collectively regulate clathrin-mediated endocytosis of the epidermal growth factor receptor. J Cell Biol 189:871–883
11. Acconcia F, Sigismund S, Polo S (2009) Ubiquitin in trafficking: the network at work. Exp Cell Res 315:1610–1618
12. Haglund K, Dikic I (2012) The role of ubiquitylation in receptor endocytosis and endosomal sorting. J Cell Sci 125:265–275
13. Aboud-Pirak E, Hurwitz E, Pirak ME, Bellot F, Schlessinger J, Sela M (1988) Efficacy of antibodies to epidermal growth factor receptor against KB carcinoma in vitro and in nude mice. J Natl Cancer Inst 80: 1605–1611
14. Penengo L, Mapelli M, Murachelli AG, Confalonieri S, Magri L, Musacchio A, Di Fiore PP, Polo S, Schneider TR (2006) Crystal structure of the ubiquitin binding domains of rabex-5 reveals two modes of interaction with ubiquitin. Cell 124: 1183–1195

Chapter 7

In Vitro Ubiquitination: Self-Ubiquitination, Chain Formation, and Substrate Ubiquitination Assays

Elena Maspero and Simona Polo

Abstract

Ubiquitination of proteins in vitro has evolved as an indispensable tool for the functional analysis of this posttranslational modification. In vitro ubiquitination is particularly helpful to study conjugation mechanisms. The efficiency of the ubiquitination reaction depends in part on the quality of the enzymes utilized. Here we introduce the assay developed in our lab to study HECT E3 ligases. It involves bacterially expressed E1, His-tagged Ube2D3 (also called UbcH5c, the best E2 for Nedd4), untagged Nedd4, and untagged ubiquitin (Ub). As tags may impair specific activity of the enzymes or even interfere with the enzymatic reaction, they should be avoided, removed, or kept to a minimal size whenever possible, unless proven to be without consequence. The protocol described here is suitable for other E3 ligases capable of forming Ub chains as pseudo-product of the enzyme reaction. It is also adapted to include substrates. In this case, substrates should be tagged and purified after the reaction is completed to allow the detection of the ubiquitinated products.

Key words Ubiquitination, In vitro assay, E3 ligase, Nedd4

1 Introduction

The development of in vitro system with purified proteins allows a direct analysis of the molecular mechanisms regulating ubiquitination of a specific protein. We developed such a system to study our favorite E3, Nedd4, an HECT ligase responsible for the ubiquitination of several endocytic proteins including eps15 and epsin [1, 2].

HECT ligases are directly involved in the substrate ubiquitination and form an intermediate thioester bond with their active Cys, before catalyzing the covalent attachment of ubiquitin to its specific substrate. The Nedd4 family comprises one member in yeast (Rsp5) and nine in humans (Nedd4, Nedd4-2, Itch, Smurf1, Smurf2, Wwp1, Wwp2, HECW1, and HECW2) [3–5]. Despite having different functions, these proteins share similar domain architecture, containing an N-terminus C2 domain, responsible for membrane binding and two to four WW domains that mediate

Rune Matthiesen (ed.), *Proteostasis: Methods and Protocols*, Methods in Molecular Biology, vol. 1449,
DOI 10.1007/978-1-4939-3756-1_7, © Springer Science+Business Media New York 2016

protein-protein interactions with substrates containing a PPxY motif, and the HECT domain at the C-terminus [3–5]. From the biological point of view, Nedd4 members are involved in cell signaling pathways that regulate cell growth and proliferation and have emerged as key regulators in several diseases from cancer to neurodegenerative disorders [3–5]. From the catalytic point of view, the majority of Nedd4 family members are K63-specific enzymes (thus not involved in proteasomal degradation) and use a sequential addition mechanism to build a chain on a substrate [6, 7]. Their enzymatic activity is tightly controlled through an autoinhibitory interaction of the C2 with the HECT domain [8, 9]. We recently provided evidence that, in Nedd4, this inhibitory close conformation can be released upon Tyr phosphorylation occurring at the C2 and at the HECT domain [10].

This chapter describes the strategies and the methods developed to study ubiquitination of Nedd4 and its activity in terms of specificity. Thanks to these assays, using a panel of Ub mutants in which lysine residues are mutated into arginine, we provided evidence that the Nedd4 family members are all K63-specific enzymes [6] that we then validated using wild-type Ub and absolute quantitation of the Ub chain types with the AQUA proteomic method [7].

2 Materials

2.1 Buffers and Solutions

1. 1 M IPTG stock solution: Dissolve IPTG in H_2O, sterile filter, and store aliquots at −20 °C.
2. 1 M Imidazole stock solution: Dissolve imidazole in H_2O and store aliquots at 4 °C.
3. 1 M DTT stock solution: Dissolve DTT in H_2O and store aliquots at −20 °C.
4. 0.1 M ATP stock solution: Dissolve ATP in H_2O, adjust the pH to 7.0, and store aliquots at −20 °C.
5. GST lysis buffer: 50 mM Na-HEPES, pH 7.5, 200 mM NaCl, 1 mM EDTA, 0.1% NP40, 5% glycerol, Protease Inhibitor Cocktail set III (Calbiochem).
6. Buffer A: 50 mM NaH_2PO_4 pH 7.8, 300 mM NaCl, 10% glycerol, 10 mM imidazole, and protease inhibitors.
7. Ub lysis buffer: 25 mM Ammonium acetate, 10 mM ß-mercaptoethanol, 10% glycerol, and protease inhibitors, pH 7.0.
8. PreScission Cleavage Buffer: 50 mM Tris–HCl, pH 7.4, 100 mM NaCl, 1 mM EDTA, 1 mM DTT, 5% glycerol.
9. Size exclusion buffer: 20 mM Tris–HCl, pH 8.0, 200 mM NaCl, 1 mM EDTA, 1 mM DTT, 5% glycerol.
10. Ubiquitination buffer: The buffer required to drive the reaction is prepared 10× concentrated and contains 250 mM Tris–HCl

pH 7.6, 50 mM $MgCl_2$, and 1 M NaCl. To the final mix freshly prepared ATP 2 mM (Sigma) and DTT 0.2 μM were added.

11. Denaturing solution: 6 M Guanidinium chloride, 20 mM Tris pH 7.5, and freshly added 1 mM PMSF and 5 mM β-mercaptoethanol.
12. RIPA buffer: 50 mM Tris–HCl, 150 mM NaCl, 1 mM EDTA, 1 % Triton, 1 % Na deoxycholate, 0.1 % SDS.
13. Laemmli buffer: Before loading on SDS-PAGE gel, samples are resuspended in 2× Laemmli buffer: 4 % SDS, 125 mM Tris pH 6.8, 20 % glycerol, 0.002 % saturated bromophenol blue, 10 % (v/v) ß-mercaptoethanol (14 M).
14. TBS-T: TBS, 0.1 % Tween 20.

2.2 Competent Cells for Protein Expression

BL21 (DE3) pLysS and Rosetta™ (DE3) pLysS cells are from Novagen.

2.3 Reagents

Glutathione Sepharose™ 4B and PreScission protease are from GE Healthcare. HisPur™ Ni-NTA Resin and Imperial Protein Stain from Life Technologies. Imidazole, DTT, and ATP from Sigma. Anti-Ub P4D1 antibody from Santa Cruz Biotechnology.

3 Methods

3.1 Protein Production

3.1.1 GST Fusion Protein

GST fusion proteins were expressed in BL21 at 18 °C for 16 h after induction with 500 μM IPTG at an OD_{600} of 0.5. Cell pellets were resuspended in GST lysis buffer. Sonicated lysates were cleared by centrifugation at 20,000 rpm = 45,000 × g for 45 min. Supernatants were incubated with 1 ml of glutathione-Sepharose beads per liter of bacterial culture. After 4 h at 4 °C, beads were washed with PBS and equilibrated in PreScission cleavage buffer. For GST-HECT^{Nedd4} production, to cleave off the GST tag, 10 units of PreScission protease per mg of substrate were incubated for 16 h at 4 °C (*see* **Note 1**). The cleaved HECT^{Nedd4} was purified onto a Superdex 200 size-exclusion chromatography column (GE Healthcare, Fig. 1).

3.1.2 His-Fusion Protein Production

His-tagged E1 enzyme Uba1 (Addgene clone #34965) was produced in Rosetta cells according to the protocol described in [11] (*see* **Note 2**).

His-tagged E2 enzyme Ube2D3 (UBCH5c) was expressed in BL21 at 18 °C for 16 h after induction with 1 mM IPTG at an OD_{600} of 0.6. Cell pellets were resuspended in Buffer A and lysed by sonication. Cell debris was removed by centrifugation and the supernatant incubated with 1 ml of HisPur Ni-NTA resin, previously washed three times with Buffer A. After 2 h at 4 °C beads were then washed three times with Buffer A, Buffer A with 1 M NaCl, and Buffer A containing 20 mM imidazole. His-fusion proteins were eluted in Buffer A containing 300 mM imidazole.

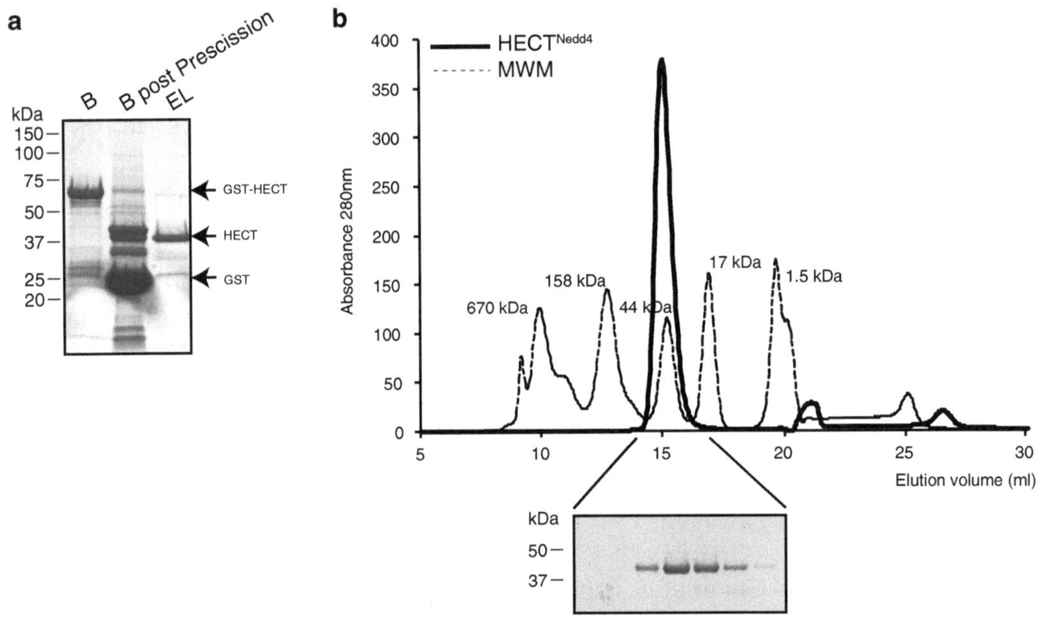

Fig. 1 Purification of GST-HECT^{Nedd4}. (**a**) PreScission cleavage of GST-HECT^{Nedd4} was controlled by loading aliquots of beads before (**b**) and after cleavage (**b** post-PreScission) and of the eluate (EL). (**b**) Chromatogram of a typical size-exclusion purification on Superdex 200 column of cleaved HECT ^{Nedd4}, purity of the protein containing fractions was assessed by SDS-PAGE and Coomassie staining

After overnight dialysis in size-exclusion buffer, Ube2D3 was purified onto a Superdex 75 size-exclusion chromatography column (GE Healthcare).

3.1.3 Untagged Ub Production

Untagged Ub WT (or mutant) was expressed in Rosetta at 18 °C for 16 h after induction with 1 mM IPTG at an OD_{600} of 0.5. Cell pellets were resuspended in Ub lysis buffer and lysed by sonication. Cell debris was removed by centrifugation and the supernatant adjusted to pH 4.5–5.0 with concentrated acetic acid. Precipitated proteins were removed by centrifugation and the supernatant containing the ubiquitin monomers was passed through a 0.45 mm PES filter. After dialysis Ub was purified onto a Superdex 75 size-exclusion chromatography column.

3.2 Protein Purification

All the proteins used during enzymatic reaction studies were purified by size-exclusion chromatography. Prior to run on a Superdex 75 or Superdex 200 size-exclusion column, the samples were concentrated using Vivaspin (of different dimension and different molecular weight cutoff according to the protein of interest) and centrifuging at 6000 × *g* in a 45 °C fixed angle rotor, at 4 °C. The concentrated sample was run onto a Superdex 75 or 200 column according to the dimension of protein to be purified; typically S200 was used for HECT protein and S75 for E2 and Ub. Gel filtration is carried out using size-exclusion buffer. The desired protein was eluted in a clear, well-isolated, peak according to its size and shape. Purity of the peak can be assayed by SDS-PAGE gel. The desired fractions were collected, pooled, and concentrated as needed using Vivaspin tubes.

3.3 In Vitro Ubiquitination Assays

We describe here two different conditions for in vitro ubiquitination assay. The first protocol is a self-ubiquitination assay in which it is possible to detect the E3 ligase activity directly on the enzyme itself or through the formation of polyubiquitin chains. It is the method of choice when specific substrate of your E3 ligase is not known. The second protocol is designed in order to investigate the activity of the E3 ligase of choice on a specific substrate. Once a robust in vitro system has been set up for the study of your E3 ligase of choice it would be possible to get information about the specificity of the enzyme introducing Ub mutant in the analysis.

3.3.1 Self-Ubiquitination Assay and Ub Chains Formation

Self-ubiquitination assay required E1 enzyme, E2 Ube2D3, HECT domain of Nedd4 produced as GST fusion protein, and Ub.

GST-HECT domain of Nedd4 not only gets ubiquitinated during the assay, but it is also able to generate free polyUb chains as pseudo-product of the enzymatic reaction. In order to investigate the nature of the Ub chain formed during the reaction ubiquitin mutants Ub can be used. Two set of mutants are available; in the first set each Lys is mutated to Arg with the exception of one position (K-only mutant), and in the second set a single Lys is substituted with Arg (K-R mutant).

Ubiquitination assay is performed in 50 μl reaction. Calculate the volume of protein stock solution needed for the assay and if necessary dilute enzymes in the ubiquitination buffer. A general protocol for self-ubiquitination assay is as follows.

1. Mix 20 nM E1, 250 nM GST-HECT^{Nedd4} on beads, 2 mM ATP, 0.2 μM DTT, and 10× ubiquitination buffer.
2. Transfer 50 μl of the common mix in a separate tube and add 1 μM Ub WT. Incubate at 37 °C for 30 min; this represents the negative control of the reaction.
3. Add to the common mix 250 nM of purified His6-tagged Ube2D3/UbcH5c and aliquot in four different tubes, each of them containing 1 μM of the different Ub mutant (i.e.,. WT, K63R). Incubate samples at 37 °C for 30 min.
4. After incubation at 37 °C for 30 min, the reaction is immediately transferred on ice and centrifuged (refrigerated) to separate the pellet containing ubiquitinated GST-HECT from free Ub chains in the supernatant.

5. Take 15 μl of the supernatant and add 2× Laemmli buffer. Run the sample on 4–20% gradient gel in order to separate the ladder of polyUb chains that are formed during the reaction.
6. Wash the pellet four times in RIPA buffer (*see* **Note 3**) before addition of 2× Laemmli buffer, heating, and loading on 8% SDS-PAGE gel. The modification by a single Ub results in 8 kDa shift in apparent molecular mass of GST-HECT. Multiple or poly-ubiquitination results in a smear higher than that.
7. Detection is performed by western blot anti-Ub (*see* Subheading 3.3.2).
8. Coomassie-stained membrane is used to show GST-fusion protein loading after western blot (*see* **Note 4**).

3.3.2 Western Blot Anti-Ub

1. Transfer proteins on a polyvinylidene fluoride (PVDF) membrane (Immobilion P, Millipore), previously activated by incubation in 100% MeOH for 5 min at RT (*see* **Note 5**).
2. After transfer (*see* **Note 6**), treat the filters in denaturing solution for 30 min at 4 °C.
3. After extensive washing in TBS-T buffer, incubate the filters overnight at 4 °C in 5% BSA (in TBS-T).
4. Incubate the filters with anti-Ub antibody diluted in TBS-T 5% BSA for 1 h at room temperature.
5. After three washes of 10 min each in TBS-T, incubate the filters with the anti-mouse horseradish peroxidase-conjugated secondary antibody diluted in TBS-T 3% BSA for 30 min at RT.
6. Wash the filters three times in TBS-T (5 min each). The bound secondary antibody is revealed using the enhanced chemiluminescence (ECL) method (Amersham).

3.3.3 Substrate Ubiquitination

Substrate ubiquitination assay required El enzyme, E2 enzyme Ube2D3, HECT domain of Nedd4 cleaved from the GST and purified, Ub, and GST fusion of the substrate of interest. Ubiquitination assay is performed in 50 μl reaction.

1. Mix 20 nM E1, 250 nM $HECT^{Nedd4}$, 300 nM GST-substrate (gamma-EnaC), 2 mM ATP, 0.2 μM DTT, 1 μM Ub WT, and 10× ubiquitination buffer.
2. Transfer 50 μl of the common mix in a separate tube and add 1 μM Ub WT. Incubate at 37 °C for 30 min; this represents the negative control of the reaction.
3. Add to the common mix 250 nM of purified His6-tagged Ube2D3/UbcH5c and incubate samples at 37 °C for 10, 30, and 60 min.
4. After incubation at 37 °C for the indicated time, the reaction is immediately transferred on ice and centrifuged (refriger-

ated) to separate to separate the pellet containing ubiquitinated GST-substrate from free Ub chains and enzymes present in the supernatant.

5. Discard the supernatant and wash the pellet four times in RIPA buffer (*see* **Note 3**) before addition of 2× Laemmli buffer, heating, and loading on 8 % SDS-PAGE gel. The modification by Ub results in 8 kDa shift in apparent molecular mass of GST-gamma EnaC.
6. Detection is performed by western blot anti-Ub (*see* Subheading 3.3.2).
7. Coomassie-stained membrane is used to show GST-fusion protein loading after western blot (*see* **Note 4**).

4 Notes

1. PreScission protease contains a noncleavable GST-tag; therefore it remains bound to the glutathione-Sepharose 4B resin.
2. We produce and purify recombinant E1 enzyme from *E. coli* [11]; alternatively E1 is commercially available from different sources.
3. If pellet is not clearly visible after centrifugation, additional GSH beads can be added to facilitate the washing step procedure.
4. Membrane is stained with Imperial Protein Stain for 30 min at room temperature. After extensive washing with water, the membrane is dried and proteins became visible.
5. Staining the filter with Ponceau solution is not recommended, as this might interfere with antibody recognition. Pre-stained molecular weight markers are used to check the transferring efficiency.
6. It is particularly important to avoid drying the membrane during these treatments. In case of drying it is possible to hydrate PVDF membrane again using MeOH for 2 min followed by extensive washing in TBS-T buffer.

Acknowledgements

The authors would like to acknowledge networking support by the Proteostasis COST Action (BM1307).

References

1. Polo S, Sigismund S, Faretta M, Guidi M, Capua MR, Bossi G, Chen H, De Camilli P, Di Fiore PP (2002) A single motif responsible for ubiquitin recognition and monoubiquitination in endocytic proteins. Nature 416:451–455
2. Woelk T, Oldrini B, Maspero E, Confalonieri S, Cavallaro E, Di Fiore PP, Polo S (2006) Molecular mechanisms of coupled monoubiquitination. Nat Cell Biol 8:1246–1254
3. Rotin D, Kumar S (2009) Physiological functions of the HECT family of ubiquitin ligases. Nat Rev Mol Cell Biol 10:398–409
4. Scheffner M, Kumar S (2014) Mammalian HECT ubiquitin-protein ligases: biological and pathophysiological aspects. Biochim Biophys Acta 1843:61–74
5. Polo S (2012) Signaling-mediated control of ubiquitin ligases in endocytosis. BMC Biol 10:25. doi:10.1186/1741-7007-10-25
6. Maspero E, Mari S, Valentini E, Musacchio A, Fish A, Pasqualato S, Polo S (2011) Structure of the HECT:ubiquitin complex and its role in ubiquitin chain elongation. EMBO Rep 12:342–349
7. Maspero E, Valentini E, Mari S, Cecatiello V, Soffientini P, Pasqualato S, Polo S (2013) Structure of a ubiquitin-loaded HECT ligase reveals the molecular basis for catalytic priming. Nat Struct Mol Biol 20:696–701
8. Mari S, Ruetalo N, Maspero E, Stoffregen MC, Pasqualato S, Polo S, Wiesner S (2014) Structural and functional framework for the autoinhibition of Nedd4-family ubiquitin ligases. Structure 22:1639–1649
9. Wiesner S, Ogunjimi AA, Wang HR, Rotin D, Sicheri F, Wrana JL, Forman-Kay JD (2007) Autoinhibition of the HECT-type ubiquitin ligase Smurf2 through its C2 domain. Cell 130:651–662
10. Persaud A, Alberts P, Mari S, Tong J, Murchie R, Maspero E, Safi F, Moran MF, Polo S, Rotin D (2014) Tyrosine phosphorylation of NEDD4 activates its ubiquitin ligase activity. Sci Signal 7:ra95
11. Berndsen CE, Wolberger C (2011) A spectrophotometric assay for conjugation of ubiquitin and ubiquitin-like proteins. Anal Biochem 418:102–110

Chapter 8

Isolation of the Ubiquitin-Proteome from Tumor Cell Lines and Primary Cells Using TUBEs

Wendy Xolalpa, Lydia Mata-Cantero, Fabienne Aillet, and Manuel S. Rodriguez

Abstract

Tandem ubiquitin-binding entities (TUBEs) act as molecular traps to isolate polyubiquitylated proteins facilitating the study of this highly reversible posttranslational modification. We provide here sample preparation and adaptations required for TUBE-based enrichment of the ubiquitin proteome from tumor cell lines or primary cells. Our protocol is suitable to identify ubiquitin substrates, enzymes involved in the ubiquitin proteasome pathway, as well as proteasome subunits by mass spectrometry. This protocol was adapted to prepare affinity columns, reduce background, and improve the protein recovery depending on the sample source and necessities.

Key words TUBEs, Ubiquitylation, Isolation, Purification, Posttranslational modifications

1 Introduction

One of the most critical steps of the entire proteomic analysis procedure is sample preparation. Obtaining a 100% representation of proteins from a biological sample certainly does not occur in practice but the most efficient methods display a high number of representative cellular proteins. Preservation of posttranslational modifications (PTMs) of proteins such as phosphorylation, ubiquitylation, or SUMOylation adds another level of complexity to the sample preparation process due to the transient and labile nature of these PTMs. The covalent attachment of one or more ubiquitin moieties to a protein substrate (known as protein ubiquitylation) implies a great diversity of conjugating and de-conjugating enzymes [1, 2], resulting in a vast repertoire of ubiquitin chains on the target proteins [3]. Consequently, this PTM is involved in numerous and crucial cellular processes [4, 5]. Different strategies have been developed to face the challenging steps of enrichment and identification of endogenous ubiquitylated proteins (including histidine

Rune Matthiesen (ed.), *Proteostasis: Methods and Protocols*, Methods in Molecular Biology, vol. 1449, DOI 10.1007/978-1-4939-3756-1_8, © Springer Science+Business Media New York 2016

pull-down and immunoaffinity purification) [6–8]. Tandem Ubiquitin Binding Entities are versatile research tools for the survey of the ubiquitin-proteasome pathway that have been successfully applied for the isolation and enrichment of polyubiquitylated proteins [9–11]. The fusion to a GST-tag allows a conventional affinity purification step by pull-down using glutathione-coupled beads. TUBEs have been demonstrated to be useful for isolation of ubiquitylated proteins from different biological sources such as cell lines, tissues, and organs [10]. Scaling up the pull-down protocol allowed the enrichment of protein samples for the study of global ubiquitylation events by mass spectrometry (MS) [12–14]. Besides the enrichment of ubiquitylated proteins in large scale, the pull-down protocol can also be suitable for the isolation of ubiquitin-interacting partners to provide a more complete view of the ubiquitin proteome.

The method described here is adapted for the isolation and identification of the ubiquitin proteome by MS. TUBEs are prepared in columns to be used as affinity matrix for capturing ubiquitin-modified proteins and interacting partners. Binding, washing, and elution steps are checked to optimize the enrichment of ubiquitylated proteins. Specific and background proteins are controlled at each step by Western blot (WB) analysis using an anti-ubiquitin antibody or gel staining in order to optimize protein recovery. The protocol should be adapted according to the source/type of biological sample since background can be increased by the presence of very abundant proteins in specific cell systems. If the aim is to isolate only ubiquitylated proteins, more stringent conditions and/or number of washes should be increased before elution to reduce the binding of nonspecific proteins. With the method described below, we have successfully isolated ubiquitylated proteins together with ubiquitin-interacting partners from tumor cell lines or primary cells such as mantle cell lymphoma (MCL) cells or red blood cells (RBCs), respectively [15]. The abundance of hemoglobin in RBCs requires specific adaptations to remove this sticky protein (indicated in detail in Subheading 4). As TUBEs are GST-fusion proteins, a GST control should be included in parallel with each sample to discriminate background from specific proteins.

The method is divided into four sections: (1) coupling TUBEs/GST to glutathione beads, (2) cell lysis and sample clearance, (3) capturing and elution of proteins, and (4) sample concentration and gel electrophoresis. Figures 1 and 2 show a scheme of the general workflow.

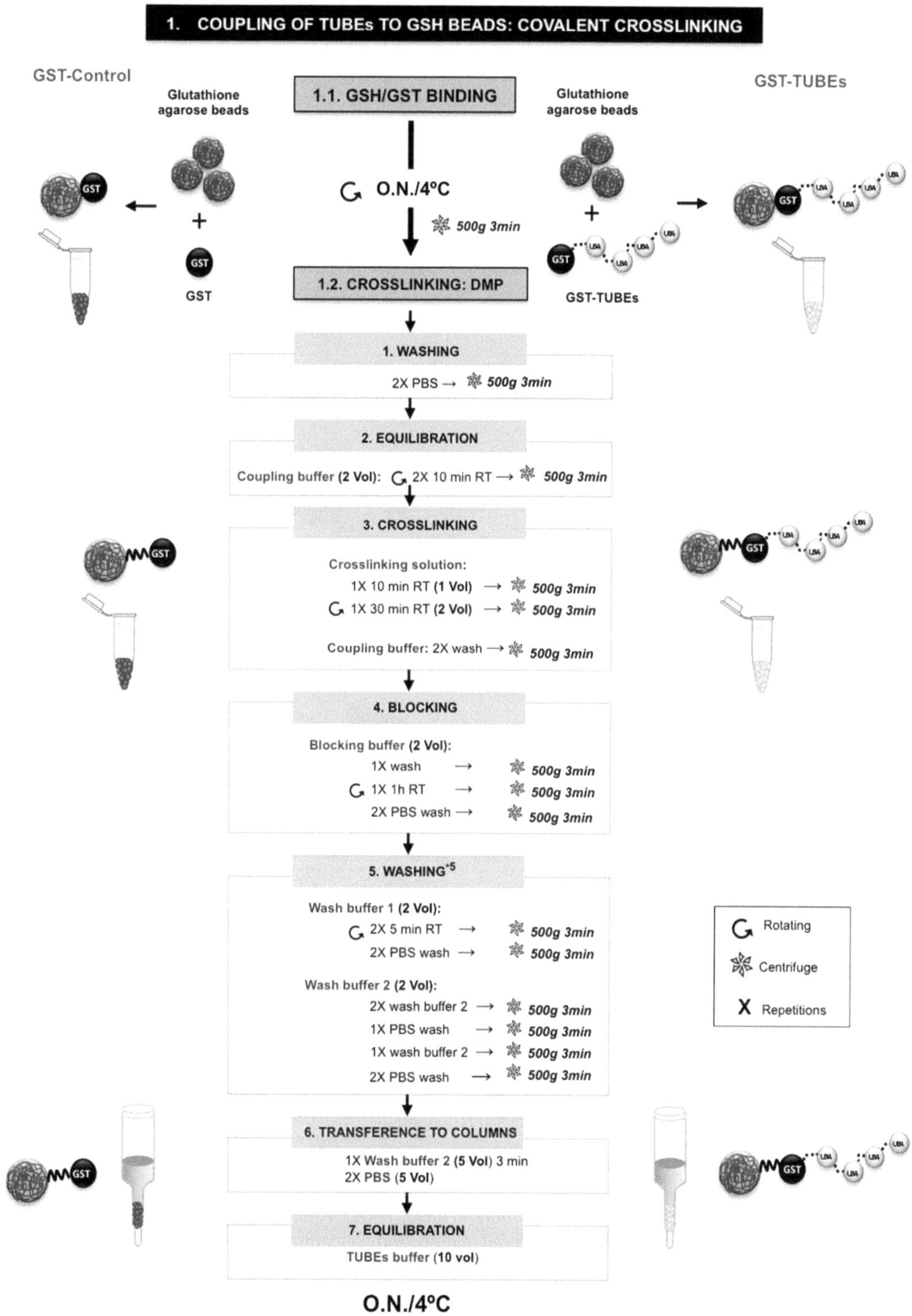

Fig. 1 Procedure to couple TUBEs/GST to GSH-agarose beads. Before covalent cross-linking using DMP, molecular traps were bound to agarose beads. After blocking active groups with ethanolamine, several washes are applied to eliminate unbound molecular traps and non-covalent interactions. Superscript numbers are associated to Subheading 4

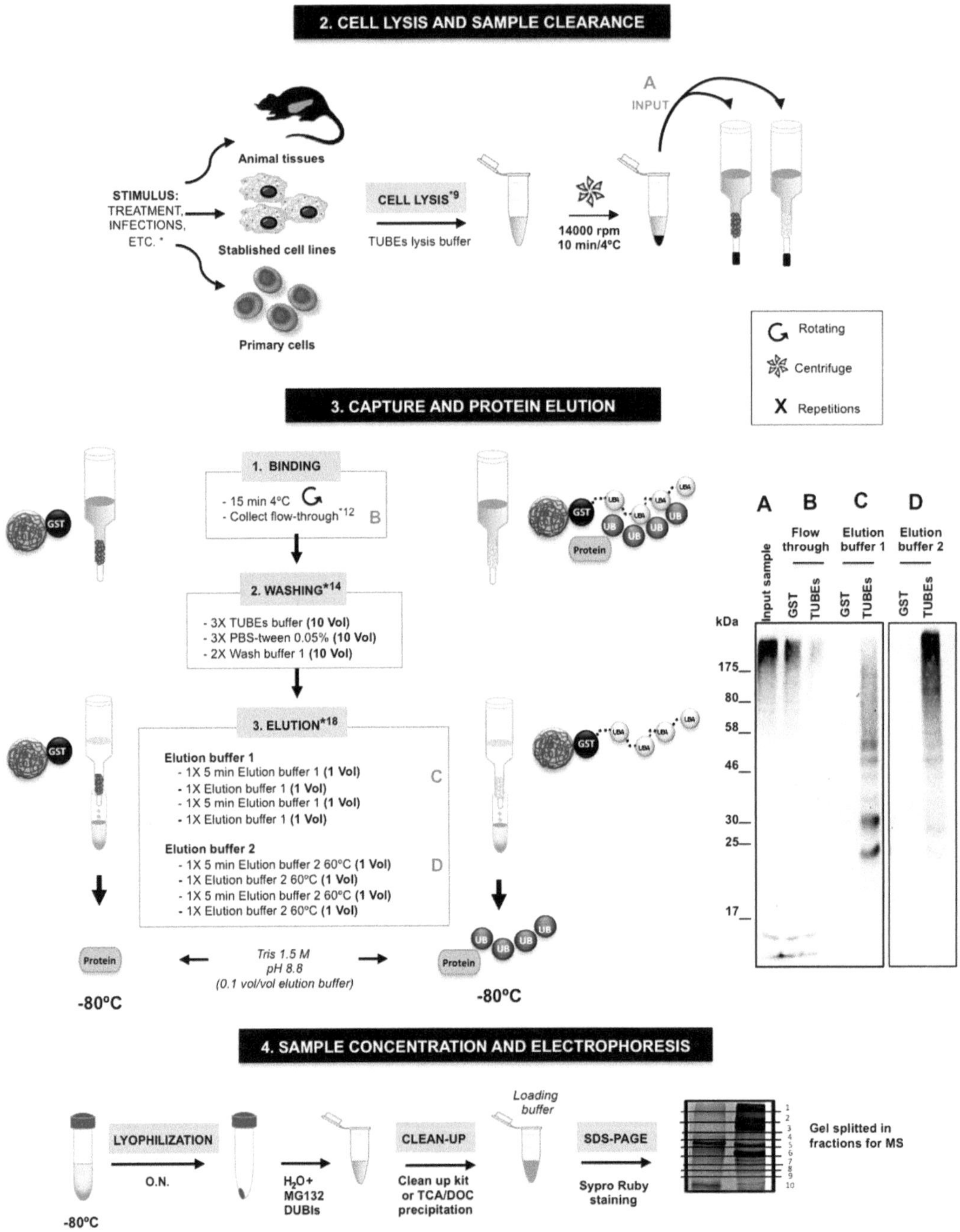

Fig. 2 Isolation of ubiquitylated proteins for MS analysis using TUBEs. After lysis of biological samples, ubiquitylated proteins and associated factors are captured, eluted, and concentrated following the illustrated procedure (Captured and eluted proteins are monitored by Western blot with an anti-ubiquitin antibody). Superscript numbers are associated to Subheading 4

2 Materials

Prepare all solutions using ultrapure water (18 M Ω cm at 25 °C) and analytical grade reagents. Solutions are filtered and stored at 4 °C (unless indicated otherwise). Filtering should be done inside a hood to avoid contaminations. Diligently follow all waste disposal regulations when disposing waste materials. Sodium azide is not added to the reagents; therefore solutions should be prepared as fresh as possible.

2.1 Coupling TUBEs/GST to Glutathione Beads

1. Glutathione beads (Biontex, Germany): Beads are previously washed twice with cold PBS and centrifuged at 2000 rpm during 15 min after each wash and kept at 4 °C until use.
2. Autoclaved PBS.
3. TUBEs and GST proteins are produced in *Escherichia coli* (C41-DE3) using a standard protocol for recombinant protein production [9] or can also be purchased through Life-Sensors Inc (Malvern, PA, USA).
4. Coupling buffer: 200 mM Borate buffer containing 3 M NaCl, pH 9. Boric acid (Sigma) is adjusted with NaOH, filtered by 0.22 μm membrane, and kept at room temperature. Do not store longer than a week as it precipitates.
5. Cross-linking solution: 50 mM Solution of dimethyl pimelimidate (DMP, Fluka) is prepared just before use. Dissolve DMP directly in coupling buffer at room temperature (RT).
6. Blocking buffer: 200 mM Ethanolamine (Sigma), pH 8.2. Carefully adjust pH with HCl in a fume hood. Filter and keep at 4 °C avoiding exposition to light by covering with aluminum foil.
7. TUBEs buffer: 20 mM Phosphate buffer pH 7.5 (Na_2HPO_4, NaH_2PO_4 from Sigma) containing 1 % Igepal (Calbiochem), 2 mM EDTA (Sigma), 50 mM sodium fluoride (Sigma), 5 mM tetra-sodium pyro-phosphate (Sigma), and 10 mM β-glycerol-2-phosphate (Sigma). Filter through 0.22 μm membrane and store at 4 °C.
8. Washing buffer 1: TUBEs buffer containing 1 M NaCl.
9. Washing buffer 2: 200 mM Glycine (Sigma) pH 2.5. Filter and store at 4 °C.
10. PBS-Tween 0.05 % (Tween 20, Sigma).
11. Empty Poly-Prep® Chromatography Columns 9 cm height (BioRad).
12. Rotating wheel for Eppendorf tubes (tube rotator).
13. Refrigerated centrifuge for Eppendorf tubes (swing rotor is optimal).

2.2 Cell Lysis and Sample Clearance

1. Tumor cell lines or primary cells are grown in appropriate medium. For MCL cells RPMI-1640 medium containing 10% fetal bovine serum, 2 mM L-glutamine, and 50 μg/mL penicillin-streptomycin were used.
2. Cold sterile PBS.
3. TUBEs lysis buffer: Supplement just before use TUBEs buffer with 1 mM PMSF (Sigma), complete mini-EDTA-free protease inhibitor cocktail (Roche), 50 μM DUB inhibitor PR-619 (Calbiochem), and 200 nM proteasome inhibitor (Bortezomib, Velcade) (*see* **Note 1**). Keep on ice until needed.
4. Round-bottom sterile polypropylene tubes (13–15 mL) suitable for sonication.
5. Sonicator probe.
6. Cold centrifuge (4 °C).
7. Wet ice.

2.3 Capturing and Elution of Proteins

1. Gravity columns pre-packed with TUBEs cross-linked beads (described in Subheading 3.1).
2. TUBEs buffer (described in Subheading 2.1).
3. Washing buffer 1: TUBEs buffer containing 1 M NaCl (prepared in Subheading 2.1).
4. PBS-Tween 0.05% (prepared in Subheading 2.1).
5. Elution buffer 1: 200 mM Glycine pH 2.5 (same as washing buffer 2 in Subheading 2.1).
6. Elution buffer 2: 1% SDS (Sigma) prepared in PBS and warmed at 60 °C before use.
7. 1.5 M Tris–HCl, pH 8.8 (filter and store at RT).
8. Rotating wheel.

2.4 Sample Concentration and Gel Electrophoresis

1. 2D-Clean up Kit (GE Healthcare).
2. Ultrapure H_2O.
3. Lyophilizer.
4. Parafilm.
5. Vortex.
6. Bath sonicator.
7. Aspiration pump.
8. Loading buffer (Laemmli buffer).
9. Reagents and buffers for SDS-PAGE.
10. Pre-cast acrylamide mini gel (1.5 mm thick) and 5-well comb.
11. Fixing solution: 10% Acetic acid and 30% ethanol. Prepared in fresh.

12. Destaining solution: 7% Acetic acid, 10% ethanol. Prepared in fresh.
13. Sypro-Ruby protein gel stain (Invitrogen).
14. Wipe Tissue (Kimberly-Clark Professional).
15. Gel imager.

3 Methods

In order to reduce contaminants, affinity chromatography is performed within a clean hood using gloves during all the procedure. The method is divided into four sections:

3.1 Coupling TUBEs/GST to Glutathione Beads: Covalent Cross-Linking

1. This step takes 2 days to be performed. If commercial TUBEs (Lifesensors) are already coupled to glutathione beads, start this procedure directly from Subheading 3.2. Never let the beads dry during any protocol step. Allow TUBEs or GST (control) binding to glutathione beads overnight (O/N) (Fig. 1). For a large-scale purification, incubate 600–700 μg of TUBEs (*see* **Note 2**) with 600–700 μL of PBS-washed glutathione beads in a 2 mL Eppendorf tube (e.g., prepare 1300 μL of a 50% slurry glutathione-beads) (*see* **Note 3**). Adjust volume to submaximal tube capacity with PBS and incubate O/N at 4 °C using a wheel for rotating incubation.
2. After TUBEs or GST incubation with glutathione beads, centrifuge at 500 × *g* for 3 min at 4 °C and remove supernatant (*see* **Note 4**).
3. Wash beads twice with PBS and centrifuge for 3 min at 500 × *g*. Discard supernatant by aspirating or remove carefully using a pipette to avoid losing beads.
4. Equilibrate beads by adding approximately 2 packed beads-volumes (Vol) of coupling buffer. Centrifuge and discard supernatant. Add 2 Vol of coupling buffer and incubate beads for 10 min by rotating at RT.
5. Before cross-linking step, centrifuge as previously, discard supernatant, and add 1 Vol of fresh 50 mM DMP dissolved in coupling buffer (cross-linking solution). Incubate for 10 min at RT, centrifuge, discard supernatant and replace by 2 Vol. of cross-linking solution. Incubate for additional 30 min by rotating at RT.
6. Wash twice with coupling buffer to remove DMP. Centrifuge and discard supernatant as previously indicated.
7. To block active amino groups, wash beads with 2 Vol of blocking buffer, centrifuge, and discard supernatant. Add another 2 Vol of blocking buffer and incubate for 1 h rotating at RT.
8. Wash twice with cold PBS.

9. To remove non-coupled proteins, wash cross-linked beads twice with 2 Vol washing cold buffer 1. Incubate beads in each washing step 5 min by rotating at RT.
10. Wash with cold PBS twice.
11. Continue removing non-coupled protein by washing twice with 2 Vol of washing buffer 2. Do not incubate; only wash by inverting tube (*see* **Note 5**).
12. Wash once with PBS and once more with washing buffer 2.
13. Wash twice again with PBS and then transfer the cross-linked TUBEs beads to an empty polypropylene column. (If the procedure does not continue on the same day, keep beads in PBS at 4 °C in the Eppendorf tube). Coupled TUBEs beads should be used as fresh as possible.
14. For the affinity chromatography step, all procedures should be done in a cold room or in a cold cupboard to ensure that all materials, buffers, and beads stay cold. Wash the beads inside the column with 5 Vol of washing buffer 2. Close column (bottom cap). Wait for 3 min and then open the bottom cap to discard the flow-through (FT).
15. Wash beads twice with 5 Vol of cold PBS (*see* **Note 6**).
16. Equilibrate beads into the column with 10 Vol of cold TUBEs buffer. Column is ready to use immediately with the lysed sample. (To check cross-linking efficacy, a sample of 10–20 μL of cross-linked beads may be analyzed by SDS-PAGE or WB anti-GST or anti-SV5) [9].

3.2 Cell Lysis and Sample Clearance

Cell lysis should be done on ice to avoid loss of protein modification and prevent other enzymatic activities. The time between the lysis and the incubation with the affinity column should be as short as possible. The following protocol has been performed with tumor cell lines or primary cells taking into account that the sample does not saturate the capacity of the coupled TUBEs beads to capture more than 80–90% of all ubiquitylated proteins present in the sample.

1. Cell culture maintenance: Mantle cell lymphoma (MCL) cells Z-138 are grown in suspension at 37 °C and 5% CO_2 humidity atmosphere. For protein extraction, around 50×10^6 cells are used for each condition or point (consider GST-bead control) (*see* **Note 7**). For a better manipulation and efficient cell lysis, each sample is splitted in two 13 mL round sterile tubes (25×10^6 cells each). Samples will be pooled later.
2. Cells are pelleted by 5-min centrifugation $300 \times g$ at RT. Wash cell pellet twice with 5 mL of cold PBS. Centrifuge and discard supernatant by aspiration (eliminate residual PBS as much as possible). Keep cell pellets on ice while performing lysis step.

3. Cell lysis (Fig. 2): Prepare supplemented TUBEs lysis buffer just before use. Add 1 mL of TUBEs lysis buffer to each 25×10^6 cell pellet (*see* **Note 8**). While keeping samples on ice, disrupt cell pellet by three pulse sonication of 30 s each, and let cool between each pulse. The lysis step should be done as quickly as possible (*see* **Note 9**). Transfer the lysates to Eppendorf tubes and clarify sample by centrifuging at $20{,}000 \times g$ during 10 min at 4 °C. Recover and pool supernatants to evaluate the input for each experimental condition. Keep a small sample for protein quantification. (Usually a lysate from 25×10^6 MCL Z-138 cell pellet resuspended in 1 mL contains about ≈2 mg/mL of total protein concentration, measured at A_{280nm}.) Therefore, approximately 4 mg of total protein is applied to the TUBEs or GST columns (*see* **Note 10**).

3.3 Capturing of Ubiquitylated Proteins by Affinity Chromatography and Elution

Ubiquitylated proteins are often interacting with other protein factors, which could increase background level. For MS purposes, longer incubation times (>1 h) increase the presence of background proteins in the sample. Due to the high TUBEs affinity for poly-ubiquitin molecules (low nanomolar range) [9], the capture of ubiquitylated proteins is very efficient in a short time (30 min or less). According to the abundance of ubiquitylated proteins in the sample, the incubation time must be optimized to preserve specific proteins and discard nonspecific ones, also present in the GST control.

1. Before starting the chromatography, do not forget to keep 50 μL of sample (input), mix with Laemmli buffer, and store at −20 °C. Analyze total ubiquitylated proteins by WB using anti-ubiquitin antibody.
2. Binding proteins to TUBEs column (Fig. 2): Apply clarified lysates directly to respective column, TUBEs or GST-control, and be sure to close bottom cap before adding the lysate. Close also top cap and allow binding of ubiquitylated proteins to the beads. Incubate during 15–30 min at 4 °C (*see* **Note 11**). For a better capture, use a wheel to keep column in rotation.
3. After the binding step, set the column into a support. To collect the flow-through fraction (FT), open first top cap and then bottom cap before collection. Keep a FT sample and mix with Laemmli buffer as it was done with the input. Keep FT fraction for WB analysis to verify that TUBEs are not saturated and ubiquitylated proteins are not being lost (*see* **Note 12**).
4. Washing unbound proteins (Fig. 2): Wash column three times with 10 Vol of TUBEs buffer. Close both sides of the column and mix by inversion. Set column in its support and discard wash flow (*see* **Note 13**).

5. Wash three times with 10 Vol PBS-Tween 0.05 %, close column, and wash as in the previous step (*see* **Note 14**).
6. Eliminating nonspecific proteins: Before elution, wash beads twice with 10 Vol of washing buffer 1 (*see* **Note 15**).
7. Eluting proteins from column (Fig. 2): Elution is performed in a sequential step: first, elute proteins with 1 Vol of elution buffer 1. Close the column, mix by flicking (do not invert), and wait for 5 min. Place column again in the support, open top and bottom caps, and collect the elution sample in a 15 mL tube (keep tube on ice to avoid ubiquitin deconjugation). Add again 1 Vol of elution buffer 1 to rinse column walls, and collect on the same tube. Add immediately to the eluted fraction 0.1 Vol of Tris 1.5 M, pH 8.8 to neutralize pH. Repeat this step once with elution buffer 1, collect and keep fraction on ice.
8. Second elution step using elution buffer 2 (pre-warmed at 60 °C): Add 1 Vol of elution buffer 2 and wait for 5 min mixing by flicking. Collect and apply 1 more Vol and collect in the same tube containing the previous eluted fraction. Repeat this step to recover tightly bound proteins (*see* **Note 16**). Neutralize eluted sample by adding Tris 1.5 M pH 8.8. Take care that the final Tris concentration does not exceed 100 mM.
9. Mix tube to homogenize eluted fractions and freeze immediately at −80 °C (*see* **Note 17**). Eluted samples can be stored at −80 °C until required MS analysis.
10. When optimizing your protocol, check also ubiquitylated proteins remaining on the beads by WB analysis. Transfer beads from column to Eppendorf tubes (with 1 Vol PBS). Drain PBS and add 300 μL of 3× Laemmli buffer.

3.4 Sample Concentration and Gel Electrophoresis

1. Lyophilization and re-constitution of samples (Fig. 2): Keep the sample frozen until lyophilization starts. Make holes in the tube caps or replace caps with parafilm (with holes) before lyophilization. To preserve frozen samples during the lyophilization process, freeze them before in liquid nitrogen (NO_2). Lyophilize samples overnight (*see* **Note 18**).
2. Re-constitute sample in 500 μL ultrapure H_2O. To avoid ubiquitin deconjugation events, sample can be supplemented with proteasome and de-ubiquitylase inhibitors, 20 μM MG132 and 50 μM PR619, respectively. Mix samples by vortexing for 15 s and/or sonicate for 5 min in a sonicator bath (*see* **Note 19**).
3. Sample cleanup: Distribute sample in 100 μL aliquots (*see* **Note 20**). Add to each aliquot 450 μL of precipitant from

2D-clean up kit and incubate tube on ice for 15 min. Add 450 μL of co-precipitant, mix the content by vortexing for 10 s, and centrifuge tubes at 15,000 × *g* for 5 min (*see* **Note 21**). A pellet should be visible in the bottom of the tube (*see* **Note 22**). Carefully aspirate the supernatant with 1 mL blue tip. Briefly spin down the tubes and aspirate remaining liquid with a 200 μL yellow tip. Add 20 μL of H_2O over the pellet followed by 1 mL of chilled wash buffer and 5 μL of wash additive. Vortex until the pellet gets in solution (*see* **Note 23**). Keep samples at −20 °C, mix tubes for 30 s, and incubate again at −20 °C. Repeat this action three times, incubate for 10–15 min between each vortex pulse, and keep samples O/N at −20 °C. Centrifuge for 5 min at 15,000 × *g*. Aspirate supernatant with a 1 mL blue tip. Spin down the tubes and aspirate the remaining liquid with a 200 μL yellow tip. Leave the pellet to dry out for 30 s keeping tubes with open caps; do not overdry pellet.

4. Resuspend the pellet in loading buffer. Add 80 μL of 1× Laemmli buffer to the first tube aliquot and pipet up and down ten times. Take the liquid and transfer it to the next tube; repeat this procedure with all the tubes to pool into one sample. Then, apply additional 50 μL of loading buffer and repeat sample transfer. The final sample volume should be of approximately 130 μL.
5. Mix protein sample by vortexing for 15 s and then boil during 5 min (repeat this twice). Keep 13 μL of sample for WB analysis; the remaining material is loaded in an SDS-PAGE gel for MS analysis. Figure 3 shows typical results from MCL in A and B or *P. falciparum*-infected red blood cells (iRBC) in C and D.
6. SDS-PAGE for mass spectrometry (Fig. 3a, c): Buy or cast gel according to the required percentage of acrylamide:bisacrylamide. After boiling samples, load the total volume of samples immediately. Load molecular weight markers leaving a well space between samples to avoid contaminations (*see* **Note 24**). Run the gel at 80 V for 10–15 min to let samples get into the stacking gel. Then increase the voltage to 125 V, for 10% gel run for about 30–60 min (*see* **Note 25**).
7. After gel electrophoresis, pry open the gel plates with the use of a clean spatula. Rinse the gel with pure MilliQ water and transfer carefully to a large glass petri dish.
8. Stain gel with Sypro-Ruby according to the manufacturer's instructions.
9. Perform de-staining step until a clean background is observed.
10. Document gel image in a digital imager.

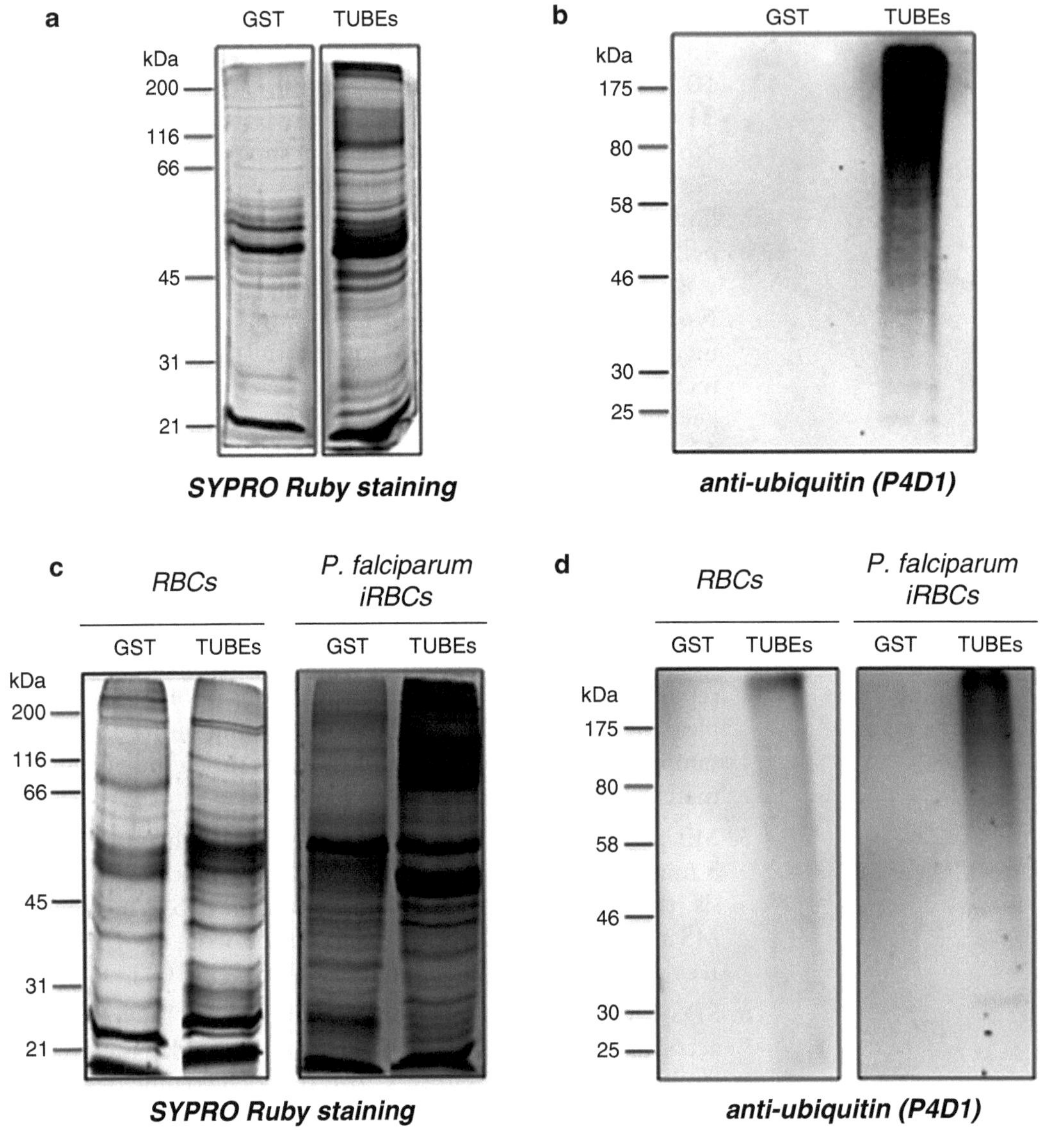

Fig. 3 Proteins captured by TUBEs for MS analysis. (**a**) SYPRO-stained gel, 10 % acrylamide (90 % of total sample from MCL cells). (**b**) WB detection of total ubiquitylated proteins (10 % of total sample from MCL cells). (**c**) SYPRO-stained gel, 10 % acrylamide (90 % of total sample from non-infected RBCs and RBCs infected (iRBC) with *Plasmodium falciparum*). (**d**) WB detection of total ubiquitylated proteins (10 % of total sample from RBC and iRBC). GST-coupled beads were used as control

4 Notes

1. Other proteasome inhibitors such as MG132 (20 μM) can be used instead of bortezomib.
2. The amount of TUBEs required to capture ubiquitylated protein from a given number of cells should be set up to avoid TUBEs saturation.

3. Add a slight excess of beads before the coupling step since after several washes a considerable amount of beads is lost.
4. During settings conditions, keep unbound fraction for WB analysis to verify that most ubiquitylated proteins were captured.
5. The pH of washing buffer 2 could alter GST binding capacity, so it is important to control incubation time with this solution.
6. Extra washes can be performed with SDS 1% preheated at 60 °C (elution buffer 2) to remove non-cross-linked proteins. It depends on the elution stringency that will be used to elute the sample.
7. Manipulate and scale the culture according to the biological sample since culture conditions are different for other cell types. The amount of cells used to isolate ubiquitylated proteins depends on the relative abundance of these proteins that will allow their optimal MS/MS detection. For example, around 2×10^9 RBCs are needed to capture ubiquitylated proteins with 700 μg of TUBEs. The number of cells should be adjusted until no ubiquitylated proteins are detected in the flow-through for a fixed amount of TUBEs.
8. During the sonication procedure, use 13 or 15 mL round-bottom tubes to avoid overflow and cell extract warming. 13 mL tubes adjust much better into sonication tip allowing an efficient lysis. Do not forget to keep tubes on ice.
9. Lysis conditions should be set up for each biological sample. By reducing the lysis time ubiquitin de-conjugation/proteasomal degradation can be limited. No more than 15 min should be used.
10. There is not a direct correlation between total protein concentration and abundance of ubiquitylated proteins. Ubiquitylation can be altered according to stimuli, cellular process, or pathology. For instance, cells treated with proteasome inhibitors can accumulate ubiquitylated proteins. If protein quantification is not possible, refer to cell number and set condition controlling by WB using anti-ubiquitin antibodies. This is the case for RBC lysates where hemoglobin interferes with standard techniques used to quantify protein concentration.
11. Do not incubate samples with beads for more than 15–30 min; otherwise the background (nonspecific proteins) will increase. Be sure to close the column tightly in both sides to avoid spills.
12. Flow-through can be passed through a new TUBEs column to avoid losing ubiquitylated proteins with weaker TUBEs affinity.

13. If desired, keep all washes to concentrate them by TCA/DOC precipitation and control unbound ubiquitylated proteins by WB analysis.
14. The number of washes and volumes may differ depending on the sample. For example higher number of washes are required when working with RBCs due to the hemoglobin content.
15. For a RBC sample, the elution step started with the washing buffer 1. After a last wash with TUBEs buffer (10 Vol), add 1 Vol of washing buffer 1 (in this case this step is considered as elution). Close the column, mix, and wait for 5 min. Open column and collect the fraction. Add 1 more Vol of washing buffer 1 and collect the elution again.
16. During elution step, the number of repetitions or rounds using elution buffer 1 and 2 can be increased to recover tightly bound proteins. For example, the elution stringency is higher for RBCs. To detach ubiquitylated proteins, 2 elution rounds with washing buffer 1, 6 rounds with elution buffer 1, and 4 rounds with elution buffer 2 are required. To set up the number of rounds required to recover most ubiquitylated proteins, all fractions are analyzed by WB anti-ubiquitin.
17. The final elution volume is around 5 mL for MCL cells. To recover most ubiquitylated proteins 2 rounds with elution buffer 1 and 2 rounds with elution buffer 2 are required. Each round includes 2 Vol of approximately 600 microlitres ul/Vol.
18. Avoid large volumes or if necessary split sample before freezing. If needed, allow samples to dry out for more than a night.
19. Final suspension volume could be 500 μL to maintain a limited number of aliquots for the cleanup step.
20. If you plan to use a kit for cleanup, scale the sample according to the manufacturer's instructions.
21. Large quantities of interfering substances can compromise protein precipitation in the sample. Therefore, precipitant and co-precipitant solutions are added in larger volumes than recommended by the kit manufacturer.
22. Sometimes the protein gets strongly attached to the Eppendorf wall and thus a pellet is not easily visible. Proceed with the cleanup protocol even if the pellet is not visible after the first centrifugation.
23. If precipitated protein sticks to the tube after washing buffer addition, carefully scratch the tube wall with a 200 μL tip. If the proteins are not well detached, final precipitation will fail.
24. Separation between samples and markers is essential to avoid protein contamination when cutting gel slides.
25. Running time may change depending on the sample, buffer, size of the gel, etc. Check that no protein runs out of the gel by following the gel migration blue front.

Acknowledgments

We thank Valérie Lang for the critical reading of this document, Gaël Roué for providing MCL cell lines, and Mikel Azkargorta and Felix Elortza for their technical support. This work was funded by the Ministerio de Economia y Competitividad, Spain grant BFU2011-28536 (MSR), Diputación Foral de Gipuzkoa (MSR and FA), and GSK OPEN-LAB foundation. LMC was supported by the GSK OPEN-LAB foundation. The authors would like to acknowledge networking support by the Proteostasis COST Action (BM1307).

References

1. Hershko A, Ciechanover A (1998) The ubiquitin system. Annu Rev Biochem 67:425–479
2. Glickman MH, Ciechanover A (2002) The ubiquitin-proteasome proteolytic pathway: destruction for the sake of construction. Physiol Rev 82(2):373–428
3. Komander D, Rape M (2012) The ubiquitin code. Annu Rev Biochem 81:203–229. doi:10.1146/annurev-biochem-060310-170328
4. Xolalpa W, Perez-Galan P, Rodríguez MS et al (2013) Targeting the ubiquitin proteasome system: beyond proteasome inhibition. Curr Pharm Des 19(22):4053–4093
5. Mata-Cantero L, Lobato-Gil S, Aillet F, Rodriguez MS (2015) The ubiquitin-proteasome system (UPS) as a cancer drug target: emerging mechanisms and therapeutics. In: Wondrak SL (ed) Stress response pathways in cancer. Springer Science + Business Media, Dordrecht, pp 225–264. doi:10.1007/978-94-017-9421-3-11
6. Hjerpe R, Rodriguez MS (2008) Efficient approaches for characterizing ubiquitinated proteins. Biochem Soc Trans 36:823–827
7. Matsumoto M, Hatakeyama S, Oyamada K et al (2005) Large-scale analysis of the human ubiquitin-related proteome. Proteomics 5(16):4145–4151
8. Schwertman P, Bezstarosti K, Laffeber C et al (2013) An immunoaffinity purification method for the proteomic analysis of ubiquitinated protein complexes. Anal Biochem 440(2):227–236
9. Hjerpe R, Aillet F, Lopitz-Otsoa F et al (2009) Efficient protection and isolation of ubiquitylated proteins using tandem ubiquitin-binding entities. EMBO Rep 10:1250–1258
10. Aillet F, Lopitz-Otsoa F, Hjerpe R et al (2012) Isolation of ubiquitylated proteins using tandem ubiquitin-binding entities. Methods Mol Biol 832:173–183.doi:10.1007/978-1-61779-474-2_12
11. Rubel CE, Schisler JC, Hamlett ED et al (2013) Diggin' on u(biquitin): a novel method for the identification of physiological E3 ubiquitin ligase substrates. Cell Biochem Biophys 67(1):127–138. doi:10.1007/s12013-013-9624-6
12. Altun M, Kramer HB, Willems LI et al (2011) Activity-based chemical proteomics accelerates inhibitor development for deubiquitylating enzymes. Chem Biol 18(11):1401–1412. doi:10.1016/j.chembiol.2011.08.018
13. Lopitz-Otsoa F, Rodriguez-Suarez E, Aillet F et al (2012) Integrative analysis of the ubiquitin proteome isolated using tandem ubiquitin binding entities (TUBEs). J Proteomics 75(10):2998–3014. doi:10.1016/j.jprot.2011.12.001
14. Shi Y, Chan DW, Jung SY et al (2011) A dataset of human endogenous ubiquitination sites. Mol Cell Proteomics 10(5):M110.002089. doi:10.1074/mcp.M110.002089
15. Mata-Cantero L, Azkargorta M et al (2016) New insights into host-parasite ubiquitin proteome dynamics in P. falciparum infected red blood cells using a TUBEs-MS approach. J Proteomics 29(139):45–59. doi:10.1016/j.jprot.2016.03.004.

Chapter 9

TUBEs-Mass Spectrometry for Identification and Analysis of the Ubiquitin-Proteome

Mikel Azkargorta, Iraide Escobes, Felix Elortza, Rune Matthiesen, and Manuel S. Rodríguez

Abstract

Mass spectrometry (MS) has become the method of choice for the large-scale analysis of protein ubiquitylation. There exist a number of proposed methods for mapping ubiquitin sites, each with different pros and cons. We present here a protocol for the MS analysis of the ubiquitin-proteome captured by TUBEs and subsequent data analysis. Using dedicated software and algorithms, specific information on the presence of ubiquitylated peptides can be obtained from the MS search results. In addition, a quantitative and functional analysis of the ubiquitylated proteins and their interacting partners helps to unravel the biological and molecular processes they are involved in.

Key words Ubiquitin, Tandem Ubiquitin Binding Entities (TUBEs), Posttranslational modification (PTM), Collision induced dissociation (CID), Mass spectrometry (MS), Gene Ontology (GO), Iodoacetamide (IAA), Chloroacetamide (CAA)

1 Introduction

Protein ubiquitylation is of paramount importance for the proper function and development of multiple cellular processes including proteolyisis, endocytosis, DNA repair, cellular localization, or activation of protein kinases [1, 2]. Its deregulation has been shown to be involved in a number of diseases, such as cancer, neurodegenerative and cardiovascular diseases, and immunological disorders, among others [3, 4].

The ubiquitin-proteome is integrated by the total ubiquitylated proteins present in the cell and their interacting partners (ubiquitin-interactome). The ubiquitin-interactome allows regulation and connection of ubiquitylated proteins with the effector functions. Large-scale analysis of protein ubiquitylation by MS has become one of the most valuable techniques to elucidate its role in physiology and pathology. However, the analysis of ubiquitylated proteins can be a daunting task because of their low

Rune Matthiesen (ed.), *Proteostasis: Methods and Protocols*, Methods in Molecular Biology, vol. 1449,
DOI 10.1007/978-1-4939-3756-1_9,

stoichiometry and their short life-span due to the action of deubiquitylation enzymes (DUBs) [5]. Therefore, protection and enrichment methods are mandatory for their analysis. A number of specific isolation methods have been developed in the last years, including the use of tandem ubiquitin-binding entities (TUBEs) [6–9], the expression of tagged ubiquitin molecules [10, 11], or even the development of anti-ubiquitin antibodies in order to pick ubiquitylated peptides [12–14].

The identification and analysis of ubiquitylated proteins by mass spectrometry involves the in vitro enzymatic digestion of the proteins of interest and the analysis of the generated peptides. Information on the mass of the peptides and their corresponding fragments is collected and contrasted with the information compiled in databases using dedicated software and algorithms (Fig. 1).

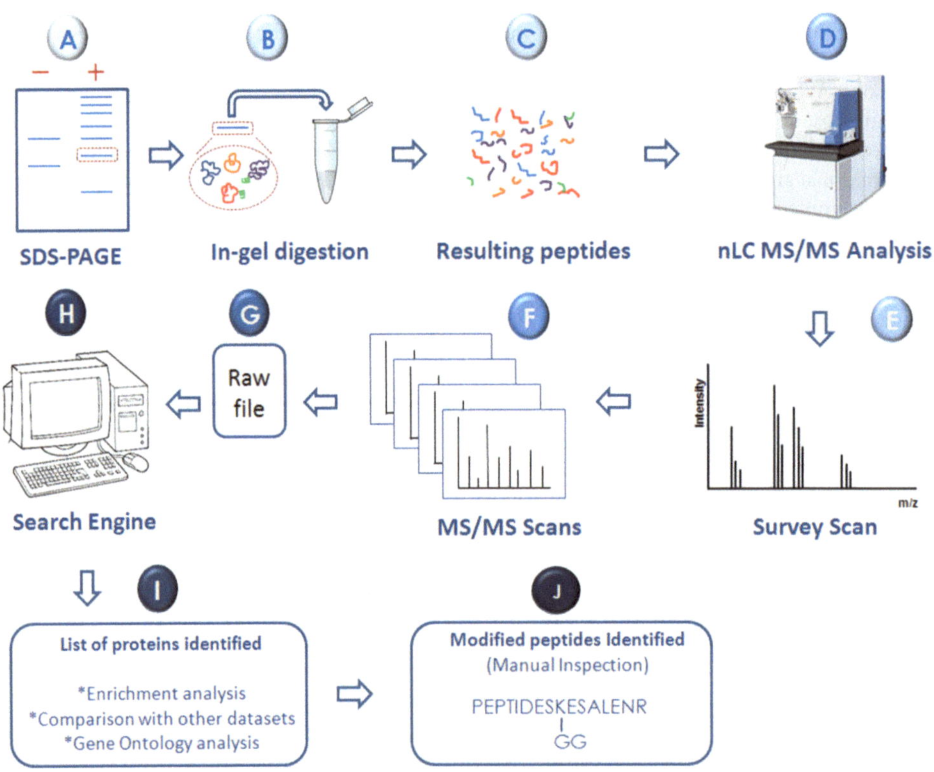

Fig. 1 Schematic overview of MS sample preparation. Processing starts from the SDS-PAGE run of the samples (**a**), followed by gel cutting and in-gel digestion of the obtained slices (**b**). The resulting peptides (**c**) are subjected to nLC-MS/MS analysis (**d**), where analysis of the peptides masses (**e**) and fragmentation patterns (**f**) is compiled. The RAW files containing this information (**g**) are loaded into the search engine (**h**), and a list of the identified proteins is obtained (**i**). Specific enrichment analysis can be carried out over this dataset, keeping only those proteins reliably enriched by TUBEs. Furthermore, this list can be compared with other datasets or subjected to functional analysis through the use of different bioinformatics tools, such as Gene Ontology (GO). Finally, modification site in a subset of identified proteins can be obtained (**j**). Manual inspection of spectra assignments is recommended in order to avoid false-positive assignments

Results provide direct information on the modified amino acids of the proteins through the identification of the GG ubiquitin signature. The tryptic cleavage of the ubiquitin sequence leaves a GG adduct attached to the substrate, increasing the mass of the peptide with 114.043 Da [15]. Due to incomplete digestion of the ubiquitin, trypsin may also leave a bigger tag comprised of LRGG, increasing the mass of the peptide with 383.228 Da [16] (Fig. 2). These mass shifts are indicative of ubiquitylation events, and are therefore used for the detection of the modified peptides in the data search step. However, identification of ubiquitylated peptides presents some drawbacks, since they are scarce in opposition to the non-modified tryptic peptides coming from the digestion of the modified protein, and are usually larger and get more charges than regular tryptic peptides [17], lowering the chances for their proper fragmentation and subsequent identification.

The experiments are typically done with at least three biological replicas, and negative controls are included for each pull down. For example, when using TUBEs pull down, beads cross-linked with GST can be used as a control. Using these negative controls, unspecifically enriched proteins can be discarded from the dataset and only those proteins more likely to be ubiquitylated (and their interacting partners) are considered for further analysis. As a starting point, this enriched dataset can be compared with other well-characterized datasets, and its functions can be outlined via a Gene Ontology term-enrichment analysis. Thus, a landscape of the molecular and biological processes these proteins are involved in can be obtained.

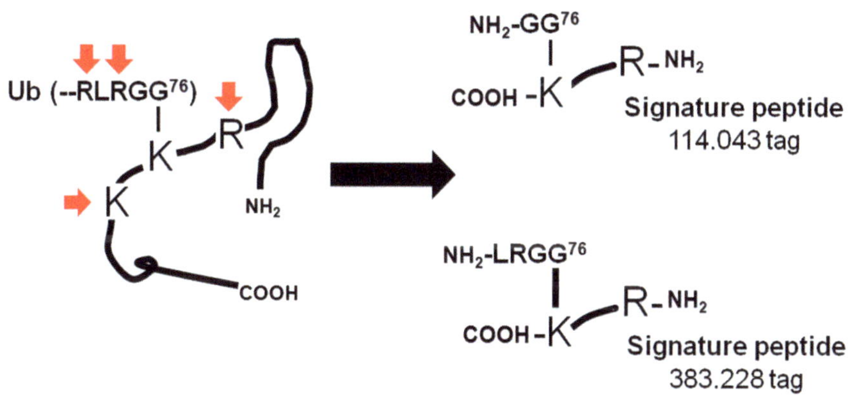

Fig. 2 Trypsin digestion of ubiquitylated proteins. Trypsin cleaves the protein sequence after K or R residues. The presence of an ubiquitin moiety attached to a K in the protein sequence usually hampers tryptic cleavage. Trypsin will cleave the available K and R in the sequence of both the protein and the ubiquitin attached to it, leaving a GG residue attached to the e-NH2 group of the ubiquitylated K. This tag can become an LRGG when trypsin fails to cut the ubiquitin sequence at the last R residue

2 Materials

2.1 Sypro Ruby Gel Staining and Image Acquisition

1. Fixing solution: 10 % Acetic acid and 30 % ethanol. Mix 100 mL acetic acid and 300 mL ethanol in a test tube, and make up to 1 L with Milli-Q water. Prepare it fresh.
2. Destaining solution: 7 % Acetic acid, 10 % ethanol. Mix 70 mL acetic acid and 100 mL ethanol in a test tube, and make up to 1 L with Milli-Q water. Prepare it fresh.
3. Wipe tissue: Precision Wipes Tissue Wipers.
4. Sypro Ruby protein gel stain, 1 L (Invitrogen).
5. Typhoon Trio Scanner, Variable Mode (GE Healthcare).

2.2 Digestion

1. 1 M Ammonium bicarbonate (AMBIC): Use Milli-Q water and store at −20 °C in 0.5 mL aliquots.
2. 100 mM AMBIC: Dilute a 0.5 mL aliquot of 1 M AMBIC to 5 mL with Milli-Q water. Prepare it fresh.
3. 50 mM AMBIC: Dilute a 0.5 mL aliquot of 1 M AMBIC to 10 mL with Milli-Q water. Prepare it fresh.
4. 1 M DTT stock: Dissolve DTT in Milli-Q water. Store at −20 °C in 10 μL aliquots.
5. 10 mM DTT: Dilute a 10 μL aliquot of 1 M DTT with 990 μL AMBIC 100 mM. Prepare it fresh.
6. 55 mM CAA: Dissolve CAA in AMBIC 100 mM. Prepare it fresh.
7. Trypsin Gold-Mass Spectrometry Grade 100 μg (Promega).
8. 1 μg/μL Trypsin Gold stock: Dissolve a vial of 100 μg trypsin gold in 0.1 ml of 50 mM acetic acid (*see* **Note 1**). Aliquots (15 μL) can be stored at −20 °C for at least 1 month.
9. 0.0125 μg/μL Trypsin Gold: Dilute the 15 μL aliquot of 1 μg/μL Trypsin Gold with 105 μL 50 mM AMBIC.
10. Trifluoroacetic acid (TFA) (Pierce).
11. 0.1 % TFA: Dissolve 0.1 mL TFA in 99.9 ml Milli-Q water.
12. Acetonitrile (ACN) (Symta).
13. SpeedVac: Rotational-Vacuum-Concentrator RVC 2–25 (Christ).
14. 50 mM Acetic acid: Dissolve 71.5 μL acetic acid in 24.28 mL Milli-Q water.
15. *E. coli* protein sample: ReadyPrep ™ *E. coli* Protein Sample 2.7 mg (Bio-Rad).

2.3 MS Analysis

1. Formic acid (FA) (Pierce).
2. 0.1 % FA: Dissolve 0.1 mL FA in 99.9 ml Milli-Q water.
3. Acetonitrile (ACN) (Symta).
4. Vials combination package (glass vial Type I) (Waters).

5. NanoAcquity UPLC System (Waters)
6. BEH C18 nanoACQUITY Column, 1.7 μm, 75 μm × 200 mm (Waters).
7. Symmetry C18 Trap Column, 5 μm, 180 μm × 20 mm (Waters).
8. Stainless steel emitters (Thermo Scientific).
9. Mass spectrometer for large-scale proteomics such as LTQ Orbitrap XL ETD Mass Spectrometer (Thermo Scientific).
10. Ultrasonic cleaning bath "Ultrasons" 6 L (J.P. Selecta).

3 Methods

Prepare all solutions using Milli-Q water (8 MΩ cm at 25 °C) and analytical grade reagents. Proceed with great care in order to avoid keratin contamination (*see* **Note 2**). Prepare and store all reagents at 4 °C (unless indicated otherwise) and follow all waste disposal regulations when disposing waste material.

3.1 Gel Staining and Image Acquisition

1. Following electrophoresis, pry the gel plates open with the use of a spatula. The gel remains on one of the glass plates. Rinse the gel with water and transfer carefully to a glass petri dish.
2. Fix the protein by the addition of 100 mL of fixing solution. Incubate for 30 min under gentle agitation, discard the solution and add 100 mL of SYPRO RUBY. Incubate overnight under agitation and in the dark (*see* **Note 3**).
3. Add 100 mL destaining solution and incubate under agitation for 30 min. Repeat this operation once. Replace the solution by 100 mL Milli-Q water, incubate for 10 min and replace the solution by 100 mL of fresh Milli-Q water and proceed to the image acquisition. Sypro Ruby images are acquired in the Typhoon Trio scanner-Variable Mode imager (GE Healthcare) using the program Typhoon scanner control v 5.0 (*see* **Note 4**).
4. Wash the scanner surface with ethanol and dry it with a wipe tissue before acquisition. Add Milli-Q water over the scanner´s surface and put the gel over the water carefully.
5. The parameters used for the acquisition are the following:
   - Acquisition mode: Fluorescence
   - Setup:
     - 610 BP 30 Deep Purple, Sypro Ruby
     - PMT: 535 V (*see* **Note 5**).
     - Laser: Blue (488) (*see* **Note 6**).
     - Sensitivity: Normal
   - Orientation: R
   - Focal plate: Platen

6. Acquire a preliminary image with a pixel size of 1000 μm and check that the selected area and laser voltage are suitable for the acquisition (*see* **Note 7**). Once the parameters are fine-tuned, acquire the image with a pixel size of 100 μm.
7. Save the image in TiFF format.
8. Once the image is acquired, remove the gel from the scanner and carefully bring it back to the Petri dish. Gels can be stored in Milli-Q water at 4 °C for at least 1 month. Clean the scanner surface with ethanol, and dry it with a wipe tissue.

3.2 Gel Cut and Digestion of the Gel Slices

Great care must be taken to avoid keratin contamination of the samples during the digestion step (*see* **Note 2**). The use of CAA instead of IAA as alkylating agent is recommended in order to avoid the generation of ubiquitylation-false positives [18].

1. Prepare a template for cutting the gel (*see* **Note 8**). Try to isolate clear bands in independent slices and keep the total number of slices limited to 10 (*see* **Note 9**) (Fig. 3a, b).
2. Print the gel image with the template at 100% of the gel image size. Put this image behind a clean glass plate, and put the gel in the upper part of this glass so that it fits the image behind it. Cut the gel following the template (*see* **Note 10**) (Fig. 3c).
3. Cut each gel slice into small pieces of approximately 1 mm^3 with a clean scalpel, and put them in a new identified Eppendorf tube with Milli-Q water (*see* **Note 11**).

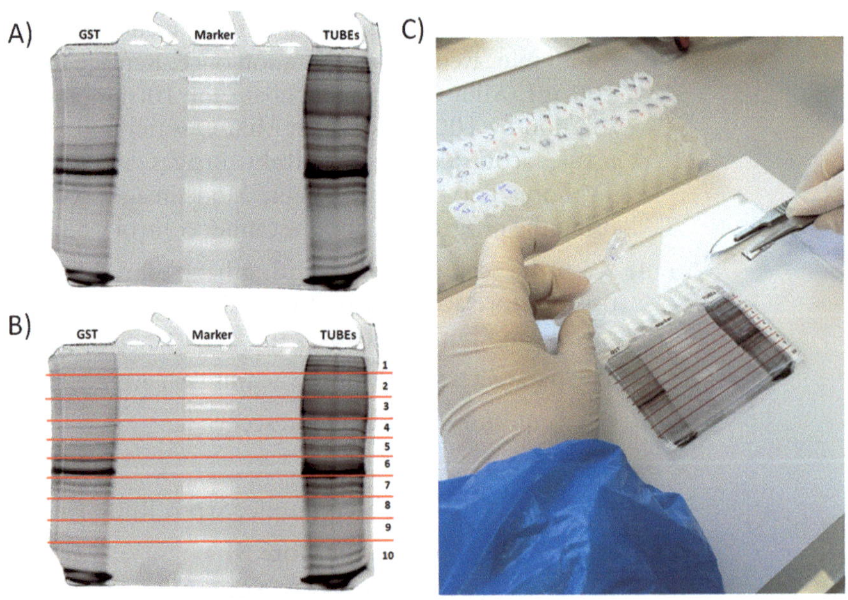

Fig. 3 Template and gel cutting. Once the gel image is acquired (**a**), a template for gel cutting can be created using different programs, such as Microsoft Powerpoint (**b**). Try to follow the band pattern of the gel and cut both the control and the sample following the same criteria to avoid variations in the pattern. Then, the template is printed at the gel size and placed behind the gel that is going to be cut (**c**). Cut the gel following the template and keep gel pieces in individual Eppendorf tubes

4. Discard Milli-Q water and add 50 μL of 50 mM AMBIC, vortex, incubate for 5 min, and discard supernatant. Repeat the procedure with 100 μL ACN (*see* **Note 12**).
5. Add 100 μL of a solution containing 10 mM DTT in 100 mM AMBIC, and incubate for 20 min at 56 °C under agitation. Discard the solution and add 55 mM CAA in 100 mM AMBIC (*see* **Note 13**). Incubate at room temperature for 30 min in the dark.
6. Wash the gel pieces adding ACN, vortex, incubate for 5 min, and discard supernatant.
7. Cover the gel with 50 μL 0.0125 μg/μL trypsin in 50 mM AMBIC. Allow the gel pieces to swell in ice for 30 min. If the gel dries out add more trypsin, and cover the gel gently.
8. Discard the trypsin supernatant and add 50 μL of AMBIC. Incubate at 37 °C overnight.
9. Add 100 μL of ACN, vortex, and incubate for 5–10 min. Put supernatants in a new microtube (one microtube per gel slice) (*see* **Note 14**).
10. Add 50 μL of 0.1 % TFA in water, vortex, and incubate for 5–10 min. Add 100 μL of ACN, vortex, and incubate for 5–10 min (*see* **Note 14**). Add the supernatants corresponding to each sample in the previously identified microtubes, and dry vacuum them in the Speed-Vac (*see* **Notes 15** and **16**).

3.3 MS Analysis of the Samples

1. Resuspend the samples in 10 μL 0.1 % FA, sonicate 5 min in the ultrasonic cleaning bath.
2. Put the resuspended samples on a vial and load the sample into the mass spectrometer (*see* **Note 17**).
3. Peptides are separated using a BEH130 C18 column, 75 μm × 200 mm, 1.7 μm coupled to a Symmetry 300 C18 UPLC Trap column, 180 μm × 20 mm, 5 μm (Waters) on a nanoACQUITY UPLC system (Waters).
4. The recommended chromatographic gradient includes the following steps (*see* **Note 18**):

Time (min)	A%	B%	Flow
0	97	3	0.3 mL/min
60	60	40	
61	15	85	
70	15	85	
72	97	3	
90	97	3	

A: FA 0.1 % in H_2O
B: FA 0.1 % in ACN

5. The MS acquisition method in the LTQ Orbitrap XL ETD includes the following parameters (*see* **Note 19**):
   - Full MS survey spectra (m/z 400 – 2000) are acquired in the orbitrap with 30,000 resolution at m/z 400.
   - Fragmentation of the six most intense precursors, with charge states equal to or greater than 2, by CID in the linear ion trap (*see* **Note 20**). Analyzed peptides are excluded from further analysis during 30 s using dynamic exclusion lists.

3.4 MS Data Analysis

As mentioned, protein identification and peptide modification assignment are carried out by searching the acquired peptide spectra in databases, such as UniProt or NCBI. Database search is carried out using different search engines, that is, algorithms that attempt to identify peptide sequences from the fragment ion spectra of the peptides in the dataset [19]. Mascot, Sequest, OMSSA, or VEMS, among others, are examples of search engines used for protein identification [20, 21]. Typical ubiquitylation tags (114.043 Da for GG, 383.228 Da for LRGG) must be considered when searching the spectra in order to find modified peptides. Furthermore, the tag LRGG gives the diagnostic ions 270.1925 (b2) and 384.2354 (b4) in MSMS which can be used to fish out potential MSMS spectra from ubiquitin modified peptides. The diagnostic ions are especially useful if MSMS spectra with high mass accuracy is available (<10 ppm).

1. Once the acquisition has ended, the generated unprocessed data are loaded into the search engine in order to identify the detected peptides and proteins.
2. Recommended search parameters include the following:
   - Carbamidomethylation of cysteines as fixed modification. Oxidation of methionines, GG (+114.043 Da) and LRGG (+383.228) modification of lysines, and protein N-terminal acetylation as variable modifications (*see* **Note 21**).
   - Peptide mass tolerance of 10 ppm and 0.5 Da fragment mass tolerance, and four missed cleavages allowed (*see* **Note 22**).
3. A decoy search is recommended in order to estimate the false discovery rate (FDR) for the samples. Once identified, selected proteins can be subjected to the functional analysis step (*see* **Note 23**).
4. Information on the presence of ubiquitylated peptides among the identified proteins can specifically be obtained by looking for those peptides carrying typical ubiquitin-modification (GG +114.043, LRGG +328.228) (*see* **Note 24**) (Fig. 4).

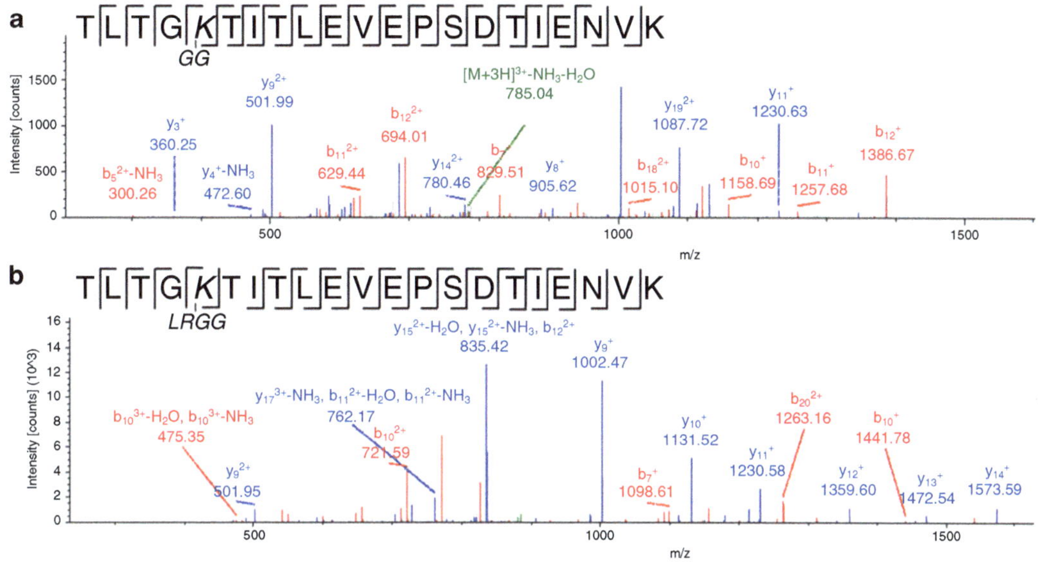

Fig. 4 Ubiquitylated peptide spectra examples. Two spectra for the ubiquitylated form of the peptide TLTGKTITLDVEPSDTIENVK, corresponding to K48 chains from polyubiquitin C (UBC_HUMAN) are provided, carrying the GG tag (**a**) or the LRGG tag (**b**) characteristic of ubiquitylation

3.5 Statistical Analysis

Once proteins are identified in both the TUBEs and negative controls (GST), it is necessary to discard unspecifically bound proteins and keep only the ubiquitylated proteins and their partners. In the example provided in this chapter, taken from Lopitz-Otsoa [22], a direct subtraction was performed. MCF7 cells treated with adriamycin were used for the characterization of the global ubiquitylation events in these cells, with the aim of pinpointing potential biomarkers and drug targets. Proteins identified in the GST controls were directly discarded from the dataset, giving a total of 643 proteins specifically bound to TUBEs. Of these, 269 were proteins consistently present in the replicates. This was the set of proteins considered as reliably enriched, and therefore further characterized in this work.

In addition to the direct subtraction of the identified proteins, relative quantitation of peptides and proteins can be carried out in order to make a more comprehensive enrichment analysis. The quantitative values can be obtained using experimental methods such as SILAC, stable isotope dimethyl labeling, tandem tags, area under the ion counts in the survey scans (XIC), or spectral counting. Matthiesen et al. [23] provides a review dealing with different MS-based quantitative methods. Then, multivariate analysis can be done in several software such as Excel, the statistical programming language R or Matlab. Below is a possible outline of the steps included in the multivariate analysis of the results.

1. Log transformation of the quantitative values. This will make the values normally distributed and lead to smaller *p*-values if a *t*-test is subsequently performed.
2. Optional normalization of the values across samples. For example, the R package "limma" supports a number of normalization procedures. References Kroll et al. and Bolstad et al. [24, 25] provide comprehensive overviews of different normalization procedures.
3. Subtraction of background values obtained from the control experiments, e.g., spectral counts from GST beads.
4. Calculate log ratios and *p* values to define difference in the level of ubiquitin modifications between experimental conditions (*see* **Note 25**). If the experiment has multiple conditions then ANOVA can be performed followed by a post-hoc test to lower the number of statistical tests.

3.6 Meta-Analysis

Frequently meta-analysis in proteomics starts out by comparing the identified proteins with proteins identified under different conditions or experimental settings. For example, in the work published by Lopitz-Otsoa et al. the set of enriched proteins was compared with the results obtained by other methods for the isolation and analysis of ubiquitylated proteins. Comparative analysis can also be done with well-known data databases such as UniProt or PhosphoSite. In this example, the Venn diagram in Fig. 5 compares TUBE-enriched proteins from Lopitz-Otsoa [22] against all O-GlcNAc-, SUMO-, and ubiquitin-annotated proteins in PhosphoSite (made with the R package "VennDiagram" which can plot Venn diagrams with up to five groups). The crosstalk between ubiquitin and different PTMs is the next level of complexity in the molecular regulation of multiple biological processes. SUMOylation and O-GlcNAcylation are known to be connected to ubiquitylation for the regulation of different functions [26–28]. Additionally, O-GlcNacylation and SUMOylation have been described to control transcription in the nucleus. In this context, the analysis of their correlation within the dataset enriched in the present analysis may be of great interest for a further characterization of these proteins.

Furthermore, proteins that are defined as significantly enriched in Subheading 3.5 above can be subjected to a functional enrichment analysis. A simple way to perform enrichment analysis is to submit the enriched protein IDs (*see* **Note 26**) to DAVID bioinformatics server [29, 30] and then export the result as text tables. The tab delimited text tables can then directly be imported in to R or Excel to produce summary graphics. For example, in Fig. 6 the ten most significant biological process categories were displayed for TUBE-enriched proteins and O-GlcNAc- and SUMO-annotated proteins in PhosphoSite.

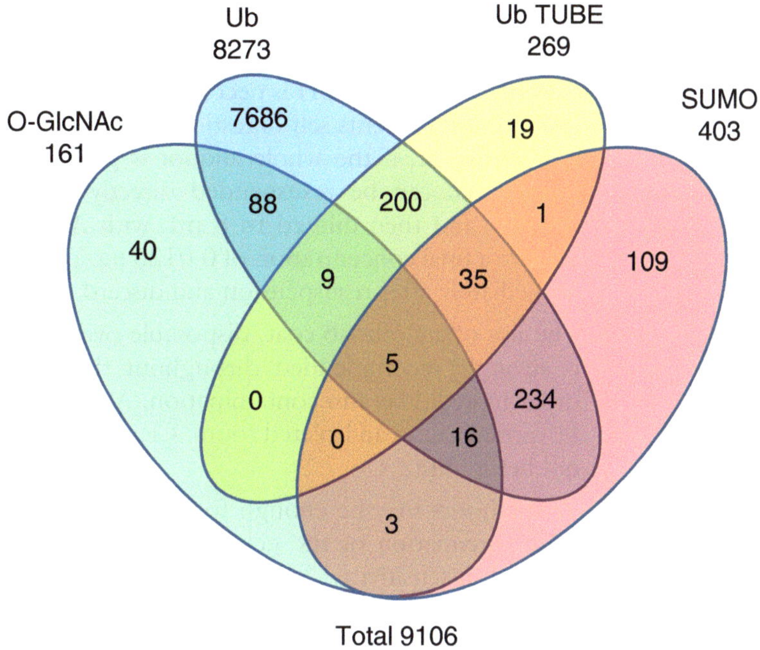

Fig. 5 Comparing all O-GlcNAc-, ubiquitin-, and SUMO-annotated proteins from PhosphoSite (Date 14-4-2015) with proteins reproducibly enriched by TUBEs and LC-MS identified by Lopitz-Otsoa et al. [22]

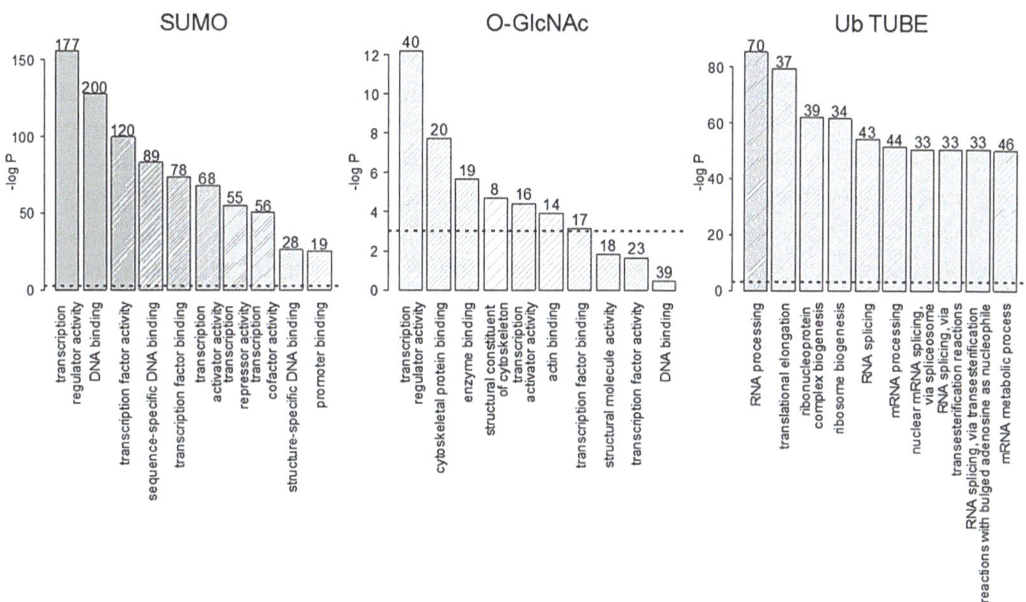

Fig. 6 Functional enrichment analysis by DAVID of the TUBEs-enriched proteins, and the SUMO- and O-GlcNAc-annotated proteins from PhosphoSite. Minus log of the FDR corrected significance of enrichment is indicated on the y-axis. The numbers on top of each bar indicate the number of proteins identified for each category

4 Notes

1. Resuspension in acid pH is necessary for the storage of trypsin, given that it prevents self-digestion events taking place in basic pH. However, if the whole aliquot is going to be used, the trypsin vial can be resuspended directly in 1 mL 50 mM AMBIC and then diluted to 8 mL with AMBIC 50 mM to achieve a final concentration of 0.0125 μg/μL. Use the trypsin immediately after resuspension and discard remnants, if any.
2. The use of a clean lab coat, disposable over-sleeves, and a cap is strongly recommended throughout the whole process in order to avoid keratin contamination. All digestion steps must be carried out in an isolated room. Clean all material with ethanol before use.
3. A few hours may be enough for protein detection, but overnight incubation of the gel in SYPRO is recommended for maximum sensitivity.
4. Switch the scanner on at least 30 min before image acquisition in order to warm up the system. The use of alternative up-to-date systems, such as the Versadoc Molecular Imager (Bio-Rad) is also a viable option for the image acquisition.
5. Laser gain values are illustrative. This value can be increased when the signal is weak, or decreased when saturated images are obtained, but it can be considered as a starting point.
6. The laser and filter setup used for this acquisition do not match with the default setup considered by the system. Therefore, a warning advice may appear when setting up the parameters. Ignore this advice and proceed with the acquisition.
7. Avoid saturation of the image and make sure that all the interesting parts of the gel are scanned before acquiring the image at high resolution. Low-resolution scans are much faster than high resolution, and therefore more suitable for the optimization of the image acquisition.
8. We used Microsoft Powerpoint, but programs intended for similar purposes can be used.
9. The number of slices may change depending on the pattern of the lane. However, keeping a reduced number of slices is a good idea in order not to increase too much the effort in the LC-MS side.
10. For a more dedicated cut, a UV transilluminator can be used. Otherwise, the entire gel lane can be cut in equal consecutive slices but the explained methodology is recommended.
11. Gel slices of a total *E. coli* extract processed in a similar way can be used in order to check the digestion process. These slices

should have a similar volume as the slices under analysis in the experiment.

12. The volumes provided are illustrative, and bigger or smaller volumes may be added. Use a volume that allows full coverage of the gel pieces.
13. Avoid the use of Iodoacetamide (IAA). IAA can artificially modify lysines with the addition of two acetamide moieties, which has the same molecular weight and chemical formula as the diglycine modification, and is completely undistinguishable by MS [18]. CAA does not provoke this effect, and therefore it is more suitable for this analysis.
14. Incubate the gel slices in ACN until they get white opaque.
15. Switch the Speed-Vac on at least 20 min before use in order to cold-up the trap.
16. Complete dryness is not recommended. Special care must be taken not overdrying the samples for a good sample recovery.
17. Sample resuspension and load may depend on the starting amount.
18. The columns and gradient are illustrative. However, a linear gradient followed by a washing step and an equilibration step are needed as part of the protocol. Length of each phase may depend on sample load, sample complexity, and/or column length among others. Adaptation to the system and optimization are therefore needed.
19. The parameters for the chromatography and the MS acquisition are illustrative. Dedicated methods and parameter optimization may be necessary for the acquisition with different equipment and samples.
20. The use of different fragmentation methods, such as ETD, has been shown to be a good alternative to CID, providing alternative fragmentation patterns. However, the method of choice for general purposes is still CID due to its ease of use and fragmentation capacity, and therefore its use is recommended.
21. If the presence of any other modification is suspected, consider it as part of the search, taking into account that it may increase search time and modify search space.
22. The parameters are typical for searching LTQ Orbitrap data. However, they should be adapted to the specific needs and characteristics of the equipment used. Four missed cleavages are allowed since the presence of ubiquitin moieties attached to the proteins is known to hamper tryptic cleavage of such residues, and therefore a high number of missed cleavages can be expected for highly modified peptides.
23. Decoy searches are recommended for complex samples. When sample complexity is low, however, its use is not recommended,

since FDR calculations might not be accurate. The threshold for protein selection may change depending on the sample. Selection of proteins with at least two peptides with a FDR < 5 % (or a Mascot *p*-value < 0.05 in the absence of an FDR estimation) or selection of the proteins with at least one peptide with a FDR < 1 % (or a Mascot *p*-value < 0.01 in the absence of an FDR estimation) are commonly used thresholds.

24. As mentioned, careful inspection of the spectra is recommended for avoiding false-positive assignments. The selection of spectra with the presence of fragments covering most of the peptide sequence and clearly assigning the modification site is recommended. Overcoming false positives is still one of the major issues when analyzing ubiquitylated proteins and peptides. The mass shift provoked by the GG addition is isobaric to many other chemical modifications, such as hydroxypropylation, asparagylation, or aspartylation, among others [17], and may therefore give rise to the detection of false positives. In addition, IAA can introduce false positives, as mentioned before. The use of high-accuracy mass spectrometers, such as the LTQ-Orbitrap XL ETD used in our approach, and the use of chloroacetamide (CAA) instead of IAA as alkylating agent during the protein digestion significantly reduce the number of false-positive assignments. However, most search engines lack robust enough tools for the unsupervised analysis and assignment of PTMs and therefore careful examination of the spectra is necessary to provide a reliable dataset of ubiquitylated peptides.
25. Published studies frequently do not provide *p*-values of enriched ubiquitin peptides because of high variance between samples. Frequently arbitrary thresholds are defined such as 1.5–2-fold enriched in a minimum number of biological replicas followed by for example validation by Western blot.
26. We find that DAVID provides a better mapping if the protein IDs and accession numbers are trimmed for version numbers. Furthermore, downloading the latest version of Gene Ontology annotation for the species of interest, and then manually map the enriched genes/proteins to Gene Ontology followed by enrichment statistics provides an even better mapping and more accurate results.

Acknowledgments

This work was funded by the Ministerio de Economia y Competitividad, Spain grant BFU2011-28536 (MSR), the Industry, Trade and Tourism Department of the Basque Government—SAIOTEK program, and Diputación Foral de Gipuzkoa (MSR). RM is funded by FCT investigator program 2012. MA, IE, and FE are part of the Proteomics Platform at CIC bioGUNE, which is member of ProteoRed-ISCIII and CIBERehd and funded by Basque Government (ETORTEK Program) and Instituto de Salud Carlos III. The authors would like to acknowledge networking support by the Proteostasis COST Action (BM1307).

References

1. Deng L, Wang C, Spencer E, Yang L, Braun A, You J, Slaughter C, Pickart C, Chen ZJ (2000) Activation of the IkappaB kinase complex by TRAF6 requires a dimeric ubiquitin-conjugating enzyme complex and a unique polyubiquitin chain. Cell 103(2):351–361
2. Wilkinson KD, Tashayev VL, O'Connor LB, Larsen CN, Kasperek E, Pickart CM (1995) Metabolism of the polyubiquitin degradation signal: structure, mechanism, and role of isopeptidase T. Biochemistry 34(44):14535–14546
3. Mata-Cantero L, Lobato-Gil S, Aillet F, Rodriguez MS (2015) The ubiquitin-proteasome system (UPS) as a cancer drug target: emerging mechanisms and therapeutics. In: Wondrak SL (ed) Stress response pathways in cancer. Springer Science+Business Media, Dordrecht, pp 225–264. doi:10.1007/978-94-017-9421-3_11
4. Xolalpa W, Perez-Galan P, Rodriguez MS, Roue G (2013) Targeting the ubiquitin proteasome system: beyond proteasome inhibition. Curr Pharm Des 19(22):4053–4093
5. Amerik AY, Hochstrasser M (2004) Mechanism and function of deubiquitinating enzymes. Biochim Biophys Acta 1695(1–3):189–207. doi:10.1016/j.bbamcr.2004.10.003
6. Aillet F, Lopitz-Otsoa F, Hjerpe R, Torres-Ramos M, Lang V, Rodriguez MS (2012) Isolation of ubiquitylated proteins using tandem ubiquitin-binding entities. Methods Mol Biol 832:173–183. doi:10.1007/978-1-61779-474-2_12
7. Hjerpe R, Aillet F, Lopitz-Otsoa F, Lang V, England P, Rodriguez MS (2009) Efficient protection and isolation of ubiquitylated proteins using tandem ubiquitin-binding entities. EMBO Rep 10(11):1250–1258. doi:10.1038/embor.2009.192
8. Hjerpe R, Rodriguez MS (2008) Efficient approaches for characterizing ubiquitinated proteins. Biochem Soc Trans 36(Pt 5):823–827. doi:10.1042/BST0360823
9. Lopitz-Otsoa F, Rodriguez MS, Aillet F (2010) Properties of natural and artificial proteins displaying multiple ubiquitin-binding domains. Biochem Soc Trans 38(Pt 1):40–45. doi:10.1042/BST0380040
10. Franco M, Seyfried NT, Brand AH, Peng J, Mayor U (2011) A novel strategy to isolate ubiquitin conjugates reveals wide role for ubiquitination during neural development. Mol Cell Proteomics 10(5):M110 002188. Doi:10.1074/mcp.M110.002188
11. Lectez B, Migotti R, Lee SY, Ramirez J, Beraza N, Mansfield B, Sutherland JD, Martinez-Chantar ML, Dittmar G, Mayor U (2014) Ubiquitin profiling in liver using a transgenic mouse with biotinylated ubiquitin. J Proteome Res 13(6):3016–3026. doi:10.1021/pr5001913
12. Sowa ME, Bennett EJ, Gygi SP, Harper JW (2009) Defining the human deubiquitinating enzyme interaction landscape. Cell 138(2):389–403. doi:10.1016/j.cell.2009.04.042
13. Udeshi ND, Mertins P, Svinkina T, Carr SA (2013) Large-scale identification of ubiquitination sites by mass spectrometry. Nat Protoc 8(10):1950–1960. doi:10.1038/nprot.2013.120
14. Xu G, Paige JS, Jaffrey SR (2010) Global analysis of lysine ubiquitination by ubiquitin remnant immunoaffinity profiling. Nat Biotechnol 28(8):868–873. doi:10.1038/nbt.1654
15. Peng J, Schwartz D, Elias JE, Thoreen CC, Cheng D, Marsischky G, Roelofs J, Finley D, Gygi SP (2003) A proteomics approach to understanding protein ubiquitination. Nat Biotechnol 21(8):921–926. doi:10.1038/nbt849
16. Shi Y, Xu P, Qin J (2011) Ubiquitinated proteome: ready for global? Mol Cell Proteomics 10(5):R110 006882. Doi:10.1074/mcp.R110.006882
17. Xu G, Jaffrey SR (2013) Proteomic identification of protein ubiquitination events. Biotechnol Genet Eng Rev 29:73–109. doi:10.1080/02648725.2013.801232
18. Nielsen ML, Vermeulen M, Bonaldi T, Cox J, Moroder L, Mann M (2008) Iodoacetamide-induced artifact mimics ubiquitination in mass spectrometry. Nat Methods 5(6):459–460. doi:10.1038/nmeth0608-459
19. Shteynberg D, Nesvizhskii AI, Moritz RL, Deutsch EW (2013) Combining results of multiple search engines in proteomics. Mol Cell Proteomics 12(9):2383–2393. doi:10.1074/mcp.R113.027797

20. Carvalho AS, Ribeiro H, Voabil P, Penque D, Jensen ON, Molina H, Matthiesen R (2014) Global mass spectrometry and transcriptomics array based drug profiling provides novel insight into glucosamine induced endoplasmic reticulum stress. Mol Cell Proteomics 13(12):3294–3307. doi:10.1074/mcp.M113.034363
21. Eng JK, Searle BC, Clauser KR, Tabb DL (2011) A face in the crowd: recognizing peptides through database search. Mol Cell Proteomics 10 (11):R111 009522. Doi:10.1074/mcp.R111.009522
22. Lopitz-Otsoa F, Rodriguez-Suarez E, Aillet F, Casado-Vela J, Lang V, Matthiesen R, Elortza F, Rodriguez MS (2012) Integrative analysis of the ubiquitin proteome isolated using tandem ubiquitin binding entities (TUBEs). J Proteomics 75(10):2998–3014. doi:10.1016/j.jprot.2011.12.001
23. Matthiesen R, Carvalho AS (2013) Methods and algorithms for quantitative proteomics by mass spectrometry. Methods Mol Biol 1007:183–217. doi:10.1007/978-1-62703-392-3_8
24. Bolstad BM, Irizarry RA, Astrand M, Speed TP (2003) A comparison of normalization methods for high density oligonucleotide array data based on variance and bias. Bioinformatics 19(2):185–193
25. Kroll TC, Wolfl S (2002) Ranking: a closer look on globalisation methods for normalisation of gene expression arrays. Nucleic Acids Res 30(11):e50
26. Aillet F, Lopitz-Otsoa F, Egana I, Hjerpe R, Fraser P, Hay RT, Rodriguez MS, Lang V (2012) Heterologous SUMO-2/3-ubiquitin chains optimize IkappaBalpha degradation and NF-kappaB activity. PLoS One 7(12):e51672. doi:10.1371/journal.pone.0051672
27. Guinez C, Mir AM, Dehennaut V, Cacan R, Harduin-Lepers A, Michalski JC, Lefebvre T (2008) Protein ubiquitination is modulated by O-GlcNAc glycosylation. FASEB J 22(8):2901–2911. doi:10.1096/fj.07-102509
28. Tatham MH, Geoffroy MC, Shen L, Plechanovova A, Hattersley N, Jaffray EG, Palvimo JJ, Hay RT (2008) RNF4 is a poly-SUMO-specific E3 ubiquitin ligase required for arsenic-induced PML degradation. Nat Cell Biol 10(5):538–546. doi:10.1038/ncb1716
29. da Huang W, Sherman BT, Lempicki RA (2009) Systematic and integrative analysis of large gene lists using DAVID bioinformatics resources. Nat Protoc 4(1):44–57. doi:10.1038/nprot.2008.211
30. da Huang W, Sherman BT, Lempicki RA (2009) Bioinformatics enrichment tools: paths toward the comprehensive functional analysis of large gene lists. Nucleic Acids Res 37(1):1–13. doi:10.1093/nar/gkn923

Chapter 10

Isolation of Ubiquitinated Proteins to High Purity from In Vivo Samples

Juanma Ramirez, Mingwei Min, Rosa Barrio, Catherine Lindon, and Ugo Mayor

Abstract

Ubiquitination pathways are widely used within eukaryotic cells. The complexity of ubiquitin signaling gives rise to a number of problems in the study of specific pathways. One problem is that not all processes regulated by ubiquitin are shared among the different cells of an organism (e.g., neurotransmitter release is only carried out in neuronal cells). Moreover, these processes are often highly temporally dynamic. It is essential therefore to use the right system for each biological question, so that we can characterize pathways specifically in the tissue or cells of interest. However, low stoichiometry, and the unstable nature of many ubiquitin conjugates, presents a technical barrier to studying this modification in vivo. Here, we describe two approaches to isolate ubiquitinated proteins to high purity. The first one favors isolation of the whole mixture of ubiquitinated material from a given tissue or cell type, generating a survey of the ubiquitome landscape for a specific condition. The second one favors the isolation of just one specific protein, in order to facilitate the characterization of its ubiquitinated fraction. In both cases, highly stringent denaturing buffers are used to minimize the presence of contaminating material in the sample.

Key words Ubiquitination, Substrates, Isolation, Denaturing conditions

1 Introduction

Ubiquitination of proteins is facilitated by the coordinated action of ubiquitin-activating E1, -conjugating E2, and -ligating E3 enzymes, and can be reversed by the so-called deubiquitinating enzymes (DUBs) [1]. Along with these ubiquitinating enzymes, proteasomal subunits, shuttling factors and other ubiquitin binding proteins regulate the fate of the ubiquitinated substrates. Altogether, nearly 1000 proteins integrate the ubiquitin proteasome system (UPS), which in addition to being the main intracellular protein degradation pathway, dynamically regulates the proteome by various other means too. Despite great advances in the mass spectrometry (MS) field, the identification of proteins regulated by the UPS is still a challenge, mostly because of the low

Rune Matthiesen (ed.), *Proteostasis: Methods and Protocols*, Methods in Molecular Biology, vol. 1449,
DOI 10.1007/978-1-4939-3756-1_10, © Springer Science+Business Media New York 2016

levels at which ubiquitin modified proteins are found within a cell. Historically, ubiquitin-binding domains (UBD) [2], ubiquitin-specific antibodies [3, 4], or epitope-tagged versions of ubiquitin [5–7] have been used in order to enrich the ubiquitinated protein fraction (ubiquitome). Here we present two methods that we recently developed: the first has allowed us to purify and enrich hundreds of ubiquitin conjugates in vivo from *Drosophila* melanogaster [7] and mouse [8] tissues, as well as from human cells [9]; the second has proved a valuable tool for validation of the ubiquitination of individual proteins and their mutants [10], for characterization of ubiquitin linkages [11], as well for identification of E3 ligases responsible for their ubiquitination [12].

1.1 The ^{bio}Ub Purification Strategy

The first successful proteomic approach ever to identify protein ubiquitination was performed in yeast [5] and based on His-tagged ubiquitin overexpression, which allowed for the use of denaturing buffers during the washing steps. A His-Ub transgenic mouse was reported [13], although no proteomics approach was published with this model. One concern using this approach would be the presence of too many endogenous histidine-rich proteins in mammals, which bind to the nickel affinity beads, resulting in excessive background for MS approaches. On the other hand, affinity pull-downs that cannot withstand denaturing conditions result in high background that can be observed by Coomassie staining [14].

In recent years, ubiquitin-remnant diGly-specific monoclonal antibodies have been used for the isolation and identification of thousands of putative ubiquitination sites in a number of systems [4, 15–19]. However, the diGly signature is also given by other ubiquitin-like proteins such as Nedd8, which is in general less abundant than ubiquitin, but whose concentration increases dramatically after the proteasome blockade commonly employed in ubiquitome studies [20]. Furthermore, the diGly approach can only isolate ubiquitinated proteins after trypsin cleavage, precluding the possibility of immunoblotting to validate them or to identify the type of ubiquitin chains (mono- or poly-) formed in vivo. The over-reliance on one single method for the identification of the substrates could also result in a significant presence of false positives [21]. Even if the actual validation of this technique is based on the identification of the diGly signature by MS, 65 % or more of the peptides isolated using diGly antibodies show no diGly signatures on them [18, 19]. One last caveat is that diGly antibodies cannot identify proteins ubiquitinated at other residues (like cysteines) in place of the canonical lysine residue [8, 22–24].

We developed the ^{bio}Ub strategy [7–9, 25] based on a short biotinylable motif [26] that is expressed N-terminally to each of several ubiquitin moieties expressed in tandem. This precursor polypeptide, which also contains the bacterial BirA enzyme, is digested by the deubiquitinating activity of endogenous DUBs, therefore allowing the BirA enzyme to specifically biotinylate all its target ubiquitins before they are attached to their substrates. Thanks to the strong affinity of the avidin-biotin interaction, very stringent washes can be applied to biotinylated material purified on avidin resins, resulting in a high enrichment of the ubiquitinated fraction in the eluted sample (Fig. 1). Furthermore, the stringent isolation and washing procedures prevent any protease and DUB activity and help to maximize the yield of purified ubiquitinated material. Additionally, the use of specific expression systems, such as the GAL4/UAS in *Drosophila* or tissue-specific/tetracycline-regulated promoters in mice, can direct the biotinylated ubiquitin (^{bio}Ub) to certain cell populations and/or stages during development, allowing the isolation of ubiquitinated material in a tissue- and time-specific manner. We have so far identified—by MS—over 4000 ubiquitinated proteins isolated using this strategy in various systems ([7, 8, 10] *and unpublished results*).

1.2 The GFP-Pulldown Strategy

In vitro strategies are routinely used to test ubiquitin substrates. In vivo validation of protein ubiquitination is more complicated: a number of pulldown approaches have been developed to test for bait/target protein ubiquitination. In most cases, however, these purifications are not carried out under denaturing conditions, leaving open the possibility that the detected ubiquitin signal arises from an interacting protein.

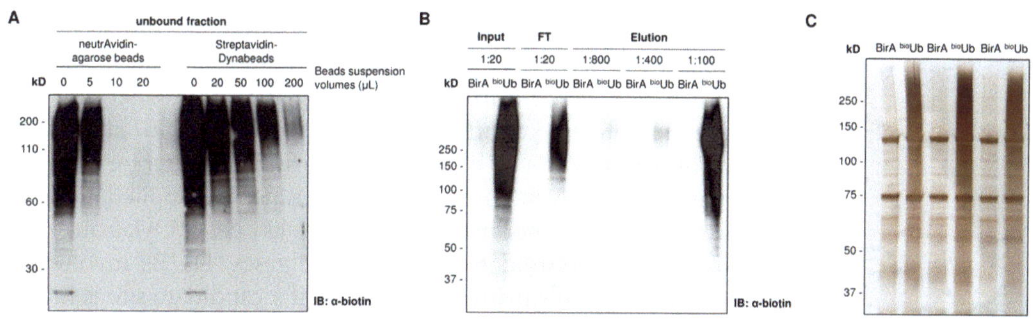

Fig. 1 Isolation of ubiquitin conjugates using the ^{bio}Ub pulldown. (**a**) Titration of bead volumes (*see* **Note 4**): Lysates from ^{bio}Ub-expressing mammalian cells were prepared as described from Subheading 3.1, **steps 1** to **7**, and incubated with the volumes of bead suspension indicated per 100 μL of lysate. 10 μL of each unbound fraction was tested by immunoblot with antibody against biotin to reveal the volume of bead suspension required for efficient pulldown at full bead capacity (between 5 and 10 μL for high-capacity NeutrAvidin-agarose (Thermoscientific) and 100 μL for streptavidin-Dynabeads (Life Technologies)). The prominent band at 25 kDa is mono-ubiquitinated histone H2A. (**b**) Monitoring of ^{bio}Ub pulldown protocol: Lysates from ^{bio}Ub-expressing *Drosophila melanogaster* embryos (^{bio}Ub) were subjected to the biotin pulldown as described in Subheading 3.1. Anti-biotin western immunoblotting to those pulldown experiments reveals a yield of approximately 25 %. Flies overexpressing only BirA were used as the control sample for the pulldown (BirA). (**c**) Eluates from ^{bio}Ub pulldown protocol: Liver lysates from ^{bio}Ub-expressing mice (^{bio}Ub) were prepared as described in Subheading 3.1. Silver staining from three independent pulldown experiments confirmed that the purification of ubiquitinated conjugates is specific to the ^{bio}Ub sample: only a few endogenously biotinylated proteins are purified from the control (BirA) mouse liver. This silver stained gel is presented here with the permission of Dr. Benoit Lectez

Single chain anti-GFP beads (Chromotek GmbH) are usually washed with non-denaturing buffers to achieve pulldowns in which protein interactions are preserved. We found, however, that these beads also allow for extremely stringent washes, giving a sensitive and quantifiable assay for ubiquitination [10, 12]. Since ubiquitination is a covalent conjugation, which can resist highly denaturing conditions, the usage of stringent washes allows for the elimination of all non-covalently bound interactions. We describe here a protocol that takes therefore full advantage of the potential of GFP nanobody technology. Capture of the GFP-tagged proteins is performed under non-denaturing conditions, with the buffer being supplemented with NEM to block deubiquitination, or post-lysis ubiquitination of substrates. Once the GFP-tagged proteins are bound to beads, highly stringent washes are applied. If improved detection of the ubiquitinated material is required, cells can be co-transfected with FLAG-tagged ubiquitin, or ubiquitin labeled with any other small tag (biotin, HA, His). Immunoblotting to reveal the ubiquitinated fraction of the purified GFP-tagged proteins can then be performed with antibodies to the tag carried on ubiquitin (Fig. 2). Observation of slower migrating ubiquitin-loaded species confirms ubiquitination of candidate substrates. Characterization of the ubiquitinated fraction can also be carried out by immunoblotting with antibodies to specific ubiquitin chain linkages [11]. The key advantage of this protocol is its simplicity and accessibility to any researcher, as well as the fact that the ubiquitin signal is not distorted by co-purifying species that would otherwise prevent a reproducible quantification of in vivo ubiquitination.

This strategy can be used to confirm and characterize the ubiquitination of a given substrate, but can also be applied to other posttranslational modifications, such as SUMOylation. The approach can be used to validate E3 ligases/DUBs involved in modifying a given substrate and to test if a candidate site is indeed modified as expected in the system of interest.

As compared to in vitro (from reconstituted components) or ex vivo (using *Xenopus* egg extracts, for example) approaches, this strategy can be used to monitor ubiquitination as it happens within the cells, from in vivo-isolated material (cells, flies, mice). It works best with co-transfection of tagged ubiquitin (since antibodies against tags tend to be more sensitive than those against ubiquitin) together with the GFP-tagged protein of interest, but we have also successfully used this assay to look at endogenous ubiquitin chains [11]. We believe this in vivo ubiquitination assay will also facilitate the identification of both ubiquitination sites and ubiquitin chain linkages by MS on a specific substrate, thanks to the purity of the sample.

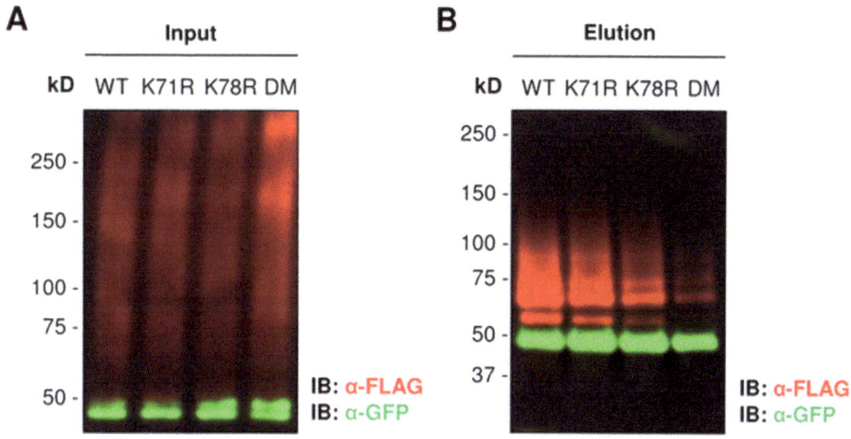

Fig. 2 Testing the in vivo ubiquitination of lysine mutants using the GFP pulldown. (**a**) *Drosophila* BG2 cells were transfected with a GFP-tagged ubiquitin substrate (wild type: WT, single lysine mutants K71R and K78R, and double lysine mutant: DM) as well as with FLAG-tagged ubiquitin. Anti-GFP (*green*) and anti-FLAG (*red*) antibodies were used to monitor the transfection levels in the whole cell extract. Both GFP-tagged protein and FLAG-tagged ubiquitin transfection efficiency were similar among samples. (**b**) Lysates from *Drosophila* BG2 cells were subjected to the GFP pulldown as described in Subheading 3.2. Anti-GFP and anti-FLAG antibodies were used to, respectively, verify that similar amounts of GFP-tagged substrate was purified for all constructs and to monitor the level of ubiquitination for each construct. While GFP levels were similar among different constructs, ubiquitination was clearly reduced in those proteins where lysines were mutated, especially in the double-lysine mutant (DM)

2 Materials

2.1 The ^{bio}Ub Purification Strategy

1. Cell lines, flies, or mice expressing either the precursor carrying the tandem modified-ubiquitin molecules plus BirA, or BirA alone as control population.
2. Two Dounce tissue grinders 7 mL (Jencons).
3. High-capacity NeutrAvidin-agarose beads (ThermoScientific).
4. PD10 desalting columns (GE Healthcare).
5. *N*-ethylmaleimide (NEM, Sigma).
6. Protease inhibitor cocktail (Roche Applied Science) at 25× concentration: 1 tablet in 2 mL of lysis/binding buffer.
7. Lysis buffer: 8 M Urea, 1% sodium dodecyl sulfate (SDS), 50 mM NEM in phosphate-buffered saline (PBS) (*see* **Note 1**).
8. Binding buffer: 3 M Urea, 1 M sodium chloride (NaCl), 0.25% SDS, 50 mM NEM in PBS.
9. Dilution buffer: 1.43 M NaCl, 1× protease inhibitor cocktail, 50 mM NEM in PBS (*see* **Note 2**).
10. Washing buffer 1 (WB1): 8 M Urea, 0.25% SDS in PBS.
11. WB2: 6 M Guanidine hydrochloride (GdnHCl) in PBS.
12. WB3: 6.4 M Urea, 1 M NaCl, 0.2% SDS in PBS.

13. WB4: 4 M Urea, 1 M NaCl, 10% isopropanol, 10% ethanol, 0.2% SDS in PBS.
14. WB5: 8 M Urea, 1% SDS in PBS.
15. WB6: 2% SDS in PBS.
16. Elution buffer (4× Laemmli SDS loading buffer): 200 mM Tris–HCl, pH 6.8, 8% SDS, 40% glycerol, 0.8 mg/mL bromophenol blue, with the addition of 100 mM dithiothreitol (DTT) prior to use.
17. Mini-column clarifying filters (Sartorious).

2.2 The GFP-Pulldown Strategy

1. Cells or tissue expressing a GFP-tagged candidate ubiquitinated protein (*see* **Note 3**).
2. Lysis buffer: 50 mM Tris–HCl pH 7.5, 150 mM NaCl, 1 mM EDTA, 0.5% Triton-X100, 1× Protease Inhibitor cocktail (Roche Applied Science), 50 mM NEM.
3. GFP Trap-A or GFP Trap-MA beads suspension (Chromotek GmbH).
4. Dilution buffer: 10 mM Tris–HCl pH 7.5, 150 mM NaCl, 0.5 mM EDTA, 1× Protease Inhibitor cocktail, 50 mM NEM.
5. Stringent wash buffer: 8 M Urea, 1% SDS in PBS.
6. Non-denaturing wash buffer: 10 mM Tris–HCl pH 7.6, 1 M NaCl, 0.5 mM EDTA, 1% (v/v) Triton-X100.
7. 1% SDS in 1× PBS.
8. Elution buffer (4× Laemmli SDS buffer): 200 mM Tris–HCl pH 6.8, 8% SDS, 40% glycerol, 0.8 mg/mL bromophenol blue, with the addition of 100 mM DTT prior to use.

3 Methods

Carry out all procedures at room temperature unless otherwise specified.

3.1 The ^{bio}Ub Purification Strategy

1. Prewash about 0.2 mL of NeutrAvidin-agarose beads suspension by resuspending in binding buffer and centrifuging 1 min at 233×*g*. Discard the supernatant and keep beads for later (*see* **Note 4**).
2. Collect ^{bio}Ub tissues in 2.5 mL of Lysis buffer+400 μL of 25× Protease Inhibitor cocktail (prepared in Lysis Buffer) (*see* **Note 5**).
3. Crush tissues using a 7 mL Dounce tissue homogenizer (*see* **Note 6**).
4. Centrifuge 1 min at 16,000×*g* and discard the pellet.
5. Centrifuge the supernatant for 5 min at 16,000×*g* at 4 °C. Repeat this step if needed.

6. Apply supernatant to a PD10 column previously equilibrated with 25 mL of binding buffer (*see* **Note 7**).
7. Collect eluate extract (3.5 mL) into 250 μL of 25× protease Inhibitor cocktail (diluted in binding buffer). Keep 1 % of the extract as input for immunoblot analysis.
8. Incubate extract with the prewashed NeutrAvidin agarose beads for 40 min at room temperature and 2 h at 4 °C (*see* **Note 8**).
9. Spin down the beads (2 min at 233×*g*) and keep the supernatant as "unbound fraction" for immunoblot analysis.
10. Wash beads in 15 mL tubes with washing buffers (WB, about 12 mL each time). Incubate 5 min with each WB and gentle rolling, then spin down the beads (2 min at 233×*g*) and discard the used buffer. The number of washes and the order in which buffers are applied is as follows: three times with WB1, three times with WB2, one time with WB3, three times with WB4, one time with WB1, one time with WB5, and three times with WB6.
11. Boil beads in 100 μL elution buffer (*see* **Note 9**) for 10 min.
12. Centrifuge boiled beads in a mini-column filter for 2 min at 16,000×*g* to recover the ubiquitin conjugates.
13. Eluted samples are typically processed for SDS gel electrophoresis and immunoblotting, or for LC-MS/MS analysis (*see* **Note 10**).

3.2 The GFP-Pulldown Strategy

We have successfully applied this basic protocol to insect cells and to mammalian cells, with different optimizations achieved where indicated. Users should optimize conditions for pulldown of their preferred GFP-tagged protein.

1. Prewash 5–15 μL per sample of GFP Trap-A or GFP Trap-MA bead suspension by collecting the beads and resuspending them in dilution buffer. GFP Trap-A beads are collected by centrifuging for 2 min at 2,700×*g* and GFP Trap-MA beads on a magnetic stand. Repeat once or twice.
2. Wash cells (or tissues) once in PBS and harvest using lysis buffer (*see* **Note 11**). We typically resuspend 3×10^6 harvested mammalian cells in 100 μL ice-cold lysis buffer and incubate on ice for 30 min with regular mixing of tube contents (*see* **Note 12**), or else scrape insect cells from 6-well plates straight into 300-500 μL of Lysis buffer per well (typically, 1×10^6 cell/mL).
3. Centrifuge lysate for 5 min at 16,000×*g* in a cold room. Collect the supernatant, which can be diluted to 0.1 % Triton X-100 (using dilution buffer) to improve binding to the beads. Remove 25 μL sample for immunoblotting as "input fraction." Mix lysate with previously washed GFP Trap beads and incubate at RT for 2 h with gentle rolling.
4. Collect beads and remove supernatant, keeping 25 μL sample as "unbound fraction" for immunoblotting.

5. Wash beads once with 1 mL of ice-cold dilution buffer.
6. Resuspend beads in 1 mL of stringent wash buffer and incubate them for 1–5 min with gentle rolling before collecting beads and discarding supernatant.
7. Wash beads three times with 1 mL of non-denaturing wash buffer, incubating for 5 min with gentle rolling each time.
8. Wash beads once with 1 mL SDS buffer, incubating for 5 min with gentle rolling.
9. Elute bound proteins from beads, by boiling at 95 °C for 10 min in 5–20 μL of elution buffer.

4 Notes

1. As solubilized urea is in equilibrium with ammonium cyanate that leads to carbamylation of amine groups in proteins, a reaction accelerated by heating, we generally use fresh urea solutions. In the case of WB3 and binding buffer, if SDS precipitates and needs to be re-dissolved, we warm up the buffer to 35 °C for no longer than 15 min.
2. Dilution buffer is only needed when ^{bio}Ub pulldown is performed from cells, to adjust the composition of cell lysates which are usually too dense to pass through the PD10 column. In this experiment, lysates are diluted with the volume of dilution buffer required to adjust Urea concentration to 3 M (as in binding buffer).
3. We have also successfully used this protocol with Venus- and YFP-tagged candidate proteins.
4. The bead volume needs to be optimized according to the amount of tissue/cells used and to the level of ^{bio}Ub expression. We usually titrate the bead volume, immunoblotting the unbound fraction to find the minimum bead volume that does not compromise the efficiency of pulldown (Fig. 1a). Similarly, the quantity of tissue/cells needs to be adjusted, as beads have a limited binding capacity. For instance, for 1 g of *Drosophila* embryos expressing ^{bio}Ub in the nervous system we typically use 0.1 mL of beads. In the case of mice, however, 0.3 mg of liver is enough to saturate a similar quantity of beads. In the case of mammalian cells we have used 1 mL of beads for 3×10^8 cells [9].
5. For 3×10^8 mammalian cells we have used 10 mL of lysis buffer.
6. For cells, the lysate should be syringed 5× through a 22G needle (to shear DNA) at this step, then diluted (*see* **Notes 2** and 7) before skipping to **step 8** of the protocol.
7. PD10 columns are used to eliminate free biotin, but also as a buffer exchange step. We equilibrate the column with binding

buffer, so the sample is exchanged into binding buffer ready to incubate with the beads. In the case of cells do not use PD10 columns; instead, add 17.5 mL of dilution buffer to 10 mL cell lysate and mix directly with 1 mL of NeutrAvidin beads.

8. Incubation has been optimized to the minimum time required for proper binding of biotinylated material to the beads. However, incubation can be alternatively performed overnight at 4 °C, although this might increase nonspecific binding.
9. The volume of elution buffer can be reduced if more concentrated sample is required.
10. We recommend to always run 10% of the eluate in a separate gel for silver staining. The observation of a significant difference in total material purified from cells expressing ^{bio}Ub versus material purified from control BirA cells, with similar levels of the endogenous biotinylated proteins (*see* Fig. 2), allow us to confirm that the whole process was performed correctly. If samples are intended for MS analysis, performing the pulldown in triplicate is recommended.
11. The number of cells and volume of beads to use should be determined empirically based on the size, expression level and ubiquitination level of the substrate. In our experience of purifying GFP-tagged proteins from mammalian cells to look for a ≤1% polyubiquitinated fraction, 5 μL GFP Trap slurry and 3×10^6 cells is required to generate a robustly quantitative signal from one immunoblot. In our experience of detecting mono- or multi-ubiquitinated fractions of GFP-tagged proteins from insect cells, 5×10^5 insect cells and 15 μL GFP Trap slurry is required. We have found GFP Trap-A and -MA beads to have a similar binding capacity for GFP-tagged proteins.
12. To extract nuclear proteins, increase Triton-X100 concentration to 1%.

Acknowledgments

While this chapter was written by the authors listed above, we would like to acknowledge other lab members who contributed to optimization of these techniques as described: Our thanks therefore to Maribel Franco, James Sutherland, Aitor Martinez, Benoit Lectez, and So Young Lee. We would also like to thank Junmin Peng and Gunnar Dittmar, without whose excellent MS support we would never have confirmed how well our ^{bio}Ub pulldown strategy was performing. The authors would like to acknowledge networking support by the Proteostasis COST Action (BM1307).

References

1. Komander D, Rape M (2012) The ubiquitin code. Annu Rev Biochem 81:203–229
2. Lopitz-Otsoa F, Rodriguez-Suarez E, Aillet F, Casado-Vela J, Lang V, Matthiesen R et al (2012) Integrative analysis of the ubiquitin proteome isolated using tandem ubiquitin binding entities (TUBEs). J Proteomics 75:2998–3014
3. Vasilescu J, Smith JC, Ethier M, Figeys D (2005) Proteomic analysis of ubiquitinated proteins from human MCF-7 breast cancer cells by immunoaffinity purification and mass spectrometry. J Proteome Res 4:2192–2200
4. Xu G, Paige JS, Jaffrey SR (2010) Global analysis of lysine ubiquitination by ubiquitin remnant immunoaffinity profiling. Nat Biotechnol 28:868–873
5. Peng J, Schwartz D, Elias JE, Thoreen CC, Cheng D, Marsischky G et al (2003) A proteomics approach to understanding protein ubiquitination. Nat Biotechnol 21:921–926
6. Greer PL, Hanayama R, Bloodgood BL, Mardinly AR, Lipton DM, Flavell SW et al (2010) The Angelman syndrome protein Ube3A regulates synapse development by ubiquitinating arc. Cell 140:704–716
7. Franco M, Seyfried NT, Brand AH, Peng J, Mayor U (2011) A novel strategy to isolate ubiquitin conjugates reveals wide role for ubiquitination during neural development. Mol Cell Proteomics 10, M110.002188
8. Lectez B, Migotti R, Lee SY, Ramirez J, Beraza N, Mansfield B et al (2014) Ubiquitin profiling in liver using a transgenic mouse with biotinylated ubiquitin. J Proteome Res 13:3016–3026
9. Min M, Mayor U, Dittmar G, Lindon C (2014) Using in vivo biotinylated ubiquitin to describe a mitotic exit ubiquitome from human cells. Mol Cell Proteomics 13:2411–2425
10. Min M, Mayor U, Lindon C (2013) Ubiquitination site preferences in anaphase promoting complex/cyclosome (APC/C) substrates. Open Biol 3:130097
11. Min M, Mevissen T, Luca MD, Komander D, Lindon C (2015) Efficient APC/C substrate degradation in cells undergoing mitotic exit depends on K11 ubiquitin linkages. Mol Biol Cell. 26:4325–32
12. Lee SY, Ramirez J, Franco M, Lectez B, Gonzalez M, Barrio R et al (2014) Ube3a, the E3 ubiquitin ligase causing Angelman syndrome and linked to autism, regulates protein homeostasis through the proteasomal shuttle Rpn10. Cell Mol Life Sci 71:2747–2758
13. Tsirigotis M, Thurig S, Dubé M, Vanderhyden BC, Zhang M, Gray DA (2001) Analysis of ubiquitination in vivo using a transgenic mouse model. Biotechniques 31:120–126, 128, 130
14. Tirard M, Hsiao H-H, Nikolov M, Urlaub H, Melchior F, Brose N (2012) In vivo localization and identification of SUMOylated proteins in the brain of His6-HA-SUMO1 knock-in mice. Proc Natl Acad Sci U S A 109: 21122–21127
15. Kim W, Bennett EJ, Huttlin EL, Guo A, Li J, Possemato A et al (2011) Systematic and quantitative assessment of the ubiquitin-modified proteome. Mol Cell 44:325–340
16. Wagner SA, Beli P, Weinert BT, Nielsen ML, Cox J, Mann M et al (2011) A proteome-wide, quantitative survey of in vivo ubiquitylation sites reveals widespread regulatory roles. Mol Cell Proteomics 10, M111.013284
17. Sarraf SA, Raman M, Guarani-Pereira V, Sowa ME, Huttlin EL, Gygi SP et al (2013) Landscape of the PARKIN-dependent ubiquitylome in response to mitochondrial depolarization. Nature 496:372–376
18. Wagner SA, Beli P, Weinert BT, Schölz C, Kelstrup CD, Young C et al (2012) Proteomic analyses reveal divergent ubiquitylation site patterns in murine tissues. Mol Cell Proteomics 11:1578–1585
19. Na CH, Jones DR, Yang Y, Wang X, Xu Y, Peng J (2012) Synaptic protein ubiquitination in rat brain revealed by antibody-based ubiquitome analysis. J Proteome Res 11: 4722–4732
20. Leidecker O, Matic I, Mahata B, Pion E, Xirodimas DP (2012) The ubiquitin E1 enzyme Ube1 mediates NEDD8 activation under diverse stress conditions. Cell Cycle Georget Tex 11:1142–1150
21. Shi Y, Xu P, Qin J (2010) Ubiquitinated proteome: ready for global? Mol Cell Proteomics 10, R110.006882
22. Williams C, van den Berg M, Sprenger RR, Distel B (2007) A conserved cysteine is essential for Pex4p-dependent ubiquitination of the peroxisomal import receptor Pex5p. J Biol Chem 282:22534–22543
23. Hensel A, Beck S, Magraoui FE, Platta HW, Girzalsky W, Erdmann R (2011) Cysteine-dependent ubiquitination of Pex18p Is linked to cargo translocation across the peroxisomal membrane. J Biol Chem 286:43495–43505
24. Wang X, Herr RA, Hansen TH (2012) Ubiquitination of substrates by esterification. Traffic Cph Den 13:19–24
25. Talamillo A, Herboso L, Pirone L, Pérez C, González M, Sánchez J et al (2013) Scavenger receptors mediate the role of SUMO and Ftz-f1 in Drosophila steroidogenesis. PLoS Genet 9:e1003473
26. Beckett D, Kovaleva E, Schatz PJ (1999) A minimal peptide substrate in biotin holoenzyme synthetase-catalyzed biotinylation. Protein Sci 8:921–929

Chapter 11

Method for the Purification of Endogenous Unanchored Polyubiquitin Chains

Daniel Scott, Jo Strachan, Varun Gopala Krishna, Barry Shaw, David J. Tooth, Mark S. Searle, Neil J. Oldham, and Rob Layfield

Abstract

Unanchored polyubiquitin chains are endogenous non-substrate linked ubiquitin polymers which have emerging roles in the control of cellular physiology. We describe an affinity purification method based on an isolated ubiquitin-binding domain, the ZnF_UBP domain of the deubiquitinating enzyme USP5, which permits the selective purification of mixtures of endogenous unanchored polyubiquitin chains that are amenable to downstream molecular analyses. Further, we present methods for detection of unanchored polyubiquitin chains in purified fractions.

Key words Ubiquitin, Unanchored polyubiquitin, Free polyubiquitin, USP5, ZnF_UBP domain, Ubiquitin-binding domain

1 Introduction

In recent years the existence and functional significance of endogenous unanchored (substrate-free) polyubiquitin chains has emerged, with chains of various linkages implicated in diverse cellular processes including, but not limited to, activation of protein kinases and regulation of the aggresome response [1–11]. To permit a detailed molecular analysis of the endogenous unanchored polyubiquitin pool we previously developed a single-step affinity purification protocol, which employs a Sepharose-coupled ubiquitin-binding domain (UBD) [12, 13]. This free ubiquitin-binding entity (FUBE) exploits the intrinsic selectivity and relatively high affinity of the ZnF_UBP domain, isolated from the human deubiquitinating enzyme USP5, for the free C-terminus of ubiquitin [14], to selectively purify unanchored (poly)ubiquitin over substrate-conjugated forms (*see* Fig. 1). Additionally the ZnF_UBP domain is specific for ubiquitin over ubiquitin-like proteins due to differences in C-terminal sequences. The FUBE displays no

Rune Matthiesen (ed.), *Proteostasis: Methods and Protocols*, Methods in Molecular Biology, vol. 1449,
DOI 10.1007/978-1-4939-3756-1_11,

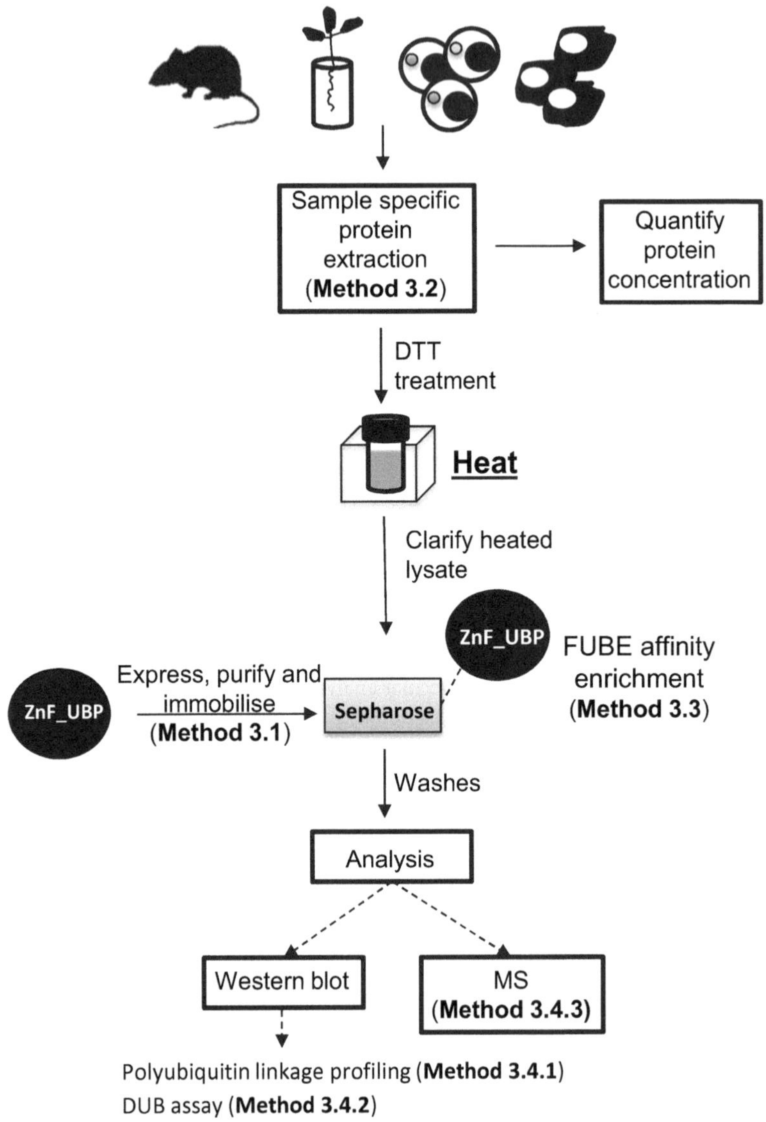

Fig. 1 FUBE work flow for the purification of endogenous unanchored polyubiquitin chains from a variety of sources

polyubiquitin-selective bias [14] and consequently captures mixtures of chains with different linkages representative of the composition within the cellular *milieu*, as well as free monoubiquitin.

Due to the high sequence conservation of ubiquitin, this method has found application in the purification of endogenous unanchored polyubiquitin chains from rat muscle and plants. Additionally endogenous unanchored polyubiquitin chains have been purified from both human and yeast cell extracts, which crucially are amenable to physiological, pharmacological, or genetic manipulation, potentially allowing further insights into the biology of unanchored polyubiquitin chains [15]. The protocols listed below outline the production and applications of the FUBE, as well as suggested techniques to probe the composition of purified unanchored polyubiquitin chain mixtures.

2 Materials

Prepare all solutions using ultrapure water and analytical grade reagents. Prepare and store all reagents at room temperature (unless indicated otherwise). Diligently follow all local waste disposal regulations when disposing waste materials.

2.1 Production of GST-ZnF_UBP

1. ZnF_UBP forward primer (5′ ACAGGATCCAAGCAGGA GGTGCAGGCATG 3′) and ZnF_UBP reverse primer (5′ GCTCGAGTTACTTGTCTGTCTTCTGCATCTTCAGC 3′).
2. pGEX-4T-1 plasmid (GE Healthcare).
3. Plasmid DNA purification: QIAquick Gel Extraction Kit (Qiagen), QIAprep Spin Miniprep Kit (Qiagen).
4. Restriction enzymes: *Bam*HI and *Xho*I, cutsmart buffer (NEB).
5. T4 DNA ligase and reaction buffer (NEB).
6. Competent cells for plasmid DNA preparation (XL10-Gold ultracompetent cells, Agilent) and protein expression (BL21 (DE3), Novagen).
7. LB medium and agar plates, containing 100 μg/mL ampicillin.
8. 1 M IPTG stock solution, dissolved in water, filter sterilized, aliquoted, and stored at −20 °C.
9. Purification buffer: 20 mM Tris (pH 7.5), 150 mM NaCl, 0.1 % (v/v) Triton X-100.
10. Thrombin cleavage buffer: 20 mM Tris (pH 8.4), 150 mM NaCl, 2.5 mM $CaCl_2$.
11. Thrombin (Sigma Aldrich) dissolved in thrombin cleavage buffer to a stock concentration of 0.5 U/μL, aliquoted, and stored at −20 °C.
12. Glutathione Sepharose 4B (GE Healthcare).
13. 5 mL Gravity flow columns (Qiagen).
14. Vivaspin 20 ultrafiltration device (MWCO 10 kDa) (Sartorius).

2.2 Covalent Immobilization of ZnF_UBP Domain to Cyanogen Bromide-Activated Sepharose 4B

1. Re-hydration buffer: 1 mM HCl.
2. Coupling buffer: 100 mM Sodium hydrogen carbonate (pH 8.3), 500 mM NaCl.
3. Sepharose blocking buffer: 1 M Ethanolamine (pH 8.0).
4. Acetate buffer: 100 mM Sodium acetate (pH 4), 500 mM NaCl.
5. Tris buffer: 100 mM Tris (pH 8.0), 500 mM NaCl.

6. Column storage buffer: 25 mM Tris (pH 7.4), 1 mM sodium azide.
7. 5 mL Gravity flow columns (Qiagen).
8. Cyanogen bromide-activated Sepharose 4B (CNBr-Sepharose, Sigma).

2.3 Preparation of Protein Extracts for Unanchored Polyubiquitin Chain Purification

1. Sample-specific lysis buffer (*see* Table 1 for buffer compositions) (*see* **Note 1**).
2. 1 M DTT stock solution.
3. Wash buffer A: 50 mM Tris (pH 7.5), 150 mM NaCl, 0.5 % (v/v) NP-40.
4. Wash buffer B: 50 mM Tris (pH 7.5), 1 mM DTT.

Table 1
Overview of sample-specific lysis protocols for the extraction of unanchored polyubiquitin chains from tissue, mammalian cells, yeast, and plant samples

Sample type	Example	Homogenizing buffer	Starting amount	Approx. total protein	Volume buffer (mL)	Lysis method
Tissue	Rat muscle	Homogenizing buffer: 50 mM Tris (pH 7.5), 150 mM NaCl, 0.5 % (v/v) NP-40, 5 mM NEM, 0.1 % (v/v) mammalian protease inhibitor cocktail (Sigma Aldrich), 20 μM MG132 (Sigma Aldrich), 0.1 % (v/v) phosphatase inhibitor (Sigma Aldrich) and 500 μM 1,10-phenanthroline (Life Sensors)	15 g	300 mg	100	Mechanical homogenizer
Cells	Human U20S cells	As above	4×10 cm dishes	15 mg	6	Sonication
Yeast	*Saccharomyces cerevisiae*	50 mM Tris (pH 7.2) 50 mM KCl; 10 μM NEM, 10 μM MG132	2 L culture	200 mg	15	Blast-freezing and grinding in liquid nitrogen
Plants	*Arabidopsis thaliana* roots	0.4 M sucrose, 10 mM Tris (pH 8.0) 5 mM β-mercaptoethanol, 0.1 mM PMSF, 1 % (v/v) plant protease inhibitor cocktail	25 g (Wet weight)	N.D.	100	

2.4 Commercially Available Ubiquitin Antibodies to Probe Purified Unanchored Polyubiquitin Samples

Antibody	Specificity	Manufacturer
VU-1	Ubiquitin	Life Sensors
P4D1	Ubiquitin	Santa Cruz
Apu-2	K48 linkage	Merck Millipore
HWA4C4	K63 linkage	Enzo Life Sciences
2A3/2E6	K11 linkage	Merck Millipore
Lub9	M1 linkage	Life Sensors

2.5 Deubiquitination Assay

1. DUB buffer: 50 mM Tris (pH 7.5), 150 mM NaCl, 10 mM DTT.
2. Full-length recombinant short isoform USP5 (Enzo Life Sciences), diluted to 10 ng/μL in DUB buffer.
3. Catalytic core of USP2 (Enzo Life Sciences), diluted to 25 ng/μL in DUB buffer.

3 Methods

3.1 Generating the FUBE

3.1.1 Cloning the ZnF_UBP Domain of USP5

1. Amplify the protein-coding sequence of the ZnF_UBP domain (residues 163–291) of USP5 from human U20S cDNA (or other mammalian cell cDNA) using ZnF_UBP forward and reverse primers, following a standard PCR method (for example GoTaq (Promega)). The PCR product contains *Bam*H1 and *Xho*1 restrictions sites at the 5′ and 3′ termini, respectively, encoded within the amplification primers.
2. Separate by DNA electrophoresis on a 1% agarose gel and extract the PCR product using the QIAquick Gel Extraction Kit (QIAGEN), elute DNA in 30 μL H_2O.
3. Double digest the PCR product and GST fusion vector (pGEX-4T-1, GE Healthcare) with *Bam*HI and *Xho*I restriction enzymes (NEB), using a standard method. Purify digested products using agarose gel electrophoresis, as above; elute DNA in 30 μL H_2O.
4. Ligate the ZnF_UBP coding sequence into the plasmid using T4 DNA ligase (NEB) according to the manufacturer's specifications.
5. Transform the ligated product into *E. coli* strains XL10-Gold ultracompetent cells (Agilent) using standard methods, and plate bacteria onto LB-ampicillin agar plates. Incubate overnight at 37 °C.
6. Select single colonies to inoculate 10 mL LB-ampicillin for growth overnight shaking at 180 r.p.m. at 37 °C.

7. Pellet cultures by centrifugation and purify plasmids using QIAprep Spin Miniprep Kit (QIAGEN), as described by the manufacturer. Confirm successful cloning of the ZnF_UBP domain by Sanger sequencing (*see* **Note 2**).

3.1.2 Over-expression and Purification of ZnF_UBP Domain

1. Transform the GST-ZnF_UBP fusion protein encoding plasmid into the *E. coli* strain BL21, using standard methods. Plate transformed bacteria onto LB agar plates, and incubate the plates overnight at 37 °C.
2. Inoculate 10 mL LB-ampicillin with a single colony, and incubate the culture overnight at 37 °C whilst shaking at 180 r.p.m.
3. Inoculate 1 L of LB-ampicillin with the 10 mL overnight culture and incubate at 37 °C while shaking at 180 r.p.m. until the optical density at 600 nm (OD_{600}) reaches 0.6.
4. Induce the overexpression of GST-ZnF_UBP fusion protein with 200 μM IPTG (final concentration) for ~16 h at 20 °C while shaking at 180 r.p.m.
5. Pellet bacterial cells by centrifugation at 3200 × *g* for 20 min at 4 °C, discard supernatant, and store the pellet at −80 °C (*see* **Note 3**).
6. Resuspend the pellet in 30 mL of purification buffer and lyse the cells by sonication on ice to avoid sample heating (30-s bursts with 30-s intervals at an amplitude of 10 μm for 5 min).
7. Clarify the lysate by centrifugation at 35,000 × *g* for 30 min at 4 °C.
8. Filter the cleared lysate using a 0.45 μm syringe filter.
9. Equilibrate 1 mL of glutathione-Sepharose 4B resin with wash buffer A in a gravity flow column.
10. Incubate the filtered lysate from **step 8**, with glutathione-Sepharose for 1 h at 4 °C with rotating.
11. Wash the Sepharose with three column volumes of purification buffer, with the final wash on a rotator for 30 min at 4 °C.
12. Wash the Sepharose with three column volumes of thrombin cleavage buffer, with the final wash on a rotator for 30 min at 4 °C.
13. To cleave the GST tag from the ZnF_UBP protein add 10 U of thrombin in 5 mL of thrombin cleavage buffer; incubate the mixture on a rotator at 4 °C overnight.
14. Collect the column flow-through containing the cleaved ZnF_UBP protein. Wash the remaining glutathione Sepharose three times with 5 mL of thrombin cleavage buffer, collecting and pooling each flow-through fraction.
15. Concentrate the sample using a Vivaspin 20 ultrafiltration device (MWCO 10 kDa), confirm the purity of thrombin cleaved ZnF_UBP by SDS PAGE, and quantify protein concentration (*see* **Note 4**).

3.1.3 Immobilization of Purified ZnF_UBP Domain onto CNBr-Sepharose

To generate the FUBE, the ZnF_UBP protein generated in Subheading 3.1.2 is covalently coupled to CNBr-Sepharose beads. Versions of the FUBE reagent are also available commercially; users should follow product specific protocols for these.

1. Hydrate the required volume of CNBr-Sepharose beads (*see* **Note 5**) for 15 min in excess re-hydration buffer.
2. Transfer the beads to a gravity column and equilibrate in ten times the bed volume in coupling buffer.
3. Incubate thrombin cleaved ZnF_UBP protein (Subheading 3.1.2), diluted into coupling buffer, with beads for 3 h at 4 °C on a rotator (*see* **Note 6**).
4. Wash then incubate the beads in ethanolamine, overnight rotating at 4 °C.
5. Wash the beads with three column volumes of coupling buffer.
6. Wash the beads alternately in a column volume of Tris buffer and acetate buffer, washing beads in each buffer four times.
7. Wash the beads with three column volumes of column storage buffer (*see* **Note 7**).

3.2 Preparation of Protein Extracts for the Purification of Unanchored Polyubiquitin Chains

Sample specific lysis is required, *see* Table 1 for details (*see* **Note 8**, which provides detail for mammalian cell lysis protocol as an exemplar).

1. Clear the lysate by centrifugation at 16,000 × *g* for 10 min at 4 °C.
2. Discard the pellet and pass supernatant through glass wool.
3. Determine protein concentration using a BCA assay, as described by the manufacturer.
4. Add DTT to the lysate to achieve a final concentration of 10 mM and incubate for 15 min at 4 °C rotating (*see* **Note 9**, not required for purification from plant material).
5. Heat lysate at 75 °C for 20 min, mixing every 5 min (*see* **Note 10**).
6. Repeat centrifugation as in **step 3**, collect supernatant and store on ice.

3.3 Affinity Purification of Unanchored Polyubiquitin Chains

1. Equilibrate the required volume of FUBE in wash buffer A in a gravity column.
2. Incubate FUBE with protein lysate prepared in Subheading 3.2 rotating at 4 °C overnight.
3. Wash the FUBE with two column volumes of wash buffer A. Perform a third final wash by rotating one column volume of wash buffer A for 15 min at 4 °C before discarding buffer.
4. Wash FUBE as in **step 3**, but using wash buffer B.

5. Re-suspend the FUBE in 1 mL wash buffer B and transfer to Eppendorf(s). Pulse centrifuge the FUBE in a microfuge. Aspirate carefully, until no liquid remains.

3.4 Visualizing the Purified Unanchored Polyubiquitin Chain Pool

The development of a range of polyubiquitin linkage-specific antibodies has permitted a simple assessment of the presence of specific linkages within a sample of purified unanchored polyubiquitin chains by western blot analysis.

3.4.1 Probing the Linkage Profile of FUBE-Purified Unanchored Polyubiquitin Chains

1. Elute FUBE-enriched polyubiquitin chains (generated in Subheading 3.3) by the addition of SDS PAGE gel loading buffer, perform western blot analysis using antibody of choice with commercial polyubiquitin ladders of different linkage as control (*see* **Note 11**).

3.4.2 Deubiquitination Assay

To confirm the unanchored nature of FUBE-purified polyubiquitin, a deubiquitination assay can be performed [16]. Here samples of FUBE-purified polyubiquitin (on beads) are incubated with full-length USP5, a deubiquitinating enzyme that selectively recognizes and disassembles unanchored polyubiquitin chains, and USP2, a broad spectrum DUB which disassembles all polyubiquitin linkages, in parallel.

1. Aliquot FUBE beads (from Subheading 3.3, **step 5**), into three aliquots. One aliquot remains as the untreated control, the other two aliquots are incubated with one bed volume of diluted USP5 or USP2. Incubate overnight with gentle agitation at 37 °C.
2. Add SDS-PAGE gel loading buffer directly to the sample mixture (beads and buffer combined) to stop the reaction.
3. Perform western blot analysis using an antibody against ubiquitin (*see* **Notes 11**).

3.4.3 Mass Spectrometry-Based Characterization of Purified Unanchored Polyubiquitin Chains

Polyubiquitin linkages present in samples following FUBE enrichment can be assessed qualitatively or more recently, quantitatively [17], by mass spectrometry. The optimal digest and analysis strategy should be empirically determined.

4 Notes

1. Store the homogenizing buffer on ice, adding inhibitors immediately prior to use.
2. The plasmid for the expression and purification of GST-tagged ZnF_UBP domain (residues 163–291) of human USP5 is available upon request from the Layfield laboratory.
3. Pause point, bacterial pellets can be stored at −80 °C for 3 months.

4. It is not essential to remove thrombin, although if required the ZnF_UBP domain can first be purified by gel filtration.
5. CNBr-Sepharose, 1 g de-hydrated Sepharose produces ~3.5 mL hydrated.
6. Thrombin cleaved ZnF_UBP coupled at a ratio of 10 mg per mL of hydrated Sepharose. Control Sepharose, where protein is omitted in binding step alongside sample, is required as a negative control.
7. Immobilised ZnF_UBP beads can be stored in column storage buffer for 3 months at 4 °C. Longer storage has not been tested.
8. Mammalian cells should be cultured under recommended conditions for their specific type. Approximately 15 mg total protein extract (concentration determined prior to DTT treatment) is recommended as a starting point for capture on 100 μL of FUBE beads; as a guide, up to 15 mg total protein can be

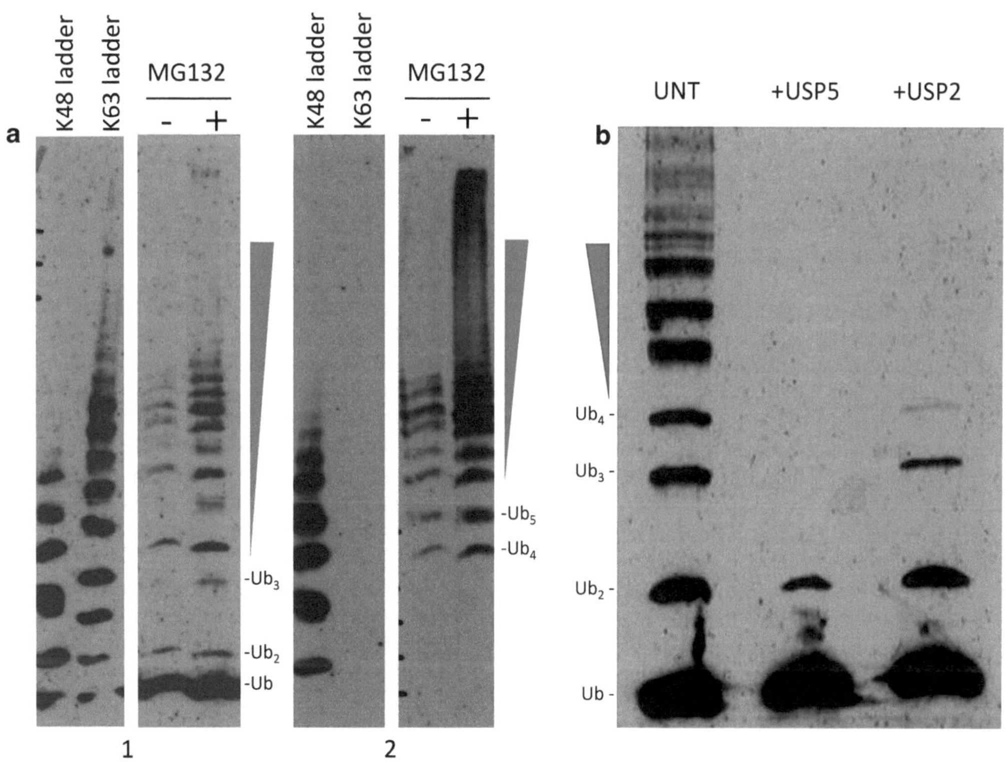

Fig. 2 Typical FUBE affinity enrichment of endogenous unanchored polyubiquitin chains from mammalian cells (in this case HEK293T). FUBE-enriched material can be western blotted and probed with a range of ubiquitin antibodies, including; anti-VU1 (no overt linkage selectivity, panel **a1** and **b**) and anti-apu2 (K48 selective, panel **a2**). Note the use of commercial polyubiquitin chain mixtures of specified linkages (K48, K63) as markers. Also note the difference in the abundance of purified unanchored polyubiquitin pool ± MG132 treatment (panel **a2**). A deubiquitination assay can be utilised to confirm the unanchored nature of purified polyubiquitin chains (panel **b**). Here FUBE-enriched material is treated with USP5 or USP2, both of which deubiquitinate the chains. UNT represents a comparable sample which is untreated with enzyme. Numbers of ubiquitin in the chains are indicated

derived from 4 × 10 cm dishes of U20S or HeLa cells at 80 % confluence [15]. Treatment of cultured cells with proteasome inhibitor (1 μM MG132) for 16 h increases the abundance of unanchored polyubiquitin chains and can be used as a positive control (*see* Fig. 2) [15].

9. DTT is added to titrate out NEM added to the homogenizing buffer (not required for plant extraction as NEM is omitted). NEM can inactivate the ZnF_UBP protein (a zinc finger); care should be taken to avoid/remove all cysteine-modifying reagents in sample buffers.
10. Sample heating is a critical step to resolve unanchored polyubiquitin chains from heat-sensitive ubiquitin-protein conjugates [15, 18, 19].
11. It is recommended to run a 5–20 % gradient gel to separate polyubiquitin chains of assorted length/linkage. Furthermore when blotting with linkage-selective antibodies, commercial purified polyubiquitin chains (Boston Biochem) of known linkages are suggested, typically 0.25 μg, as controls (*see* Fig. 2). For recommended dilutions/protocol and buffers to be used with antibodies see the manufacturer's guidelines.

References

1. Xia ZP et al (2009) Direct activation of protein kinases by unanchored polyubiquitin chains. Nature 461(7260):114–119
2. Nanduri P et al (2015) Chaperone-mediated 26S proteasome remodeling facilitates free K63 ubiquitin chain production and aggresome clearance. J Biol Chem 290(15):9455–9464
3. Ouyang H et al (2012) Protein aggregates are recruited to aggresome by histone deacetylase 6 via unanchored ubiquitin C termini. J Biol Chem 287(4):2317–2327
4. Banerjee I et al (2014) Influenza A virus uses the aggresome processing machinery for host cell entry. Science 346(6208):473–477
5. Hao R et al (2013) Proteasomes activate aggresome disassembly and clearance by producing unanchored ubiquitin chains. Mol Cell 51(6):819–828
6. Rajsbaum R et al (2014) Unanchored K48-linked polyubiquitin synthesized by the E3-ubiquitin ligase TRIM6 stimulates the interferon-IKKepsilon kinase-mediated antiviral response. Immunity 40(6):880–895
7. Pertel T et al (2011) TRIM5 is an innate immune sensor for the retrovirus capsid lattice. Nature 472(7343):361–365
8. Miranzo-Navarro D, Magor KE (2014) Activation of duck RIG-I by TRIM25 is independent of anchored ubiquitin. PLoS One 9(1):e86968
9. Dayal S et al (2009) Suppression of the deubiquitinating enzyme USP5 causes the accumulation of unanchored polyubiquitin and the activation of p53. J Biol Chem 284(8): 5030–5041
10. Amerik AY et al (1997) In vivo disassembly of free polyubiquitin chains by yeast Ubp14 modulates rates of protein degradation by the proteasome. EMBO J 16(16):4826–4838
11. Braten O et al (2012) Generation of free ubiquitin chains is up-regulated in stress and facilitated by the HECT domain ubiquitin ligases UFD4 and HUL5. Biochem J 444(3):611–617
12. Strachan J et al (2012) Insights into the molecular composition of endogenous unanchored polyubiquitin chains. J Proteome Res 11(3): 1969–1980
13. Scott D et al (2015) Ubiquitin-binding domains: mechanisms of ubiquitin recognition and use as tools to investigate ubiquitin-modified proteomes. Proteomics 15(5–6):844–861
14. Reyes-Turcu FE et al (2006) The ubiquitin binding domain ZnFUBP recognizes the C-terminal diglycine motif of unanchored ubiquitin. Cell 124(6):1197–1208
15. Strachan J et al (2013) Broad utility of an affinity-enrichment strategy for unanchored polyubiquitin chains. J Proteomics Bioinform S7: 001. http://dx.doi.org/10.4172/jpb.S7–001

16. Komander D et al (2009) Molecular discrimination of structurally equivalent Lys 63-linked and linear polyubiquitin chains. EMBO Rep 10(5):466–473
17. Peng J et al (2003) A proteomics approach to understanding protein ubiquitination. Nat Biotechnol 21(8):921–926
18. Zeng W et al (2010) Reconstitution of the RIG-I pathway reveals a signaling role of unanchored polyubiquitin chains in innate immunity. Cell 141(2):315–330
19. Morimoto D et al (2015) The unexpected role of polyubiquitin chains in the formation of fibrillar aggregates. Nat Commun 6:6116

Chapter 12

Fluorescent Tools for In Vivo Studies on the Ubiquitin-Proteasome System

Olli Matilainen, Sweta Jha, and Carina I. Holmberg

Abstract

The ubiquitin-proteasome system (UPS) plays a key role in maintaining proteostasis by degrading most of the cellular proteins. Traditionally, UPS activity is studied in vitro, in yeast, or in mammalian cell cultures by using short-lived GFP-based UPS reporters. Here, we present protocols for two fluorescent tools facilitating real-time imaging of UPS activity in living animals. We have generated transgenic *Caenorhabditis elegans* (*C. elegans*) expressing a photoconvertible UbG76V-Dendra2 UPS reporter, which permits measurement of reporter degradation by the proteasome independently of reporter protein synthesis, and a fluorescent polyubiquitin-binding reporter for detection of the endogenous pool of Lys48-linked polyubiquitinated proteasomal substrates. These reporter systems facilitate cell- and tissue-specific analysis of UPS activity especially in young adult animals, but can also be used for studies during development, aging, and for example stress conditions.

Key words Ubiquitin-proteasome system (UPS), UbG76V-Dendra2, Polyubiquitin-binding reporter, Photoconversion, Live imaging, Proteostasis, *C. elegans*

1 Introduction

The ubiquitin-proteasome system (UPS) is one of the main safeguards of protein homeostasis by degrading cellular proteins including short-lived regulators, unfolded and damaged proteins. In UPS, the substrate is polyubiquitinated through the actions of ubiquitin-activating (E1), -conjugating (E2), and -ligating (E3) enzymes. The polyubiquitinated substrate is then recognized and degraded by the proteasome. The proteasome is a large (over 2.5 megadaltons) multisubunit protein complex consisting of a barrel-shaped 20S core particle capped from one or both ends with 19S regulatory particles or alternative activators [1]. Dysfunctions of the UPS are associated with severe proteotoxic conditions such as age-related neurodegenerative diseases and some cancers [2]. In addition, changes in proteasomal degradation have been detected in aging organisms including humans and *C. elegans* [3–5].

Rune Matthiesen (ed.), *Proteostasis: Methods and Protocols*, Methods in Molecular Biology, vol. 1449,
DOI 10.1007/978-1-4939-3756-1_12, © Springer Science+Business Media New York 2016

Development of short-lived GFP-based UPS reporters for human cell culture studies has provided key tools for investigations on UPS-mediated proteolysis [6, 7]. More recently, the nematode *C. elegans* has started to be utilized as a multicellular model system to unravel functions of the UPS [5, 8, 9]. *C. elegans* is a widely used model organism in biomedical research and, due to its short life-span and conserved signaling pathways, it is also a popular model for aging studies. Moreover, the transparent body of *C. elegans* makes it highly suitable for live imaging. We have created two fluorescent tools to study UPS-mediated protein degradation in *C. elegans*: a photoconvertible UPS reporter and a fluorescent polyubiquitin-binding reporter [5, 10].

Our photoconvertible fluorescent UPS reporter (Fig. 1a) is based on the coral fluorescent Dendra2 protein, which can be irreversibly photoconverted from green-to-red fluorescent form by using 405 nm or 488 nm wavelength [11]. We have tagged Dendra2 with the uncleavable ubiquitin molecule UbG76V turning it into an ubiquitin fusion degradation (UFD) substrate [12, 13]. This N-terminal ubiquitin form cannot be removed by deubiquitinases and functions as an anchor for polyubiquitin chains, thereby targeting the complete fusion protein for degradation by the proteasome. UbG76V-Dendra2 is expressed under tissue-specific promoters in *C. elegans*, which enables UPS activity studies in different cell types and tissues (Fig. 1b). Photoconversion of UbG76V-Dendra2 from a green-to-red state enables quantification of proteasomal turnover rate for the subset of photoconverted reporter proteins, thus avoiding the effect of newly synthesized reporter proteins on the experimental outcome. In comparison, constitutively fluorescent GFP-based UPS reporters require expression of a second fluorescent protein (e.g., mRFP, mCherry), preferably in the same tissue of the animal, to distinguish whether the observed effect is due to changes in degradation rate or rate of protein synthesis.

The stability of ubiquitin-tagged UPS reporters is not only affected by proteasome activity, but also by the activity of upstream UPS components such as E3 ligase(s). We therefore used an alternative approach to design another UPS reporter for live imaging of the cellular pool of endogenous Lys48-linked polyubiquitinated proteins in *C. elegans*. This polyubiquitin-binding reporter has two ubiquitin-interacting motif (UIM) domains derived from the *C. elegans* proteasome subunit RPN-10 fused to the N-terminus of the commercially available ZsProSensor-1 fluorescent reporter (Fig. 1c) [10]. ZsProSensor-1 composes of the fluorescent ZsGreen protein couple in the C-terminus to the mouse ornithine decarboxylase (MODC) domain, leading to a short-lived fusion protein targeted for ubiquitin-independent degradation by the proteasome. Accordingly, expression of ZsProSensor-1 in *C. elegans* intestinal cells did not result in detectable reporter fluorescence. The UIM-domains capture Lys48-linked polyubiquitinated

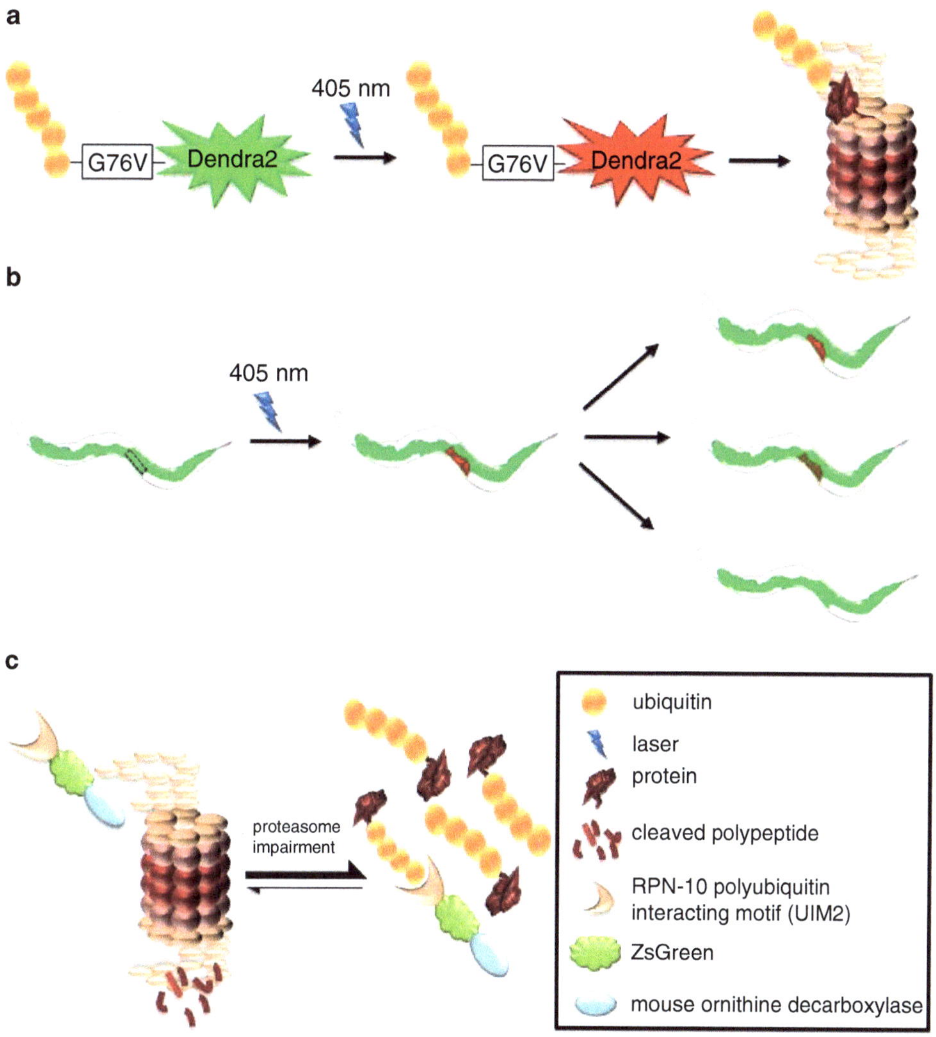

Fig. 1 In vivo UPS reporter systems. (**a**) UPS reporter for direct measurement of proteasome-mediated protein degradation. The UbG76V-Dendra2 UPS reporter can be irreversibly photoconverted from a green to a red fluorescent state by exposure to intense light (e.g., 405 nm). The UbG76V-Dendra2 is degraded by the proteasome in a polyubiquitin-dependent manner. (**b**) Animal model reporting on cell or tissue-specific UPS activity. Transgenic *C. elegans* expressing UbG76V-Dendra2 in for example intestinal cells can be exposed to single-cell photoconversion. Normal, enhanced or impaired proteasomal degradation of the photoconverted reporter can be measured independently of new reporter synthesis in a living animal. (**c**) Animal model for detection of the cellular pool of endogenous polyubiquitinated proteins. The polyubiquitin-binding reporter is stabilized upon binding to Lys48-linked polyubiquitinated substrates in the cell. The polyubiquitin-binding reporter is targeted for ubiquitin-independent proteasomal degradation via the mouse ornithine decarboxylase degradation domain (MODC domain) fused to the fluorescent ZsGreen protein. The reporter binds polyubiquitinated endogenous substrates via the ubiquitin-interacting motif (UIM) domains of the *C. elegans* proteasome subunit RPN-10. Impaired proteasomal degradation results in accumulation of polyubiquitinated proteins and increased reporter fluorescence

endogenous substrates in the cell, thus stabilizing the reporter and leading to fluorescent worms. As a readout, increased fluorescence can be interpreted as an accumulation of polyubiquitinated proteins due to impaired proteasome-mediated degradation. Polyubiquitin-binding domains have previously been used as tools for capturing the cellular pool of polyubiquitin chains from mammalian cell lysates [14] and for visualizing polyubiquitinated proteins in mammalian cells [15].

The experimental procedures and equipment requirements differ between the above described reporters, as UbG76V-Dendra2 reporter worms should preferably be individually imaged with confocal microscope, whereas the polyubiquitin reporter worms can be imaged with standard fluorescence microscope. This protocol description focuses on providing key points in the UbG76V-Dendra2 reporter animal analysis, as well as a brief description of imaging of the polyubiquitin reporter worms. By using these complementing in vivo UPS reporter systems, we have been able to start unravelling cell-type- and aging-specific changes in UPS activities in *C. elegans*, as well as identified tissue-specific regulatory mechanisms of the UPS [5, 10, 16].

2 Materials

2.1 Transgenic C. elegans

Transgenic *C. elegans* strains expressing UbG76V-Dendra2 or the polyubiquitin-binding reporter in body wall muscle cells, neurons, or intestinal cells [5, 10, 16].

2.2 Agarose (Fischer Scientific)

3–5 % melted agarose in H_2O.

2.3 Glass Slides and Cover Slips (Thermo Scientific)

1 mm × 26 mm × 76 mm (thickness, lenght, width) glass slides and 0.13–0.16 mm × 20 mm × 20 mm (thickness, lenght, width) cover slips.

2.4 Levamisole Hydrochloride (Sigma-Aldrich)

0.5–1 mM in M9 buffer (22 mM KH_2PO_4, 41 mM Na_2HPO_4, 8.5 mM NaCl, and 19 mM NH_4Cl). If worms are not paralyzed, the concentration can be carefully increased.

2.5 Microscopes

For UbG76V-Dendra2 *C. elegans* imaging, we preferably like to use motorized Zeiss Axio Observer Z1 inverted confocal microscope with LSM 5 Live line scanner and LSM AIM software Rel. 4.2. 518F immersion oil (Zeiss) is used with 63× objective. However, by optimizing photoconversion, also other confocal microscopes can be used. Live imaging of polyubiquitin-binding reporter strains can be performed with Zeiss Axioplan microscope or any other equivalent fluorescent microscope.

3 Methods

3.1 Agarose Pads

Use 3–5 % melted agarose to prepare thick agarose pads for worm mounting. Thick agarose pads can be prepared by placing spacers, e.g., two glass slides of 1 mm on top of each other, on each side of the sample slide to which ~400 μl melted agarose is added. A glass slide is then temporarily placed on the melted agarose and the spacers to generate a flattened 1 mm thick agarose pad. The top glass slide is then removed and the pad is placed in a humidified box for immediate use.

3.2 Transgenic *C. elegans* Maintenance

1. Transgenic *C. elegans* strains are grown under standard conditions [17] at 20 °C (*see* **Note 1**). Worms can be imaged at any time during their life. However, UbG76V-Dendra2 strains tend to loose fluorescence upon aging, and therefore it is recommended to image them as young adult worms when possible.

3.3 Live Imaging

1. For imaging, worms are mounted on agarose pads with a drop of 1 mM levamisole and topped with a cover slip. It is recommendable to use a 1 mm thick agarose pad, as it functions as a cushion, and the worm does not flatten when the cover slip is placed on top of it. The mounted worms should be used for immediate imaging.
2. At its unconverted state, Dendra2 has excitation and emission maxima at 490 and 507 nm, respectively, and at its photoconverted state, the excitation and emission maxima are at 553 and 573 nm, respectively. After the worm has been localized under the microscope, fast scanning with the GFP channel should be used to identify and focus the cell for photoconversion. Images are acquired with 63× 1.4 NA plan-apochromat objective and 518F immersion oil. Before the photoconversion, one scan should be taken and saved with both green and red channels to serve as a "before photoconversion" time point. For photoconversion we use 405 nm diode laser. Dendra2 can also be photoconverted, but less efficiently, with 488 nm laser. After selecting the target cell, photoconversions are done by using 25–50 iterations (scanning speed 1.6 μs per pixel) with 100 % laser output. In addition to single cells, whole tissues and worms can also be photoconverted and images acquired with 10× 0.45 NA plan-apochromat objective (*see* **Notes 2** and **3**).
3. After acquiring the image "after photoconversion," the worm should be removed from the pad to recover on NGM-agar plate before later time-point imaging. If the worm will be used in time-lapse imaging with short intervals, it can be kept on the agarose pad for 2–3 h. In this case, it is good to check that the worm does not dry out. This can be prevented by sealing the

cover slip with agarose. For measurements of UbG76V-Dendra2 proteasomal degradation in the dorsorectal ganglion, we have imaged the worm every 10 min for up to 3 h, while imaging of the UPS reporter in dopaminergic neurons, body wall muscle cells, and intestinal cells has been performed at 3, 6, 12, and 24 h after photoconversion due to slower degradation rates. For time-lapse imaging manual focusing is required at each time-point, as the worm may slightly twist or subside in the agarose during microscopy (*see* **Notes 4** and **5**).

4. We have used LSM AIM software Rel. 4.2 to analyze signal intensities. For analysis, we have quantified similar pixel amount at each image from different time points. To calculate the relative intensities, the absolute values of fluorescence before and right after photoconversion are set as 100% for the green and red fluorescence signals, respectively.
5. For analysis of polyubiquitin-binding reporter worms, the worms are immobilized with levamisole on similar agarose pads as UbG76V-Dendra2 worms, and imaged with Zeiss Axioplan microscope, or any other equivalent fluorescent microscope. Alternatively, worms can be immobilized with levamisole on a foodless agar growth plate, and imaged by using standard fluorescent stereomicroscope used for normal fluorescence worm maintenance. Fluorescent intensities can be quantified by using ImageJ.

4 Notes

1. Detailed information on worm maintenance and generation of transgenic animals are described in the online published WormBook at www.wormbook.org.
2. Depending on the cell-type of expression, the short-lived UbG76V-Dendra2 may not be visible in all cells of the same tissue in the worm. Especially extrachromosomal arrays give mosaic expression, which can be avoided by transgene integration. For example, an extrachromosomal array with intestinal expression does not usually show UbG76V-Dendra2 fluorescence in every intestinal cell. In addition, in worms with stronger UPS activity than in the wild-type (N2) background, such as the long-lived *daf-2(e1370)* mutants, it may be difficult to find young adults with fluorescent intestinal cells. Worms exhibiting mosaic reporter expression can also be taken advantage of by investigating if variation in reporter expression levels, as reflected by differences in fluorescence intensity, affects its degradation rate. For photoconversion, it is recommended to pick worms with fluorescence in at least two cells, because depending on the orientation of the worm, the other cell may

not come into clear focus. This is affected by the position of the gonad: if the gonad is above the cell being photoconverted, it hampers focusing. It is very important to keep the same lateral orientation of the worm during photoconversion and imaging at the later timepoint(s), i.e., when the worm previously exposed to photoconverison is placed back on the agarose pad after a recovery period in the normal growth plate. If the photoconverted cell resides on the other side of the worm compared to it its position during photoconversion, it will reduce the fluorescence intensity due to the lack of focus. The lateral orientation is convenient to check, e.g., from the position of the vulva (left vs. right) in relation to head of the worm.

3. At the start of photoconversion, when searching for the worm on the pad under the confocal microscope, it is recommended to avoid using the fluorescence light source and use only transmitted light. Especially with high magnification objectives (40× and 63×) the fluorescent light easily photoconverts the whole worm. Scanning speed and number of iterations for the photoconversion step should be determined separately for each microscope. In our Dendra2 experiments, we have used Zeiss LSM 5 Duo confocal microscope and its fast line-scanner LSM 5 Live. We have noticed that confocal microscopes with slow scanning speed are not suitable for the photoconversion experiments described here.
4. It is important to notice that lasers on confocal microscopes can heat up when they are kept on for several hours. Laser heating increases the signal intensity and can affect the results. For example, it is common that the imaging lasers produce a stronger signal in the afternoon compared to the signal right after the photoconversion in the morning, if the lasers have been on the whole day. This problem can be avoided by keeping the lasers on for a while before starting the photoconversion.
5. It is always important to choose healthy looking age-synchronized worms for experiments.

Acknowledgement

We thank the *Caenorhabditis* Genetics Center, which is funded by NIH Office of Research Infrastructure Programs (P40 OD010440), for strains and the Biomedicum Imaging Unit, University of Helsinki, for imaging services. This work was supported by grants to C.I.H. from the Academy of Finland (131408 and 259797), University of Helsinki Funds, International Human Frontier Science Program Foundation, and the Sigrid Juselius Foundation. The authors would like to acknowledge networking support by the Proteostasis COST Action (BM1307).

References

1. Finley D (2009) Recognition and processing of ubiquitin-protein conjugates by the proteasome. Annu Rev Biochem 78:477–513
2. Dahlmann B (2007) Role of proteasomes in disease. BMC Biochem 8(Suppl 1):S3
3. Chondrogianni N, Gonos ES (2005) Proteasome dysfunction in mammalian aging: steps and factors involved. Exp Gerontol 40:931–938
4. Chondrogianni N, Voutetakis K, Kapetanou M, Delitsikou V, Papaevgeniou N, Sakellari M, Lefaki M, Filippopoulou K, Gonos ES (2014) Proteasome activation: an innovative promising approach for delaying aging and retarding age-related diseases. Ageing Res Rev 23:37–55
5. Hamer G, Matilainen O, Holmberg CI (2010) A photoconvertible reporter of the ubiquitin-proteasome system in vivo. Nat Methods 7:473–478
6. Dantuma NP, Lindsten K, Glas R, Jellne M, Masucci MG (2000) Short-lived green fluorescent proteins for quantifying ubiquitin/proteasome-dependent proteolysis in living cells. Nat Biotechnol 18:538–543
7. Bence NF, Sampat RM, Kopito RR (2001) Impairment of the ubiquitin-proteasome system by protein aggregation. Science 292:1552–1555
8. Liu G, Rogers J, Murphy CT, Rongo C (2011) EGF signalling activates the ubiquitin proteasome system to modulate C. elegans lifespan. EMBO J 30:2990–3003
9. Segref A, Torres S, Hoppe T (2011) A screenable in vivo assay to study proteostasis networks in caenorhabditis elegans. Genetics 187:1235–1240
10. Matilainen O, Arpalahti L, Rantanen V, Hautaniemi S, Holmberg CI (2013) Insulin/IGF-1 signaling regulates proteasome activity through the deubiquitinating enzyme UBH-4. Cell Rep 3:1980–1995
11. Chudakov DM, Lukyanov S, Lukyanov KA (2007) Tracking intracellular protein movements using photoswitchable fluorescent proteins PS-CFP2 and Dendra2. Nat Protoc 2:2024–2032
12. Johnson ES, Bartel B, Seufert W, Varshavsky A (1992) Ubiquitin as a degradation signal. EMBO J 11:497–505
13. Johnson ES, Ma PC, Ota IM, Varshavsky A (1995) A proteolytic pathway that recognizes ubiquitin as a degradation signal. J Biol Chem 270:17442–17456
14. Bennett EJ, Shaler TA, Woodman B, Ryu KY, Zaitseva TS, Becker CH, Bates GP, Schulman H, Kopito RR (2007) Global changes to the ubiquitin system in huntington's disease. Nature 448:704–708
15. Sims JJ, Scavone F, Cooper EM, Kane LA, Youle RJ, Boeke JD, Cohen RE (2012) Polyubiquitin-sensor proteins reveal localization and linkage-type dependence of cellular ubiquitin signaling. Nat Methods 9:303–309
16. Li X, Matilainen O, Jin C, Glover-Cutter KM, Holmberg CI, Blackwell TK (2011) Specific SKN-1/nrf stress responses to perturbations in translation elongation and proteasome activity. PLoS Genet 7:e1002119
17. Brenner S (1974) The genetics of caenorhabditis elegans. Genetics 77:71–94

Chapter 13

Bimolecular Fluorescence Complementation to Assay the Interactions of Ubiquitylation Enzymes in Living Yeast Cells

Ewa Blaszczak, Claude Prigent, and Gwenaël Rabut

Abstract

Ubiquitylation is a versatile posttranslational protein modification catalyzed through the concerted action of ubiquitin-conjugating enzymes (E2s) and ubiquitin ligases (E3s). These enzymes form transient complexes with each other and their modification substrates and determine the nature of the ubiquitin signals attached to their substrates. One challenge in the field of protein ubiquitylation is thus to identify the E2–E3 pairs that function in the cell. In this chapter, we describe the use of bimolecular fluorescence complementation to assay E2–E3 interactions in living cells, using budding yeast as a model organism.

Key words Ubiquitin, Ubiquitin-conjugating enzyme, Ubiquitin ligase, Protein–protein interactions, Protein-fragment complementation assay, BiFC, Living cell, *Saccharomyces cerevisiae*, Microscopy, Linear unmixing

1 Introduction

Conjugation of the small protein ubiquitin to other cellular proteins, a process termed ubiquitylation, regulates the homeostasis and activity of thousands of proteins in eukaryotic cells [1–3]. It is achieved through a hierarchical network of enzymes that comprises ~30 ubiquitin-conjugating enzymes (E2s) and more than 600 known or putative ubiquitin ligases (E3s) in human cells [4, 5]. In this network, E2s carry activated ubiquitin, while E3s allow the transfer of ubiquitin from E2s to substrate proteins. E2s and E3s can also conjugate ubiquitin to ubiquitin moieties already attached to substrate proteins, which leads to the assembly of polymeric ubiquitin chains. In ubiquitin chains, any of the seven lysine residues of ubiquitin or its N-terminus can be modified by a subsequent ubiquitin. Substrate proteins can thus be modified by mono-ubiquitin moieties or by various types of poly-ubiquitin chains that can be complex and contain heterogeneous

Rune Matthiesen (ed.), *Proteostasis: Methods and Protocols*, Methods in Molecular Biology, vol. 1449,
DOI 10.1007/978-1-4939-3756-1_13, © Springer Science+Business Media New York 2016

ubiquitin-ubiquitin linkages [6]. It is now well established that the nature of the ubiquitin modification attached to a substrate protein encode distinct molecular signals that trigger different responses in the cell. Deciphering how this ubiquitin code is written by E2s and E3s and interpreted by the cell machinery is thus a central question in the field [7].

Structural and biochemical studies have revealed many details on the interaction and catalytic mechanism of individual E2s and E3s, but an important challenge is to understand how these enzymes operate at a network level in living cells. For instance, when investigating the activity of a given E3, it is critical to exhaustively describe the range of E2s that can function with this E3. This is not easily done, since we are currently not able to accurately predict which E2s and E3s can interact with each other and conventional biochemical methods such as immunoprecipitation often do not succeed to capture E2–E3 interactions due to their low affinity. Yeast two-hybrid approaches are able to detect weak interactions and have been used with some success to systematically assay the human E2–E3 interactome [8, 9]. However, these screens did not identify E2 partners for numerous E3s, which may in part be due to the fact that many E3s function as heterodimers or as large protein complexes that are not reconstituted in a yeast two-hybrid assay. For instance, E2 partners of the human BRCA1-BARD1 heterodimeric E3 complex could only be identified by yeast two-hybrid when using a bait construct consisting of the catalytic domains of BRCA1 and BARD1 fused in a single polypeptide that folds into a correct E3 structure [10]. To overcome this limitation, we recently introduced the use of bimolecular fluorescence complementation (BiFC) as a mean to assay E2–E3 interactions in their native cellular context [11]. BiFC is a protein-fragment complementation assay where two proteins of interest, here an E2 and an E3, are fused to complementary N- and C-terminal fragments of a fluorescent protein reporter (reviewed in [12–16]). Upon E2–E3 interaction, the fragments of the fluorescent protein are brought into close proximity, allowing them to fold and to reconstitute an active fluorescent protein, which can then be detected using fluorescence microscopy (Fig. 1).

In this chapter, we describe critical aspects on the design of BiFC experiments and present imaging conditions and image processing steps for sensitive detection and quantification of BiFC complex formation in budding yeast (protocols describing how to implement BiFC experiment in other model organisms have been described elsewhere, *see* for instance [17–19] and **Note 1**). The sensitivity of fluorescence microscopy experiments in yeast is limited by the background fluorescence (autofluorescence) of the cells that hinders the detection of weak fluorescence signals of interest. This is particularly an issue in BiFC experiments as only a fraction of the fusion proteins form BiFC complexes. The fluorescence

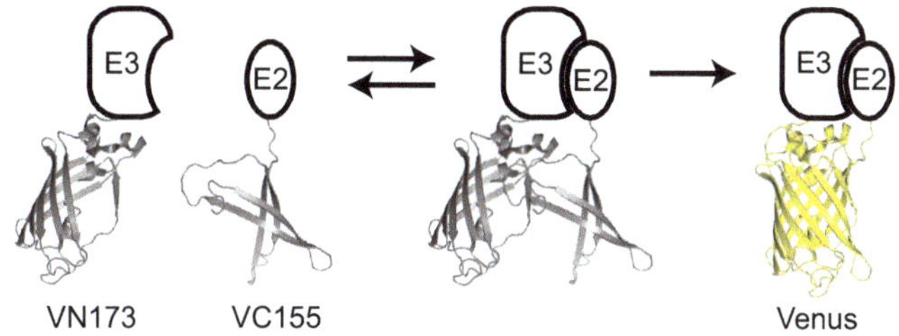

Fig. 1 *Principle of BiFC to image E2–E3 interactions.* E2s and E3s of interest are tagged with complementary fragments of a fluorescent protein (e.g., VN173 and VC155). Upon E2–E3 interaction, the fragments are brought in close proximity which allows irreversible reconstitution of the fluorescent protein (e.g. Venus)

intensities produced by BiFC complexes are thus typically less than 10% of the fluorescence intensity that would be produced by the corresponding proteins fused to an intact fluorescent protein [14]. In the method section, we therefore first describe how to cultivate yeast cells to minimize cell autofluorescence and how to setup imaging conditions to lower its contribution in the images. We then describe an image processing workflow to digitally subtract autofluorescence from BiFC images and quantify BiFC signals in single cells (Fig. 2). Overall this method enables sensitive visualization and quantification of E2–E3 interactions in budding yeast.

1.1 Critical Considerations and Design of BiFC Experiments in Yeast

1.1.1 Advantages and Limitations of BiFC

Excellent reviews have described in details the characteristics, advantages, and limitations of BiFC [12–16]. In addition to the ease with which it can be implemented, an important asset of BiFC over other methods used to monitor protein–protein interactions (PPIs) in living cells is its ability to detect very weak PPIs, with dissociation constants up to 1 mM [20, 21]. BiFC is thus perfectly suited to reveal E2–E3 interactions that have dissociation constants in the micromolar range [4]. This ability of BiFC to detect weak interactions originates from the fact that the reconstitution of a fluorescent protein from its complementary fragments is essentially irreversible (Fig. 1). This property has been documented in vitro and in vivo with several fluorescent proteins, including the widely used variant of the yellow fluorescent protein Venus (see [22] and references therein). BiFC thus acts as a trap that captures PPIs. Inevitably, it can also capture nonspecific protein–protein collisions that occur randomly in the cell, leading to false positive fluorescence. This caveat of BiFC is particularly problematical when proteins are highly expressed or locally concentrated as this leads to higher collision frequencies [23]. BiFC is therefore a valuable method to investigate E2–E3 interactions in the context of living cells, but adequate controls (see below) and independent assays are required to demonstrate that the detected interactions are indeed specific and biologically meaningful.

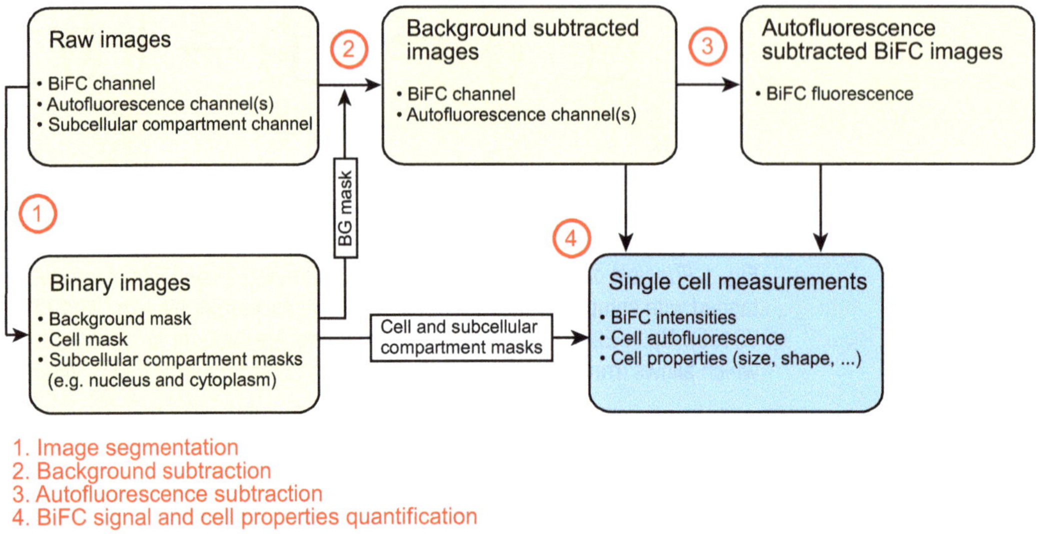

Fig. 2 *Scheme of the image processing workflow*. The image processing procedure described in this chapter involves four steps: Image segmentation (1), background subtraction (2), autofluorescence subtraction (3), and BiFC signal and cell property quantification (4). (1) Image segmentation produces binary images (*see* Fig. 3) that are then used in the subsequent image processing steps and for fluorescence quantification. (2) Background subtraction is performed using background mask produced during image segmentation. This step is required to be able to perform quantitative measurements of BiFC signals. (3) Autofluorescence subtraction reduces the contribution of autofluorescence in BiFC channel images, which improves the quality of the BiFC images (*see* Fig. 4) and the quantification of BiFC signals. (4) BiFC signal is quantified in single cells using the autofluorescence subtracted BiFC channel image and subcellular compartment masks. The quantification of other cell properties can also improve the analysis of BiFC signals. For instance, quantifying cell autofluorescence is useful to eliminate dead (strongly autofluorescent) cells, while measuring cell size can enable to distinguish bud and mother cells

Another limitation of BiFC is the slow maturation of fluorescent proteins. In budding yeast, the half-life of Venus maturation has been estimated to be ~15 min [24]. This creates a delay between the time when the fusion proteins interact with each other and the time when the complex actually becomes fluorescent. Both the delay and the irreversible nature of fluorescent protein reconstitution limit the use of BiFC to investigate temporal changes in E2–E3 interactions. Since interactions are not observed in real time, care should also be taken in the interpretation of fluorescence localizations. What is observed in BiFC images is the localization of trapped BiFC complexes, which may not always correspond to the site where the interaction of the two proteins takes place.

1.1.2 Choice of Fluorescent Protein Fragments

Numerous fluorescent proteins have been used in BiFC assays (reviewed in [15, 16]). In yeast as in other organisms, the Venus fluorescent protein is most widely used because its fragments produce the highest level of BiFC fluorescence [25]. It is commonly split at residues 173 and 155 to produce overlapping N-terminal and C-terminal fragments (VN173 and VC155, respectively), yet other efficient fragment combinations have been described (see for instance [26, 27]). As aforementioned, Venus fragments are prone to self-assembly. Multiple attempts have been made to improve the specificity of Venus-based BiFC [22, 27–30], but many of the proposed solutions also reduce the intensity of specific BiFC fluorescence and have not been tested in yeast. As long as optimized fragments have not been clearly established in yeast, we suggest using the VN173 and VC155 fragments for which most tools are currently available. These tools notably include plasmids for one-step PCR-mediated fusion of endogenous genes with VN173 or VC155 [31] (these plasmids are available from the EUROSCARF, http://web.uni-frankfurt.de/fb15/mikro/euroscarf/data/Huh.html), but also a collection of 5809 VN173-tagged yeast strains that comprises most yeast E2s and E3s [32] (these strains are commercially available as single strains or as the whole collection from the Korean Biotech company Bioneer, http://eng.bioneer.com/products/YeastGenome/VN-FusionLibrary-overview.aspx). Note that it is possible to introduce the A206K mutation in VC155 to prevent dimerization of the reconstituted Venus protein [26].

1.1.3 Construction of Yeast Strains for BiFC Experiments

The design of the fusion proteins is an essential step in BiFC experiments. Clearly, the localization and interaction of the two protein partners should not be impaired by the fluorescent protein fragments. In addition, the fluorescent proteins fragments should be positioned in such a way that, upon interaction of the two partners, they can meet and reconstitute the reporter fluorescent protein. These criteria are often tested empirically by fusing the fragments to either end of the investigated proteins. Since many E3s are large multi-domain proteins, we suggest tagging them first at the end which is the closest to their catalytic domain (i.e. the C-terminus for most E3s). This should help to position the fluorescent protein fragment in proximity to any potential interacting E2. Note that tagging E3s may impair their catalytic activity without necessarily disturbing E2 interactions. For instance, C-terminal tags inactivate HECT E3s [33, 34] because they impair the positioning of residues of the E3 C-terminal tail that are involved in catalysis [35] but that do not participate in E2 recruitment [36, 37]. Yeast E2s are small proteins and may successfully be tagged at either end, with the exception of Ubc6 and Ubc7 that have to be tagged N-terminally (Ubc6 C-terminus contains a transmembrane domain and faces the lumen of the endoplasmic reticulum [38], while Ubc7 C-terminus is involved in the interaction with its partner Cue1 [39]).

While performing BiFC experiments in yeast, it is best to replace the endogenous genes with their tagged versions. This ensures that the tagged proteins are expressed at physiological concentrations and that there is no competition between the tagged and untagged proteins. Yet, we observed that several E2s endogenously tagged with VC155 are significantly less expressed than the wild-type proteins (unpublished results). We have not examined the reason for this, but it may partly be due to poor folding of the VC155 fragment [14].

BiFC assays in yeast are particularly well suited for large scale analysis of PPIs. It is therefore advantageous to construct the E2 and E3 tagged strains in a genetic background compatible with high-throughput yeast manipulation. The strains we use carry the can1::STE2pr-spHIS5 and lyp1::STE3pr-HPH markers to allow automatic strain crossing and selection of the haploid progeny of either MATa or MATalpha mating type [11]. Protocols for high-throughput yeast manipulation have been described in details elsewhere [40]. In addition, we recommend including in the constructed strains a marker of a subcellular compartment fused to a red fluorescent protein (e.g., we used Rpn7-tDimer2 as a nuclear marker [11]). This enables to get precise information on the possible subcellular localization of the interaction, but also helps to achieve robust and sensitive measurements of BiFC fluorescence intensities.

1.1.4 Negative Controls

One of the challenges while performing BiFC experiments is the identification of appropriate negative controls to distinguish *bona fide* interactions from nonspecific self-assembly of the fluorescent protein fragments. Ideally, one should replace one of the two binding partners with a version mutated in its interaction surface [15]. The mutant protein should be fused to the fluorescent protein fragment in the same way as the wild-type protein and should display the same expression level and subcellular localization. Designing such mutants of E2s or E3s may not always be straightforward. In some instances, E3s engage multiple contacts with their E2s that involve not only the E3 catalytic domain, but also another region of the E3 or an auxiliary subunit [41]. In the case where E2 or E3 interaction mutants cannot be easily designed, it is possible to perform competition experiments by overexpressing an untagged version of one of the binding partners [15]. Importantly, the use of fluorescent protein fragments unfused, or fused to an irrelevant protein, is not a suitable negative control because the efficiency of non-specific self-assembly of fluorescent protein fragments is influenced by the nature of the proteins they are fused to [42]. In addition, such constructs are unlikely to be expressed at the same level and to have the same subcellular localization as the original fusion protein.

Importantly, the biological significance of specific PPIs identified by BiFC should be established using fully independent assays. BiFC may reveal indirect or enzyme-substrate interactions (see for instance [43]). Furthermore, some E2s and E3s can interact *via* their catalytic domains without triggering ubiquitylation (see for instance [10, 44]). E2–E3 interactions revealed by BiFC therefore need to be carefully characterized by independent in vivo and in vitro experiments to determine their nature and functional relevance.

2 Materials

2.1 Yeast Cultures

1. YPD plates: 1% (w/v) yeast extract, 2% (w/v) peptone, 2% (w/v) dextrose, 2% (w/v) agar in dH_2O. Autoclave at 121 °C for 15 min. Cool down to 55 °C before pouring the plates.
2. YPD + Ade medium: 1% (w/v) yeast extract, 2% (w/v) peptone, 2% (w/v) dextrose, 20 mg/L adenine hemisulfate in dH_2O. Autoclave at 121 °C for 15 min.
3. Sterile 14 mL round-bottom culture tubes or sterile U-shaped 2 mL 96-deepwell plates (to be sealed using sterile air-permeable sealing films for cell culture).

2.2 Microscopy

1. 10× Low-fluorescence nitrogen base: 5 g $(NH_4)_2SO_4$, 1 g KH_2PO_4, 0.5 g $MgSO_4$, 0.1 g NaCl, 0.1 g Ca_2Cl, 0.5 mg H_3BO_4, 0.04 mg $CuSO_4$, 0.1 mg KI, 0.2 mg $FeCl_3$, 0.4 mg $MnSO_4$, 0.2 mg Na_2MoO_4, 0.4 mg $ZnSO_4$, 2 μg biotin, 0.4 mg calcium pantothenate, 2 mg inositol, 0.4 mg niacin, 0.2 mg PABA, 0.4 mg pyridoxine HCl, 0.4 mg thiamine in 100 mL dH_2O. Autoclave at 121 °C.
2. 10× Amino acids: 20 mg Adenine hemisulfate, 20 mg Uracil, 20 mg L-Histidine HCl, 30 mg L-lysine HCl, 60 mg L-leucine, 20 mg L-methionine, 20 mg L-tryptophan in 100 mL dH_2O. Filter sterilize.
3. Low-fluorescence medium (LFM): 2 g Dextrose, 10 mL 10× low-fluorescence nitrogen base, 10 mL 10× amino acids in 100 mL dH_2O. Filter sterilize.
4. 8-Well coverglass imaging chambers (e.g., Nunc™ Lab-Tek™ II chambers, Thermo Scientific, USA) or 96-well coverglass imaging plates (e.g., Imaging Plates CG, ZellKontakt GmbH, Germany).
5. Inverted epifluorescence or confocal microscope equipped with suitable filters and objectives (*see* **Note 2** and Subheading 3). For high-throughput BiFC experiments, the microscope should be equipped with a XYZ motorized stage and a 96-well plate holder.

2.3 Image Processing

1. Image processing software (e.g., ImageJ, Fiji or CellProfiler).

3 Methods

3.1 Cell Preparation for Microscopy

1. *Day 1*: Inoculate YPD agar plates with the yeast strains of interest and incubate them overnight at 30 °C. Include positive, negative and no-BiFC control strains (*see* **Note 3**).
2. *Day 2, morning*: Inoculate 1 mL liquid YPD + Ade cultures at an OD_{600} of ~0.2 using the freshly grown cells (*see* **Note 4**).

Depending on the number of strains to analyze, the cultures can be grown in individual sterile 14 mL round-bottom tubes or U-shaped 2 mL 96-deepwell plates sealed with an air-permeable sealing. Cultivate under constant agitation at 25 °C in a shaking incubator (*see* **Note 5**).

3. *Day 2, evening*: Use 100 μL of each culture to measure their OD_{600} and dilute them to an OD_{600} of 0.001–0.005 in 1 mL YPD + Ade (*see* **Note 6**). Cultivate overnight under constant agitation at 25 °C in a shaking incubator.
4. *Day 3, morning*: Harvest the cells from the overnight cultures by centrifugation at 3000 × *g* for 3 min and resuspend them in 300 μL of liquid LFM medium prewarmed to 25 °C. Use 100 μL to measure the OD_{600} and use the rest of the cells to inoculate 0.5 mL LFM cultures at an OD_{600} of 0.3 in individual tubes or 96-deepwell plates. Incubate the cultures under agitation for at least 3 h at 25 °C (*see* **Note 7**).
5. *Day 3, afternoon*: Microscopy can be performed in 8-well coverglass chambers or 96 well coverglass plates, depending on the number of strains to analyze. For 8-well chambers, place 200 μL of each culture in the wells and then add in each well 300 μL of LFM medium prewarmed to 25 °C. For 96-well plates, place 80 μL of each culture in the wells and then add in each well 120 μL of prewarmed LFM medium. Let the cells settle to the bottom of the wells for 30 min before proceeding with imaging (*see* **Note 8**).

3.2 Image Acquisition

1. *Objective lens*: The choice of the objective is critical. To maximize the amount of fluorescence collected from the cells and obtain a good horizontal resolution, choose a high numerical aperture (NA) and high magnification objective. Avoid objectives designed for phase contrast and remove differential interference contrast phase plates and prisms from the optical path because they would significantly reduce transmission. Objectives with a correction collar are convenient to correct small variations in cover glass thickness and achieve maximum image quality. Note that when imaging yeast strains in 96-well plates it is more convenient to use water or glycerol rather than oil as the immersion medium. We use a Leica HC PL APO 63×/1.20 W motCORR CS2 objective.
2. *BiFC channel*: This channel collects the light emitted by BiFC complexes but also by cell autofluorescence. To lower the contribution of autofluorescence, design image acquisition settings that maximize the ratio of the light collected from BiFC fluorescence over autofluorescence. This is typically achieved using a narrow bandpass filter around the emission peak of the fluorescent protein. Excitation should also be performed using a narrow passband at the excitation peak of the fluorescent

protein. We typically use a 514 nm excitation laser and a 525–538 nm bandpass emission filter to image Venus BiFC (Venus excitation and emission peaks are 515 and 528 nm, respectively). The image acquisition settings need to be optimized for each microscope using positive and no-BiFC control strains (*see* **Note 3**).

3. *Autofluorescence channel(s)*: To be able to digitally subtract autofluorescence from BiFC channel images, it is necessary to record independent images of the cell autofluorescence. The image acquisition settings for those images should be designed to maximize the ratio of the light collected from autofluorescence over BiFC fluorescence (*see* **Note 9**). Excitation may be performed using a passband away from the excitation peak of the fluorescent protein. In addition, to achieve accurate autofluorescence subtraction, it can be beneficial to define several autofluorescence channels. In our experiments, we typically use two autofluorescence channels, acquired with 458 and 514 nm excitation lasers and 500–540 and 480–505 nm bandpass emission filters, respectively (*see* Fig. 4). Importantly, the primary autofluorescence channel will be used for segmentation of the cells (see below and Fig. 3). It should have a good signal-to-noise ratio and should enable to clearly recognize the contour of the cells.
4. *Subcellular compartment channel*: We recommend acquiring images of a subcellular compartment of interest stained with a protein marker fused to a red fluorescent protein. For instance, we routinely use Rpn7-tDimer2 as a nuclear marker (Fig. 3). We acquire these images simultaneously with the BiFC channel images using a 561 nm excitation laser and a 580–630 nm bandpass emission filter. These images need to have a sufficiently good signal to noise ratio to enable segmentation (see below).
5. *Confocal-specific settings:* Pixel size and pinhole diameter need to be carefully adjusted as these parameters strongly influences the quality of the images. Larger pixels yield brighter images with better signal-to-noise ratios. For sensitive quantification of weak BiFC signals, it is therefore beneficial to increase pixel size, even if this is at the cost of a reduced spatial resolution. We routinely use 0.25 mm wide pixels. Similarly, opening the pinhole allows more light to reach the photodetector and yields brighter images.

3.3 Image Processing and BiFC Signal Quantification

Image processing is used to digitally subtract autofluorescence from BiFC channel images and to produce quantitative BiFC measurements in single cells. The workflow of the image processing steps is schematized in Fig. 2. It can be automatized using macros or plug-ins in ImageJ and Fiji (*see* **Note 10**), or pipelines in CellProfiler.

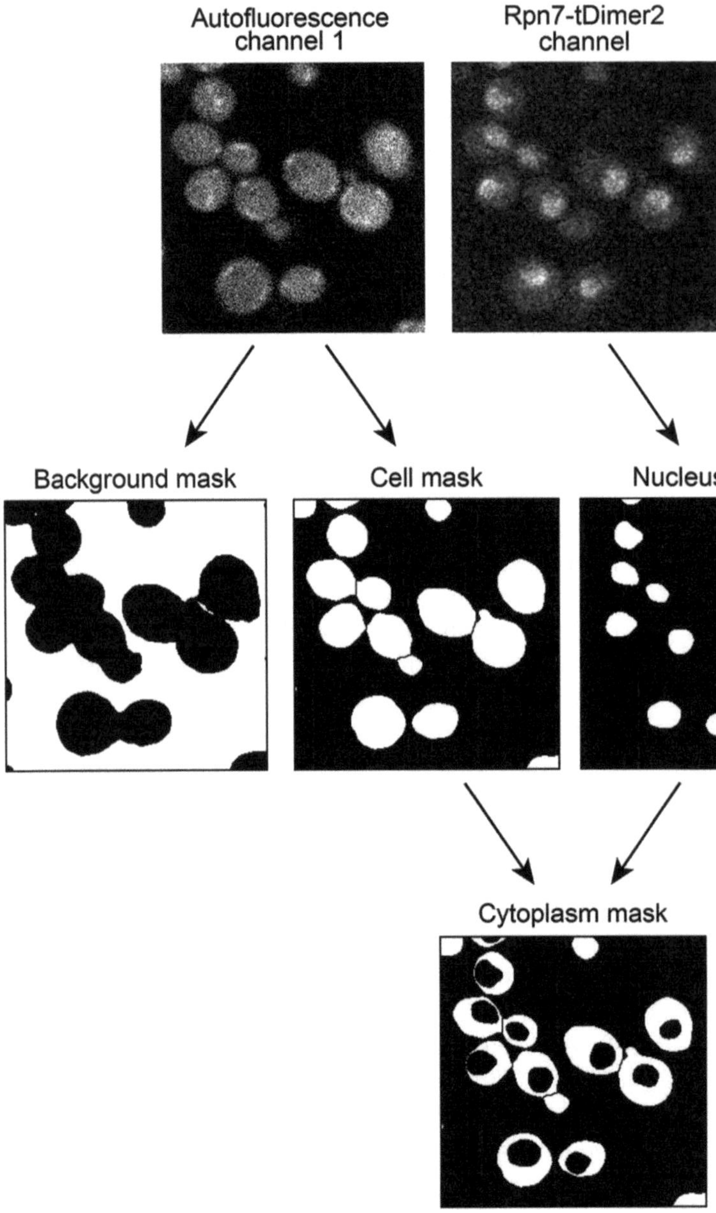

Fig. 3 *Image segmentation*. The image segmentation procedure described in this chapter uses the raw images acquired with the primary autofluorescence channel and with the subcellular compartment channel to produce four types of binary images. The autofluorescence image is first processed to select either the lower intensity pixels, which produces a binary image of the background pixels (background mask), or the higher-intensity pixels, which produces a binary image of the cell pixels (cell mask). Similarly, the subcellular compartment image (here Rpn7-tDimer2) is processed to select the higher-intensity pixels, which produces a binary image of the subcellular compartment pixels (e.g., nucleus mask). Combining this image with the cell mask enables to produce a binary image of the rest of the cell (e.g., cytoplasm mask)

Fig. 4 (continued) were cultivated and imaged as indicated in Subheading 3. The background subtracted BiFC and autofluorescence channel images are shown in the *left panel* and the BiFC images produced after autofluorescence subtraction are shown in the *right panel*. These images were then further processed using the PureDenoise Plugin for ImageJ [52] to reduce pixel noise and improve BiFC signal visualization. The interaction between VC155-Ubc6 and Asi3-VN173 (*top row*) produces a BiFC signal at the nuclear rim that can be easily detected in the background-subtracted image and that is improved after autofluorescence subtraction. In contrast, the interaction between VC155-Ubc6 and Asi1-VN173 (*bottom row*) produces a BiFC signal that is barely detectable without autofluorescence subtraction

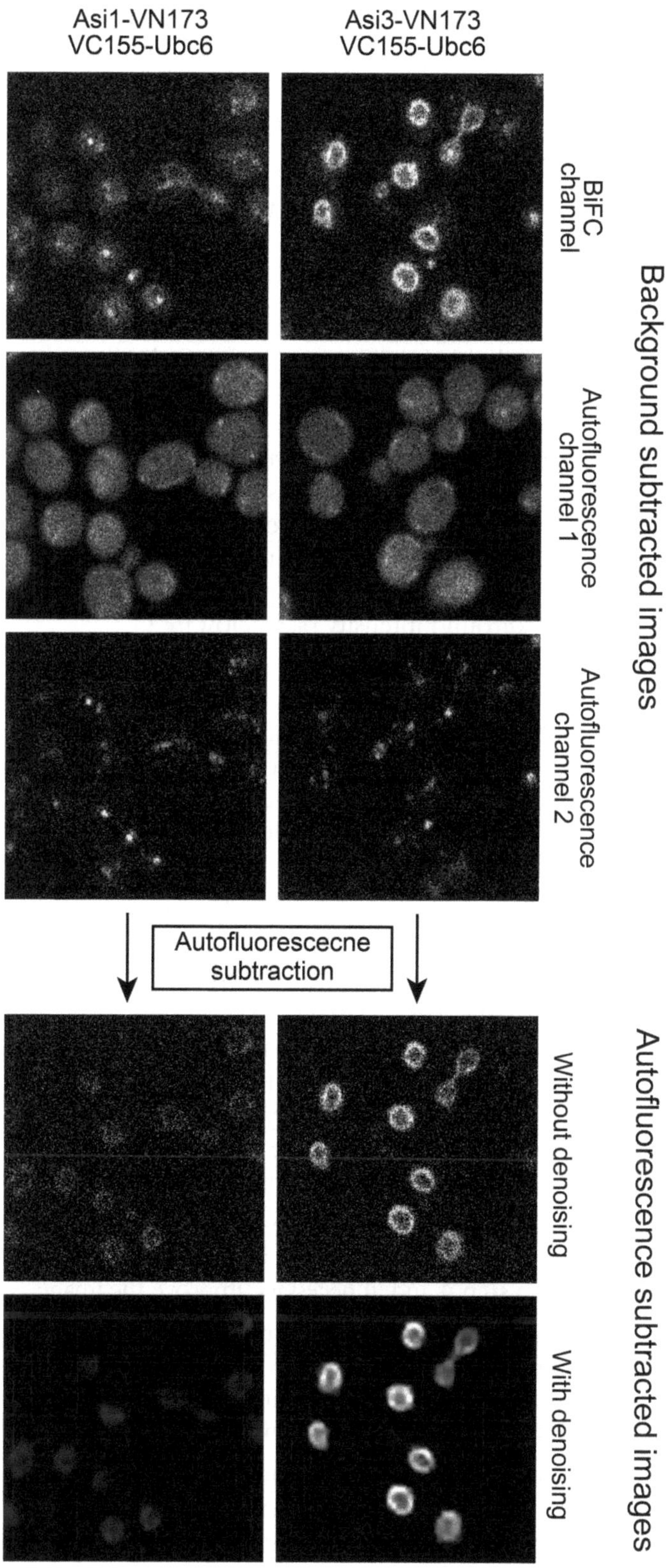

Fig. 4 *Autofluorescence subtraction improves the detection of weak BiFC signals*. This figure illustrates how autofluorescence subtraction enables to improve the quality of BiFC images and the detection of weak BiFC signals. Haploid yeast cells expressing the E2 Ubc6 tagged with VC (VC155-Ubc6) and the inner nuclear membrane localized E3s Asi3 or Asi1 tagged with VN (Asi3-VN173, *top row*, and Asi1-VN173, *bottom row*) from their endogenous chromosomal loci

3.3.1 Image Segmentation

The first step in image processing is segmentation. The procedure described here produces 4 binary images (*see* Fig. 3) that are then used to select the pixels to include in fluorescence measurements.

1. Open the image acquired with the primary autofluorescence channel.
2. Apply a spatial filter to remove pixel noise and small objects in this image (*see* **Note 11**). Duplicate the filtered image.
3. Threshold the filtered image: Set the lower threshold to the minimum pixel intensity of the image and adjust the upper threshold value to produce a binary image corresponding to background regions of the image field (*see* **Note 12**). The selected threshold value should be sufficiently low to ensure that the selected pixels do not contain any fluorescence from cell edges. Divide the resulting image with 255 (*see* **Note 13**). This step produces a binary mask that will be used to quantify background intensity (see below).
4. Threshold the duplicated filtered image: Set the upper threshold to the maximum pixel intensity of the image and adjust the lower threshold value to produce a binary image corresponding to the cells (*see* **Note 12**). The threshold value selected here should be higher than the threshold used in **step 3** and enable to nicely delineate the contour of individual cells.
5. *Optional step*: Improve the binary image produced in **step 4** by applying morphological operators. For instance, performing an erosion followed by dilation smooths objects and removes isolated pixels.
6. Apply a watershed transformation. This operation is essential to individualize cells that are touching each other and that could not be separated by thresholding. This step produces a binary image that will be used to identify individual cells for BiFC fluorescence quantification (see below).
7. Open the image acquired with the subcellular compartment channel. Apply a spatial filter as in **step 1**, threshold the filtered image as in **step 4** and, if necessary, improve the image as in **step 5**.
8. Divide the subcellular compartment binary image produced in **step** 7 by 255 (*see* **Note 13**). This produces the binary mask that will be used to quantify fluorescence signals in this compartment (e.g., the nucleus) (see below).
9. Invert the subcellular compartment binary image produced in **step** 7, multiply it with the binary image produced in **step 6** and divide the resulting image by 255 (*see* **Note 13**). This produces the binary mask that will be used to quantify fluorescence signals in the rest of the cell (e.g., the cytoplasm) (see below).

3.3.2 Background Subtraction

Background must be subtracted from BiFC and autofluorescence channel images before further processing (*see* **Note 14**).

1. Open the image acquired with the BiFC channel. Convert it to a 32-bit float image (*see* **Note 15**).
2. Multiply the image with the background binary mask produced in **step 3** of image segmentation. This produces an image where all pixel values are set to zero except background pixels.
3. Measure the integrated density (i.e., the sum of all pixel values) in this image and divide it with the integrated density of the binary mask. This operation calculates the average intensity of background pixels.
4. Subtract the average background intensity from all pixel values in the BiFC channel image.
5. Repeat **steps 1–4** with the image(s) acquired with the autofluorescence channel(s).

3.3.3 Autofluorescence Subtraction

This step aims to remove autofluorescence signals from the BiFC channel image, which significantly improves the detection and quantification of weak BiFC signals (Fig. 4). To this end, the autofluorescence channel images are rescaled and subtracted from the BiFC channel images. The identification of a correct rescaling factor for each autofluorescence channel is done empirically using images of no-BiFC control cells (*see* **Note 3**). Once such factors have been identified, they can be applied to all other images acquired in identical conditions.

1. Open the background subtracted BiFC channel and autofluorescence channel images of no-BiFC control cells.
2. Multiply each autofluorescence channel image by a separate rescaling factor. The initial value of each rescaling factor can be set arbitrarily, for instance to a value of 0.1.
3. Subtract each rescaled autofluorescence channel image from the BiFC channel image.
4. Examine the quality of the autofluorescence subtraction and repeat **steps 2** and **3** until a correct rescaling factor has been identified for each autofluorescence channel. The quality of the subtraction can be evaluated in several ways. Visual inspection of the subtracted image gives a qualitative impression of the efficiency of the subtraction and enables to readily identify a range of possible rescaling factors. To objectively fine tune each rescaling factor, it is possible to measure the mean and standard deviation of all pixel intensities in the subtracted image. A perfectly well-subtracted image should have a mean pixel intensity of zero. A positive mean indicates that a rescaling factor is too small, while a negative mean indicates that a rescaling factor is too large. In addition, the standard deviation

of pixel intensities in the entire image should be as low as possible and should be equal to the standard deviation of pixel intensities in background regions. Therefore, correct rescaling factors can be identified by minimizing both the absolute value of the mean and the standard deviation of all pixel intensities in the subtracted image.

5. Once each rescaling factor have been identified using images of no-BiFC control cells, they can be applied to subtract autofluorescence in other images acquired in identical conditions by repeating **steps 1–3** with these images.

3.3.4 BiFC Fluorescence and Cell Property Quantification

1. Open the subtracted BiFC fluorescence image.
2. Open the subcellular compartment mask produced in **step 8** of image segmentation. Multiply it with the subtracted BiFC fluorescence image. This produces an image where all pixel values are set to zero, except for pixels from the imaged subcellular compartment.
3. Open the mask produced in **step 9** of image segmentation. Multiply it with the subtracted BiFC fluorescence image. This produces an image where only the pixels corresponding to the imaged subcellular compartment are set to 0.
4. Open the binary image of the cells produced in **step 6** of image segmentation.
5. Perform a particle analysis in this image to define regions of interest (ROIs) corresponding to the cells that will be used in fluorescence quantification. Exclude cells that are touching the image edges or that are not circular. Set minimum and maximum pixel size areas to exclude too small cells and abnormally large cells or cell aggregates.
6. For each ROI, measure the integrated density in the image produced in **step 2** and divide it with the integrated density of the corresponding binary mask. This operation calculates the average BiFC fluorescence intensity in the subcellular compartment of each selected cell.
7. Repeat the operations described in **step 6** using the image produced in **step 3** and the corresponding binary mask. This calculates the BiFC fluorescence intensity in the rest of the cells.
8. Repeat **steps 1–7** with background subtracted images of the primary autofluorescence channel. This enables to identify cells that display an abnormally high or low fluorescence (e.g., as dead cells or out of focus cells, respectively) and to eliminate them in further analysis. It is also interesting to measure other cell properties such as size and shape parameters to be able to relate differences in BiFC intensities with different cell types.
9. To be able to compare BiFC fluorescence intensities measured in different experiments we recommend standardizing the

measured intensities such that BiFC signals measured in no-BiFC control cells have a mean of zero (which would be the case if background and autofluorescence subtraction were perfect) and a standard deviation of one. This operation is possible when a sufficient number of no-BiFC control cells are included in the analysis to precisely estimate these values.

10. Represent the standardized BiFC fluorescence intensities using a scientific graph plotting software.

4 Notes

1. To fully benefit from the capacity of BiFC to assay protein–protein interactions in near physiological conditions, we recommend performing BiFC experiments using cells that originate from the same organism as the investigated proteins. Still, it is possible to use BiFC in yeast to assay the interaction of heterologous proteins, as this is done for instance in yeast two hybrid assays. Plasmids that can be used to express heterologous proteins in yeast for BiFC experiments have for instance been described in [45].
2. BiFC images can be acquired with epifluorescence or confocal microscopes. In general, using a confocal microscope is not beneficial for yeast imaging, because there is no significant out-of-focus fluorescence [46]. Most protocols for live cell imaging of yeast therefore use epifluorescence (see for instance [47]). However, modern confocal microscope can be equipped with tunable band filters or spectral detectors. This offers a great flexibility in the selection of the emission passband and can be advantageous to define optimal BiFC imaging conditions and enable efficient autofluorescence subtraction. We currently perform our BiFC experiments using a Leica TCS SP8 confocal microscope.
3. Positive control strains are isogenic strains expressing fusion proteins known to produce a well-detectable BiFC signal (see for instance in [11]). They are used to verify the overall quality of the imaging procedure. no-BiFC control strains are isogenic strains that cannot produce any BiFC fluorescence, for instance strains that only express one of the two putative interaction partners. They are used to define parameters for autofluorescence subtraction, to verify its efficiency and to standardize the BiFC fluorescence measurements (*see* Subheading 3.3.4). no-BiFC control cells must be included in every BiFC experiment. They should not be confused with negative control strains that are isogenic strains designed to assay the specificity of PPIs detected by BiFC.
4. Many common laboratory strains (e.g., W303) are mutated in the *ADE2* gene. When grown in conventional YPD, these strains accumulate phosphoribosylaminoimidazole, an inter-

mediate in the adenine biosynthesis pathway, which is converted in the vacuole into a red pigment that strongly interferes with fluorescence microscopy. This can be minimized by supplementing the growth medium with 20 μg/mL extra adenine or by using ADE^+ strains (e.g., BY4741).

5. Protein folding and maturation of fluorescent proteins is temperature dependent [48]. Although Venus has been optimized for expression in mammalian cells at 37 °C, we observed that growing cells at 25 °C rather than 30 °C yields brighter BiFC fluorescence. Similar observations have been made with YFP [25, 49]. Performing the entire experiment at 20–25 °C also simplifies the imaging step since it is not necessary to use a microscope stage temperature controller.
6. To reduce yeast autofluorescence and avoid cell cycle synchronization it is best to keep cells actively growing (OD_{600} below 2) for several generations prior to imaging. To achieve this, overnight yeast cultures need to be inoculated at a low density so that they are not overgrown in the next morning. The exact OD_{600} at which the cultures are inoculated needs to be determined according to each strain generation time, which is ~2 h for wild-type haploid laboratory strains when cultivated at 25 °C.
7. Yeast imaging is performed in LFM medium [50], which does not contain riboflavin and folic acid and is therefore less autofluorescent than minimal media prepared from complete yeast nitrogen base (YPD is highly autofluorescent and must be avoided in fluorescence microscopy). Yeast should be cultivated in LFM medium a few hours prior to imaging.
8. In this protocol, yeast cells are imaged unattached, settled down on the glass cover slips. For best results, cells should be neither too scarce nor too dense. We suggest using a density of $\sim 2 \times 10^4$ cells per square millimeter, which usually corresponds to ~2 μL of cells at an OD_{600} of 0.5. 8-Well chambers and 96-well plates have well surfaces of ~70 and ~30 mm^2, respectively.
9. The image processing procedure for autofluorescence subtraction described in this chapter works well if the autofluorescence channel images contain minimal bleed-through from BiFC fluorescence. If this is not the case, it is possible to perform a more sophisticated linear unmixing procedure which enables to separate and quantify overlapping fluorescence signals [51].
10. An example macro showing how the image processing steps described in this section can be automatized in ImageJ is available at https://github.com/grabut/BiFCanalysis
11. Several spatial filters can be used for image denoising. The Gaussian Blur and FFT Bandpass filters perform very well to remove pixel noise and small objects but they smooth edges. A

median filter better preserves edges. It is also possible to use more sophisticated algorithms such as anisotropic diffusion (http://fiji.sc/Anisotropic_Diffusion_2D) and non-local means filtering (http://fiji.sc/Non_Local_Means_Denoise).

12. Thresholding is a critical step in image processing as it strongly influences the final results. Initially, we suggest performing this step manually, using interactive selection of threshold values and visual inspection of the resulting binary images. However, when analyzing large series of images acquired under similar conditions, more robust results can be obtained using automatic thresholding procedures that are not affected by subjective selection of threshold values. Identification of a suitable automatic thresholding algorithm is not always easy. The Otsu and Mixture-of-Gaussian thresholding methods are commonly used in fluorescence microscopy.
13. Binary images produced by thresholding in ImageJ and Fiji have only two pixel values, 0 and 255, that represent black and white on an 8-bit scale. To be used as masks in image calculations, they need to be divided by 255 to have pixel values of 0 and 1. However, binary image operations and commands in ImageJ and Fiji (e.g., the "Analyze Particles") require binary images with pixel values of 0 and 255.
14. The background subtraction procedure described here assumes that the background intensity is evenly distributed in the imaging field. If this is not the case, more sophisticated procedures are required. For instance, if uneven background is due to uneven illumination, a flat-field correction should be applied [46].
15. In 32-bit float images, pixels can be assigned negative values, which is best for image processing and quantification (no pixel information is lost during background and autofluorescence subtraction).

Acknowledgements

This work received funding from the ANR (grant ANR-12-JSV8-0003-001) and Biosit. G.R. was supported by INSERM and E.B. received fellowships from the Ministère de la Recherche et de l'Enseignement Supérieur and La Ligue Contre le Cancer. Confocal microscopy was performed at the Microscopy Rennes Imaging Center (MRic) facility. The authors also would like to acknowledge networking support by the Proteostasis COST Action (BM1307).

References

1. Grabbe C, Husnjak K, Dikic I (2011) The spatial and temporal organization of ubiquitin networks. Nat Rev Mol Cell Biol 12:295–307
2. Rabut G (2012) Introduction to the pervasive role of ubiquitin-dependent protein degradation in cell regulation. Semin Cell Dev Biol 23:481
3. Clague MJ, Heride C, Urbé S (2015) The demographics of the ubiquitin system. Trends Cell Biol 25:417–426
4. Ye Y, Rape M (2009) Building ubiquitin chains: E2 enzymes at work. Nat Rev Mol Cell Biol 10:755–764
5. Li W, Bengtson MH, Ulbrich A et al (2008) Genome-wide and functional annotation of human E3 ubiquitin ligases identifies MULAN, a mitochondrial E3 that regulates the organelle's dynamics and signaling. PLoS One 3:e1487
6. Kulathu Y, Komander D (2012) Atypical ubiquitylation—the unexplored world of polyubiquitin beyond Lys48 and Lys63 linkages. Nat Rev Mol Cell Biol 13:508–523
7. Komander D, Rape M (2012) The ubiquitin code. Annu Rev Biochem 81:203–229
8. Markson G, Kiel C, Hyde R et al (2009) Analysis of the human E2 ubiquitin conjugating enzyme protein interaction network. Genome Res 19:1905–1911
9. van Wijk SJL, de Vries SJ, Kemmeren P et al (2009) A comprehensive framework of E2-RING E3 interactions of the human ubiquitin-proteasome system. Mol Syst Biol 5:295
10. Christensen DE, Brzovic PS, Klevit RE (2007) E2-BRCA1 RING interactions dictate synthesis of mono- or specific polyubiquitin chain linkages. Nat Struct Mol Biol 14:941–948
11. Khmelinskii A, Blaszczak E, Pantazopoulou M et al (2014) Protein quality control at the inner nuclear membrane. Nature 516:410–413
12. Kerppola TK (2008) Bimolecular fluorescence complementation: visualization of molecular interactions in living cells. In: Sullivan KF (ed) Methods in cell biology. Academic Press, New York, pp 431–470
13. Shyu YJ, Hu C-D (2008) Fluorescence complementation: an emerging tool for biological research. Trends Biotechnol 26:622–630
14. Kerppola TK (2009) Visualization of molecular interactions using bimolecular fluorescence complementation analysis: characteristics of protein fragment complementation. Chem Soc Rev 38:2876–2886
15. Kodama Y, Hu C-D (2012) Bimolecular fluorescence complementation (BiFC): a 5-year update and future perspectives. Biotechniques 53:285–298
16. Miller KE, Kim Y, Huh W-K et al (2015) Bimolecular fluorescence complementation (BiFC) analysis: advances and recent applications for genome-wide interaction studies. J Mol Biol 427:2039–2055
17. Kerppola TK (2013) Bimolecular fluorescence complementation (BiFC) analysis of protein interactions in live cells. Cold Spring Harb Protoc 2013:727–731
18. Weber-Boyvat M, Li S, Skarp K-P et al (2015) Bimolecular fluorescence complementation (BiFC) technique in yeast Saccharomyces cerevisiae and mammalian cells. Methods Mol Biol 1270:277–288
19. Schütze K, Harter K, Chaban C (2009) Bimolecular fluorescence complementation (BiFC) to study protein-protein interactions in living plant cells. Methods Mol Biol. 479:189–202. doi:10.1007/978-1-59745-289-2_12. PMID: 19083187
20. Morell M, Espargaró A, Avilés FX et al (2007) Detection of transient protein-protein interactions by bimolecular fluorescence complementation: the Abl-SH3 case. Proteomics 7:1023–1036
21. Magliery TJ, Wilson CGM, Pan W et al (2005) Detecting protein-protein interactions with a green fluorescent protein fragment reassembly trap: scope and mechanism. J Am Chem Soc 127:146–157
22. Kodama Y, Hu C-D (2010) An improved bimolecular fluorescence complementation assay with a high signal-to-noise ratio. Biotechniques 49:793–805
23. Levy ED, Kowarzyk J, Michnick SW (2014) High-resolution mapping of protein concentration reveals principles of proteome architecture and adaptation. Cell Rep 7:1333–1340
24. Ball DA, Lux MW, Adames NR et al (2014) Adaptive imaging cytometry to estimate parameters of gene networks models in systems and synthetic biology. PLoS One 9:e107087
25. Shyu YJ, Liu H, Deng X et al (2006) Identification of new fluorescent protein fragments for bimolecular fluorescence complementation analysis under physiological conditions. Biotechniques 40:61–66
26. Manderson EN, Malleshaiah M, Michnick SW (2008) A novel genetic screen implicates Elm1 in the inactivation of the yeast transcription factor SBF. PLoS One 3:e1500
27. Ohashi K, Kiuchi T, Shoji K et al (2012) Visualization of cofilin-actin and Ras-Raf interactions by bimolecular fluorescence complementation assays using a new pair of split Venus fragments. Biotechniques 52:45–50

28. Saka Y, Hagemann AI, Piepenburg O et al (2007) Nuclear accumulation of Smad complexes occurs only after the midblastula transition in Xenopus. Development 134:4209–4218
29. Nakagawa C, Inahata K, Nishimura S et al (2011) Improvement of a Venus-based bimolecular fluorescence complementation assay to visualize bFos-bJun interaction in living cells. Biosci Biotechnol Biochem 75:1399–1401
30. Lin J, Wang N, Li Y et al (2011) LEC-BiFC: a new method for rapid assay of protein interaction. Biotech Histochem 86:272–279
31. Sung M-K, Huh W-K (2007) Bimolecular fluorescence complementation analysis system for in vivo detection of protein-protein interaction in Saccharomyces cerevisiae. Yeast 24:767–775
32. Sung M-K, Lim G, Yi D-G et al (2013) Genome-wide bimolecular fluorescence complementation analysis of SUMO interactome in yeast. Genome Res 23:736–746
33. Salvat C, Wang G, Dastur A et al (2004) The -4 phenylalanine is required for substrate ubiquitination catalyzed by HECT ubiquitin ligases. J Biol Chem 279:18935–18943
34. Kühnle S, Mothes B, Matentzoglu K et al (2013) Role of the ubiquitin ligase E6AP/UBE3A in controlling levels of the synaptic protein Arc. Proc Natl Acad Sci U S A 110:8888–8893
35. Maspero E, Valentini E, Mari S et al (2013) Structure of a ubiquitin-loaded HECT ligase reveals the molecular basis for catalytic priming. Nat Struct Mol Biol 20:696–701
36. Huang L, Kinnucan E, Wang G et al (1999) Structure of an E6AP-UbcH7 complex: insights into ubiquitination by the E2-E3 enzyme cascade. Science 286:1321–1326
37. Kamadurai HB, Souphron J, Scott DC et al (2009) Insights into ubiquitin transfer cascades from a structure of a UbcH5B approximately ubiquitin-HECT(NEDD4L) complex. Mol Cell 36:1095–1102
38. Sommer T, Jentsch S (1993) A protein translocation defect linked to ubiquitin conjugation at the endoplasmic reticulum. Nature 365:176–179
39. Metzger MB, Liang Y-H, Das R et al (2013) A structurally unique E2-binding domain activates ubiquitination by the ERAD E2, Ubc7p, through multiple mechanisms. Mol Cell 50:516–527
40. Tong AHY, Boone C (2007) 16 High-throughput strain construction and systematic synthetic lethal screening in. In: Stansfield I, Stark MJR (eds) Methods in microbiology. Academic Press, New York, pp 369–707
41. Berndsen CE, Wolberger C (2014) New insights into ubiquitin E3 ligase mechanism. Nat Struct Mol Biol 21:301–307
42. Robida AM, Kerppola TK (2009) Bimolecular fluorescence complementation analysis of inducible protein interactions: effects of factors affecting protein folding on fluorescent protein fragment association. J Mol Biol 394:391–409
43. Blondel M, Bach S, Bamps S et al (2005) Degradation of Hof1 by SCF(Grr1) is important for actomyosin contraction during cytokinesis in yeast. EMBO J 24:1440–1452
44. Huang A, de Jong RN, Wienk H et al (2009) E2-c-Cbl recognition is necessary but not sufficient for ubiquitination activity. J Mol Biol 385:507–519
45. Malleshaiah MK, Shahrezaei V, Swain PS et al (2010) The scaffold protein Ste5 directly controls a switch-like mating decision in yeast. Nature 465:101–105
46. Waters JC (2009) Accuracy and precision in quantitative fluorescence microscopy. J Cell Biol 185:1135–1148
47. Rines DR, Thomann D, Dorn JF et al (2011) Live cell imaging of yeast. Cold Spring Harb Protoc 2011
48. Shaner NC, Steinbach PA, Tsien RY (2005) A guide to choosing fluorescent proteins. Nat Methods 2:905–909
49. Horstman A, Tonaco IAN, Boutilier K et al (2014) A cautionary note on the use of split-YFP/BiFC in plant protein-protein interaction studies. Int J Mol Sci 15:9628–9643
50. Sheff MA, Thorn KS (2004) Optimized cassettes for fluorescent protein tagging in Saccharomyces cerevisiae. Yeast 21:661–670
51. Zimmermann T (2005) Spectral imaging and linear unmixing in light microscopy. In: Microscopy techniques. Springer, Heidelberg, Berlin, pp 245–265
52. Luisier F, Vonesch C, Blu T et al (2010) Fast interscale wavelet denoising of Poisson-corrupted images. Signal Process 90:415–427

Chapter 14

Monitoring Ubiquitin-Coated Bacteria via Confocal Microscopy

Marie Lork, Mieke Delvaeye, Amanda Gonçalves, Evelien Van Hamme, and Rudi Beyaert

Abstract

Salmonella is a gram-negative facultative intracellular pathogen that is capable of infecting a variety of hosts. Inside host cells, most *Salmonella* bacteria reside and replicate within *Salmonella*-containing vacuoles. They use virulence proteins to manipulate the host cell machinery for their own benefit and hijack the host cytoskeleton to travel toward the perinuclear area. However, a fraction of bacteria escapes into the cytosol where they get decorated with a dense layer of polyubiquitin, which labels the bacteria for clearance by autophagy. More specifically, autophagy receptor proteins recognize the ubiquitinated bacteria and deliver them to autophagosomes, which subsequently fuse to lysosomes. Here, we describe methods used to infect HeLa cells with *Salmonella* bacteria and to detect their ubiquitination via immunofluorescence and laser scanning confocal microscopy.

Key words *Salmonella*, Bacteria, Ubiquitination, Autophagy, Confocal microscopy, Imaging, Immunofluorescence, Host–pathogen interaction, Bacterial clearance

1 Introduction

Salmonella has traditionally been characterized as a vacuolar pathogen since most *Salmonella* bacteria reside in the so-called *Salmonella*-containing vacuole (SCV), a membrane bound compartment derived from the endocytic pathway [1, 2]. Recent work indicates that depending on the cell type, *Salmonella* can occupy different niches within the cell. In epithelial cells, next to the vacuolar bacteria, a considerable proportion of the total bacteria population has been found to inhabit the cytosol, where they hyper-replicate [3]. However, once the bacteria reach the cytosol they are exposed to the host ubiquitin system [4]. The bacteria are decorated with a dense layer of polyubiquitin chains, which peaks at 4 h post infection [5]. This ubiquitin coat is recognized by autophagy receptors, NDP52, OPTN, and p62, which can bind to ubiquitin and members of the autophagosomal protein

Rune Matthiesen (ed.), *Proteostasis: Methods and Protocols*, Methods in Molecular Biology, vol. 1449,
DOI 10.1007/978-1-4939-3756-1_14, © Springer Science+Business Media New York 2016

microtubule-associated protein 1A/1B-light chain 3 (LC3) family, leading to targeting of bacteria into autophagosomes and their subsequent degradation in autophagolysosomes [6–8]. Ubiquitination of bacteria is also necessary for the direct recruitment of other important components of the autophagy machinery, including the Atg16L1 complex, the ULK1 complex, and ATG9L1 [9]. Knockdown of autophagy receptors leads to enhanced bacterial replication, indicating that autophagy restricts the growth of cytosolic bacteria and is necessary for their clearance.

It is still unclear whether the formation of a ubiquitin coat on the surface of intracellular bacteria involves the ubiquitination of bacterial surface proteins or the ubiquitination of host cell proteins that bind the bacteria upon membrane rupture. Either way, it is likely that diverse E3 ubiquitin ligases are involved in the recognition and targeting of cytosolic bacteria. Huett et al. reported that the E3 ubiquitin ligase LRSAM1 contributes to the ubiquitination of invading bacteria [10]. However, there is still a substantial amount of bacteria ubiquitinated in LRSAM1-deficient cells, indicating that other E3 ligases are also involved. Moreover, it has been reported that the ubiquitin coat surrounding *Salmonella* contains at least M1- and K63-linked polyubiquitin chains [11], while in vitro ubiquitination of *Salmonella* with LRSAM1 primarily produces K6- and K27-linked polyubiquitin [10]. Clearly, determination of the molecular machinery involved in the ubiquitination of invading bacteria and their subsequent clearance by autophagy still needs significant research. In this context, monitoring and quantifying ubiquitinated intracellular bacteria by confocal microscopy is very valuable (Fig. 1). Here we describe methods for the infection of HeLa cells with *Salmonella* bacteria, immunofluorescence staining with anti-ubiquitin, cellular imaging via confocal microscopy, and image processing.

2 Materials

2.1 Human Cell Culture and Bacterial Infection

1. Culture medium for HeLa cells (human cervical carcinoma cell line): Dulbecco's Modified Eagle Medium (DMEM) (Gibco/BRL, Bethesda, MD), supplemented with 10% fetal calf serum, 0.4 mM sodium pyruvate, 2 mM L-glutamine, 0.1 mM nonessential amino acids.
2. Trypsin/EDTA buffer: 400 ml 0.04% EDTA (1 mM EDTA, 0.1 mM NaCl, 2.5 mM KCl, 7 mM $Na_2HPO_4 \cdot 12H_2O$, and 2 mM KH_2PO_4) and 100 ml trypsin solution (0.1 mM Trypsin, 0.1 M NaCl, 2.5 mM KCl, 66 mM Tris–HCl, 0.7 mM Na_2PO_4, and 1% phenol red), pH 7.6. Store at 4 °C.
3. 10× Phosphate-buffered saline (PBS): 95.44 g Dulbecco's Phosphate Buffered Saline (Lonza, Verviers, Belgium) in 1 l water. Store at 4 °C.

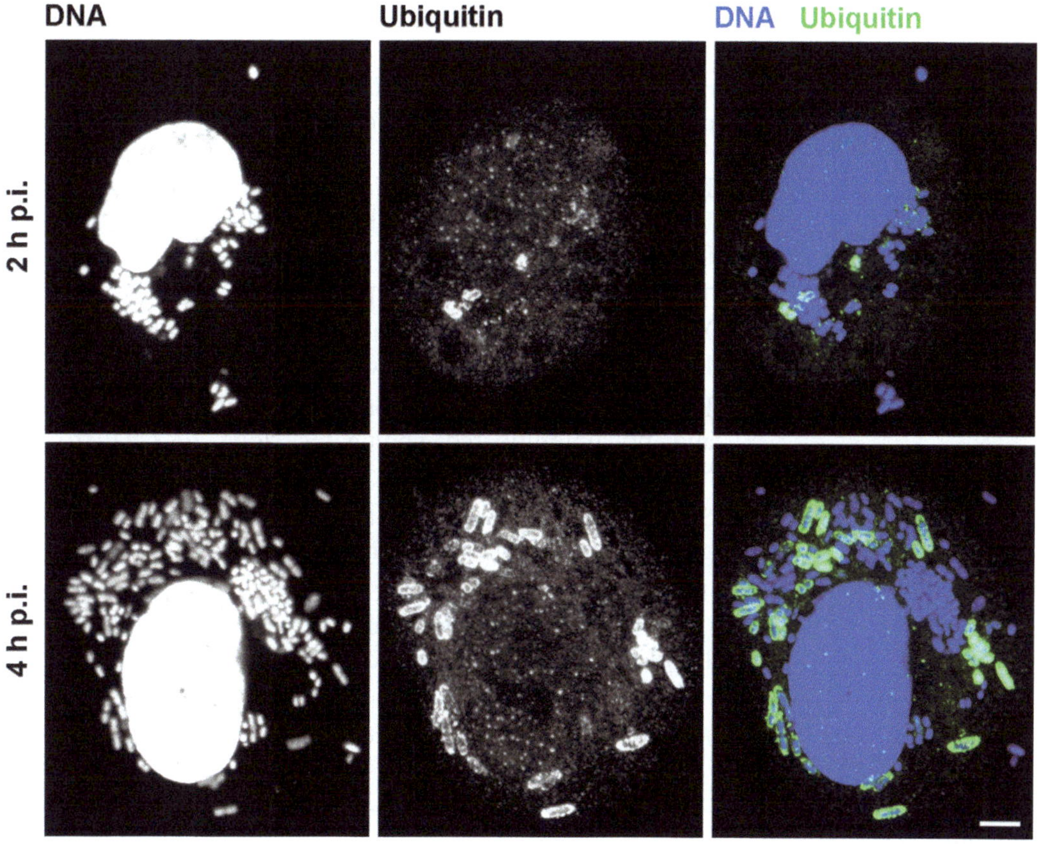

Fig. 1 Confocal microscopy of ubiquitination of intracellular *Salmonella* bacteria. Infected HeLa cells were fixed at indicated time points after infection (hours p.i.). *Left panels* show cellular and bacterial DNA stained with DAPI, *middle panels* show anti-ubiquitin visualized with Dylight488, and *right panels* show the overlays of both channels. All images are maximum intensity projections of confocal Z-slices acquired with a step size of 0.5 μm. Scale bar, 5 μm

4. 8-Well μ-Slides (ibidi, Munich, Germany).
5. Luria Broth (LB) medium: 1% bactotryptone, 0.5% bacto-yeast, 1% NaCl; autoclave; store at room temperature or 4 °C.
6. LB agar plates: Prepare LB medium as above, add 15 g agar/L before autoclaving. After autoclaving, cool to approximately 55 °C, and pour into petri dishes; let harden, then invert and store at 4 °C.
7. Pathogenic *Salmonella enterica* ssp. enterica strain LMG3264 was purchased from the Belgian Coordinated Collections of Microorganisms (BCCM)-LMG Bacteria Collection (Ghent University, Belgium).
8. Gentamycin (Gibco/BRL, Bethesda, MD).

2.2 Immunofluorescence Staining

1. Fixation buffer: 4% paraformaldehyde (PFA) in PBS (*see* **Note 1**). For 1 l of 4% PFA, heat up 800 ml PBS in a glass beaker to

approximately 60 °C while stirring. Do not let the solution boil. Add 40 g of PFA powder to the heated PBS solution. Slowly raise the pH by adding 1 N NaOH drop wise until the solution clears. Once the PFA is dissolved, let the solution cool down and filter through a 0.45 μm filter. Adjust the volume to 1 l with 1× PBS. Recheck the pH and adjust with small amounts of diluted HCl to approximately 6.9. The solution can be aliquoted and frozen or stored at 2–8 °C for up to 1 month.

2. TBS-TX (25 mM Tris–HCl pH 8, 150 mM NaCl, 0.1 % Triton X-100, 1 % bovine serum albumin (BSA)). Triton X-100 and BSA are freshly added prior to use.
3. Primary antibody: mouse monoclonal antibody against mono- and polyubiquitinated conjugates clone FK2 (Enzo Life Sciences, Farmingdale, NY).
4. Secondary antibody: DyLight 488 secondary antibody conjugates (Thermo Fisher Scientific, Waltham, MA).
5. 4′,6-Diamidino-2-phenylindole (DAPI) FluoroPure™ grade (Thermo Fisher Scientific, Waltham, MA). A 500 μM stock of DAPI is prepared in deionized water and aliquots are stored at −20 °C.
6. Mounting medium: 1 % *N*-propyl-gallate in glycerol.

3 Methods

3.1 Seeding of HeLa Cells

1. Plate 10^4 HeLa cells per well in 200 μl DMEM in 8-well μ-Slides (*see* **Notes 2** and **3**).
2. Grow cells overnight in a humidified atmosphere at 37 °C and 5 % CO_2.

3.2 Infection with Salmonella

1. Two days prior to infection, streak bacteria from frozen glycerol stocks onto a fresh LB agar plate and incubate overnight at 37 °C (*see* **Notes 4–6**).
2. Inoculate a single bacterial colony into 5 ml LB medium and incubate in a shaking incubator overnight at 37 °C (*see* **Note 7**).
3. Subculture the bacterial suspension 1:33 in fresh LB medium and incubate in a shaking incubator at 37 °C for 2.5–3.5 h until they reach an OD_{600} of 0.7–1.
4. Prior to infection refresh HeLa cells with new DMEM.
5. Infect cells with 20 μl *Salmonella* subculture and incubate for 15 min at 37 °C/5 % CO_2.
6. After infection remove extracellular bacteria by washing twice with warm PBS and add DMEM containing 100 μg/ml gentamycin (*see* **Note 8**).

7. After 1 h, change the medium to DMEM containing 20 μg/ml gentamycin.
8. Collect samples at appropriate time points post infection.

3.3 Immunofluorescence Staining

1. At appropriate time points wash cells twice with PBS (*see* **Notes 8** and **9**).
2. Fix cells with fixation buffer for 20 min at room temperature (RT) to insure cellular preservation while imaging (*see* **Notes 10** and **11**).
3. Wash fixed cells twice with PBS (*see* **Note 12**).
4. Permeabilize the cells by incubating the slides for 10 min in TBS-TX.
5. Dilute the primary antibody 1:200 in TBS-TX and incubate the slides with primary antibody overnight at 4 °C (*see* **Note 13**).
6. Wash the slides three times for 5 min and once for 30 min with TBS-TX (*see* **Note 14**).
7. Dilute the secondary antibody 1:500 in TBS-TX and incubate the slides with secondary antibody for 2 h at RT protected from light.
8. Wash the slides three times for 5 min and once for 30 min with PBS.
9. Incubate the slides with 500 nM DAPI in PBS for 20 min at RT protected from light in order to visualize the DNA present in the HeLa cells as well as the *Salmonella* bacteria.
10. Rinse three times with PBS.
11. Remove PBS and mount the slides with 200 μl mounting medium per well to prevent photo bleaching (*see* **Notes 15** and **16**).

3.4 Laser Scanning Confocal Microscopy and Image Processing

1. Images are acquired using a Leica TCS SP5 AOBS confocal system (Leica, Manheim, Germany), with a 63× HCX PLApo 1.4 oil-immersion objective, with a format of 1024 × 1024, line average of 4, and zoom of 3 (pixel size: 70.6 nm by 70.6 nm). Bright-field images are acquired with the HeNe 543 nm laser line, while ubiquitin labeled with DyLight488 is imaged with Ar laser excitation at 488 nm, and nucleic acids with diode laser excitation at 405 nm. *Z*-stacks acquired for visualization purposes are imaged with axial resolution of 0.5 μm, while the stacks acquired for image analysis are imaged at nyquist Z resolution (step size 0.13 μm).
2. Image analysis is performed with Volocity 3D Image Analysis Software (Perkin Elmer Life Sciences). A protocol can be developed where the bacteria are segmented as objects with fluorescence intensity above background and of predicted size. In the same protocol, a second measurement can be performed to segment for ubiquitin. Finally, an "intersect" algorithm is

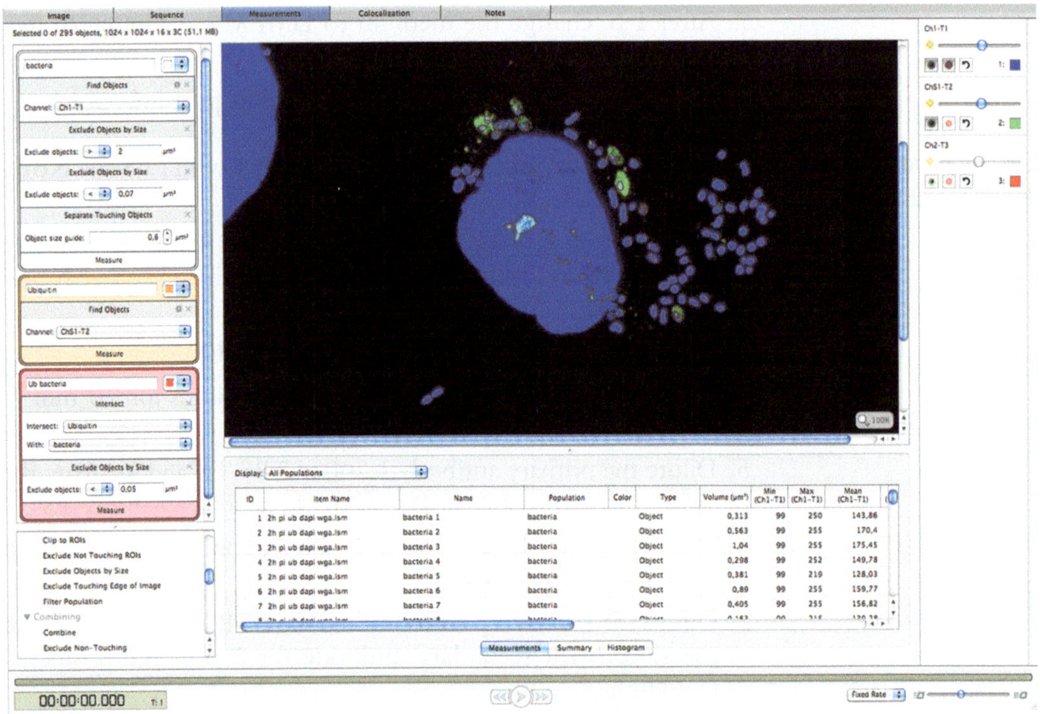

Fig. 2 Image analysis to identify ubiquitin-coated *Salmonella* bacteria. A protocol was developed in Volocity software to segment bacteria and ubiquitin in order to identify ubiquitin-coated bacteria. On the left side the protocol is shown, on the right side the result of the segmentation is visible in the image. The boundaries of the segmented areas are shown in a different color for each of the three populations

applied to identify ubiquitin-coated bacteria (Fig. 2). Additionally, the number of nuclei can be determined to give an overall count of the cells in the field of view. This can be useful to determine the average number of bacteria per cell and the number of ubiquitin-coated bacteria per cell.

4 Notes

1. PFA is toxic. Gloves should be worn and solutions should be made in the fume hood.
2. The cells need to be sub-confluent and nicely stretched out at the time of infection to enable an adequate analysis. As different cells in different laboratories tend to grow at different rates, it is advisable to determine the optimal cell concentration of the used cell line.
3. Up to five 8-well μ-Slides can be stored in one 14 cm petri dish for easy handling and transportation.
4. *Salmonella enterica* is classified as a biosafety level 2 (BL2) organism. All handling of these bacteria must be performed in a BL2 facility.

5. Submerge an inoculation needle into the frozen glycerol stock and apply several streaks on an LB plate. Use a fresh inoculation needle and streak through the applied lines. Repeat several times in order to obtain single bacteria colonies.
6. The LB plate containing bacterial colonies can be stored at 4 °C for several weeks. For new experiment, a single colony should be streaked on a fresh LB plate.
7. We use 50 ml Falcon tubes to grow overnight cultures.
8. For all washing steps remove the liquid carefully with a pipette from the corner of the well. Avoid scratching the monolayer with the tip of the pipette. When adding new liquid, pipette carefully against the wall of the chamber to avoid detachment of the cells.
9. Do not let your cells dry at any point during the staining.
10. Once the slides have been fixed in fixation buffer, all following steps may be performed outside the BL2 facility.
11. Do not exceed the fixing time as over-fixation can lead to artifacts.
12. Fixed cells can be stored in PBS at 4 °C for approximately 2 weeks without reducing the staining efficiency.
13. Always include a negative staining control without primary antibody using only the secondary antibody to exclude non-specific binding of the secondary antibody.
14. Extensive washing is required to reduce background.
15. Cut the tip off for pipetting the mounting medium.
16. Once stained and covered with mounting medium, the fluorescent staining will stay for up to 1 month if you have a strong signal and if stored at 4 °C in the dark.

Acknowledgement

M.L. holds a predoctoral fellowship from the Fund for Scientific Research Flanders (FWO). Work in the authors' lab is supported by grants from the FWO, "Belgian Foundation Against Cancer", "Agency for Innovation by Science and Technology", "Interuniversity Attraction Poles", "Concerted Research Actions", and "Group-ID Multidisciplinary research partnership" from Ghent University, VIB. The authors would like to acknowledge networking support by the Proteostasis COST Action (BM1307).

References

1. Drecktrah D, Knodler LA, Howe D, Steele-Mortimer O (2007) *Salmonella* trafficking is defined by continuous dynamic interactions with the endolysosomal system. Traffic 8(3):212–225. doi:10.1111/j.1600-0854.2006.00529.x
2. Steele-Mortimer O, Meresse S, Gorvel JP, Toh BH, Finlay BB (1999) Biogenesis of *Salmonella* typhimurium-containing vacuoles in epithelial cells involves interactions with the early endocytic pathway. Cell Microbiol 1(1):33–49

3. Malik-Kale P, Winfree S, Steele-Mortimer O (2012) The bimodal lifestyle of intracellular *Salmonella* in epithelial cells: replication in the cytosol obscures defects in vacuolar replication. PloS One 7(6):e38732. doi:10.1371/journal.pone.0038732
4. Perrin AJ, Jiang X, Birmingham CL, So NS, Brumell JH (2004) Recognition of bacteria in the cytosol of Mammalian cells by the ubiquitin system. Curr Biol 14(9):806–811. doi:10.1016/j.cub.2004.04.033
5. Birmingham CL, Smith AC, Bakowski MA, Yoshimori T, Brumell JH (2006) Autophagy controls *Salmonella* infection in response to damage to the *Salmonella*-containing vacuole. J Biol Chem 281(16):11374–11383. doi:10.1074/jbc.M509157200
6. Thurston TL, Ryzhakov G, Bloor S, von Muhlinen N, Randow F (2009) The TBK1 adaptor and autophagy receptor NDP52 restricts the proliferation of ubiquitin-coated bacteria. Nat Immunol 10(11):1215–1221. doi:10.1038/ni.1800
7. Wild P, Farhan H, McEwan DG, Wagner S, Rogov VV, Brady NR, Richter B, Korac J, Waidmann O, Choudhary C, Dotsch V, Bumann D, Dikic I (2011) Phosphorylation of the autophagy receptor optineurin restricts *Salmonella* growth. Science 333(6039):228–233. doi:10.1126/science.1205405
8. Zheng YT, Shahnazari S, Brech A, Lamark T, Johansen T, Brumell JH (2009) The adaptor protein p62/SQSTM1 targets invading bacteria to the autophagy pathway. J Immunol 183(9):5909–5916. doi:10.4049/jimmunol.0900441
9. Fujita N, Morita E, Itoh T, Tanaka A, Nakaoka M, Osada Y, Umemoto T, Saitoh T, Nakatogawa H, Kobayashi S, Haraguchi T, Guan JL, Iwai K, Tokunaga F, Saito K, Ishibashi K, Akira S, Fukuda M, Noda T, Yoshimori T (2013) Recruitment of the autophagic machinery to endosomes during infection is mediated by ubiquitin. J Cell Biol 203(1):115–128. doi:10.1083/jcb.201304188
10. Huett A, Heath RJ, Begun J, Sassi SO, Baxt LA, Vyas JM, Goldberg MB, Xavier RJ (2012) The LRR and RING domain protein LRSAM1 is an E3 ligase crucial for ubiquitin-dependent autophagy of intracellular Salmonella Typhimurium. Cell Host Microbe 12(6):778–790. doi:10.1016/j.chom.2012.10.019
11. van Wijk SJ, Fiskin E, Putyrski M, Pampaloni F, Hou J, Wild P, Kensche T, Grecco HE, Bastiaens P, Dikic I (2012) Fluorescence-based sensors to monitor localization and functions of linear and K63-linked ubiquitin chains in cells. Mol Cell 47(5):797–809. doi:10.1016/j.molcel.2012.06.017

Chapter 15

Detection and Analysis of Cell Cycle-Associated APC/C-Mediated Cellular Ubiquitylation In Vitro and In Vivo

Cesyen Cedeño, Esther La Monaca, Mara Esposito, and Gustavo J. Gutierrez

Abstract

The anaphase-promoting complex or cyclosome (APC/C) is one of the major orchestrators of the cell division cycle in mammalian cells. The APC/C acts as a ubiquitin ligase that triggers sequential ubiquitylation of a significant number of substrates which will be eventually degraded by proteasomes during major transitions of the cell cycle. In this chapter, we present accessible methodologies to assess both in in vitro conditions and in cellular systems ubiquitylation reactions mediated by the APC/C. In addition, we also describe techniques to evidence the changes in protein stability provoked by modulation of the activity of the APC/C. Finally, specific methods to analyze interactors or posttranslational modifications of particular APC/C subunits are also discussed. Given the crucial role played by the APC/C in the regulation of the cell cycle, this review only focuses on its action and effects in actively proliferating cells.

Key words Anaphase-promoting complex or cyclosome (APC/C), Cell cycle, Ubiquitin, Ubiquitylation, Phosphorylation, Cdc27/APC3, Cdh1, Cdc20

1 Introduction

Progression through the cell division cycle in mammalian cells is an intricate process exquisitely regulated by posttranslational modifications such as phosphorylation and ubiquitylation [1, 2]. In particular, inactivation of essential cell cycle regulators is achieved through their ubiquitylation and proteasome-mediated degradation. Ubiquitylation mechanisms associated with proteolysis often involve the formation of a polyubiquitin chain covalently attached to the substrates. Those chains are generated via the repeated and sequential action of three types of enzymes called ubiquitin-activating enzymes (E1), ubiquitin-conjugating enzymes (E2), and ubiquitin ligases (E3). Chains of ubiquitin can be formed through lysines 6, 11, 27, 29, 33, 48, or 63 or at the amino terminal methionine of ubiquitin [3]. The anaphase-promoting complex or cyclosome (APC/C) is arguably the most important ubiquitin ligase

Rune Matthiesen (ed.), *Proteostasis: Methods and Protocols*, Methods in Molecular Biology, vol. 1449,
DOI 10.1007/978-1-4939-3756-1_15, © Springer Science+Business Media New York 2016

(E3) acting during and orchestrating the mammalian cell cycle [4]. The APC/C is a giant multisubunit complex of about 1.2 MDa. The core of the APC/C is composed of at least 19 subunits (14 distinct polypeptides) in mammalian cells [5, 6]. In addition to its core complex, the APC/C also associates to substrate-binding/activatory regulatory subunits, namely Cdc20 or Cdh1. The APC/C belongs to the family of cullin-RING (Really Interesting New Gene) finger E3 ubiquitin ligases and it is capable of catalyzing ubiquitylation of a myriad of substrates that modulate major cell cycle processes including DNA replication, chromosomal segregation, progression through and exit from mitosis, maintenance of G1 phase of the cell cycle and cytokinesis [7, 8]. Most APC/C substrates are essential regulatory components of the cell cycle, such as geminin, cyclin A, securin, cyclin B1, Aurora A, and Polo-like kinase 1 (Plk-1).

The APC/C can mediate protein ubiquitylation with the help of several ubiquitin-conjugating enzymes: UbcH5A (Ube2D1), UbcH5B (Ube2D2/Ubc4), and in particular UbcH10 (Ube2C/UbcX/E2C/Vihar) are responsible for the initial or "priming" ubiquitylation activities of the APC/C, i.e., catalyzing the conjugation of the first ubiquitin moiety(ies) into putative substrates, while it has been found that another E2 termed Ube2S rather confers the APC/C ability to extend the length of the ubiquitin chains attached into its substrates, including generating branched ubiquitin chains [9–13].

Remarkably, the complexity of the APC/C also applies to the diverse topology of ubiquitin chains that can generate into its substrates. The APC/C is indeed capable of forming ubiquitin chains using lysines 11, 48, and 63 of ubiquitin although the emerging view related to the mammalian APC/C indicates that polyubiquitylation via lysine 11 is predominant at least during the cell cycle-related APC/C activation in conjunction with the activity of Ube2S [14].

The APC/C targets substrates possessing short linear recognition sequences such as the D-[RXXLXXXX(N)] and KEN-[KENXXX(N)] boxes using a mechanism involving a highly processive initial reaction followed by multiple encounters with the substrates and slower rates reactions that are favored by the presence of the "primed" ubiquitylated sites [15].

Finally, it is worth noting that a growing body of evidence indicates that the APC/C also plays important roles in highly differentiated and specialized cells such as neurons [16]. Methods to analyze and detect the activity of the APC/C in this particular cellular context will nonetheless not be discussed in this review.

2 Materials

2.1 In Vitro Analysis Methods of APC/C-Mediated Ubiquitylation

1. E1: Ube1 (*UBA1*) can be purchased from different sources including Boston Biochem or in-house produced from insect cells using a baculovirus-infection-based system.
2. E2: UbcH5A or B, UbcH10, Ube2S can be purchased from different sources such as Boston Biochem. They can also be produced in and purified from bacteria with different tags (6xHistidine, Glutathione S-transferase (GST), or Maltose-Binding Protein (MBP)).
3. E3: the APC/C can be immunoprecipitated from mammalian cells (as described in Subheading 3 below). A handful group of laboratories have been able to produce and purify recombinant APC/C using insect cells infected with baculovirus-based systems [17–21]. These protocols are however not discussed in this review.
4. Anti-Cdc27/APC3 antibodies: AF3.1 clone from Santa Cruz Biotechnology.
5. Cdc20: it can be produced and purified as a tagged full-length protein from insect cells. Alternatively, it can be in vitro transcribed-translated in rabbit reticulocyte lysates and employed to activate the APC/C as a programmed extract.
6. Cdh1: it can be produced and purified as a tagged full-length protein from insect cells. Alternatively, it can be in vitro transcribed-translated in rabbit reticulocyte lysates and employed to activate the APC/C as a programmed extract [22].
7. APC/C substrates: many substrates [geminin, securin, cyclin A, cyclin B1, the N-terminus of cyclin B1 (comprising the D-boxes), Cdc20 or Cdh1 themselves, Aurora A, or Plk-1] can be used to test the ubiquitylation activity of the APC/C. In this review, we propose to generate the substrates radiolabeled in rabbit reticulocyte extracts from a vector under the control of a T7 promoter. Alternatively, substrates can also be fluorescently labeled [21].
8. TnT® T7 Coupled Reticulocyte Lysate System (Promega).
9. L-[35S]-methionine (Perkin Elmer).
10. Micro Bio-spin 6 columns (Bio-Rad).
11. Adenosine triphosphate (ATP) (Sigma Aldrich).
12. Ubiquitin: can be purchased untagged or tagged from different sources or it can also be produced and purified from bacteria using pET or pGEX vectors.
13. Ubiquitylation buffer: 25 mM Tris–HCl (pH = 7.5), 50 mM NaCl and 10 mM $MgCl_2$ supplemented with an ATP regenerating system.

14. ATP regenerating system: 1.25 mM ATP, 1.25 mM $MgCl_2$, 1.9 mM Creatine Phosphate and 6.25 μg/ml Creatine Phosphokinase (components purchased from Sigma Aldrich).
15. Bovine serum albumin (BSA) (Sigma Aldrich).
16. Polyacrylamide gel electrophoresis equipment and ad-hoc SDS-PAGE solutions.
17. Gel staining solution: 10% acetic acid, 40% methanol and Coomassie blue (Carl Roth) in milli-Q graded water.
18. Gel destaining solution: 10% acetic acid and 40% methanol in milli-Q graded water.
19. Gel drying system.
20. Phosphorimaging screen (GE Healthcare).
21. Personal Molecular Imager™ (PMI™) system (Phosphorimager) (Bio-Rad).
22. Image Lab™ software or Quantity One 1-D analysis software (Bio-Rad).
23. Reagents for immunoblotting and detection using infrared technology.
24. Odyssey scanner and software (LI-COR Biosciences).
25. Quikchange Site Directed Mutagenesis kit (Agilent Technologies).

2.2 In Vivo Analysis Methods of APC/C-Mediated Ubiquitylation

1. Nocodazole (Sigma Aldrich).
2. Thymidine (Sigma Aldrich).
3. Cycloheximide (Sigma Aldrich).
4. Phosphate-buffered saline (PBS): 137 mM NaCl, 2.7 mM KCl, 4.3 mM Na_2HPO_4 and 1.5 mM KH_2PO_4.
5. Lysis Buffer: 50 mM Tris–HCl (pH = 7.5), 150 mM NaCl, 2 mM $MgCl_2$, 0.5% NP-40, 1 mM DTT and 10% glycerol, supplemented with a proteases-phosphatases inhibitors cocktail (Roche Diagnostics).
6. Nitrogen decompression buffer: 20 mM Tris–HCl (pH = 7.5), 5 mM KCl, 1.5 mM $MgCl_2$ and 1 mM DTT, supplemented with a proteases-phosphatases inhibitors cocktail (Roche Diagnostics).
7. Protein A/G PLUS-agarose (Santa Cruz Biotechnology).
8. Protein A sepharose beads (Bio-Rad).
9. Protein G agarose resin (Roche Diagnostics).
10. Proteasome inhibitor MG-132 (Selleckchem).
11. Ubiquitin aldehyde (Boston Biochem).
12. 4× SDS sample buffer: 250 mM Tris–HCl (pH = 6.7), 8% SDS, 40% glycerol, 0.4 M dithiothreitol (it can be replaced by β-mercaptoethanol) and 0.02% bromophenol blue.

13. Emi1: vector to transfect mammalian cells.
14. proTAME (tosyl-L-arginine methyl ester) (Boston Biochem).
15. Human cell lines: HeLa S3, HEK-293T, HFF-1 or other cells of interest (ATCC).
16. Tissue culture reagents, materials and media.
17. Immunoprecipitation buffer: 20 mM Tris–HCl (pH = 8.0), 150 mM NaCl, 1 % NP-40 and 2 mM EDTA, supplemented with a proteases-phosphatases inhibitors cocktail (Roche Diagnostics) and MG-132 (at a final concentration of 50 μM)
18. Bacterially expressed and purified recombinant proteins (GST or 6xHistidine-tagged).
19. Binding buffer: 50 mM Tris–HCl (pH = 7.5), 150 mM NaCl, 0.5 % NP-40, 5 mM EGTA, 5 mM EDTA and 2 mM $MgCl_2$, supplemented with a proteases-phosphatases inhibitors cocktail (Roche Diagnostics).
20. Nickel Xpure agarose resin (Bio-Connect).
21. DNA mini-prep and Maxi-prep purification kits.

3 Methods

3.1 General Methods

3.1.1 In Vitro Production of Unlabeled or [^{35}S]-Methionine-Labeled Proteins

Purified plasmids preparations (at least DNA mini-prep quality graded) encoding known substrates of the APC/C (see Materials) are transcribed-translated in vitro in the presence of L-[^{35}S]-methionine in rabbit reticulocyte lysates following manufacturer's instructions, then loaded on pre-equilibrated (20 mM Tris–HCl, pH = 7.4) Micro Bio-spin 6 columns in order to eliminate non-incorporated radioactive methionine.

3.1.2 Cell Cycle Synchronization Protocols

To properly assess the effects of the APC/C in the ubiquitin-triggered degradation of substrates along the cell cycle, mastering common methods used to synchronize cell lines during their division cycle is required. There are several specialized reviews which have addressed specific methodologies to synchronize cells along the cell cycle [23]. Those methodologies are cell-type-specific and in all cases need to be verified either by fluorescence-activated cell sorting (FACS) monitoring DNA content (by propidium iodide labeling for example) or by microscopy or biochemical analyses (enzymatic assays and/or Western-blotting) visualizing known cell cycle markers [22, 24–26]. Here, we will briefly summarize the most widely used and simple methods of cell cycle synchronization:

Nocodazole Block

Cells are grown in the presence of 330 nM nocodazole for 18 h. After that incubation, a significant fraction of cells will arrest in prometaphase. Release from the arrest can be accomplished by 1–2 washes (depending on the attachment displayed by the cells) with PBS followed by growth in fresh medium.

Double-Thymidine Block

Cells are grown in the presence of 2 mM thymidine for 18 h, washed twice with PBS and released into thymidine-free media for 6–8 h, and finally grown again for 12 h in the presence of 2 mM thymidine. Under these conditions, cells are arrested in late G1 phase. A final release can be accomplished by 1–2 washes with PBS followed by growth in fresh medium.

Thymidine-Nocodazole Block

Cells are treated for 18–24 h with media containing 2 mM thymidine. After two washes and a release into fresh media for 6 h, cells are treated with 330 nM nocodazole for 10–12 h. Using this protocol, a population of cells in G1 can be obtained by washing 1–2 times with PBS and releasing the nocodazole-arrested cells into fresh media for 3 h.

3.1.3 Cells Extracts Preparation

From each of the protocols described above, it is possible to produce cell lysates that will allow to: (1) semi-purify the APC/C in order to perform in vitro ubiquitylation assays; or (2) perform in vitro degradation assays monitoring the capacity of the APC/C to instigate ubiquitin-mediated substrates degradation.

In the case of cells synchronized using a thymidine-nocodazole block for example, cells are first synchronized at prometaphase after the consecutive treatment, then washed (1–2 times with PBS) and finally released for 3 h into fresh media. After the short release, a significant fraction of cells are found in G1 phase of the cell cycle. Extracts from cells at G1 phase can be used to obtain a semi-purified and activatable APC/C. After cell synchronization, cells are subsequently harvested and lysed in a buffer containing 50 mM Tris–HCl (pH = 7.5), 150 mM NaCl, 2 mM $MgCl_2$, 0.5 % NP-40, 1 mM DTT, and 10% glycerol, supplemented with proteases-phosphatases inhibitors cocktail (*see* **Note 1** for an alternative lysis/cell disruption method).

3.2 APC/C-Catalyzed In Vitro Substrate Ubiquitylation

The in vitro APC/C-mediated ubiquitylation assay uses ^{35}S-methionine radiolabeled proteins as putative substrates and semi-purified APC/C from extracts of G1-synchronized HeLa S3 cells.

1. One ml of lysate from G1-synchronized cells in a single Eppendorf tube is used in order to perform ten ubiquitylation reactions.
2. One ml of lysate from G1-synchronized cells is incubated with 10 μg of antibodies raised against human Cdc27/APC3 during 4 h at 4 °C with constant rotation.
3. The mixture is incubated with beads capable of binding the anti-Cdc27/APC3 antibodies during 2 h at 4 °C with constant rotation (*see* **Note 2**).

4. The mixture is then centrifuged at 4 °C for 5 min at 1,600 × *g*.
5. Precipitates are washed at least three times with the appropriate buffer used to prepare the lysate but without detergents.
6. Ubiquitylation reactions are started by mixing in a maximum volume of 20 μl the following reagents: 5–10 μl of beads containing the immunoprecipitated and semi-purified APC/C, 1–2 μl of in vitro transcribed-translated and radiolabeled substrate, 100 nM Ubel (E1), 2 μM UbcH10/Ube2C (and/or 0.1 μM Ube2S) depending on the E2 of interest, 20 mM ATP, 1.5 mg/ml ubiquitin, 10 mM DTT and 2 mg/ml BSA in a ubiquitylation buffer including 25 mM Tris–HCl (pH = 7.5), 50 mM NaCl and 10 mM $MgCl_2$, supplemented with an ATP regenerating system (1.25 mM ATP, 1.25 mM $MgCl_2$, 1.9 mM Creatine Phosphate and 6.25 μg/ml Creatine Phosphokinase) (see Note 3).
7. In the case of elution of the APC/C from the antibody-beads reagent, a total of 20 nM of semi-purified APC/C is used per reaction. Elution of the APC/C can be eventually performed using a competing peptide (see Note 4).
8. According to the particular APC/C which wants to be analyzed, the reaction must be supplemented with 1–2 nM of recombinant Cdc20 or Cdh1. Alternatively, 1–2 μl of rabbit reticulocyte extracts (depending on the efficiency of accumulation of the translated protein in the extract) programmed to in vitro transcribe-translate unlabeled Cdc20 or Cdh1 can also be employed.
9. The reaction is incubated at 30 °C with gentle shaking if the APC/C was not eluted from the agarose or sepharose beads after immunoprecipitation. It is recommended to perform a kinetic involving several time points (0, 15, 30, 60, 90, and 120 min) when attempting for the first time an APC/C-instigated in vitro ubiquitylation reaction.
10. The ubiquitylation reaction is stopped by addition of 4× sample buffer.
11. Samples are boiled at 95 °C for 5 min and centrifuged at 1,600 × *g* for 1 min.
12. Samples are loaded in an SDS-PAGE (we typically use 10–12 % Laemmli gels) and separated by applying a current of 160 V constant.
13. After electrophoresis, the gel is immediately dried on top of a Whatman paper with the help of a gel drying system (Bio-Rad).
14. Dried gels are exposed overnight using a Phosphorimaging screen (GE Healthcare Life Sciences) and scanned in a Phosphorimager (Bio-Rad).
15. Signal can be appropriately visualized and quantified with the help of ad-hoc software (such as Quantity One and/or Image Lab).

3.3 APC/C-Triggered In Vitro Substrate Degradation

In order to evidence cell cycle-dependent degradation mediated by the APC/C using mammalian extracts in vitro, the following procedure can be utilized:

1. Concentrated lysates (at least 10 μg/μl) are prepared from cell cycle-synchronized cells at the different phases of the cell cycle (as described above) (*see* **Note 5**).
2. Functional cell extracts are supplemented with an ATP regenerating system, 10 μM ubiquitin and 0.1 mg/ml cycloheximide to block protein synthesis in the extract.
3. Five μl of concentrated extracts are mixed with a 1/10 volume of radiolabeled substrate from a reticulocyte lysate in vitro transcription-translation reaction.
4. Extracts can eventually be supplemented with Cdc20 or Cdh1 (wild-type or mutant proteins).
5. Incubation is performed at 30 °C according to a kinetic including several time points between 0 and 3 h. The reaction is scaled-up accordingly to the number of time points.
6. Fractions are analyzed by SDS-PAGE.
7. Gels are fixed and stained with a solution containing 10% acetic acid, 40% methanol, and Coomassie blue staining. Destaining can be achieved by incubating the gels in the same solution in the absence of Coomassie blue.
8. Stained gels showing equal loading of total extracts are dried in Whatman papers with the help of a gel drying system.
9. Dried gels are visualized and analyzed with the help of a Phosphorimager.
10. Rates of degradation for a particular substrate can be obtained by fitting the intensity of the visualized bands to an exponential decay function, and eventually subtracting the rate of nonspecific degradation in the presence of 50 μM MG-132 to inhibit cellular proteasomes.
11. In order to confirm that the degradation observed for a particular substrate in the concentrated cell extracts is due to the APC/C, several experimental conditions can be tested: (a) addition of recombinant Emi1 (produced and purified from bacteria) to the extract should block degradation [27, 28]; (b) mutagenesis of linear motifs present in the substrates (such as D- or KEN-boxes) which are able to directly interact with the substrate adaptors/activators of the APC/C (Cdc20 and/or Cdh1) or potentially other subunits of the APC/C core, should also abrogate degradation [29, 30]; (c) blockade of the APC/C by chemical inhibitors such as TAME should also decrease or abrogate rates of substrates degradation [31]; and finally (d) extracts lacking appropriate levels of Cdc20 or Cdh1 should be less efficient instigating ubiquitylation and proteasomal degradation of APC/C substrates (*see* **Note 6**).

3.4 APC/C-Instigated In Vivo Substrate Ubiquitylation

Although most methodological approaches related to the study of the APC/C often utilize in vitro experiments in which the complexity of the cellular systems is reduced, it could also be beneficial to employ a cell line system in order to quickly be able to screen mutations performed at the level of the substrate but also at the level of the conjugating ubiquitins. Indeed, in this protocol, we propose the use of hemagglutinin (HA)-tagged ubiquitin either as a wild-type construct or mutated in one, several, or all possible residues (1 methionine and 7 lysines) used in ubiquitin to form a polymeric chain. On the other hand, the putative APC/C substrate will be also tagged (with either a FLAG- or a myc-tag) allowing the analysis of ubiquitylation of substrates in cells by a simple immunoprecipitation procedure as follows:

1. Tissue culture cell lines (such as HeLa S3, HEK-293T, or HFF-1) are grown in ad-hoc culture media. Typically, at least 1 × 10 cm dish of confluent cells are needed for each immunoprecipitation.
2. Cells are first washed once with PBS and harvested by trypsinization and centrifugation at room temperature at 1,600 × *g* for 5 min.
3. Cell pellets are washed once with PBS and subsequently kept on ice.
4. Lysis is performed in immunoprecipitation buffer (20 mM Tris–HCl, pH = 8.0; 150 mM NaCl; 1% Nonidet P-40 (NP-40) and 2 mM EDTA) supplemented with a proteases-phosphatases inhibitors cocktail (Roche) and 50 μM MG-132 in order to enhance visualization of ubiquitylated substrates. Lysis is performed on ice for not more than 20 min in a volume of ~1 ml per 10 cm dish of confluent cells.
5. Lysates are centrifuged at 16,000 × *g* for 20 min at 4 °C. The supernatant is transferred to pre-chilled Eppendorf tubes while pellets are either discarded or saved at −80 °C.
6. Cleared supernatants are incubated in the presence of ~1 μg antibodies directed against the tag of the APC/C substrate of interest. We routinely use anti-c-myc antibodies chemically conjugated to agarose beads (SC-40 AC from Santa Cruz Biotechnology) or anti-FLAG M2 affinity gel (Sigma Aldrich) in order to immunoprecipitate myc- or FLAG-tagged substrates, respectively. Incubation is performed for 6 h at 4 °C with constant rotation.
7. Immunoprecipitates are collected via centrifugation at 1,600 × *g* for 5 min at 4 °C, washed three times with immunoprecipitation buffer and one time with PBS at 4 °C.
8. The experiment is stopped by adding 20 μl of 2× sample buffer to the immunoprecipitated beads slurry.

9. Samples are boiled for 5 min at 95 °C and analyzed by SDS-PAGE and Western-blotting.
10. We suggest to perform first a Western-blot using an anti-HA polyclonal (rabbit) antibody in order to detect putative ubiquitylated proteins that will migrate at higher molecular weights compared to the weight of the unmodified substrate. By using antibodies raised in different species for both immunoprecipitation and Western-blotting, the idea is to reduce to a bare minimum the detection of the heavy and light chains of the immunoglobulins from the immunoprecipitating antibodies. The chemically conjugated commercially available antibody-beads (usually) display reduced background in this particular configuration. Subsequently, a re-blot should be performed using anti-tag substrate or anti-substrate antibodies in order to confirm immunoprecipitation of the right target. This is of crucial importance especially in cases when ubiquitylation patterns will be qualitatively and/or quantitatively compared under different experimental conditions such as mutations in the substrate and/or the conjugating ubiquitins.
11. Western-blotting is performed following classical methodologies. In our experience, we favor detection of bands using infrared fluorescently labeled secondary antibodies with the help of an Odyssey device from LI-COR Biosciences.
12. If ubiquitylation of a substrate is detected, inhibition of the APC/C by proTAME or by downregulating APC/C subunits can be done in order to confirm specificity of the reaction.

3.5 APC/C-Mediated In Vivo Substrate Degradation

The following protocol relies on the rationale that tempering with the protein levels of different key subunits of the APC/C should influence the cellular's ability to control the stability of a putative substrate of the APC/C via its proteasomal-mediated degradation.

Given the cell cycle-dependent activation of the APC/C, it is plausible that some of the effects observed will be better evidenced when cells are synchronized at specific cell cycle stages employing the protocols mentioned above.

3.5.1 Downregulation and Overexpression of APC/C Co-activators

Cdh1

Downregulation of Cdh1 can be achieved by using the pSUPER-Cdh1 construct that encodes for a synthetic siRNA against Cdh1: 5′-UGAGAAGUCUCCCAGUCAGTT-3′. The pSUPER empty plasmid is used as a negative control [22]. On the other hand, overexpression of Cdh1 can be accomplished with the help of the pCMV-myc-Cdh1 vector that expresses a myc-tagged Cdh1 protein in cells [22].

Cdc20

Downregulation of Cdc20 can be successfully achieved using siRNA reagents commercialized by Dharmacon (GE Healthcare) such as the ON-TARGETplus CDC20 siRNA SMART pool. On the other hand, overexpression of Cdc20 can be accomplished with the help of the pCMV-myc-Cdc20 vector expressing a myc-tagged Cdc20 protein in cells [22].

3.5.2 Downregulation of APC/C Subunits

Depletion of specific APC/C subunits can be performed with 100 pmol siGENOME Smartpool siRNA commercialized by Dharmacon (GE Healthcare) per well of a 6-well dish using lipofectamine RNAiMax (Life Technologies) as transfection reagent. Efficiency of the siRNAs targeting the subunits APC1, APC3, APC4, APC5, APC6, and APC8 of the APC/C have been recently reported [32].

Although downregulation or overexpression of subunits of the APC/C can readily alter the steady-state levels of the APC/C substrates, it is nevertheless highly recommended to perform careful calculations of the substrates half-life in the presence of cycloheximide during a kinetic of several hours including an adequate number of time points.

Overexpression of APC/C subunits can be achieved by transfecting cells with lipofectamine 3000 (Life Technologies) using untagged or tagged plasmids. Expression plasmids for the different APC/C subunits can be purchased from sources such as GeneCopoeia or OriGene.

3.5.3 Biochemical Inhibition of the APC/C

In order to block APC/C-instigated degradation in cells, several inhibitors can be used:

1. Emi1, a protein that is capable of inhibiting the APC/C, can be overexpressed in cells using the pCS2+-myc-Emi1 vector (generated by Peter K. Jackson, Genentech).
2. Overexpression of peptides comprising tandems of D-box motifs can also be used as a way to block APC/C-mediated ubiquitylation and proteasomal degradation of APC/C substrates in cells.
3. More recently exciting advances have been achieved by the group of Randall W. King in their efforts to chemically inhibit the APC/C. Currently, proTAME is a permeable prodrug that is converted inside the cells into its active compound (TAME) and that is capable of prematurely stopping efficient ubiquitylation of APC/C substrates. The drug is commercially available. More recently, King's group has reported another inhibitor (named APC inhibitor or apcin) that binds to Cdc20 and disrupts the ubiquitylation of D-box containing substrates [33, 34].

3.6 Binding Assays Using APC/C Subunits

In the following protocol, the APC/C subunits can be used either as baits or as putative interaction partners for virtually any other protein of interest:

Bacterially expressed and purified recombinant proteins, either GST- or 6xHistidine-tagged (1–5 μg) are pre-bound to 10 μl of glutathione sepharose beads (GE Healthcare) or Nickel-NTA agarose beads (Qiagen) respectively, by incubating for 2 h at 4 °C on a rotating-wheel, followed by three washes with binding buffer (50 mM Tris–HCl, pH = 7.5; 150 mM NaCl; 0.5 % NP-40; 5 mM EGTA; 5 mM EDTA; and 2 mM $MgCl_2$, supplemented with a proteases-phosphatases inhibitors cocktail (Roche)). Bait-bead complexes are then mixed with 5–10 μl of reticulocyte lysate (which has been previously programmed to express the protein of interest, labeled with ^{35}S-methionine) in a total volume of around 750 μl of binding buffer. After incubation for 4–6 h at 4 °C on a rotating-wheel, the beads are gently centrifuged, washed three times with binding buffer and resuspended in 25 μl of 2× sample buffer. The beads are finally boiled for 5 min and subjected to SDS-PAGE, followed by Coomassie staining of the gel in order to verify equal amounts of recombinant pulled-down proteins in every sample. Gels are finally dried and visualized/analyzed with the help of a Phosphorimager (Bio-Rad).

3.7 Regulation of APC/C Subunits by Posttranslational Modifications

Many subunits of the APC/C have been reported to be phosphorylated by several kinases. In particular, the substrate adaptors/activators Cdc20 and Cdh1 are the target of a complex regulation of their function by multisite phosphorylation [22].

To study phosphorylation of Cdc20 and/or Cdh1, in vitro kinase reactions can be performed with *bona fide* kinases (obtained as recombinant proteins or immunoprecipitated from cells) and recombinantly purified Cdc20 or Cdh1. Given the difficulties to produce those adaptors from bacteria, they should be either produced using the baculovirus-insect cells system or as chopped tagged proteins in bacterial systems. We have successfully produced mutant proteins of Cdh1 (deleted of its N- or C-terminus) in bacteria by tagging them with an MBP solubility-helping tag (encoded by the *MalE* gene). Those proteins were purified by standard protocols using amylase magnetic beads (New England Biolabs).

In vitro kinase assays can be performed in the so-called histone H1 kinase buffer (50 mM Tris–HCl pH = 7.5; 20 mM EGTA; 10 mM $MgCl_2$; 1 mM β-glycerophosphate; and 1 mM DTT) that especially works for kinases of the CDK family but that can be adopted (or adapted) for other kinases.

3.7.1 In Vitro Kinase Assays

1. Mix the kinase reaction in pre-chilled Eppendorf tubes on ice. A typical reaction includes in a total volume of 10–20 μl: 1–2 μg of the phosphorylation reaction substrate, 50 μM of unlabeled ATP, the kinase of interest (in suspension or bound to beads after immunoprecipitation), 0.1–1 μl of 32P-γ-ATP from a 10 μCi/μl stock and a sufficient volume of kinase buffer.
2. Incubate the reaction at 30 °C during 30–60 min with mild shaking if beads are present in the reaction.
3. Stop the reaction by adding 4× sample buffer.

4. Boil the samples at 95 °C for 5 min.
5. Subject the reaction to SDS-PAGE analysis (see Note 7).
6. Stain and destain the gel in order to visualize the reaction.
7. Dry the gel, visualize and analyze with the help of a Phosphorimager. The time of exposure using a Phosphorimaging plate can be gauged based on the emission detected with a Geiger counter. Time of exposure can vary from minutes to days.

4 Notes

1. Alternatively, in order to better preserve enzymatic activities from the extracts, harvested cells can be preferably disrupted by nitrogen decompression using a buffer composed of 20 mM Tris–HCl (pH = 7.5), 5 mM KCl, 1.5 mM $MgCl_2$, and 1 mM DTT, supplemented of a proteases-phosphatases inhibitors cocktail.
2. We have successfully used protein A/G agarose beads (Santa Cruz Biotechnology), protein A sepharose beads (Bio-Rad), or protein G agarose resins (Roche). The incubation with the beads (an amount equivalent to 50–100 μl of compacted slurry) is performed during 2 h at 4 °C with constant rotation. It is important that enough volume of mixture is present in the Eppendorf tube to guarantee a proper mixing of the beads during the incubation.
3. Accessorily, 50 μM MG-132 (a proteasome inhibitor) and 3 μM Ubiquitin aldehyde (Ubal, a deubiquitinating enzymes inhibitor) can be used depending on the particular experimental condition considered in order to detect enhanced substrate ubiquitylation.
4. In the case of the AF3.1 antibody (Santa Cruz Biotechnology), the competing peptide corresponds to amino acids 814–823 of human Cdc27/APC3.
5. In order to detect APC/C-Cdc20-mediated ubiquitylation and subsequent proteasomal degradation, extracts from cells transitioning during metaphase-anaphase are required. On the other hand, in order to detect APC/C-Cdh1-mediated ubiquitylation and subsequent proteasomal degradation, extracts from cells exiting mitosis and in early G1 phase are needed.
6. Concentrated extracts can be immunodepleted from APC/C subunits by sequential rounds of immunoprecipitation. Alternatively, relevant subunits of the APC/C involved in substrates' modification can be initially downregulated in the cells by means of shRNA or siRNA technologies.
7. Of note, the non-incorporated 32P-γ-ATP migrates slightly ahead than the bromophenol blue present in the sample buffer

forming the migration front of the gel. If the gel is long enough, the non-incorporated 32P-γ-ATP can be kept inside the gel and discarded as a solid waste, avoiding contamination of the gel running buffer and other laboratory plastic ware.

Acknowledgements

Research performed at the laboratory of Pathophysiological Cell Signaling is funded by the following bodies: FWO (G0C7514N grant), BELSPO (IAP-VII/07 program), VUB Research Council (new PI grant), and Innoviris (Brains Back to Brussels program to GJG). CC thanks financial support from the European Union's Seventh Framework Program (FP7) 2007–2013 under grant agreement no. 264257. The authors would also like to acknowledge networking support by the Proteostasis COST Action (BM1307).

References

1. Teixeira LK, Reed SI (2013) Ubiquitin ligases and cell cycle control. Annu Rev Biochem 82:387–414. doi:10.1146/annurev-biochem-060410-105307
2. Gutierrez GJ, Ronai Z (2006) Ubiquitin and SUMO systems in the regulation of mitotic checkpoints. Trends Biochem Sci 31(6):324–332. doi:10.1016/j.tibs.2006.04.001, S0968-0004(06)00095-8 [pii]
3. Mocciaro A, Rape M (2012) Emerging regulatory mechanisms in ubiquitin-dependent cell cycle control. J Cell Sci 125(Pt 2):255–263. doi:10.1242/jcs.091199
4. Peters JM (2006) The anaphase promoting complex/cyclosome: a machine designed to destroy. Nat Rev Mol Cell Biol 7(9):644–656. doi:10.1038/nrm1988
5. Chang L, Barford D (2014) Insights into the anaphase-promoting complex: a molecular machine that regulates mitosis. Curr Opin Struct Biol 29:1–9. doi:10.1016/j.sbi.2014.08.003
6. Barford D (2015) Understanding the structural basis for controlling chromosome division. Philos Trans A Math Phys Eng Sci 373(2036). doi: 10.1098/rsta.2013.0392
7. Pines J (2011) Cubism and the cell cycle: the many faces of the APC/C. Nat Rev Mol Cell Biol 12(7):427–438. doi:10.1038/nrm3132
8. Lu D, Hsiao JY, Davey NE, Van Voorhis VA, Foster SA, Tang C, Morgan DO (2014) Multiple mechanisms determine the order of APC/C substrate degradation in mitosis. J Cell Biol 207(1):23–39. doi:10.1083/jcb.201402041
9. Matsumoto ML, Wickliffe KE, Dong KC, Yu C, Bosanac I, Bustos D, Phu L, Kirkpatrick DS, Hymowitz SG, Rape M, Kelley RF, Dixit VM (2010) K11-linked polyubiquitination in cell cycle control revealed by a K11 linkage-specific antibody. Mol Cell 39(3):477–484. doi:10.1016/j.molcel.2010.07.001
10. Meyer HJ, Rape M (2011) Processive ubiquitin chain formation by the anaphase-promoting complex. Semin Cell Dev Biol 22(6):544–550. doi:10.1016/j.semcdb.2011.03.009
11. Song L, Rape M (2011) Substrate-specific regulation of ubiquitination by the anaphase-promoting complex. Cell Cycle 10(1):52–56
12. Williamson A, Banerjee S, Zhu X, Philipp I, Iavarone AT, Rape M (2011) Regulation of ubiquitin chain initiation to control the timing of substrate degradation. Mol Cell 42(6):744–757. doi:10.1016/j.molcel.2011.04.022
13. Kelly A, Wickliffe KE, Song L, Fedrigo I, Rape M (2014) Ubiquitin chain elongation requires E3-dependent tracking of the emerging conjugate. Mol Cell 56(2):232–245. doi:10.1016/j.molcel.2014.09.010
14. Meyer HJ, Rape M (2014) Enhanced protein degradation by branched ubiquitin chains. Cell 157(4):910–921. doi:10.1016/j.cell.2014.03.037
15. Lu Y, Wang W, Kirschner MW (2015) Specificity of the anaphase-promoting complex: a single-molecule study. Science 348(6231):1248737. doi:10.1126/science.1248737
16. Puram SV, Bonni A (2011) Novel functions for the anaphase-promoting complex in neurobiology. Semin Cell Dev Biol 22(6):586–594. doi:10.1016/j.semcdb.2011.03.006

17. Zhang Z, Yang J, Kong EH, Chao WC, Morris EP, da Fonseca PC, Barford D (2013) Recombinant expression, reconstitution and structure of human anaphase-promoting complex (APC/C). Biochem J 449(2):365–371. doi:10.1042/BJ20121374
18. Chang L, Zhang Z, Yang J, McLaughlin SH, Barford D (2014) Molecular architecture and mechanism of the anaphase-promoting complex. Nature 513(7518):388–393. doi:10.1038/nature13543
19. Frye JJ, Brown NG, Petzold G, Watson ER, Grace CR, Nourse A, Jarvis MA, Kriwacki RW, Peters JM, Stark H, Schulman BA (2013) Electron microscopy structure of human APC/C(CDH1)-EMI1 reveals multimodal mechanism of E3 ligase shutdown. Nat Struct Mol Biol 20(7):827–835. doi:10.1038/nsmb.2593
20. Brown NG, Watson ER, Weissmann F, Jarvis MA, VanderLinden R, Grace CR, Frye JJ, Qiao R, Dube P, Petzold G, Cho SE, Alsharif O, Bao J, Davidson IF, Zheng JJ, Nourse A, Kurinov I, Peters JM, Stark H, Schulman BA (2014) Mechanism of polyubiquitination by human anaphase-promoting complex: RING repurposing for ubiquitin chain assembly. Mol Cell 56(2):246–260. doi:10.1016/j.molcel.2014.09.009
21. Brown NG, VanderLinden R, Watson ER, Qiao R, Grace CR, Yamaguchi M, Weissmann F, Frye JJ, Dube P, Ei Cho S, Actis ML, Rodrigues P, Fujii N, Peters JM, Stark H, Schulman BA (2015) RING E3 mechanism for ubiquitin ligation to a disordered substrate visualized for human anaphase-promoting complex. Proc Natl Acad Sci U S A 112(17):5272–5279. doi:10.1073/pnas.1504161112
22. Gutierrez GJ, Tsuji T, Chen M, Jiang W, Ronai ZA (2010) Interplay between Cdh1 and JNK activity during the cell cycle. Nat Cell Biol 12(7):686–695, doi:ncb2071 [pii] 10.1038/ncb2071
23. Rosner M, Schipany K, Hengstschläger M (2013) Merging high-quality biochemical fractionation with a refined flow cytometry approach to monitor nucleocytoplasmic protein expression throughout the unperturbed mammalian cell cycle. Nat Protoc 8(3):602–626. doi:10.1038/nprot.2013.011
24. Gutierrez GJ, Tsuji T, Cross JV, Davis RJ, Templeton DJ, Jiang W, Ronai ZA (2010) JNK-mediated phosphorylation of Cdc25C regulates cell cycle entry and G(2)/M DNA damage checkpoint. J Biol Chem 285(19):14217–14228, doi:M110.121848 [pii] 10.1074/jbc.M110.121848
25. Chen M, Gutierrez GJ, Ronai ZA (2011) Ubiquitin-recognition protein Ufd1 couples the endoplasmic reticulum (ER) stress response to cell cycle control. Proc Natl Acad Sci U S A 108(22):9119–9124, doi:1100028108 [pii] 10.1073/pnas.1100028108
26. Chen M, Gutierrez GJ, Ronai ZA (2012) The anaphase-promoting complex or cyclosome supports cell survival in response to endoplasmic reticulum stress. PLoS One 7(4):e35520, doi:PONE-D-11-23151 [pii] 10.1371/journal.pone.0035520
27. Reimann JD, Freed E, Hsu JY, Kramer ER, Peters JM, Jackson PK (2001) Emi1 is a mitotic regulator that interacts with Cdc20 and inhibits the anaphase promoting complex. Cell 105(5):645–655
28. Wang W, Kirschner MW (2013) Emi1 preferentially inhibits ubiquitin chain elongation by the anaphase-promoting complex. Nat Cell Biol 15(7):797–806. doi:10.1038/ncb2755
29. Pfleger CM, Kirschner MW (2000) The KEN box: an APC recognition signal distinct from the D box targeted by Cdh1. Genes Dev 14(6):655–665
30. Pfleger CM, Lee E, Kirschner MW (2001) Substrate recognition by the Cdc20 and Cdh1 components of the anaphase-promoting complex. Genes Dev 15(18):2396–2407. doi:10.1101/gad.918201
31. Zeng X, Sigoillot F, Gaur S, Choi S, Pfaff KL, Oh DC, Hathaway N, Dimova N, Cuny GD, King RW (2010) Pharmacologic inhibition of the anaphase-promoting complex induces a spindle checkpoint-dependent mitotic arrest in the absence of spindle damage. Cancer Cell 18(4):382–395. doi:10.1016/j.ccr.2010.08.010
32. Clark E, Spector DH (2015) Studies on the contribution of human cytomegalovirus UL21a and UL97 to viral growth and inactivation of the anaphase promoting complex/cyclosome (APC/C) E3 ubiquitin ligase reveal a unique cellular mechanism for down-modulation of the APC/C subunits APC1, APC4, and APC5. J Virol. doi:10.1128/JVI.00403-15
33. Zeng X, King RW (2012) An APC/C inhibitor stabilizes cyclin B1 by prematurely terminating ubiquitination. Nat Chem Biol 8(4):383–392. doi:10.1038/nchembio.801
34. Sackton KL, Dimova N, Zeng X, Tian W, Zhang M, Sackton TB, Meaders J, Pfaff KL, Sigoillot F, Yu H, Luo X, King RW (2014) Synergistic blockade of mitotic exit by two chemical inhibitors of the APC/C. Nature 514(7524):646–649. doi:10.1038/nature13660

Chapter 16

Detection and Analysis of SUMOylation Substrates In Vitro and In Vivo

Cesyen Cedeño, Esther La Monaca, Mara Esposito, and Gustavo J. Gutierrez

Abstract

SUMOylation is a widely used protein posttranslational mechanism capable of regulating substrates localization, stability, and/or activity. Identification and characterization of *bona fide* SUMO substrates is a laborious task but its discovery can shed light to exquisite and crucial regulatory signaling events occurring within the cell. Experiments performed in the SUMOylation field often demand a good understanding of the putative substrate's function and necessitate a solid knowledge regarding both in vitro and in vivo approaches. This contribution offers a simplified view into some of the most common experiments performed in biochemical and cell biological research of the SUMO pathway in mammalian systems. It also summarizes and updates well established protocols and tricks in order to improve the likelihood to obtain reliable and reproducible results.

Key words SUMO pathway, SUMO substrates, SUMOylation

1 Introduction

Small ubiquitin-like modifier (SUMO, originally named Sentrin) is a protein composed of about 97 residues (depending of the isoform considered) after proteolytic processing of its C-terminal tail mediated by SUMO-specific proteases. SUMO is present in eukaryotic genomes as several isoforms with, in many cases, overlapping functions [1].

SUMO-activating enzyme (SAE) mediates the attachment of SUMO into substrates in a similar fashion as ubiquitin-activating enzymes (E1s) in the ubiquitylation reaction activate ubiquitin. However, in the case of SUMO, SAE is not a single protein but a heterodimer composed of SAE1 and SAE2 [2]. Phylogenetic conservation analyses of the heterodimer show that regions of the ubiquitin E1s are split into SAE1 and SAE2 in the SUMO pathway. The mechanistic details concerning adenylation and thio-esther bond formation are conserved among both processes.

Rune Matthiesen (ed.), *Proteostasis: Methods and Protocols*, Methods in Molecular Biology, vol. 1449, DOI 10.1007/978-1-4939-3756-1_16, © Springer Science+Business Media New York 2016

In the SUMOylation cascade, there is a unique SUMO-conjugating enzyme (E2) responsible for the transfer of SUMO into the substrate, termed Ubc9. The existence of a handful of SUMO ligases (E3s), directly responsible for mediating the conjugation of SUMO moieties into substrates, have been well established in the literature (such as the members of the PIAS family); however, it is also clear that Ubc9 can act at least in in vitro reactions as both E2 and E3 components during SUMO substrates' modification [3, 4].

Upon SUMOylation, several substrates' functions can be affected and there is no obvious rule to predict how SUMOylation will alter the structural and/or functional properties of a substrate. Remarkably, SUMOylation is a reversible mechanism since deSUMOylation can be achieved by SUMO proteases or SENtrin-specific proteases (SENPs) [5].

From many studies aimed to analyze the cellular SUMO proteome, it has become apparent that SUMOylation substrates often possess a SUMOylation consensus site in which a lysine (to which SUMO gets attached) is located in the middle of a linear motif defined by ΨXKE (where Ψ is a hydrophobic acid, X is any amino acid, and E is a glutamic acid or negatively charged residue). In addition, SUMOylation can also occur at different lysines which are not necessarily found within the context of the SUMOylation consensus sites of a given substrate.

The SUMOylation pathway also displays a great level of diversity and versatility. SUMOylated substrates can indeed be modified by mono-SUMOylation but also by poly- or multi-SUMOylation. Remarkably, SUMO isoforms have different preferences toward one or the other modifications, with SUMO1 being more prone to only conjugate once while SUMO2/3 rather generates poly-SUMOylated species, at least in mammals. Poly-SUMOylation occurs when the reaction employs another SUMO protein (which has been previously attached to a substrate) as the substrate. Among the three major SUMO isoforms in mammalians, only SUMO2 and 3 can form SUMO chains due to the presence of a SUMO consensus site in their primary sequences. SUMO1 lacks this site and therefore cannot form chains.

Finally, SUMOylated species can be recognized by other proteins via electrostatic interactions with SUMO. This process is mediated by the so-called SIMs or SUMO interacting motifs which are involved in the regulation of many different signaling pathways [6–8].

This chapter provides an updated insight into different protocols and experimental techniques that aim to identify and characterize SUMO substrates. Our contribution includes a standard research workflow (*see* Fig. 1) in which in vitro preliminary observations are subsequently confirmed and complemented with in vivo experiments.

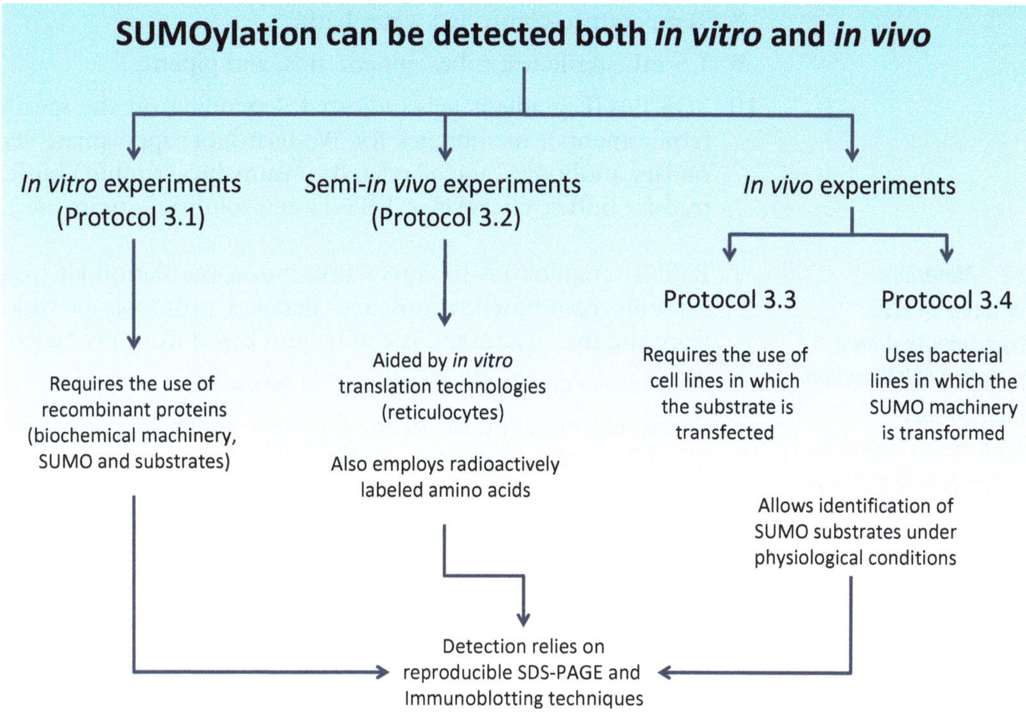

Fig. 1 Graphical workflow for detection of SUMOylation substrates

2 Materials

2.1 Materials for In Vitro Experiments Aimed to Detect SUMOylation of Recombinantly Purified Substrates

1. Recombinant E1 heterodimer or SUMO-activating enzyme (SAE): It can be purchased from Enzo Life Sciences Cat No. BML-UW9330.
2. Recombinant E2 or SUMO-conjugating enzyme: Enzo Life Sciences supplies an untagged Ubc9 (Cat No. ALX-201-046, *see* **Note 1**).
3. SUMO isoforms: SUMO 1, 2, or 3: They can be purchased from several sources including Boston Biochem.
4. SUMO-specific primary antibodies: anti-SUMO antibodies specific for each isoform are commercially available from different providers.
5. ATP regenerating system: 1.25 mM ATP, 1.25 mM $MgCl_2$, 1.9 mM Creatine Phosphate, and 6.25 μg/mL Creatine Phosphokinase (all components can be purchased from Sigma Aldrich).
6. 10× SUMO buffer (500 mM Tris–HCl pH = 7.5, 50 mM $MgCl_2$, supplemented with 10 mM ATP).
7. Purified substrate: produced and purified from recombinant organisms (such as *E. coli*). The substrate can be tagged in order to facilitate its detection by Western-blotting analysis or non-tagged if primary antibodies are available.

8. Temperature-controlled water bath.
9. 1.5 mL sterile test tubes, pipette tips, and pipettes.
10. SDS-PAGE gradient gels (adjusted depending on the specific requirements), membranes for Western-blot, appropriate secondary antibodies, and all related consumables (running buffer, transfer buffer, filter paper, PBS-Tween solution, casein, etc.).

2.2 Materials for Semi-In Vivo Experiments Aimed to Detect SUMOylation of Substrates Produced in Rabbit Reticulocytes Extracts

1. Rabbit reticulocytes in vitro transcription-translation kit (read carefully recommendations and detailed protocols provided from the manufacturer). It can be purchased from Promega.
2. ^{35}S-Methionine or ^{35}S-Cysteine (*see* **Note 2**).
3. Plasmids containing the gene of interest under the control of a T7, T3, or SP6 promoter (*see* **Note 2**).
4. *N*-ethylmaleimide, NEM (it acts as a SUMO protease (SENP) or deSUMOylase inhibitor) [9].
5. 1.5 mL sterile test tubes, pipette tips, and pipettes.
6. 15% SDS-PAGE gels (the acrylamide concentration can be adjusted depending on the specific requirements), and all related consumables (running buffer, Western-blot detection reagents, etc.).
7. All personal protecting equipment and devices needed in order to prevent and/or eliminate contaminations with radioactive material.

2.3 Materials for In Vivo Experiments Aimed to Detect SUMOylated Proteins in Cells

1. Cell lines transfected with the gene(s) of interest.
2. Plasmids for SUMO pathway proteins (SAE1/SAE2, Ubc9, and the SUMO isoforms). pCDNA3.1 plasmids for all components are available.
3. Agarose beads with immobilized coating of protein A/G.
4. Suitable primary anti-SUMO antibodies.
5. Substrate-specific antibodies and/or tag-specific antibodies (these antibodies can be used either for detection or for co-immunoprecipitation experiments).
6. Agarose glutathione beads.
7. SUMO Interacting Motif traps (SIMtraps).
8. *N*-ethylmaleimide, NEM (it acts as a SUMO protease or deSUMOylase inhibitor).
9. DMEM media, fetal bovine serum, 6-well plates or 10 cm dishes, and all needed tissue culture reagents.
10. 1.5 mL sterile test tubes, pipette tips, and pipettes.
11. 15% SDS-PAGE gels (acrylamide concentration can be adjusted depending on the specific requirements), membranes for Western-blotting, appropriate secondary antibodies, and all related consumables (running buffer, transfer buffer, filter paper, PBS-Tween solution, casein, etc.).

2.4 Materials for Semi-In Vivo Experiments Aimed to Detect SUMOylated Proteins in E. coli

1. Competent *Escherichia coli* BL21 (DE3).
2. Suitable bicistronic or tricistronic plasmids containing SAE1, SAE2, E2, and/or E3 enzymes as well as SUMO isoforms of interest (it is recommended to obtain these constructs from original sources reported in previous studies) [4, 10].
3. Antibiotics, LB media, LB agar plates, small culture tubes (with loose caps) and 1.5 mL sterile test tubes, pipette tips, and pipettes.
4. Plasmids containing the wild-type substrate of interest and also mutated (lysine-to-arginine) in its putative SUMOylation sites.
5. Suitable anti-SUMO-specific primary antibodies.
6. Substrate-specific antibodies and/or tag-specific antibodies for Western-blot applications.
7. 15% SDS-PAGE gels (acrylamide concentration can be adjusted depending on the specific requirements), membranes for Western-blot, appropriate secondary antibodies, and all related consumables (running buffer, transfer buffer, filter paper, PBS-Tween solution, casein, etc.).
8. Commassie-staining solution and distaining solution.
9. Bradford reagent and suitable protein standards.

3 Methods

3.1 In Vitro Experiments Aimed to Detect SUMOylated Substrates

1. Prepare and mix recombinant proteins (SAE1, SAE2, Ubc9, and SUMO isoforms) according to manufacturer's instructions in a 1.5 mL test tube. Add ATP regenerating system if desired (*see* **Note 3**).
2. Add the substrate protein and adjust the volume up to 30 μL with SUMO buffer and deionized water. Mix/homogenize properly.
3. Incubate the sealed test tube at 37 °C for 1 h. It is recommended to prepare reactions in which the different components of the SUMO machinery are individually removed as well as preparing a reaction without substrate as negative controls.
4. Upon completion, the reaction is stopped by addition of 10 μL of 4× SDS-PAGE sample buffer and boiled at 95 °C for 5 min.
5. Spin down all samples and load 20 μL of each assay into two 15% SDS-PAGE gels.
6. Perform SDS-PAGE adjusting time and voltage. One of the gels will be subsequently used for Western-blot analysis while the other will be Coomassie-stained.
7. Transfer the first gel onto a suitable membrane and block according to traditional methods for Western-blot. Incubate with primary antibodies against the specific substrate (or tag of the substrate). It is recommended to re-blot the membrane using

antibodies against the specific SUMO isoform(s) employed. Develop the membrane using adequate secondary antibodies.

8. A complete analysis of this experiment should reveal in first instance whether SUMOylation of the substrate takes place or not by evidencing a shift of ~15–20 kDa in the apparent weight of the substrate per SUMO moiety conjugated (*see* **Note 4**).
9. In the second gel, it is also possible to visualize by Coomassie-staining the SUMOylation reactions although it is not always obvious to detect it given the substoichiometric amounts of substrates that are usually SUMOylated in the in vitro reactions. In addition, there is a higher risk of ambiguity related to the exact nature of putative SUMOylated bands observed.

3.2 Semi-In Vivo Experiments Aimed to Detect SUMOylated Proteins Using Rabbit Reticulocyte Extracts

3.2.1 Production of the Radiolabeled Substrate for SUMOylation

1. Thaw the components of a rabbit reticulocyte lysates kit on ice 5–10 min prior to starting the assay. In parallel, place 1.5 mL test tubes on ice and set the temperature of a thermal bath at 37 °C.
2. Organize all personal protective equipment according to radioactive material safety protocols.
3. For a 10 μL reaction, add 5 μL of rabbit reticulocyte lysates into pre-chilled tubes and label them properly depending on the type of samples or controls to be tested.
4. Add the T7, T3, or SP6 promoter-based DNA plasmid containing the gene of interest. Use concentrated and pure DNA (>500 ng/μL) in order to dilute as little as possible the reticulocyte extracts and mix gently, always keeping the samples on ice. The volume of DNA added should not exceed 2 μL.
5. Extract enough ^{35}S-Methionine or ^{35}S-Cysteine from the shielded radioactive vial following safety protocols and transfer it to a designated tube kept on ice.
6. Add 2 μL of radiolabeled amino acid (according to the manufacturer's instructions) to the mixture still on ice, mix/homogenize and incubate the reaction for 90 min at 30 °C. Use a radiation-proof shield in front of the thermoblocker in order to contain radioactive emissions.
7. Stop the reaction by cooling on ice and store the sample at −80 °C. Take an aliquot of the reaction (0.5–2 μL) and add SDS-PAGE sample buffer and heat at 95 °C for 5 min avoiding spills or that the lids will "pop" due to the boiling. Spin down each sample before further handling.
8. Prepare an SDS-PAGE system in order to run the samples always behind a radiation-proof shield. Load samples and run using the appropriate settings for voltage and time.
9. Stop the run before the migration front of the gel escapes from the SDS-PAGE system in order to minimize radioactivity in the contaminated liquid waste.

10. Consider nonetheless the running buffer as radioactive waste and dispose it accordingly. Glass plates (if applicable), electrodes, and tank can be washed and reused only if they are decontaminated following protocols designed to handle contaminated radioactive materials (please check local regulations with the safety officer of your institution).
11. The gel (eventually after Coomassie-staining) is vacuum- and heat-dried using an appropriate device and the correct protections: (a) plastic wrap to avoid contact of the gel with the surfaces inside the gel drier; and (b) radiation-proof shields.
12. Expose the dried gel using a Phosphorimaging screen overnight at room temperature. Visualize and analyze the results with the help of a Phosphorimager and ad hoc software. The gel must be treated as radioactive solid waste once the analysis is done.
13. Analysis of the position of the radioactive band should reveal efficient transcription-translation of the substrate in vitro.

3.2.2 SUMOylation Assay Using the Radiolabeled Substrate

1. To 1–2 μL of rabbit reticulocyte lysates programmed to produce the radiolabeled substrate of interest, add a given amount (50–200 ng) of SUMO isoforms (mainly SUMO1, 2, or 3); 200 nM SAE1/SAE2; 500 nM Ubc9; and required volumes of 10× SUMO buffer and ATP regenerating system into a final volume reaction of 15–20 μL. Always keep in mind not to dilute too much the reactions.
2. Incubate the reaction at 37 °C for the desired time. We recommend a kinetic including time zero, 15, 30, 45, 60, and 120 min.
3. Analyze the reaction by SDS-PAGE as indicated above (Subheading 3.2.1). In this case, the radioactive bands detected by the Phosphorimager should show (usually) a major band corresponding to the molecular weight of the substrate while its SUMOylated form(s) should display apparent higher molecular weight(s) of 15–20 kDa per SUMO moiety conjugated.
4. A complete analysis using this method should include reactions in which the lysine-to-arginine mutants (generated by DNA mutagenesis) are also included.

3.3 Methodology for In Vivo Experiments Aimed to Detect SUMOylated Substrates

1. Obtain a cell line (transiently or stably) expressing or overexpressing the protein of interest. Protocols and suitable cell lines will depend on the nature of the given pathway that is scrutinized.
2. Prepare a lysis buffer according to the specific experimental requirements of the substrate and cell line utilized. We suggest a buffer with the following composition: salts 0–1 M, ionic detergent (if applicable): 0.01–0.5 %, non-ionic detergent (if applicable): 0.1–1 %, divalent cations: 0–10 mM, EDTA: 0–5 mM, pH = 7–8. As an example, we propose the following buffer: 100 mM Tris–HCl, 5 mM EDTA, 1 mM TCEP, 0.5 % NP-40, supplemented with Complete tablets (Roche Diagnostics) and 1 mM NEM.

3. Select a plate or set of wells containing at least 10^6 cells (transfected and non-transfected as negative controls), strip them out using trypsin-EDTA in the case of adherent cells. Wash the cells with PBS once and resuspend them in 500 μL of lysis buffer and incubate on ice for 20 min.
4. Centrifuge the lysates at 16,000 × *g* at 4 °C for 20 min. The pellet contains cellular debris and insoluble material. Save the supernatant and collect it into pre-chilled 1.5 mL test tubes kept on ice.
5. Measure the total protein concentration of the supernatant by performing a Bradford assay and incubate identical amounts (usually at least 1 mg) of lysates produced from transfected and non-transfected cells with 1 μg of an antibody directed against the substrate of interest (tag or protein substrate itself). Incubate at 4 °C for 4–6 h with constant rotation.
6. Prepare agarose-protein A and/or G beads (according to the specific antibody used) by washing and equilibrating them in the appropriate buffer used for co-immunoprecipitation (see details above). Add protein A and/or G beads (15 μL of slurry) to the lysate previously incubated with the immunoprecipitating antibody and incubate again for 1–2 h at 4 °C with constant rotation.
7. Centrifuge for 5 min at 1,600 × *g* at 4 °C to separate supernatant from pellet (containing the beads). Wash the pellet (beads + antibodies + immunoprecipitated proteins) three times with 1 mL of buffer, always performing very mild centrifugation steps (1,600 × *g* at 4 °C for 5 min).
8. Add 15 μL 2× of SDS-PAGE loading buffer into the tube and boil for 5 min at 95 °C. Load samples onto an SDS-acrylamide gel. Set the proper voltage and time and perform SDS-PAGE.
9. Transfer the gel into a suitable membrane (usually nitrocellulose or PVDF) and block the membrane according to traditional methods for Western-blotting. Incubate with primary antibodies directed against different SUMO isoforms (mainly SUMO1, 2, or 3). Develop the membrane using appropriate secondary antibody taking into consideration that IgGs from the antibody used for immunoprecipitation were also transferred to the same membrane. A careful choice of antibodies can avoid cross-reactions and interference between antibodies (*see* **Note 5**).
10. The result of this Western-blot will complement any in vitro assay previously done with the same substrate. This experiment provides evidence that the endogenous cellular machinery can SUMOylate the substrate of interest. In the same way as indicated before, the SUMOylation site can be confirmed or identified if several conditions using candidate lysine-to-arginine mutants are employed.
11. SUMOylation efficiency in cells can be "boosted" by co-overexpressing Ubc9 together with the substrate of interest.

On the other hand, a dominant-negative (or activity-dead) mutant of Ubc9 (mutated in the Cysteine used for SUMO esterification) should reduce the efficiency of SUMOylation of the substrate by the cellular endogenous machinery.

3.3.1 Optional Protocol

1. From **step 5**, it is possible to perform a pull-down experiment rather than an immunoprecipitation. This protocol takes advantage of the ability of some short linear motifs to bind SUMO isoforms [6]. SUMO-interacting motifs (SIM) are conserved linear elements that can be placed in tandem along an engineered sequence to create an artificial polypeptide called SUMO-trap [11]. In this particular protocol, SUMO-traps are tagged with GST (but other tags such as V5 also exist).
2. Wash and equilibrate GSH beads.
3. Mix SUMO-trap and lysates prior to incubation with GSH beads. Incubation of SUMO-traps with the lysate should be adjusted depending on the substrate since selective attachment of SUMO1, 2, or 3 may affect interaction with SUMO-traps (*see* **Note 6**). In general, incubation should be carried out at 4 °C for several hours.
4. Wash beads with lysis buffer three times and recover the beads (and save the supernatants that can also be analyzed by SDS-PAGE to confirm depletion of SUMOylated species by the SUMO-traps) after centrifuging the mixture at 4 °C for 5 min at 1,600 × *g*.
5. Elute bound material to SUMO-traps using a small volume of lysis buffer supplemented with 25 μM freshly dissolved Glutathione (check pH of the buffer when preparing it). Centrifuge at 3000 × *g* at 4 °C for 3 min and recover supernatant in a new pre-chilled test tube.
6. Proceed with the regular (previous) protocol from **step 8**.

3.4 Methodology for Semi-In Vivo Experiments Aimed to Detect SUMOylated Proteins in E. coli

1. Transform SAE1/SAE2 or fusion construct into *E. coli* BL21 (DE3) cells. Obtain positive colonies and proceed sequentially with another transformation using Ubc9 DNA plasmid [4, 10]. Plate and select using antibiotics after each transformation.
2. Generate competent cells from the positive colonies above and split into three different new transformations, one for each SUMO isoform. Prepare competent cells of all the bacterial lines.
3. Finally, transform the gene of interest or its lysine-to-arginine mutants in different reactions, plate and select them using the appropriate antibiotics. Keep in mind that a negative control should be included in which an empty plasmid (the backbone plasmid of the substrate) should be transformed in parallel.
4. Select a single colony and grow it on 3 mL LB media supplemented with the adequate antibiotics overnight at 37 °C with shaking.

5. Transfer 250 μL of each culture into fresh media up to a final volume of 5 mL. Supplement with the proper antibiotics and grow until the O.D. at 600 nm reaches between 0.6 and 1.0.
6. Induce using the corresponding compounds (IPTG, arabinose, etc.).
7. Collect the pellet after 2 h of expression and lyse the cells by sonication using the lysis buffer described in Subheading 3.3, **step 2** but without detergents.
8. Measure total protein concentration by Bradford and prepare samples of ~30 μL of each lysate with the same amount of proteins (60–80 μg).
9. Add 10 μL of 4× SDS-PAGE sample buffer into each tube and boil for 5 min at 95 °C. Split samples in two.
10. Load samples onto two different SDS-PAGE gels. Set the proper voltage and time and perform SDS-PAGE.
11. Transfer each gel into a suitable membrane and block according to traditional methods for Western-blotting. Incubate one membrane with primary antibodies against different SUMO isoforms (mainly SUMO1, 2, or 3). The second membrane could be blotted using a substrate-specific antibody.
12. Process the membranes using the corresponding secondary antibodies (*see* **Note** 7).

4 Notes

1. It is also possible to use a fusion protein containing the functional units of both SAE1 and SAE2 [12].
2. Check how many methionines or cysteines are encoded within the sequence of the substrate as it can determine the feasibility of detecting a proper signal and which radiolabeled amino acid(s) to employ.
3. Protein concentrations can be adapted but is recommended to use a molar ratio of 1:2:20 of E1 heterodimer:Ubc9:SUMO isoforms at the nM–μM range. Substrates can be added as a 10× molar excess compared to the total amount of SUMO in the test tube.
4. Consensus SUMOylation sites can be inferred using different SUMO predictors available on the Internet. Based on this information, it is also possible to produce lysine-to-arginine mutants at particular locations on the substrates. Quikchange mutagenesis can be performed to disrupt the SUMOylation site(s). Including this control in the protocol will offer an invaluable test of the specificity of the SUMOylation reaction, in addition to allowing the identification of the modification site(s) in the substrate.

5. It is also possible to overexpress, in the cells transfected with the putative SUMOylatable substrate, the different (tagged) SUMO isoforms using mammalian plasmids suitable for transfection.
6. SIM-traps can be produced *in-house*, requested to other laboratories under MTAs, or purchased from providers such as Ubiquitin-Proteasome Biotechnologies (Cat. No. J4410) or Boston Biochem (Cat. No. AM-200).
7. The results from these blots will provide evidence that the SUMO machinery can SUMOylate the substrate under physiological conditions in a heterologous system. As suggested, the SUMO site can be confirmed if several bacterial clones are created using candidate lysine-to-arginine mutants. The system is advantageous in the sense that large amounts of SUMOylated substrate can be produced and purified in order to perform further biophysical characterizations. There is also a well-described way of performing this protocol using the plant-specific SUMO machinery [13].

Acknowledgements

Research performed at the laboratory of Pathophysiological Cell Signaling is funded by the following bodies: FWO (G0C7514N grant), BELSPO (IAP-VII/07 program), VUB Research Council (new PI grant), and Innoviris (Brains Back to Brussels program to GJG). CC thanks financial support from the European Union's Seventh Framework Program (FP7) 2007–2013 under grant agreement no. 264257. The authors would also like to acknowledge networking support by the Proteostasis COST Action (BM1307).

References

1. Hay RT (2013) Decoding the SUMO signal. Biochem Soc Trans 41:463–473
2. Tang Z, Hecker CM, Scheschonka A et al (2008) Protein interactions in the sumoylation cascade—lessons from X-ray structures. FEBS J 275:3003–3015
3. Wang J, Taherbhoy AM, Hunt HW et al (2010) Crystal structure of UBA2ufd-Ubc9: insights into E1-E2 interactions in sumo pathways. PLoS One 5:e15805
4. O'Brien SP, DeLisa MP (2012) Functional reconstitution of a tunable E3-dependent sumoylation pathway in Escherichia coli. PLoS One 7:e38671
5. Hickey CM, Wilson NR, Hochstrasser M (2012) Function and regulation of SUMO proteases. Nat Rev Mol Cell Biol 13:755–766
6. Song J, Durrin LK, Wilkinson TA et al (2004) Identification of a SUMO-binding motif that recognizes SUMO-modified proteins. Proc Natl Acad Sci U S A 101:14373–14378
7. Namanja AT, Li YJ, Su Y et al (2012) Insights into high affinity small ubiquitin-like modifier (SUMO) recognition by SUMO-interacting motifs (SIMs) revealed by a combination of NMR and peptide array analysis. J Biol Chem 287:3231–3240
8. Kerscher O (2007) SUMO junction-what's your function? New insights through SUMO-interacting motifs. EMBO Rep 8:550–555
9. Da Silva-Ferrada E, Lopitz-Otsoa F, Lang V et al (2012) Strategies to identify recognition signals and targets of SUMOylation. Biochem Res Int 2012:875148

10. Mencía M, De Lorenzo V (2004) Functional transplantation of the sumoylation machinery into Escherichia coli. Protein Expr Purif 37:409–418
11. Da Silva-Ferrada E, Xolalpa W, Lang V et al (2013) Analysis of SUMOylated proteins using SUMO-traps. Sci Rep 3:1690
12. Uchimura Y, Nakao M, Saitoh H (2004) Generation of SUMO-1 modified proteins in E. coli: towards understanding the biochemistry/structural biology of the SUMO-1 pathway. FEBS Lett 564:85–90
13. Okada S, Nagabuchi M, Takamura Y et al (2009) Reconstitution of Arabidopsis thaliana SUMO pathways in E. coli: functional evaluation of SUMO machinery proteins and mapping of SUMOylation sites by mass spectrometry. Plant Cell Physiol 50:1049–1061

Chapter 17

Detection of Protein–Protein Interactions and Posttranslational Modifications Using the Proximity Ligation Assay: Application to the Study of the SUMO Pathway

Marko Ristic, Frédérique Brockly, Marc Piechaczyk, and Guillaume Bossis

Abstract

The detection of protein–protein interactions by imaging techniques often requires the overexpression of the proteins of interest tagged with fluorescent molecules, which can affect their biological properties and, subsequently, flaw experiment interpretations. The recent development of the proximity ligation assays (PLA) technology allows easy visualization of endogenous protein–protein interactions at the single molecule level. PLA relies on the use of combinations of antibodies coupled to complementary oligonucleotides that are amplified and revealed with a fluorescent probe, each spot representing a single protein–protein interaction. Another application of this technique is the detection of proteins posttranslational modifications to monitor their localization and dynamics in situ. Here, we describe the use of PLA to detect protein SUMOylation, a posttranslational modification related to ubiquitination, as well as interaction of SUMOylated substrates with other proteins, using both adherent and suspension cells.

Key words Proximity ligation assay, SUMOylation, Protein–protein interaction

1 Introduction

SUMOylation involves the covalent conjugation of the ubiquitin-related modifiers SUMO-1, -2, or -3 via isopeptide bond formation on target protein lysines. It has appeared in the recent years that SUMOylation plays a role as important as that of phosphorylation in the control of protein function and fate. Among others, SUMOylation has been involved in the regulation of cellular processes such as replication, transcription, DNA damage repair, protein stability, and localization [1].

The study of SUMOylation is a challenging task due to the low steady-state abundance of SUMOylated proteins (typically 0.1–1 % of a given target is SUMOylated at steady state) [2]. Moreover,

Rune Matthiesen (ed.), *Proteostasis: Methods and Protocols*, Methods in Molecular Biology, vol. 1449,
DOI 10.1007/978-1-4939-3756-1_17, © Springer Science+Business Media New York 2016

localizing SUMOylated proteins using cellular imaging techniques has been hampered by the lack of specific tools. In particular, antibodies strictly specific for the SUMOylated forms of particular proteins (i.e., comparable to phosphoprotein-specific antibodies) are still lacking. To circumvent this limitation, we have recently generated the first antibody of this kind to specifically recognize the SUMOylated form of the oncogenic c-Fos transcription factor [3], which is one of the best-studied components of the dimeric AP-1 transcriptional complex. Using this antibody, we could clearly demonstrate by Chromatin Immunoprecipitation (ChIP) the binding of SUMOylated c-Fos to its target promoters during transcriptional activation. However, we could not analyze the intranuclear localization of SUMOylated c-Fos (hereafter called c-Fos-SUMO) using this antibody in regular immunofluorescence assays, most probably because of its low abundance and, maybe, the poor accessibility of the epitope. This underlines the necessity to resort to other approaches to address this issue.

Proximity ligation assay (PLA) is a technology that enables visualizing and localizing single proteins (in particular, in the case of low abundance proteins that are not detectable with immunofluorescence techniques), protein–protein interactions, and protein posttranslational modifications in fixed cells [4]. It relies, first, on the use of two antibodies from different species, directed to (1) the same protein for visualization of individual proteins, (2) two different proteins for detection of protein–protein interactions, and (3) a protein and a posttranslational modifier when studying posttranslation modification. Another essential requirement is the use of two secondary antibodies, which, on the one hand, are capable of recognizing the two species type of primary antibodies and, on the other hand, are coupled to specific oligonucleotides (they are called PLA probes PLUS and MINUS). When the two primary antibodies are in close proximity (i.e., within <40 nm), the addition of an oligonucleotide complementary to those conjugated to the secondary antibodies allows a circle of DNA to form. Then, a rolling circle polymerization process can be used to amplify the hybridized probes, which are detected using a specific fluorescent oligonucleotidic probe. PLA signals are analyzed using a fluorescence microscope and appear as individualized fluorescent spots, each spot representing a single interaction between the PLUS and MINUS probes (Fig. 1)

Here, we describe a detailed protocol to perform PLA to detect, not only SUMOylated proteins, but also interaction of SUMOylated proteins with specific partners. In particular, we apply this method to detect the SUMOylation of RanGAP1, the most abundant SUMO-1 target whose SUMOylation is required for its binding to the nuclear pore complex [5] as well as that of the c-Fos transcription factor [3, 6]. In addition, we use PLA to show that SUMOylated proteins are found associated with histone marks characteristic of both repressed and active chromatin. Protocols are provided for both adherent and suspension cells.

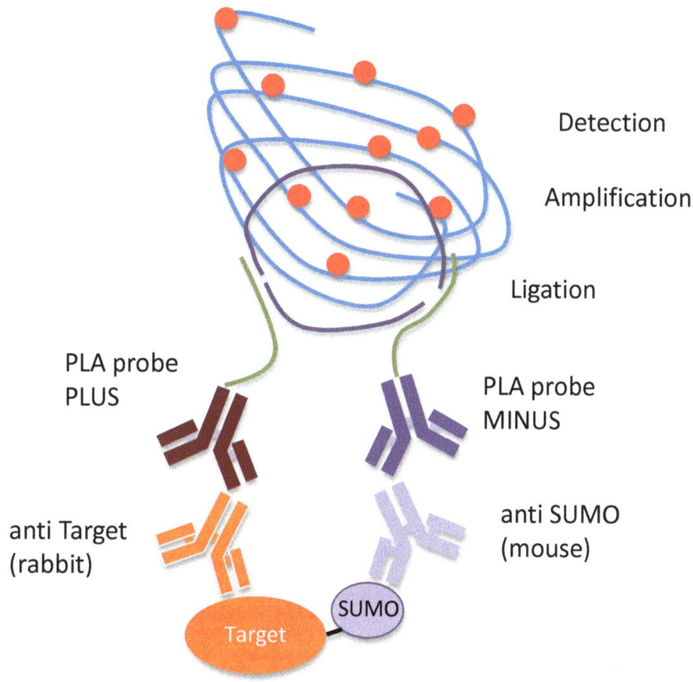

Fig. 1 Proximity ligation assay to detect protein SUMOylation. Two primary antibodies, generated in different species, are used. One recognizes the protein of interest and the other SUMO-1 or SUMO-2. Secondary antibodies specific for the primary antibodies species and coupled to oligonucleotide probes (PLA probes PLUS and MINUS) are then used. The PLA probes are then ligated and circularized. The circular DNA is then amplified upon a rolling circle polymerization. A fluorescently labeled oligonucleotide is then used as a detection probe

2 Material

All buffers are prepared extemporaneously. The volumes of buffers vary according to the number of samples.

1. PBSi is a phosphate-balanced saline (PBS) buffer (137 mM NaCl, 2.7 mM KCl, 10 mM sodium phosphate buffer, pH 7.4) containing various inhibitors: 10 mM *N*-methyl-maleimide (NEM) (*see* **Note 1**), 1 μg/ml Aprotinin, 1 μg/ml Pepstatin, and 1 μg/ml Leupeptin.
2. For studying cells in suspension, the CSK-0.1 % buffer can be used (*see* **Note 4**). It contains: 0.1 % Triton X-100, 300 mM sucrose, 100 mM NaCl, 3 mM $MgCl_2$, 1 mM EDTA, 10 mM Tris–HCl pH 7.4, 10 mM NEM, 1 μg/ml Aprotinin, 1 μg/ml Pepstatin, and 1 μg/ml Leupeptin.
3. For studying adherent cells, the CSK-0.5 % buffer can be used (*see* **Note 4**). It contains: 0.5 % Triton X-100, 300 mM sucrose, 100 mM NaCl, 3 mM $MgCl_2$, 1 mM EDTA, 10 mM Tris–HCl pH 7.4, 10 mM NEM, 1 μg/ml APL.

4. 4% Paraformaldehyde (PFA) in PBSi.
5. 10 mM Glycine in PBS.
6. 0.5% Triton X-100 in PBS.
7. 10% Fetal Calf Serum in PBS.
8. 5% Fetal Calf Serum in PBS complemented with 100 mM NaCl (final concentration of NaCl: 237 mM).
9. Wash Buffer A: 25 mM Tris–HCl pH 7.4, 100 mM NaCl, 0.05% Tween 20.
10. Wash Buffer B: 200 mM Tris–HCl pH 7.4, 100 mM NaCl.
11. ProLong Diamond Antifade Mountant with DAPI (Life Technologies: P36962).
12. Liquid Blocker Super Pap Pen (e.g., Life Technologies: 00-8899)
13. Coverglas forceps.
14. 76×26 mm Microscope slides (e.g., Knittel Glass StarFrost).
15. 12 mm-diameter Cover glasses.
16. Diagnostic slides (10-well slides with Teflon, e.g., Thermo Scientific, ER-208B-CE24) for studying adherent cells.
17. Cytospin, Cytofunnel, and Cytoclips (Thermo Scientific, 3120110) for studying cells in suspension.

2.1 PLA Kit

1. PLA PLUS and MINUS probes (Olink) are distributed by Sigma-Aldrich. They are available for different species (goat, mouse, rabbit). The PLUS probe needs to be combined with a MINUS probe. Alternatively, primary antibodies can be directly coupled to PLUS or MINUS probes using Duolink Probemaker kits (Sigma-Aldrich, DUO92909 for PLUS kit and DUO92910 for MINUS kit).
2. Duolink In Situ Detection Reagents (Olink) are distributed by Sigma-Aldrich. The Duolink kit contains all reagents for the ligation of probes, their amplification, and their detection. It exists in different colors (green, orange, red, far red) and for brightfield.

3 Methods

All steps are carried out at room temperature unless otherwise specified. The protocol is described for both suspension and adherent cells.

3.1 Preparation of Cells

3.1.1 Adherent Cells

1. Add 50 μl of polylysine (*see* **Note 2**) to each well of 10-well diagnostic slides (*see* **Note 3**) for 10 min. Remove polylysine and let slides dry.
2. Seed the cells on the slides in 50 μl of culture medium and incubate at 37 °C for 24 h. Typically, seed 5×10^3 cells per well.

3. Remove the medium and incubate with CSK-0.5 % (*see* **Note 4**) at room temperature for 10 min.
4. Remove the CSK-0.5 % buffer and replace by 50 μl of 4 % paraformaldehyde (PFA) (*see* **Note 5**) in PBSi. Incubate for 15 min.
5. Wash with 10 mM Glycine in PBS for 10 min.
6. Permeabilize cells with 0.1 % Triton X-100 in PBS for 5 min (*see* **Note 6**).
7. Wash twice with PBS.

3.1.2 Suspension Cells

1. For easier handling, cells are cytospined on coverslips rather than on regular microscope slides. First, the opening of the Cytofunnels should be marked on microscope slides. To this aim, place a glass slide into the cytospin cassette (Cytoclips) together with the plastic Cytofunnels and mark the position of its opening with a pen.
2. Remove the microscope slide, add a drop of water on the marked circle and cover with a coverslip (the water is used to stick the coverslip). Remove the excess of water.
3. Cover the glass slide with a cardboard filter in which you make a hole facing the one in the Cytofunnel. Assemble the glass slide (with the coverslip) covered with the filter and the Cytofunnel in the Cytoclip.
4. Resuspend the cells at a concentration of 6×10^6/ml in PBSi.
5. Transfer 70 μl of the cell suspension in the Cytofunnel and spin at 1200 rpm for 2 min.
6. Place the coverslip with the cells on 50 μl of CSK-0.1 % for 1 min (do not exceed this time, as the cells will detach) [4]. The buffer is deposited on a parafilm and the coverslip is flipped on the drop.
7. Transfer the coverslip on 50 μl of 4 % paraformaldehyde, in PBSi and incubate for 15 min.
8. Transfer the coverslip in a 12-well plate and wash with PBS containing 10 mM Glycine for 5 min. Repeat the washing step.
9. Permeabilize the cells by placing the coverslip on 50 μl of 0.5 % Triton X-100 in PBS for 5 min.
10. Wash once with PBS. Remove the remaining PBS with a vacuum pump and place the coverslip, cells on the top, on a parafilm and wait that the remaining PBS on the coverslip forms a drop where cells are located. Make a circle on the edge of this drop with the Liquid Blocker pen. This step is performed to minimize the study area and, thereby, to use lower amounts of the PLA kit.

3.2 Proximity Ligation Assay

All further steps are common to adherent and suspension cells.

1. Incubate cells in the blocking buffer (10% fetal calf serum-containing PBS (*see* **Note 7**)) at 37 °C for 30 min.

2. Mix both primary antibodies (*see* **Note 8**) from different species (dilutions from 1/100 to 1/500, depending on the antibody) in 5% fetal calf serum-containing PBS complemented with 100 mM NaCl. A control should be made where only one of the primary antibody is used. To prove the specificity of the signal, another control should be made where at least one of the target proteins is suppressed, for example using RNA interference.
3. Add 25 μl of antibody solution onto cells. Incubate at 37 °C for 2 h in a humidity chamber.
4. Wash 2 × 5 min with Buffer A (*see* **Note 9**).
5. Dilute PLA probes (1/5) (*see* **Note 10**) in the buffer provided in the kit. Use one MINUS- and one PLUS probe, each one of them specific for a different antibody species.
6. Add 20 μl on the cells and incubate at 37 °C for 1 h in a humidity chamber.
7. Wash 2 × 5 min with Buffer A (*see* **Note 9**).
8. Prepare the ligation mix (Ligation Buffer 1/5, Ligase 1/40 in H_2O).
9. Add 20 μl on the cells and incubate at 37 °C for 30 min in a humidity chamber.
10. Wash 2 × 5 min with Buffer A (*see* **Note 9**).
11. Prepare the amplification mix (Amplification Buffer 1/5, Polymerase 1/80 in H_2O)
12. Add 20 μl onto cells and incubate at 37 °C for 100 min in a humidity chamber.
13. Wash 2 × 5 min with Buffer B (*see* **Note 9**).
14. Dry the slides/coverslips in the dark and mount them using a DAPI-containing mounting medium such as the Duolink In Situ Mounting medium (Olink) or the Prolong Diamond (Life Technologies).

3.3 Imaging and Image Analysis

PLA signals appear as bright fluorescent spots in epifluorescence microscope analyses. Each spot corresponds to a single interaction between the two antibodies used in the assay. As the spots can be on different focal plans, it is advised to use confocal microscope analysis and to scan all Z plans. All images should be acquired using the same settings. PLA signals can be quantified as the number of PLA spots per cell. This analysis can be performed using the Duolink Image Tool.

3.4 Application of PLA to the Study of SUMOylation

3.4.1 Detection of Protein SUMOylation

PLA can be used to analyze posttranslational modifications of proteins, in particular SUMOylation. An antibody to both SUMO and the target protein should be used. We routinely use the 21C7 [7] and the 8A2 [8] monoclonal antibodies to detect SUMO-1 and SUMO-2, respectively. The hybridomas for both of these mouse monoclonals are available through the DHSB (Developmental Studies Hybridoma Bank, http://dshb.biology.uiowa.edu/). RanGAP is the most prominent SUMO-1 target and its SUMOylation is required for its targeting to the nuclear pore complex [5, 7]. Using PLA with both RanGAP and SUMO-1 antibodies, we could confirm the close proximity between RanGAP and SUMO-1 at the nuclear periphery, likely representing its SUMOylated form (Fig. 2). However, it could also represent the binding of RanGAP to a SUMOylated protein. Unfortunately, mutating the SUMO acceptor sites to induce loss of the PLA signal could not be considered to prove formally that the PLA signal corresponded to SUMOylated RanGAP, as SUMOylation is required for proper localization of RanGAP at the pore. PLA has also been used to detect the SUMOylation of a GFP fusion of the ZBTB1 protein. Importantly in this case, the mutation of the acceptor lysines abolished the PLA signal, which confirmed that it actually corresponded to the SUMOylated form of ZBTB1 and not on interaction of ZBTB1 with a SUMOylated partner [9].

An alternative choice to detect SUMOylated proteins is the use of antibodies specific for the SUMOylated form of a given protein, as we described for c-Fos [3]. As mentioned above, the low abundance of c-Fos itself, as well as that of its SUMOylated form (c-Fos-SUMO <1 % of total c-Fos) in living cells and the poor accessibility of the targeted epitope (T-shaped peptide containing the C-terminus of SUMO bound to the c-Fos SUMOylation domain; see [3]) made regular immunofluorescence studies unlikely to be successful. To circumvent this limitation, we used PLA with, on one side, an anti-c-Fos-SUMO antibody and, on the other side, anti-SUMO-1 or -2 antibodies to amplify the signal and detect the SUMOylated form of c-Fos (Fig. 3). Thanks to its amplification step, PLA allowed the visualization of endogenously SUMOylated c-Fos [3].

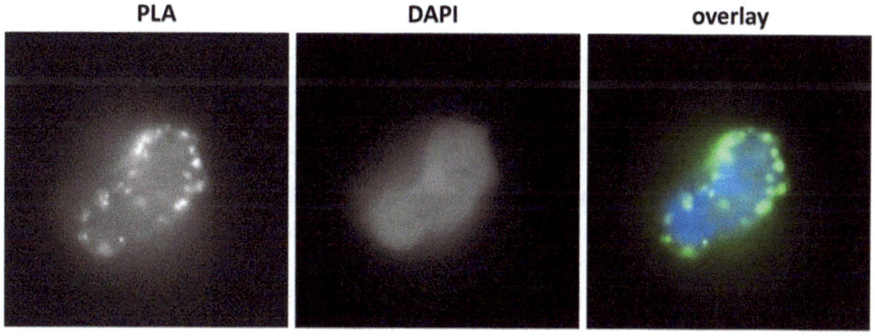

anti-RanGAP and anti-SUMO-1

Fig. 2 In situ detection of the SUMOylation of RanGAP. The human Acute Myeloid Leukemia cell line HL-60 was used in the PLA protocol for suspension cells. PLA was carried out using a goat anti-RanGAP antibody [15], the mouse monoclonal 21C7 SUMO-1 antibody [7] and the Duolink In Situ detection kit with green fluorochrome. Images were acquired with a Leica DM6000 microscope

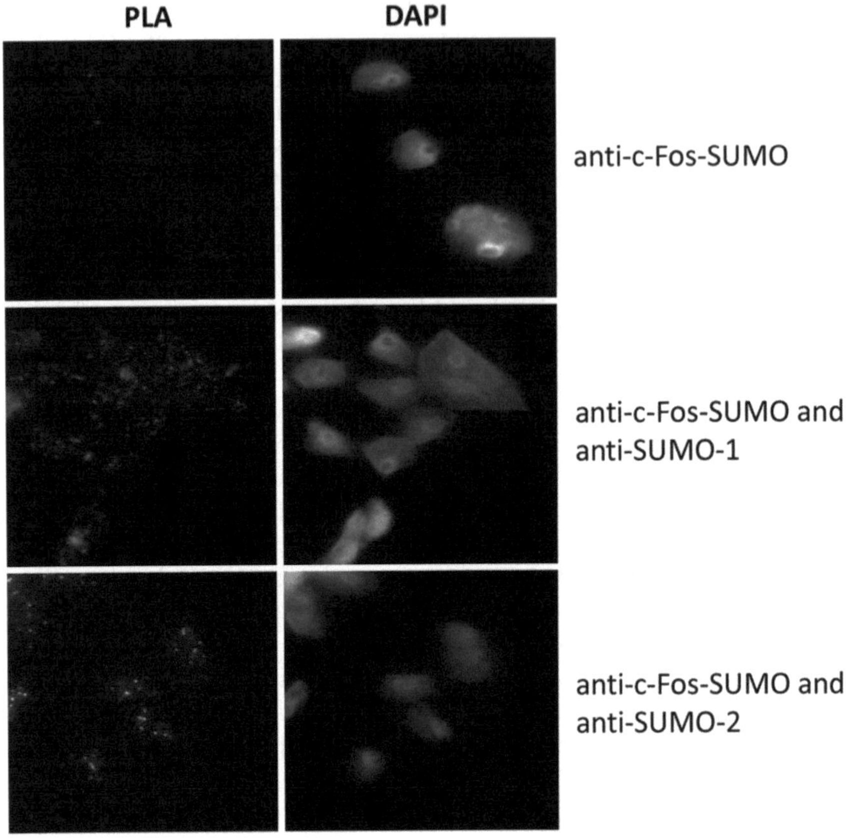

Fig. 3 In situ detection of the SUMOylated form of c-Fos. The human cell line 293T was used in the PLA protocol for adherent cells. PLA was carried out using a rabbit anti-c-Fos-SUMO antibody [3] alone or with either the mouse monoclonal 21C7 SUMO-1 antibody [7] or the 8A2 SUMO-2 antibody [8] and the Duolink In Situ detection kit with green fluorochrome. Images were acquired with a Leica DM6000 microscope

3.4.2 SUMOylated c-Fos Is Found in Actively Transcribed Chromatin

As mentioned previously, PLA's main application is the in situ detection of protein–protein interactions. We have used this technique to address whether SUMOylated c-Fos can be present on actively transcribed genes. To this aim, we carried out PLA using an antibody directed to c-Fos-SUMO and a second antibody directed to a histone mark, such as $H3K4me_3$, that is associated with transcriptionally active genes. The presence of PLA spots confirmed that SUMOylated c-Fos can be found on actively transcribed chromatin (Fig. 4). The same was observed between SUMOylated c-Fos and RNA Polymerase II [3]. These results contrasted with the idea broadly accepted at that time that SUMOylation of transcription factors is essentially associated with transcription repression. Complementary experiments, however, demonstrated that one essential role of SUMOylation is to limits c-Fos transcriptional activity, most probably to avoid the deleterious effects of the overexpression of c-Fos target genes [3].

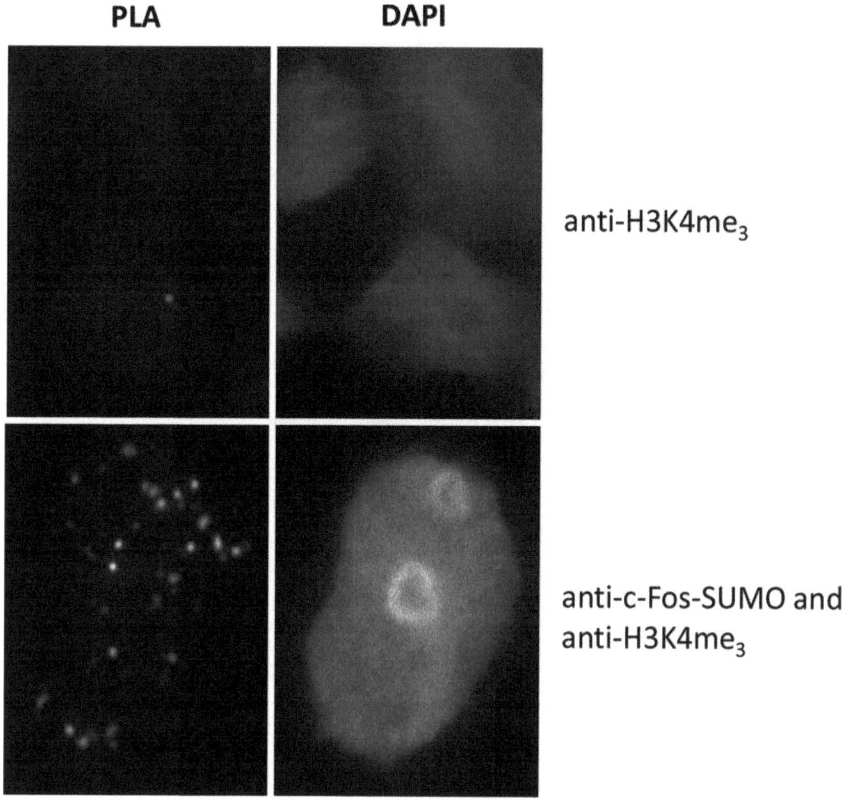

Fig. 4 In situ detection of the interaction between the SUMOylated form of c-Fos and the active transcription-associated histone mark H3K4me$_3$. The human cell line 293T was used in the PLA protocol for adherent cells. PLA was carried out using a rabbit anti-c-Fos-SUMO antibody [3] and a mouse monoclonal antibody to H$_3$K$_4$me$_3$ (Diagenode, Mab-152050) and the Duolink In Situ detection kit with green fluorochrome. Images were acquired with a Leica DM6000 microscope

3.4.3 Localization of SUMOylated Proteins on Active of Repressed Chromatin

As mentioned above, SUMOylation has long and principally been associated with heterochromatin and repressed genes [10, 11]. However, recent genome-wide analyses have shown that SUMOylated proteins are also highly enrich on actively transcribed genes [12–14]. To visualize the association of SUMOylated proteins with active or repressed chromatin, we carried out PLA using both SUMO-1 or SUMO-2 antibodies together with antibodies to histone marks associated with repressed (H3K27me$_3$) or active chromatin (H3K4me$_3$). This confirmed that SUMOylated proteins are found both on active and repressed genes (Fig. 5).

In conclusion, PLA is a powerful method to study SUMOylation. Although it requires the use of well-defined controls, it allows the visualization of endogenous protein–protein interactions and to study their dynamic regulation, in particular by stresses or physiological signaling.

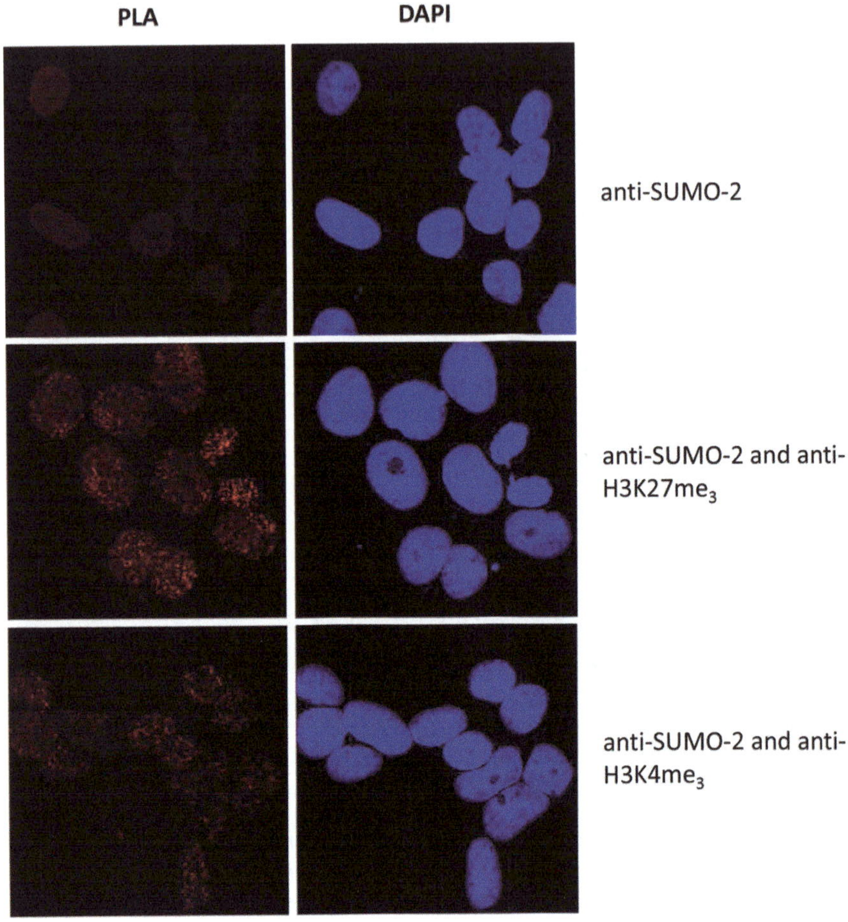

Fig. 5 In situ detection of the interaction between SUMOylated proteins and the repressed transcription-associated histone mark $H3K27me_3$ or the active transcription-associated histone mark $H_3K_4me_3$. The human cell line 293T was used in the PLA protocol for adherent cells. PLA was carried out using the mouse 8A2 SUMO-2 antibody and a rabbit antibody to $H3K4me_3$ (Millipore, 04-745) or a rabbit antibody to $H3K27me_3$ (Millipore, C5200603) and the Duolink In Situ detection kit with red fluorochrome. Images were acquired with a Zeiss Axioplan2/LSM 510 META confocal microscope

4 Notes

1. 1 M NEM stock solutions should be prepared fresh in DMSO. NEM is highly toxic and should be handled with care.
2. Polylysine coating is optional for cells that adhere strongly to glass slides. However, we find that most cell lines attach better on polylysine.
3. These slides are used to minimize the surface required for the reaction and, therefore, to use minimum amounts of the Duolink kit. The protocol can also be carried out on regular coverslips.

4. CSK buffer is used to remove soluble proteins and keep proteins associated with DNA. If the protein of interest does not bind to DNA or specific structures, this step should be skipped.
5. Alternative fixation methods can be used, depending on the cell type and the antigen studied.
6. This step can be skipped if the CSK buffer is used in **step 3**.
7. Other blocking buffers can be used that might work better, depending on the antibodies. A blocking buffer is also provided with the PLA probes kit.
8. The primary antibodies to be used for PLA must be highly specific. If an antibody shows background in other applications (for example, immunofluorescence or immunoblotting), it might also give false-positive signals in PLA.
9. For suspension cells, place the coverslips in 12-well plates and wash with 2 ml of the various buffers. For adherent cells grown on 10-well slides, put the slides on a histology tray and wash with 100 ml of washing buffers. After washes, the Teflon should be dried with a vacuum pump.
10. PLA probes could be diluted down to 1/15 without any loss in efficiency for most antibody couples we tested. The optimal concentration of probes can be tested for your own application.

Acknowledgments

MP's laboratory is an "Equipe Labellisée" of the Ligue Nationale contre le Cancer. This work was also supported by the CNRS, the Association pour la Recherche sur le Cancer (ARC), the INCA (PLBIO 2012-105 and 2013-1-MELA-05-1), the Marie-Curie Initial Training Network "UPStream," and the Region Languedoc Roussillon (programme "Chercheur d'Avenir"). The authors would like to acknowledge networking support by the Proteostasis COST Action (BM1307).

References

1. Flotho A, Melchior F (2013) Sumoylation: a regulatory protein modification in health and disease. Annu Rev Biochem 82:357–385
2. Hay RT (2005) SUMO: a history of modification. Mol Cell 18:1–12
3. Tempé D, Vives E, Brockly F et al (2014) SUMOylation of the inducible (c-Fos:c-Jun)/AP-1 transcription complex occurs on target promoters to limit transcriptional activation. Oncogene 33:921–927
4. Soderberg O, Gullberg M, Jarvius M et al (2006) Direct observation of individual endogenous protein complexes in situ by proximity ligation. Nat Meth 3:995–1000
5. Mahajan R, Delphin C, Guan T et al (1997) A small ubiquitin-related polypeptide involved in targeting RanGAP1 to nuclear pore complex protein RanBP2. Cell 88:97–107
6. Bossis G, Malnou CE, Farras R et al (2005) Down-regulation of c-Fos/c-Jun AP-1 dimer

activity by sumoylation. Mol Cell Biol 25: 6964–6979
7. Matunis MJ, Coutavas E, Blobel G (1996) A novel ubiquitin-like modification modulates the partitioning of the Ran-GTPase-activating protein RanGAP1 between the cytosol and the nuclear pore complex. J Cell Biol 135:1457–1470
8. Zhang X-D, Goeres J, Zhang H et al (2008) SUMO-2/3 modification and binding regulate the association of CENP-E with kinetochores and progression through mitosis. Mol Cell 29:729–741
9. Matic I, Schimmel J, Hendriks IA et al (2010) Site-specific identification of SUMO-2 targets in cells reveals an inverted SUMOylation motif and a hydrophobic cluster SUMOylation motif. Mol Cell 39:641–652
10. Garcia-Dominguez M, Reyes JC (2009) SUMO association with repressor complexes, emerging routes for transcriptional control. Biochim Biophys Acta 1789:451–459
11. Gill G (2005) Something about SUMO inhibits transcription. Curr Opin Genet Dev 15:536–541
12. Neyret-Kahn H, Benhamed M, Ye T et al (2013) Sumoylation at chromatin governs coordinated repression of a transcriptional program essential for cell growth and proliferation. Genome Res 23:1563–1579
13. Paakinaho V, Kaikkonen S, Makkonen H et al (2014) SUMOylation regulates the chromatin occupancy and anti-proliferative gene programs of glucocorticoid receptor. Nucleic Acids Res 42(3):1575–1592
14. Rosonina E, Duncan SM, Manley JL (2010) SUMO functions in constitutive transcription and during activation of inducible genes in yeast. Genes Dev 24:1242–1252
15. Bossis G, Melchior F (2006) Regulation of SUMOylation by reversible oxidation of SUMO conjugating enzymes. Mol Cell 21:349–357

Chapter 18

Dissecting SUMO Dynamics by Mass Spectrometry

Krzysztof Drabikowski and Michał Dadlez

Abstract

Protein modification by SUMO proteins is one of the key posttranslational modifications in eukaryotes. Here, we describe a workflow to analyze SUMO dynamics in response to different stimuli, purify SUMO conjugates, and analyze the changes in SUMOylation level in organisms, tissues, or cell culture. We present a protocol for lysis in denaturing conditions that is compatible with downstream IMAC and antibody affinity purification, followed by mass spectrometry and data analysis.

Key words SUMO, Ubiquitin like proteins, Posttranslational modifications, Label-free proteomics, Stress response

1 Introduction

Small Ubiquitin-related MOdifier (SUMO) proteins are covalently and reversibly coupled to several intracellular protein targets, modulating protein–protein, and protein–DNA interactions. SUMOylation regulates DNA-repair, transcription, chromatin organization, protein complex assembly, protein trafficking, and homeostasis [1]. Several SUMO targets belong to signaling pathways involved in cancer, neurodegenerative, and heat diseases [2]. SUMO modification is highly dynamic throughout cell cycle [3] and in response to several stress stimuli [4].

Here, we describe a workflow to analyze by mass spectrometry the dynamics of SUMO modifications in organisms, tissues, or cell culture. Different stimuli induce changes in pattern of SUMO modifications at different time frames [3, 5–7] (Drabikowski et al. submitted). Thus, we first analyze the kinetics of changes in SUMO pattern by western blot. Time of response to stimuli by SUMO depends on the stimulus and ranges from minutes to hours. Based on the level and pattern of modification on assessed by the western blot, we choose appropriate time points for proteomics analysis.

Classical approaches to isolate SUMO conjugates are based on purification in denaturing conditions in high concentration of urea

Rune Matthiesen (ed.), *Proteostasis: Methods and Protocols*, Methods in Molecular Biology, vol. 1449,
DOI 10.1007/978-1-4939-3756-1_18,

[7, 8] of poly His-tagged SUMO proteins or in SDS [9] of endogenous SUMO. SUMO modification is very dynamic and unstable due to high activity of de-SUMOylating enzymes. Furthermore, SUMO modifies large macromolecular complexes making it difficult to isolate all SUMO conjugates in mild extraction conditions. On the other hand, denaturing conditions are not fully compatible with downstream antibody affinity purification and require protein renaturation or dilution of the denaturing agent. Our approach is based on denaturing the samples in SDS and precipitating excess of SDS on ice followed by downstream exchange of SDS to sarcosyl (sodium lauroyl sarcosinate). This protocol was proposed by Schlager et al. for purification of overexpressed his-tagged proteins from inclusion bodies in *E. coli* [10]. We have applied this procedure to isolate conjugates his-tagged SUMO from the nematode *Caenorhabditis elegans* (Drabikowski et al. submitted). Denaturing lysis of cells or tissues, followed by precipitation of SDS is also compatible with downstream antibody purification without the need to renature or to dilute the sample. For higher specificity, it is possible to combine IMAC purification with second step of antibody affinity purification. Next, we prepare the samples for mass spectrometry analysis using the filter assisted sample prep (FASP) protocol [11].

Modern mass spectrometers are very sensitive and protein identification without negative controls may generate false positive results. Especially, use of agarose or sepharose IMAC support results in unspecific protein purification. Therefore, we identify in parallel proteins from assayed and control samples. When transgenic cells or animals are used, empty vector transfected cells or wild-type animals are a good control. When samples from non-transgenic animals or cells are analyzed, immunoprecipitaion without primary antibody or unrelated antibody control can be used. For label-free protein quantification, we use at least three biological replicates, the more the better. Two technical replicates increase confidence of protein identification.

This protocol is also applicable to other ubiquitin like proteins, isolated from cells, tissues, and organisms.

2 Materials

2.1 Lysate Preparation and SUMO Conjugate Purification

1. Lysis buffer. 50 mM $NaPO_4$ pH 7.5, 300 mM NaCl 1 % SDS, 20 mM DTT (*see* **Note 1**).
2. IMAC equilibration buffer: 50 mM $NaPO_4$ pH 7.5, 150 mM NaCl, 20 mM imidazlole, 0.1 % sarcosyl.
3. IMAC wash buffer: 50 mM $NaPO_4$ pH 7.5, 150 mM NaCl, 50 mM imidazlole, 0.1 % sarcosyl.

4. IMAC elution buffer: 50 mM $NaPO_4$ pH 7.5, 150 mM NaCl, 300 mM imidazlole, 0.1 % sarcosyl.
5. Antibody affinity wash buffer: 50 mM $NaPO_4$ pH 7.5, 150 mM NaCl, 0.1 % sarcosyl.
6. Antibody affinity elution buffer: 100 mM glycine pH 2.5.
7. IMAC metal affinity resin.
8. Anti-tag or anti-SUMO antibody cross-linked to beads (*see* **Notes 2** and **3**).

2.2 FASP

1. Vivacon 500 ultrafiltration unit MWCO 30000 cat no VNO1H22. Sartorius.
2. UA: 8 M urea in H_2O, 1 ml for sample.
3. DTT solution: 50 mM DTT in UA.
4. IAA solution: 50 mM iodoacetamide in UA.
5. ABC buffer: 50 mM NH_4HCO_3 in water.
6. Proteomics grade trypsin.

2.3 Equipment

1. SDS-PAGE and western blot setup.
2. Sonicator.
3. Direct Detect® Spectrometer (EMD Millipore).
4. Nano LC–MS/MS LTQ Orbitrap or Q-Exactive Orbitrap (Thermo) mass spectrometer.

2.4 Data Analysis

1. Maxquant software [12] http://www.maxquant.org.
2. Perseus software http://www.perseus-framework.org.

3 Methods

Carry out all procedures at room temperature unless otherwise specified.

3.1 Test Dynamics of SUMO Modification Changes

Time of response to stimuli by SUMO depends on the stimulus and ranges from minutes to hours. To choose optimal time points for proteomics identifications, start with several time points, for example, 10 min, 30 min, 1, 2, 4, and 8 h. Collect samples and analyze by western blot with anti-SUMO or anti-tag antibody (if the SUMO protein is tagged). If the resolution of the initial time point setup is not satisfactory, analyze more time points.

3.2 Lysate Preparation

1. Pulverize the tissue in liquid nitrogen in a mortar. Depending on the designed downstream purification steps use 0.5–10 g of tissue (*see* **Note 5**).

2. Place the pulverized tissue/worms/... into boiling lysis buffer, at least 20 volumes of lysis buffer per one volume of pulverized material.
3. When isolating SUMO conjugates from cells in culture, wash the cells with PBS and pour 2 ml of boiling lysis buffer per 10 cm dish. Transfer the lysate to a test tube.
4. Sonicate immediately. Avoid foaming of the SDS since foam formation decreases sonication efficiency.
5. Centrifuge the lysate for 30 min at 24,000 × *g*.
6. Transfer the lysate to new tubes.
7. Incubate on ice for 30 min. The lysate should appear as a milky slurry. Centrifuge 30 min at 10,000 × *g* at 0–4 °C (*see* **Note 6**).
8. Filter the clear lysate through a 45 μm filter (*see* **Note 7**).

3.3 IMAC Purification

1. Add imidazole to 20 μM final concentration to the cleared lysate.
2. Equilibrate the IMAC beads with IMAC equilibration buffer.
3. Apply the cleared lysate to the IMAC column.
4. Wash with at least 20 bed volumes with IMAC wash buffer.
5. Elute with IMAC elution buffer.
6. Measure the protein concentration with Direct Detect® Spectrometer or other applicable method.

3.4 Antibody Affinity Purification (See Notes 1–4 and 12)

1. Pre-elute the unbound antibody from the beads with 100 mM glycine pH 2.5.
2. Equilibrate the beads/column with the antibody purification wash buffer.
3. Apply the lysate or eluate from IMAC purification and incubate O/N at 4 °C (*see* **Note 8**).
4. Wash with at least 20 column volume of antibody purification wash buffer or if using magnetic beads at least five times 5 min with the wash buffer.
5. Elute with the antibody purification elution buffer.
6. Measure the protein concentration with Direct Detect® Spectrometer or other applicable method.

3.5 Filter Aided Sample Prep (FASP)

1. Measure the concentration of purified proteins.
2. Take 10–50 μg of purified protein per sample. Reduce with 50 mM DTT at 55 °C for 30 min.
3. Dilute the sample 1:8 in UA.
4. Apply to Vivacon filter units, centrifuge 14,000 × *g* for 15 min (*see* **Note 9**).
5. Add 200 μg 8 M urea to filter unit and centrifuge at 14,000 × *g* for 15 min. repeat this step.

6. Add 100 µl 50 µM IAA solution. Mix and incubate in the dark at room temperature for 30 min.
7. Centrifuge at 14,000 × *g* for 15 min. repeat this step twice.
8. Add 200 µl ABC buffer to the filter unit and centrifuge at 14,000 × *g* for 15 min. repeat this step twice.
9. Add 50 µl ABC with trypsin (enzyme to protein ration 1:100), mix for 1 min.
10. Incubate the units in a wet chamber at 37 °C for 4–16 h (*see* **Note 10**).
11. Transfer the filter unit to new collection tube, centrifuge at 14,000 × *g* for 15 min.
12. Add 50 µl ABC to the filter unit and centrifuge at 14,000 × *g* for 15 min.
13. Measure the concentration of eluted peptides (*see* **Note 11**).

3.6 Mass Spectrometry

Detail description of the mass spectrometry analysis depends on exact instrument setup available and is beyond the scope of this chapter. For label-free proteomics quantifications, precise protein quantification is essential. We routinely apply 1 µg of peptides sample per run. Overloading the instrument decreases the dynamic range and disturbs the quantification. To avoid falls positive identification by carry over proteins, run the control samples before the experimental ones. For efficient comparison and matching between runs we run all samples consecutively on one instrument.

3.7 Data Analysis

1. Load the LC–MS/MS .raw files to Maxquant (Raw files/load/…).
2. Define "parameter group" if comparing different experiments. Define "Experiment": Control 1, Control 2…. Sample 1, Sample 2. Leave "Fraction" unchanged. If you define more than one "parameter group," define "group specific parameters" for all groups.
3. Setup "Modifications."
4. Setup "Label-free quantification." Click off "Fast LFQ."
5. Check/global parameters/general "Match between the runs," to align.
6. Select database. Global parameters/global/Fasta files/add files.
7. Select chemical modifications for protein quantification (the same as Subheading 3.7, **step 3**)/Global parameters/protein quantification.
8. Press "Start" to run the program.
9. Open/combined/txt/proteinGroups.txt file in Excel or other spreadsheet program.
10. Remove rows containing proteins identified in more than one control sample identified by 2 or more peptides.

11. Load the edited proteinGroups.txt into Perseus.
12. Select "LFQ intensity" for the "Expression" tab.
13. Select "only identified by site" and "reverse" in "Categorical annotation" tab.
14. Select "ID," "Proteins," "Peptides," "Unique Peptides," and "iBAQ" in "Numerical annotation" tab.
15. Select "Fasta header," in "Textual annotation" tab.
16. Remove proteins identified only by one peptide (/Filter rows/ Filter rows based on numerical/expression column). Setup x: peptides. Setup Relation 1 $x <= 1$, setup filer mode: Reduce matrix.
17. Remove proteins for which no unique peptide was identified (/Filter rows/Filter rows based on numerical/expression column). Setup x: Unique peptides. Setup Relation 1 $x < 1$, setup filer mode: Reduce matrix.
18. Perform logarithmic transformation ($\log_2$).
19. Replace the missing LFQ values with normal distribution. (/Imputation/Replace missing values with normal distribution).
20. Divide the data sets into groups, for example control, samples. Annotate rows/categorical annotation rows.
21. Calculate the relative protein enrichment and analyze the statistics to test the significance between category groups (for comparing two conditions/Tests/two-sample tests/*t*-test, for multiple conditions/Tests/multiple sample tests/Anova). Use "Permutation based FDR" set number of randomizations to 2500, uncheck "$\log_{10}$."
22. In the "output" data matrix, the significant differences will be marked with "+" for in "*T*-test significance" or "Anova significance" columns.

4 Notes

1. Since lysis is performed with boiling SDS, there is no need to use inhibitors of de-SUMOylating enzymes as *N*-Ethylmaleimide.
2. Magnetic beads tend to give less background than agarose or sepharose beads.
3. Cross-linking antibodies to the beads and washing of unbound antibodies is essential because eluted antibodies interfere with precise determination of protein quantity used for proteomics identification.
4. Use of camelid single chain antibodies, for example anti-GFP monobody [13], gives less background antibody signal in the sample than classical antibodies.

5. Efficiency of protein extraction depends on success in tissue pulverization. It is worth to monitor level of pulverization by taking a small part of the powder and inspect it under dissecting microscope.
6. For downstream steps of binding to the affinity columns, it is essential to remove the free SDS from the sample. After centrifugation, the SDS pellet should be approximately 10% volume of the lysate. If the pellet is small repeat the precipitation and centrifugation. If the repeated centrifugation does not yield more SDS precipitate, dilute the lysate with more lysis buffer. To avoid SDS resolubilisation, keep the lysate at 0 °C at all times.
7. If the filter clogs fast, it is most probably due to insufficient removal of insoluble debris. Repeat Subheading 3.2, **step 5**.
8. If antibody affinity purification is the second step of purification, dilute 10× the eluate from IMAC purification with antibody purification wash buffer. High concentration of imidazole interferes with antibody binding.
9. After each centrifugation step in FASP, there should be no visible liquid on top of the spin column. If some liquid remain, repeat centrifugation.
10. The volume of the digestion mixture is very small and prone to drying. Thus, it is important to have tight wet chamber and preincubate it at 37 °C for 1 h prior to applying the samples.
11. If measuring the protein/peptide concentration in ABC buffer in the Direct Detect®Spectrometer, perform several rounds of drying the samples or leave the Direct Detect® card to dry overnight to allow ammonium bicarbonate to fully evaporate. Ammonium bicarbonate is detected at the same wavelength as peptide bond.
12. Two step purification using IMAC followed by antibody affinity purification increases specificity and sensitivity of SUMO target identification but at the cost of reproducibility between biological replicates. Thus, for label-free quantification, we suggest to perform single step IMAC or antibody purification, including the control samples.

Acknowledgments

The work was supported by National Science Centre (Narodowe Centrum Nauki) grant DEC 1/Z/2011/01/M/NZ2/02997. The authors would like to acknowledge networking support by the Proteostasis COST Action (BM1307).

References

1. Flotho A, Melchior F (2013) Sumoylation: a regulatory protein modification in health and disease. Annu Rev Biochem 82:357–385
2. Yang XJ, Chiang CM (2013) Sumoylation in gene regulation, human disease, and therapeutic action. F1000Prime Rep 5:45
3. Schimmel J et al (2014) Uncovering SUMOylation dynamics during cell-cycle progression reveals FoxM1 as a key mitotic SUMO target protein. Mol Cell 53:1053–1066
4. Impens F, Radoshevich L, Cossart P, Ribet D (2014) Mapping of SUMO sites and analysis of SUMOylation changes induced by external stimuli. Proc Natl Acad Sci U S A 111:12432–12437
5. Psakhye I, Jentsch S (2012) Protein group modification and synergy in the SUMO pathway as exemplified in DNA repair. Cell 151:807–820
6. Golebiowski F et al (2009) System-wide changes to SUMO modifications in response to heat shock. Sci Signal 2:ra24
7. Hendriks IA et al (2014) Uncovering global SUMOylation signaling networks in a site-specific manner. Nat Struct Mol Biol 21:927–936
8. Denison C et al (2005) A proteomic strategy for gaining insights into protein sumoylation in yeast. Mol Cell Proteomics 4:246–254
9. Becker J et al (2013) Detecting endogenous SUMO targets in mammalian cells and tissues. Nat Struct Mol Biol 20:525–531
10. Schlager B, Straessle A, Hafen E (2012) Use of anionic denaturing detergents to purify insoluble proteins after overexpression. BMC Biotechnol 12:95
11. Wiśniewski JR, Zougman A, Nagaraj N, Mann M (2009) Universal sample preparation method for proteome analysis. Nat Methods 6:359–362
12. Cox J, Mann M (2008) MaxQuant enables high peptide identification rates, individualized p.p.b.-range mass accuracies and proteome-wide protein quantification. Nat Biotechnol 26:1367–1372
13. Płociński P et al (2014) Identification of protein partners in mycobacteria using a single-step affinity purification method. PLoS One 9:e91380

Chapter 19

Isolation of Lysosomes from Mammalian Tissues and Cultured Cells

Carmen Aguado, Eva Pérez-Jiménez, Marcos Lahuerta, and Erwin Knecht

Abstract

Lysosomes participate within the cells in the degradation of organelles, macromolecules, and a wide variety of substrates. In any study on specific roles of lysosomes, both under physiological and pathological conditions, it is advisable to include methods that allow their reproducible and reliable isolation. However, purification of lysosomes is a difficult task, particularly in the case of cultured cells. This is mainly because of the heterogeneity of these organelles, along with their low number and high fragility. Also, isolation methods, while disrupting plasma membranes, have to preserve the integrity of lysosomes, as the breakdown of their membranes releases enzymes that could damage all cell organelles, including themselves. The protocols described below have been routinely used in our laboratory for the specific isolation of lysosomes from rat liver, NIH/3T3, and other cultured cells, but can be adapted to other mammalian tissues or cell lines.

Key words Lysosomes, Liver, Cultured cells, Subcellular fractionation, Differential and gradient centrifugation

1 Introduction

Lysosomes are a group of organelles with varying sizes, forms, content, and densities. Lysosomes enclose a wide variety of acid hydrolases, including proteases (cathepsins), lipases, glucosidases, and nucleases [1, 2]. Macromolecules and other substrates reach the lysosomes by diverse mechanisms, including endocytosis, crinophagy, and different kinds of autophagy (macroautophagy, microautophagy, and chaperone-mediated autophagy) [3]. Once inside the lysosomes, the sequestered materials are degraded and their building blocks are recycled. Lysosomes are involved in a large variety of cell processes, including differentiation, development, aging and cell death, and they also play an important role in many pathological disorders, such as cancer and neurological diseases.

Rune Matthiesen (ed.), *Proteostasis: Methods and Protocols*, Methods in Molecular Biology, vol. 1449,
DOI 10.1007/978-1-4939-3756-1_19, © Springer Science+Business Media New York 2016

Since the discovery of lysosomes 60 years ago [4] and their initial implication in the so-called lysosomal storage diseases [5, 6], the studies on these organelles, which usually include their isolation, have grown exponentially. For example, isolation of lysosomes followed by proteomic analysis has been used to identify new lysosomal membrane proteins (e.g., [7–9]). These and other studies where isolation of lysosomes was employed have been useful to progress our knowledge on the physiology and pathology of lysosomes.

Subcellular fractionation has allowed researchers to study the characteristics and function of different cellular components. Lysosomes were first identified by de Duve when trying to find the structure where glucose 6-phosphatase was localized [10]. This was done by differential centrifugation of liver homogenates to obtain a light mitochondrial (LM) fraction, followed by a sucrose gradient that separated lysosomes from other cell components in this fraction by their different density. Because of the above-mentioned heterogeneity of lysosomes, purification of a pure fraction of lysosomes is a difficult task and it is common to find these fractions contaminated, mainly with mitochondria and peroxisomes but also with other cell components. This caveat was first overcome by the injection of detergent Triton WR-1339 [11, 12]. The detergent is selectively taken up by lysosomes, reducing the lysosomal density and therefore allowing a better separation from mitochondria. Other strategy for achieving a better separation, for example, is the loading of lysosomes with colloidal gold [13, 14], producing an increase in the lysosomal density. These methods are currently in use (e.g., [7]) and allow a better purification of the lysosomal fraction. However, they imply a change in the lysosomal composition, as they are loaded with external agents and this could affect the lysosomal function. In fact, the organelles purified with both procedures are tritosomes and aurosomes, which are not strictly lysosomes.

The extraction of pure "intact" lysosomes is therefore a critical step for researchers in this field. The isolation method has been improved by the use of gradients of Percoll, Metrizamide, Nycodenz, and other reagents (e.g., [9, 15]). Although these methods have been widely used for the isolation of lysosomes from animal tissues, this is much more arduous when using cultured cells, particularly because the difficulty of disrupting their plasma membranes without affecting the lysosomal membranes. Anyway, there are several reports in the literature for the isolation of lysosomes or their subpopulations from different sources (e.g., [7, 14, 16–19]). Here, we describe the protocols that we have followed for many years in our laboratory to isolate these organelles.

2 Materials

2.1 Isolation of Lysosomes from Rat Liver

Prepare all solutions using ultrapure water and centrifugation or analytical grade reagents. Solutions must be freshly made or stored at −20 °C, unless otherwise stated. All experiments involving animals should be conducted in compliance with approved Institutional Animal Use Committee protocols.

1. Male Wistar rats weighing 200–250 g starved during 16–24 h (only with water *ad libitum*).
2. Dissection instruments (scissors, clamps, and tweezers).
3. Glassware: 250 mL Erlenmeyer flask, 250 mL beakers, and 100 mL graduated cylinder.
4. Gauze (two layers) or cheesecloth.
5. Thomas Pestle Tissue Grinder homogenizer (55-mL, Thomas Scientific, catalog number 3431E55, Swedesboro, NJ, USA) attached to an IKA RW-20 digital dual-range mixer (Cole-Parmer, catalog number EW-50705-00, Vernon Hills, IL, USA).
6. Refractometer (Carl Zeiss, Jena, Germany).
7. Cold finger (10 × 150 mm test tube filled with ice).
8. Pasteur pipettes (plastic and glass) and pipette rubber bulbs.
9. Homogenization medium: 0.3 M sucrose. This solution can be prepared the previous day and stored at 4 °C.
10. 85.6% Metrizamide in water at pH 7.0. Preparation of this solution is extremely tedious (*see* **Note 1**).
11. Heraeus centrifuge (Biofuge 28RS) equipped with a 3745 rotor or equivalent (Hanau, Germany) and 50-mL polycarbonate tubes.
12. Sorvall centrifuge (Evolution RL) equipped with a SA-300 rotor or equivalent (Thermo Fisher Scientific, Waltham, MS, USA) and 50-mL polycarbonate tubes.
13. Ultracentrifuge (Optima XL-100K) equipped with a SW40 Ti rotor (5–6 g of liver, *see* **Note 2**) and 14 × 95 mm Ultra-Clear tubes (catalog number 344060) (Beckman Coulter, Brea, CA, USA).

2.2 Isolation of Lysosomes from Cultured Cells

1. Polystyrene square cell culture dishes (Nunc, Thermo Fisher Scientific, catalog number 166508). Culture area: 500 cm^2.
2. Heraeus centrifuge (Biofuge 28RS) equipped with a 3745 rotor or equivalent and 50-mL polycarbonate tubes.
3. Thomas Pestle Tissue Grinder, 10-mL (Thomas Scientific, catalog number 3431E45).
4. Polyethylene cell lifters (Corning Incorporated, catalog number 3008, NY, USA).

5. Parr nitrogen bomb (Parr, model 4639, IL, USA) (*see* **Note 3**).
6. Ultracentrifuge: Optima XL-100K equipped with a SW40 Ti and a 70.1 Ti rotors Optima MAX-130 equipped with a TLA-100 and a TLA-55 rotors. Ultracentrifuge tubes: 14 × 95 mm Ultra-Clear (catalog number 344060), 16 × 76 mm thickwall polycarbonate (catalog number 355630), 11 × 39 mm microcentrifuge polyallomer (catalog number 357448), and 7 × 20 mm thickwall polyallomer (catalog number 343621) (Beckman Coulter).
7. Krebs-Henseleit (KH) medium: 118.4 mM NaCl, 4.75 mM KCl, 1.19 mM KH_2PO_4, 2.54 mM $MgSO_4$, 2.44 mM $CaCl_2{\cdot}2H_2O$, 28.6 mM $NaHCO_3$, 10 mM glucose, containing 10 mM HEPES, pH 7.4.
8. Phosphate buffered saline (PBS) medium: 150 mM NaCl, 2.7 mM KCl, 1.4 mM KH_2PO_4, 10 mM Na_2HPO_4, pH 7.4.
9. Percoll/Metrizamide solutions and 0.25 M sucrose (*see* **Note 4**).
10. Homogenization buffer (HB) 10×: 2.5 M sucrose, 100 mM HEPES, 10 mM EDTA, pH 7.3.
11. Pasteur pipettes (plastic and glass) and pipette rubber bulbs.

3 Methods

3.1 Isolation of Lysosomes from Rat Liver (Adapted from Wattiaux et al. [15])

Once the liver is obtained all the steps should be performed on ice or at 4 °C to prevent lysosomes from becoming damaged by released enzymes.

1. Obtain liver from a starved Wistar rat. Remove skin and fat tissue to get the liver as clean as possible and weigh the liver.
2. Wash extensively with cold 0.3 M sucrose in a precooled beaker and cut into small pieces with the aid of scissors to facilitate homogenization.
3. Homogenize in 3 volumes of 0.3 M sucrose per gram of tissue by 8 strokes at 500 rpm in a homogenizer.
4. Add 4 additional volumes of 0.3 M sucrose and filter the homogenate through double gauze without pressing it and collect in a precooled Erlenmeyer flask. Separate 60 μL of homogenate in an Eppendorf tube for determination of specific activities of enzymes.
5. Split the homogenate into two 50-mL polycarbonate tubes (use 0.3 M sucrose to equilibrate) and centrifuge at 4800 × *g* for 5 min at 4 °C in a Heraeus 3745 rotor.
6. After centrifugation, collect the supernatant into clean 50-mL polycarbonate tubes (*see* **Note 5**).
7. Centrifuge at 17,000 × *g* for 10 min at 4 °C in a Heraeus 3745 rotor.

8. Remove the supernatant (*see* **Note 6**). Resuspend the pellet from this second centrifugation with a "cold finger" and wash it with 0.3 M sucrose (3.5 volumes per gram of liver).
9. Centrifuge as in #7. Remove the supernatant.
10. Resuspend the pellet with a "cold finger" in 1 mL of 0.3 M sucrose. This is an LM fraction.
11. Place the LM fraction at the bottom of an Ultra-Clear ultracentrifuge tube and add 3.39 g of 85.6% Metrizamide to (adjust final Metrizamide concentration to 57%). Mix gently by vortexing (*see* **Note 7**).
12. Generate a discontinuous Metrizamide gradient by carefully and consecutively overlaying the following Metrizamide solutions on top of the LM fraction mixed with Metrizamide (Fig. 1):
    - 2.13 mL of 32.8% Metrizamide.
    - 3.50 mL of 26.3% Metrizamide.
    - 3.85 mL of 19.8% Metrizamide.

    Fill the tube up to about 2–3 mm from the border with 0.3 M sucrose, to prevent collapse during centrifugation.
13. Centrifuge the gradient at 141,000 × *g* for 90 min at 4 °C in a SW40 Ti rotor (*see* **Note 8**).

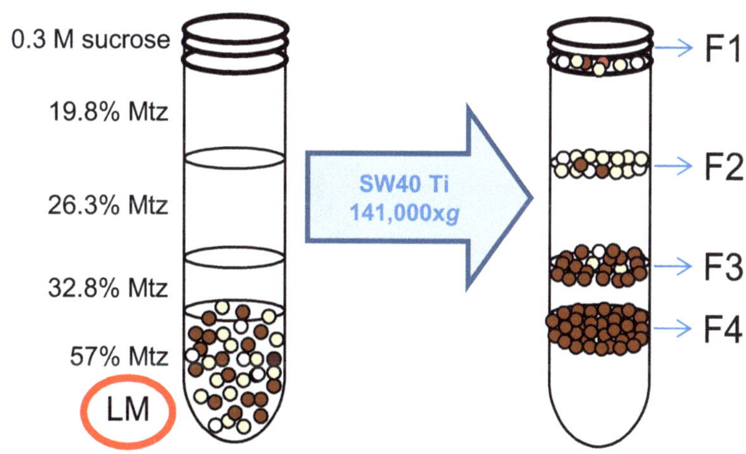

Fig. 1 Isolation of lysosomes from rat liver. Once the light mitochondrial (LM) fraction is obtained, it is loaded on the bottom of an Ultra-Clear ultracentrifuge tube and adjusted to 57 % Metrizamide (Mtz). Three layers are overlaid on top to create a discontinuous Mtz gradient and a 0.3 M sucrose layer is finally added. After 90 min centrifugation at the indicated conditions, lysosomes are located at the 0.3 M sucrose/19.8 % Mtz and 19.8 %/26.3 % Mtz interfaces (F1 and F2). F3 and F4 are enriched in mitochondria. *Whitish circles* represent lysosomes and *brown circles* mitochondria

14. After ultracentrifugation, four bands (F1 to F4, from top to bottom) are distinguished in the gradient interfaces (*see* Fig. 1). With a glass Pasteur pipette collect the two upper whitish bands (F1 + F2), which contain lysosomes (approximately 2–3 mL).
15. Mix with 10 volumes of ice cold 0.3 M sucrose in 50-mL polycarbonate tubes and centrifuge at 37,000 × *g* for 10 min at 4 °C (to remove Metrizamide) in a Sorvall centrifuge using a SA-300 rotor. This pellet contains the lysosomes. Resuspend them with a blunt Pasteur pipette (*see* **Note 9**) in 30 mL of ice cold 0.3 M sucrose and wash two times (Fig. 2a).
16. Resuspend the last sediment as above in 250–300 μL of the solution suitable for your following experiments (*see* **Note 10**).
17. Separate 40 μL for quantification of protein and enzymatic activities (*see* **Note 11**).

3.2 Isolation of Lysosomes from Cultured Cells (Adapted from Storrie and Madden [16])

For optimal isolation use 4–5 square cell culture dishes per condition (*see* **Note 12**). After washing the cells with 70 mL PBS, treat them for 30 min to 4 h at 37 °C with 70 mL KH medium/dish (*see* **Note 13**).

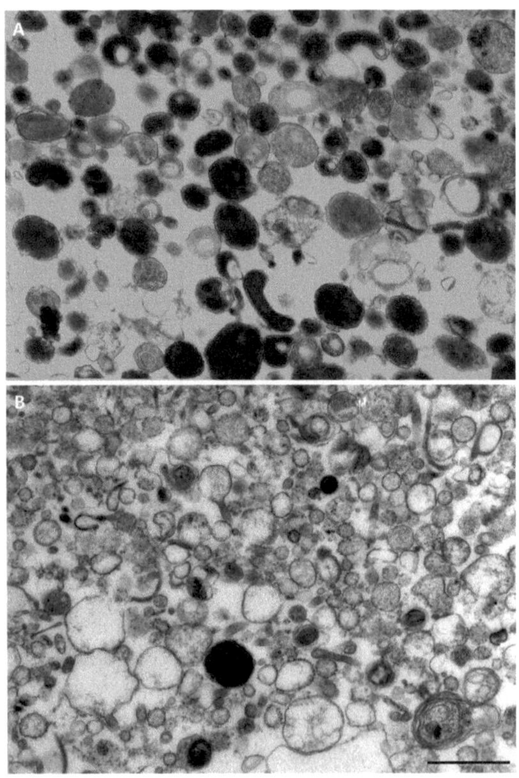

Fig. 2 Purified lysosomes. Representative electron micrographs of lysosomes isolated from rat liver (**a**) and NIH/3T3 cells (**b**). Bar: 1 μm

1. Remove the KH medium from the plates and carefully detach the cells with a cell lifter in cold PBS (2 mL/ plate). Collect the cells into 50-mL tubes with a plastic Pasteur pipette.
2. Centrifuge cells at 800 × *g* for 5 min at 4 °C in a Heraeus 3745 rotor.
3. Wash with 0.25 M sucrose (25–30 mL). Centrifuge as in #2.
4. Resuspend the pellet in 3 mL of 0.25 M sucrose. Place the cell suspension in the precooled chamber of the nitrogen bomb. Slowly, open the valve and allow the pressure to reach 35 psi (2.41 Bar) in 1 min. Keep pressure at 35 psi for a total of 7 min with occasional shaking (optimize conditions according to cell type, *see* **Note 3**).
5. Further homogenize the cells on ice using a precooled 10-mL homogenizer and 8 strokes of its Teflon pestle. Collect the homogenate and wash the homogenizer with 1 mL of 0.25 M sucrose, to get a final volume of about 4 mL. Separate 60 μL of homogenate in an Eppendorf tube for determination of specific activities of enzymes.
6. Centrifuge at 2500 × *g* for 15 min at 4 °C in a Heraeus centrifuge using a 3745 rotor (*see* **Note 14**).
7. In the meantime, prepare the gradient in an Ultra-Clear tube (14 × 95 mm) by adding the following solutions from bottom to top (Fig. 3a):
   - 35 % Metrizamide in 0.25 M sucrose: 2.6 mL.
   - 17 % Metrizamide in 0.25 M sucrose: 2.6 mL.
   - 6 % Percoll in 0.25 M sucrose: 3.9 mL.
8. Add the supernatant from #6 (PNS, approximately 3.5 mL) on top of the discontinuous gradient (Fig. 3a).
9. Centrifuge at 70,000 × *g* for 35 min at 4 °C in a SW40 Ti rotor. LM fraction appears in the 17 % Metrizamide/6 % Percoll interface (*see* Fig. 3a).
10. Collect the LM band, approximately 1.4 mL (*see* **Note 15**). Put at the bottom of an Ultra-Clear tube (14 × 95 mm) with 1.1 mL of 80 % Metrizamide and mix gently by vortexing. Add the following solutions from bottom to top (Fig. 3b):
    - 17 % Metrizamide in 0.25 M sucrose: 2.6 mL.
    - 5 % Metrizamide in 0.25 M sucrose: 2.6 mL.
    - 0.25 M sucrose up to 3 mm from the border, approximately 5 mL.
11. Centrifuge at 70,000 × *g* for 35 min at 4 °C in a SW40 Ti rotor.
12. Collect the lysosomes, approximately 1.2 mL (second band from bottom, Fig. 3b). Dilute with 3–4 volumes of 0.25 M sucrose

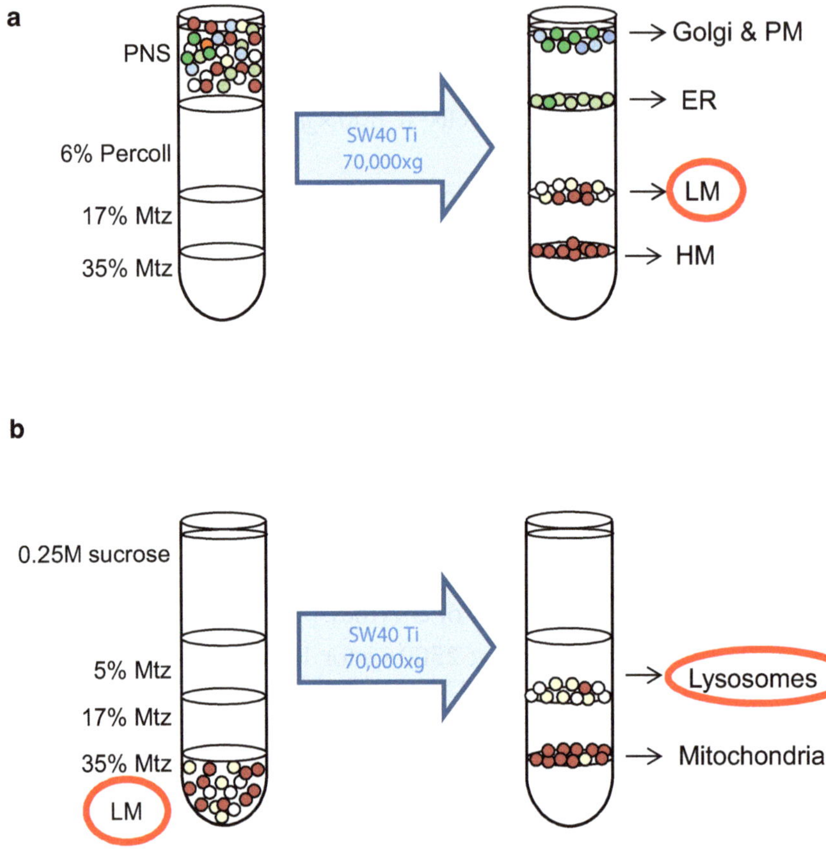

Fig. 3 Isolation of lysosomes from cultured cells. (**a**) Once the postnuclear supernatant (PNS) is obtained, it is layered on top of a Percoll/Metrizamide (Mtz) gradient. After a 35 min centrifugation at the indicated conditions, the light mitochondrial (LM) fraction is recovered from the 17 % Mtz/6 % Percoll interface. (**b**) This fraction is loaded on the bottom of an Ultra-Clear ultracentrifuge tube and adjusted to 35 % Mtz. Three layers are overlaid on top to create a discontinuous Mtz gradient. After a 35 min centrifugation at the indicated conditions, lysosomes are located at the 17 %/5 % Mtz interface. *Green* and *blue circles* represent Golgi complex and plasma membrane (PM); light *green circles* represent endoplasmic reticulum (ER); *whitish circles* represent lysosomes and *brown circles* represent mitochondria (light and heavy (HM))

and centrifuge at 100,000×*g* for 30 min at 4 °C in a 70.1 Ti rotor using thickwall polycarbonate tubes. Repeat this step.

13. Using a blunt Pasteur pipette, resuspend the lysosomal pellet in 150–200 μL of the solution suitable for your following experiments (*see* **Note 10**).
14. Separate 20 μL at −80 °C for protein and enzymatic activities measurements (*see* **Note 16**).

An alternative method to avoid the use of Metrizamide requires modification in the following specific steps of the previous procedure:

1. Wash the cells with 1× HB (about 33 mL/tube) instead of 0.25 M sucrose.

2. Resuspend the pellet in 3 mL of 1× HB and homogenize as in the method above (*see* **Note 17**).
3. Prepare the gradient by adding the following solutions from bottom to top:
   - 10× HB: 1.2 mL.
   - 17.5 % Percoll in 1× HB: 8.5 mL.
4. Centrifuge at 67,000 × *g* for 30 min at 4 °C in a 70.1 Ti rotor.
5. Remove the first 1 mL and take 9 different fractions of 1.2 mL in separate Eppendorf tubes. Lysosomes are mainly located in the first three fractions, but it is advisable when using other cell types to recover and analyze the other fractions too. Pool fractions 1–3 in a thickwall polycarbonate tube.
6. Centrifuge at 100,000 × *g* for 30 min at 4 °C in a 70.1 Ti rotor (*see* **Note 18**). Continue with #12 of the main protocol.

4 Notes

1. Preparation of Metrizamide solutions. The stock solution (85.6 % Metrizamide) should be dissolved in the dark. To avoid the formation of clumps of undissolved material, start with half of the final volume of water and add small amounts of Metrizamide while stirring. Do not add more Metrizamide until the former has been fully dissolved. Adjust the pH of the solution to 7.0 with 0.01 M NaOH while stirring (since this solution has an elevated density, wait a while for an accurate pH measurement). Using a refractometer, adjust more precisely the concentration of the different solutions by adding water at pH 7.0 or stock solution to bring them to the corresponding refractive index (see below). Store all solutions at −20 °C. The volumes needed for two samples are shown below:

	85.6 % Metrizamide (mL)	Water pH 7.0 (mL)	Refractive index (20 °C)
85.6 %	5.00	–	1.4710
32.8 %	1.92 (2.80 g)	3.08	1.3854
26.3 %	2.46 (3.58 g)	5.54	1.3763
19.8 %	1.85 (2.71 g)	6.15	1.3643

Nycodenz, more easily available and less expensive than Metrizamide, could probably replace the use of Metrizamide in a similar protocol [9], but further analyses are recommended.

2. For larger amounts of tissue (about 20 g) use the SW28 rotor, adjusting the volumes of the different Metrizamide layers in the gradient.
3. Parr nitrogen bomb is a cell disruptor by nitrogen decompression. It preserves organelle integrity in cultured cells better than other methods. Conditions provided in this method are optimized for NIH/3T3 cells. Time and nitrogen pressure conditions should be optimized for other cell lines.
4. Percoll/Metrizamide solutions.

   (a) 80 % Metrizamide in 0.25 M sucrose pH 7.2. All Metrizamide solutions are prepared in 0.25 M sucrose pH 7.2.

   Preparation: put 50 mL of 0.25 M sucrose in a glass beaker and slowly add 80 g of Metrizamide. Adjust the pH with 0.01 N NaOH and add 0.25 M sucrose to a final volume of 100 mL. Remember (*see* **Note 1**) to carry out these preparations in the dark. The volumes needed for two samples are shown below:

	80 % Metrizamide (mL)	Percoll (mL)	0.25 M sucrose (mL)
35 % Metrizamide	2.40	–	3.10
17 % Metrizamide	2.34	–	8.66
15 % Metrizamide	0.34	–	5.16
6 % Percoll	–	0.51	8.00

   (b) 17.5 % Percoll in 1× HB: 18 mL are needed for two samples. Mix 3.15 mL of 100% Percoll, 1.8 mL of 10× HB and 13.05 mL of water.

5. Take care to avoid collecting portions of the white sediment from this first centrifugation, which consists of erythrocytes, nuclei, large mitochondria, and some cell debris.
6. The supernatants from centrifugation at 17,000 ×*g* for 10 min contain cytosol, microsomes, etc.
7. To collect the LM fraction, Pasteur pipettes or other wide mouth pipettes are better than micropipette tips to reduce damage to lysosomal membranes.
8. It is advisable to carry out all centrifugations of gradients without brake and with low acceleration.
9. To prepare the blunt Pasteur pipette, heat the edge of a glass pipette to make a glass drop at its end.
10. Lysosomes can be used for different experiments and they should be treated accordingly. For example, for *chaperone-mediated autophagy studies*, freshly isolated lysosomes should be immediately resuspended, after washing, in 10 mM MOPS/0.3 M sucrose

pH 7.0 and incubated with the appropriate substrates and reagents as described [18, 20]. For *proteomic studies of lysosomal components* (e.g., [8]), it is advisable to incubate the lysosomes for 30 min at 37 °C in an iso-osmotic medium (250–300 mOsmol/L) to allow degradation of the cytoplasmic material sequestered by the lysosomes. For *isolation of lysosomal membranes* follow this procedure:

- Resuspend lysosomes in 200 μL of water plus protease inhibitors (1 mM phenylmethylsulfonyl fluoride, 100 μM leupeptin, 10 μM pepstatin A, 2 mM EDTA, 2 mM dithiothreitol).
- Disrupt lysosomes by 6–8 freeze-thaw cycles and homogenize.
- Centrifuge at 130,000 × *g* for 10 min at 4 °C in an Optima MAX-130 equipped with a TLA-100 rotor or equivalent.
- Wash the membranes three times with 100 μL of water plus protease inhibitors to remove residual components from the lysosomal matrix. Centrifuge as above.
- Resuspend in 100 μL of water plus protease inhibitors (see above) and analyze.

11. To calculate the yield and the degree of purification of this lysosomal fraction relative to the original homogenate, we routinely use β-*N*-acetyl-glucosaminidase and β-hexosaminidase as lysosomal marker enzymes. In our hands, yield and degree of purification are 6–10% and (70–80)-fold. For experiments with intact lysosomes, calculate lysosomes latency by measuring the activity of a lysosomal enzyme in the presence or not of 0.15% Triton X-100. Preparations with latency values below 95% are discarded for those experiments. Protein concentration can be evaluated using a Micro BCA protein assay kit (Thermo Scientific) or an equivalent procedure.
12. We use 90% confluent NIH/3T3 cells, about 8×10^7cells/plate.
13. KH medium or equivalent starvation media are used to increase the cellular lysosomal mass.
14. The supernatant and the pellet from this centrifugation are the postnuclear supernatant (PNS) and pellet (PNP), respectively. Resuspend the PNP in 4 mL of 0.25 M sucrose and reserve an aliquot for enzymatic activity measurements.
15. To avoid the contamination of the lysosomal fraction with other cellular components, remove the two upper bands with a Pasteur pipette.
16. Calculate yield and degree of purification of the lysosomal fraction as in **Note 11**. In our hands, yield and degree of purification are 3–5% and 20- to 30-fold, respectively.

17. The addition of 10 mM potassium phosphate pH 7.4 alters the sedimentation rate of mitochondria without affecting the lysosomes, improving the purity of the lysosomal fraction [21].
18. If the sediments of this centrifugation are loose, centrifuge at 70,000 × *g* for 30 min at 4 °C using an Optima MAX-130 equipped with a TLA-55 rotor or equivalent.

Acknowledgements

Current work in the laboratory is supported by grants from the Spanish Ministry of Economy and Competitivity (BFU2011-22630 and SAF2014-54604-C3-2-R), Generalitat Valenciana (Prometeo 2009/051), and Instituto de Salud Carlos III (ACCI2015 action from CIBER on Rare Diseases). The authors would like to acknowledge networking support by the Proteostasis COST Action (BM1307).

References

1. Saftig P (ed) (2005) Lysosomes. Landes Bioscience, Georgetown, TX, USA
2. Wartosch L, Bright NA, Luzio JP (2015) Lysosomes. Curr Biol 25(8):R315–R316
3. Knecht E, Aguado C, Cárcel J, Esteban I, Esteve JM, Ghislat G, Moruno JF, Vidal JM, Saez R (2009) Intracellular protein degradation in mammalian cells: recent developments. Cell Mol Life Sci 66(15):2427–2443
4. De Duve C, Pressman BC, Gianetto R, Wattiaux R, Appelmans F (1955) Tissue fractionation studies. 6. Intracellular distribution patterns of enzymes in rat-liver tissue. Biochem J 60(4):604–617
5. Hers HG, van Hoof F (eds) (1973) Lysosomes and storage diseases. Academic, New York, NY, USA
6. Parenti G, Andria G, Ballabio A (2015) Lysosomal storage diseases: from pathophysiology to therapy. Annu Rev Med 66:471–486
7. Zhang H, Fan X, Bagshaw R, Mahuran DJ, Callahan JW (2008) Purification and proteomic analysis of lysosomal integral membrane proteins. Methods Mol Biol 432:229–241
8. Ghislat G, Aguado C, Knecht E (2012) Annexin A5 stimulates autophagy and inhibits endocytosis. J Cell Sci 125(Pt 1):92–107
9. Chapel A, Kieffer-Jaquinod S, Sagne C, Verdon Q, Ivaldi C, Mellal M, Thirion J, Jadot M, Bruley C, Garin J, Gasnier B, Journet A (2013) An extended proteome map of the lysosomal membrane reveals novel potential transporters. Mol Cell Proteomics 12(6):1572–1588
10. de Duve C (2005) The lysosome turns fifty. Nat Cell Biol 7(9):847–849
11. Leighton F, Poole B, Beaufay H, Baudhuin P, Coffey JW, Fowler S, De Duve C (1968) The large-scale separation of peroxisomes, mitochondria, and lysosomes from the livers of rats injected with triton WR-1339. Improved isolation procedures, automated analysis, biochemical and morphological properties of fractions. J Cell Biol 37(2):482–513
12. Horvat A, Baxandall J, Touster O (1969) The isolation of lysosomes from Ehrlich ascites tumor cells following pretreatment of mice with Triton WR-1339. J Cell Biol 42(2):469–479
13. Henning R, Plattner H (1974) Isolation of rat liver lysosomes by loading with colloidal gold. Biochim Biophys Acta 354(1):114–120
14. Berg TO, Fengsrud M, Stromhaug PE, Berg T, Seglen PO (1998) Isolation and characterization of rat liver amphisomes. Evidence for fusion of autophagosomes with both early and late endosomes. J Biol Chem 273(34):21883–21892
15. Wattiaux R, Wattiaux-De Coninck S, Ronveaux-dupal MF, Dubois F (1978) Isolation of rat liver lysosomes by isopycnic centrifugation in a metrizamide gradient. J Cell Biol 78(2):349–368
16. Storrie B, Madden EA (1990) Isolation of subcellular organelles. Methods Enzymol 182:203–225

17. Graham JM (2001) Purification of a crude mitochondrial fraction by density-gradient centrifugation. Curr Protoc Cell Biol Chapter 3:Unit 3 4
18. Kaushik S, Cuervo AM (2008) Chaperone-mediated autophagy. Methods Mol Biol 445:227–244
19. Seglen PO, Brinchmann MF (2010) Purification of autophagosomes from rat hepatocytes. Autophagy 6(4):542–547
20. Patel B, Cuervo AM (2015) Methods to study chaperone-mediated autophagy. Methods 75:133–140
21. Lardeux B, Gouhot B, Forestier M (1983) Improved recovery of rat liver fractions enriched in lysosomes by specific alteration of the sedimentation properties of mitochondria. Anal Biochem 131(1):160–165

Chapter 20

Analysis of Relevant Parameters for Autophagic Flux Using HeLa Cells Expressing EGFP-LC3

Sandra Muñoz-Braceras and Ricardo Escalante

Abstract

Macroautophagy (called just autophagy hereafter) is an intracellular degradation machinery essential for cell survival under stress conditions and for the maintenance of cellular homeostasis. The hallmark of autophagy is the formation of double membrane vesicles that engulf cytoplasmic material. These vesicles, called autophagosomes, mature by fusion with endosomes and lysosomes that allows the degradation of the cargo. Autophagy is a dynamic process regulated at multiple steps. Assessment of autophagy is not trivial because the number autophagosomes might not necessarily reflect the real level of autophagic degradation, the so-called autophagic flux. Here, we describe an optimized protocol for the analysis of relevant parameters of autophagic flux using HeLa cells stably expressing EGFP-LC3. These cells are a convenient tool to determine the influence of the downregulation or overexpression of specific proteins in the autophagic flux as well as the analysis of autophagy-modulating compounds. Western blot analysis of relevant parameters, such as the levels of EGFP-LC3, free EGFP generated by autophagic degradation and endogenous LC3-I–II are analyzed in the presence and absence of the autophagic inhibitor chloroquine.

Key words Autophagy, Autophagosome, LC3, HeLa

1 Introduction

Autophagy is a degradative pathway in which cytoplasmic constituents are degraded in the lysosomes. The process comprises the formation of structures, called phagophores, which sequester the cytoplasmic cargo and elongate to form closed double membrane vesicles, the autophagosomes. Subsequently, they fuse with compartments of the endolysosomal pathway to form autolysosomes, where the inner membrane of the vesicle and its content are degraded [1]. It occurs constitutively at basal levels but it is further induced in response to starvation and to other circumstances that could threaten cellular homeostasis, such as the presence of defective organelles or accumulated protein aggregates [2, 3].

Although numerous proteins participate in the constant formation and degradation of autophagic structures [4–6], only

Rune Matthiesen (ed.), *Proteostasis: Methods and Protocols*, Methods in Molecular Biology, vol. 1449,
DOI 10.1007/978-1-4939-3756-1_20, © Springer Science+Business Media New York 2016

Atg8—and its homologs in other eukaryotes—remains associated to all those structures (phagophores, autophagosomes, and autolysosomes). Precisely because of its association to autophagic membranes until the end of the process, Atg8 is the quintessential protein for monitoring autophagy. In mammals, one of its homologs, microtubule-associated protein 1 light chain 3 (LC3), is the most widely used as an autophagic marker [7].

Membrane association of LC3 requires the posttranslational modification of the protein. Newly synthesized LC3 (proLC3) is cleaved by Atg4 at Gly120 at the C-terminus to form LC3-I, which is diffusely localized in the cytoplasm [8]. During autophagosome formation, LC3-I is conjugated to phosphatidylethanolamine in an ubiquitin-like reaction and the lipidated LC3 form (LC3-II) binds to the outer and the inner membranes of autophagosomes [9–11]. After autophagosome–lysosome fusion, outer membrane-bound LC3-II is again cleaved by Atg4 and dissociated from the membrane [12]. On the contrary, inner membrane-bound LC3-II is degraded within the autolysosome [8], *see* Fig. 1 for a schematic representation.

Most current assays to monitor autophagy are based on LC3 conversion and degradation. LC3 conversion can be traced by microscopy and immunoblotting techniques as LC3-II presents a punctate localization instead of a diffuse pattern and migrates faster

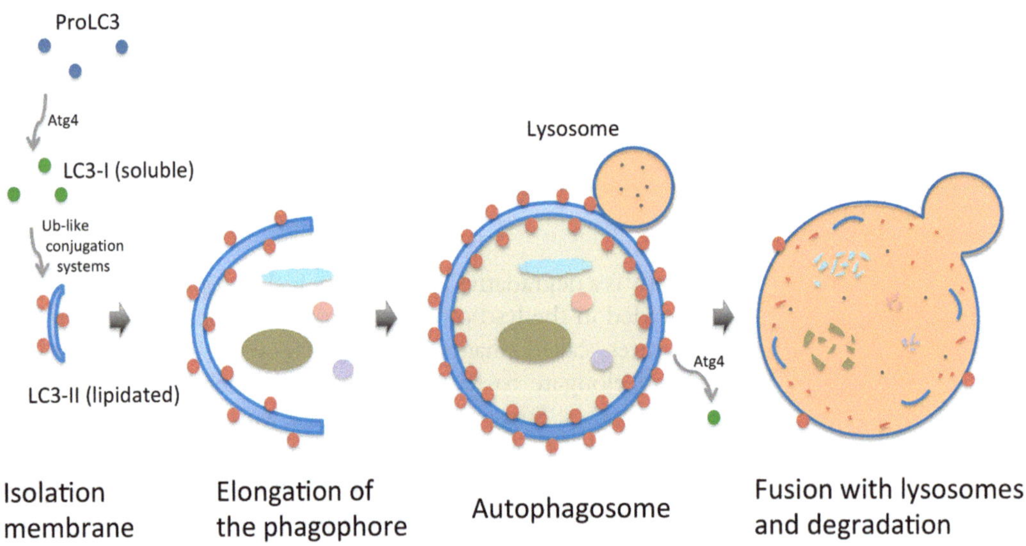

Fig. 1 Simplified representation of LC3 lipidation and turnover during autophagy. Cellular proLC3 is first cleaved by the protease Atg4 at the C-terminus to form LC3-I, also known as the soluble form. Upon autophagy induction, LC3-I is conjugated to the membrane of the nascent autophagosome and remains attached to both sides of the double membrane as the vesicle elongates, engulfs the cargo, and closes up. At that point, the outer LC3 is cleaved and recycled, the autophagosome fuses with lysosomes, and the internal LC3 is the degraded together with the inner membrane and the cargo. For cells expressing GFP-LC3, the fused protein suffers the same process as the endogenous LC3. The degradation of LC3 and GFP-LC3 in the autophagosome is the base for the methods described here, as their levels and localization change in the course of autophagic flux

than LC3-I on SDS-PAGE due to its hydrophobicity. LC3-II puncta and LC3-II levels correlate with the number of autophagic structures present in the cell but they do not inform about the actual autophagic activity. The reason is that autophagy is a dynamic process, in which the number of autophagosomes at a certain moment depends on their formation and degradation. This is referred to as autophagic flux and can be conveniently inferred from the assessment of LC3-II degradation in the lysosomes [13]. For that, LC3-II accumulation is monitored by immunoblotting after the addition of compounds that inhibit lysosomal degradation such as chloroquine.

The GFP-tagged version of LC3 at the N-terminus (GFP-LC3) is widely used to visualize autophagosomes. Following autophagy induction, GFP-LC3 is lipidated and associates to forming autophagosomes, resulting in a punctate pattern of the marker. Non-autophagic LC3 puncta can also occur due to the aggregation of the overexpressed GFP-LC3 [14] but this artifact is prevented in cell lines stably expressing GFP-LC3 at moderate levels. We would like to emphasize that not only the fluorescence of the lipidated membrane-bound form (GFP-LC3-II) can be used to monitor autophagy, but that the fluorescence of the soluble form (GFP-LC3-I) can also be informative of the autophagic process. In particular, when GFP-LC3 is overexpressed, GFP-LC3-I is diffusely located in the nucleus but translocates to the cytoplasm upon autophagy induction [15]. Thus, translocation of the diffuse fluorescence reflects autophagic activity. Differences in intensity of the diffuse fluorescence pattern can give a clue about the rate of LC3 conversion and degradation. However, it should be noted that differences of intensity can be due to different expression levels of the marker, so this is only applicable when using samples expressing a relatively homogeneous level of GFP-LC3, which is achieved in cell lines stably expressing the marker.

In addition, the potential of GFP-LC3 to monitor autophagy goes beyond its use as a fluorescent label. The fusion protein undergoes the same conversion and degradation of endogenous protein, but GFP is more resistant to lysosomal proteases than LC3. Thus, the appearance of free GFP also serves as an indicator of autophagic degradation [16]. We have found that most of the published studies limit the use of GFP-LC3 in immunoblotting to the detection of the free GFP band and that lysosomal inhibitors are not always added to properly characterize autophagic flux. However, as GFP-LC3 overexpression does not affect autophagic activity [17], we consider that the detection of both forms of the overexpressed and endogenous protein (GFP-LC3-I/II and LC3-I/II) together with the detection of the free GFP fragment, in the presence and absence of lysosomal inhibitors, provides a more complete view of the autophagic flux in a given experimental condition.

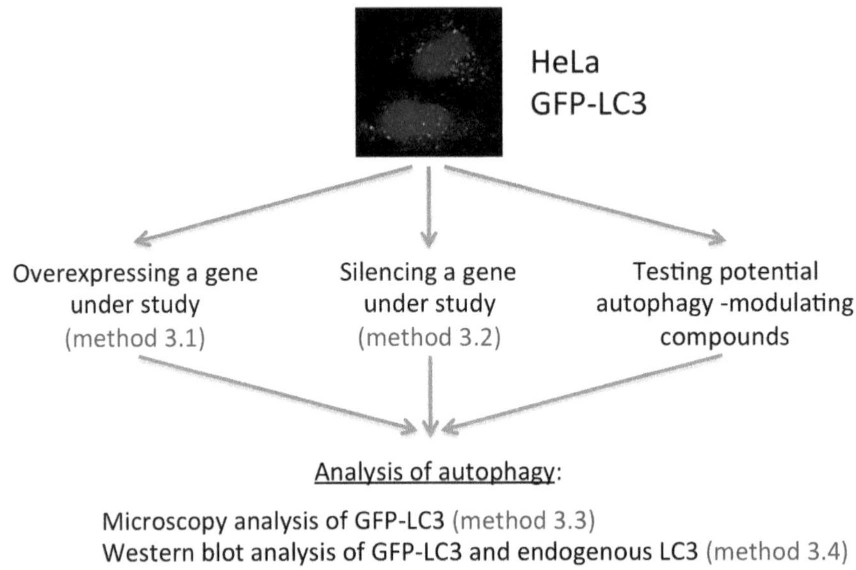

Fig. 2 Schematic representation of the methods described in this chapter. HeLa cells stably expressing the autophagosome marker GFP-LC3 can be used to determine autophagy changes upon overexpression (Subheading 3.1) or downregulation (Subheading 3.2) of genes under study as well as for testing the effect of compounds on autophagy (for example, in preclinical and high-throughput screenings). The effect of these alterations in autophagy can then be monitored by microscopy analysis (Subheading 3.3) or western blot to determine autophagic flux using the change in the levels of LC3-I and II (or GFP-LC3-I, II) and free GFP fragments upon autophagy inhibition with chloroquine (Subheading 3.4)

We describe in this chapter a series of protocols to efficiently assess autophagy using HeLa cells stably expressing EGFP-LC3 and a combination of fluorescence microscopy and immunoblotting procedures. These methods complement each other, which helps to interpret changes in the autophagic activity upon downregulation or overexpression of your favorite protein. This protocol can easily be adapted to the analysis of autophagy-modulating compounds. Figure 2 shows a diagram of the different methods described in this chapter.

2 Materials

2.1 Reagents

1. HeLa cells stably expressing EGFP-LC3 were kindly provided by Aviva M Tolkovsky (John Van Geest Center for Brain Repair, Cambridge, UK) and described previously [18].
2. Complete cell culture medium: DMEM (Dulbecco's Modified Eagle's Medium) (Sigma-Aldrich) supplemented with 10% FBS (Fetal Bovine Serum) (Gibco) and 1× penicillin-streptomycin (Gibco).

3. Starvation cell culture medium: EBSS, Earle's Balanced Salt Solution (Sigma-Aldrich).
4. PBS (phosphate-buffered saline) 1×: 133 mM NaCl, 8 mM Na_2HPO_4, 2 mM KH_2PO_4; pH 7.4.
5. TrypLE Express Enzyme (Gibco) for detachment of adherent cells.
6. Lipofectamine 2000 (Invitrogen).
7. Lipofectamine RNAiMAX (Invitrogen).
8. Opti-MEM (Invitrogen).
9. Silencer Select siRNA, target and negative control (Ambion).
10. 4 % Paraformaldehyde (Merk) in PBS. Prepare the solution in a hood, heating it to approximately 60 °C and stirring it gently. 1 M NaOH is added drop by drop to clear the solution, but taking care that the pH is maintained around 7.4.
11. 100 mM Glycine (Carlo Erba) in PBS.
12. DAPI (4′,6-Diamidino-2-Phenylindole, Dihydrochloride) (Molecular Probes).
13. ProLongGold antifade mountant (Molecular Probes).
14. Chloroquine diphosphate salt (Sigma-Aldrich) stock solution at 1 mM, prepared with deionized distilled water (ddH_2O).
15. RIPA lysis buffer: 50 mM Tris–HCl pH 8, 1 % NP-40, 0.1 % SDS, 0.5 % sodium deoxycholate, 150 mM NaCl, 2 mM EDTA. Protease inhibitor cocktail (Sigma-Aldrich) is added to the lysis buffer at a 1:100 dilution just before use.
16. Pierce BCA protein assay kit (Thermo Scientific).
17. Sample loading buffer 5×: 250 mM Tris–HCl pH 6.8, 10 % SDS, 25 % glycerol, 0.04 % bromophenol blue, 100 mM dithiothreitol (DTT).
18. SDS-polyacrylamide resolving gel (14 %). Recipe for 10 ml (enough to prepare a 1.5 mm thick mini-protean gel): 3.85 ml ddH_2O, 3.5 ml 40 % acrylamide/bis-acrylamide solution 37.5:1 (Bio-Rad), 2.5 ml Tris buffer 1.5 M pH 8.8, 100 μl SDS 10 %, 50 μl ammonium persulfate (APS) 10 % in ddH_2O (Bio-Rad, 161-0700), 5 μl *N,N,N′,N′*-Tetramethylethylenediamine (TEMED) (Sigma-Aldrich).
19. SDS-polyacrylamide stacking gel (4 %). Recipe for 4 ml (enough to prepare a 1.5 mm thick mini-protean gel): 3.04 ml ddH_2O, 0.4 ml 40 % acrylamide/bis-acrylamide solution 37.5:1, 0.5 ml Tris buffer 1 M pH 6.8, 40 μl SDS 10 %, 20 μl APS 10 %, 4 μl TEMED.
20. SDS running buffer: 25 mM Tris Base, 192 mM glycine, 0.1 % SDS in ddH_2O.
21. Transfer buffer: 25 mM Tris Base, 192 mM glycine, 20 % methanol in ddH_2O.

22. Tris buffered saline containing tween (TBS-T): 0.136 mM NaCl, 20 mM Tris Base, 0.05% Tween-20 (Sigma-Aldrich). The pH should be adjusted to 7.4 with HCl.
23. Skim milk (Sigma-Aldrich).
24. Primary antibodies: anti-LC3 (Cell Signaling, 2775), anti-GFP (Sigma-Aldrich, G1544), anti-GAPDH antibody (Enzo LifeSciences, ADI-CSA-335).
25. Secondary antibodies are horseradish peroxidase (HRP)-conjugated antibodies: goat anti-rabbit IgG-HRP (Santa Cruz Biotechnology, sc-2004), goat anti-mouse IgG-HRP (Santa Cruz Biotechnology, sc-2005).
26. Amersham ECL Western Blotting detection reagent (GE Healthcare).

2.2 Other Materials and Equipment

1. Falcon 100 mm TC-treated polystyrene cell culture dishes (Corning), 6-well and 24-well clear TC-treated polystyrene multiwell cell culture plates (Corning).
2. CO_2 incubator.
3. 12 mm coverslips (Heinz Herenz) and microscope slides (VWR).
4. Inverted Zeiss LSM 710 laser confocal microscope (Zeiss) equipped with a Plan-Apochromat 63×/1.40 NA oil-immersion objective. ZEN2009 acquisition software and ImageJ processing software.
5. Mini-PROTEAN Electrophoresis System and Mini Trans-Blot Electrophoretic Transfer Cell (Bio-Rad).
6. BioTrace Polyvinylidene fluoride (PVDF) transfer membrane (Pall Life Sciences).
7. CURIX RP2 Plus film (Agfa) and an X-ray film processing machine. A digital imaging system (GE Healthcare ImageQuant LAS4000) can be used instead.

3 Methods

3.1 DNA Transfection

1. The day before, detach the adherent cells on culture with the aid of a dissociating solution, such as TrypLE Express Enzyme, centrifuge the cells, and seed approximately 2×10^6 cells in a 100 mm dish to become 80% confluent at the time of transfection. Incubate the cells in complete cell culture medium at 37 °C in a CO_2 incubator.
2. Prior to transfection, replace the growing medium with 8 ml of prewarmed complete medium without antibiotics because the presence of antibiotics during transfection may increase cell death.

3. For each sample, dilute 40 μl Lipofectamine 2000 in 1 ml of serum-free medium Opti-MEM, mix gently by pipetting and incubate at room temperature for 5 min.
4. Dilute 20 μg of plasmid DNA in 1 ml Opti-MEM. Make the equivalent dilution with the same amount of a control empty vector.
5. Add the Lipofectamine 2000 dilution to the DNA dilution (2 ml of final volume for each sample), mix by pipetting, and incubate at room temperature for 20 min.
6. Add the 2 ml of the solution with the DNA-Lipofectamine 2000 complexes to the cells slowly and move the dish to evenly distribute the solution.
7. Incubate the cells with the DNA-Lipofectamine 2000 complexes at 37 °C in a CO_2 incubator for 4 h to minimize toxicity, aspirate the medium, and add fresh complete medium.
8. Incubate for about 24–30 h to allow the cells to recover from transfection. The cells are now ready for the subsequent autophagy experiments (*see* Subheadings 3.3 and 3.4) (*see* **Note 1**).

3.2 siRNA Reverse Transfection

1. The day of the transfection, detach the adherent cells on culture, count, centrifuge the cells, and dilute approximately 1×10^6 cells in 8 ml of prewarmed complete medium without antibiotics. Keep the cells in the falcon until **step 6** or alternatively **step 1** can be performed during the 20 min incubation described in **step 4**.
2. To prepare the complexes, for each sample, dilute 35 μl Lipofectamine RNAiMAX in 1 ml Opti-MEM, mix gently by pipetting, and incubate at room temperature for 5 min.
3. Dilute 100 pmol (10 μl of a 10 μM solution) of siRNA in 1 ml Opti-MEM. Make the equivalent dilution with the same amount of a negative control siRNA.
4. Add the Lipofectamine RNAiMAX dilution to the siRNA dilution (2 ml of final volume for each sample), mix by pipetting, and incubate at room temperature for 20 min.
5. Add the siRNA-Lipofectamine RNAiMAX complexes to the 100 mm dish and move back and forth to cover the surface of the dish with the mixture.
6. Add the 8 ml dilution of the cells to the dish containing the 2 ml of siRNA-Lipofectamine RNAiMAX complexes (the final siRNA concentration is 10 nM). Rock the dish gently.
7. Incubate at 37 °C in a CO_2 incubator for 48 h. Changing the medium is not necessary.
8. Repeat the siRNA reverse transfection from **steps 1–6** (*see* **Note 2**) using again 1×10^6 cells and the same volumes of every reagent for each transfection.

9. Three days after the second transfection the cells can be used for the subsequent autophagy experiments (*see* Subheadings 3.3 and 3.4).

3.3 Microscopy Analysis of GFP-LC3 Fluorescence

1. The day before the experiment, seed approximately 40,000 transfected cells on sterilized coverslips in 24-well plates. The culture should be 60–70% confluent at the time of the experiment (*see* **Note 3**).
2. For starvation treatment to induce autophagy, wash the cells twice with 1 ml of prewarmed PBS and once again with 1 ml of prewarmed starvation medium EBSS. Aspirate off the medium and add 1 ml of EBSS. Similarly, aspirate off the medium, wash the cells, and add 1 ml of prewarmed fresh complete medium (DMEM with 10% FBS) to the samples in which basal autophagy will be analyzed.
3. Incubate the cells at 37 °C and 5% CO_2 for 2 h (*see* **Note 4**).
4. Discard the medium and rinse the cells with PBS.
5. Fix the cells with 4% paraformaldehyde in PBS, pH 7.4, for 15 min at room temperature (*see* **Note 5**).
6. Wash the cells three times with PBS.
7. To quench the possible fluorescence signal from free aldehyde groups in paraformaldehyde, it can be convenient to incubate with 100 mM glycine in PBS for 30 min at room temperature.
8. At this point, permeabilization and incubation with primary and fluorescent-secondary antibodies can also be performed if the detection of additional proteins is desired. A nuclear dye such as DAPI can also be employed to stain nuclei (*see* **Note 6**).
9. Mount the samples using an antifade reagent such as ProLongGold. Put 4 μl of ProLongGold onto a microscope slide and with the aid of a forceps, place the coverslip onto the drop with the cells facing down and avoiding air bubble trapping. Leave the mounted samples in the dark at room temperature until the ProLongGold reagent is dry and then store them in the dark at 4 °C until visualization.
10. Observe the cells under the microscope. We routinely use a 63× objective. Higher magnification (100×) can be used to visualize the ring-shape of the autophagosomes.
11. Acquire the necessary captures along the *z*-axis to image the whole cell. Then, perform the montage to obtain the maximum intensity *z*-projection of the stack (*see* **Note 7**). Figure 3a shows images of the control sample of a representative experiment.
12. Count the puncta (*see* **Notes 8** and **9**). Figure 3b illustrates an example of quantification of the puncta observed in the control samples of independent experiments.

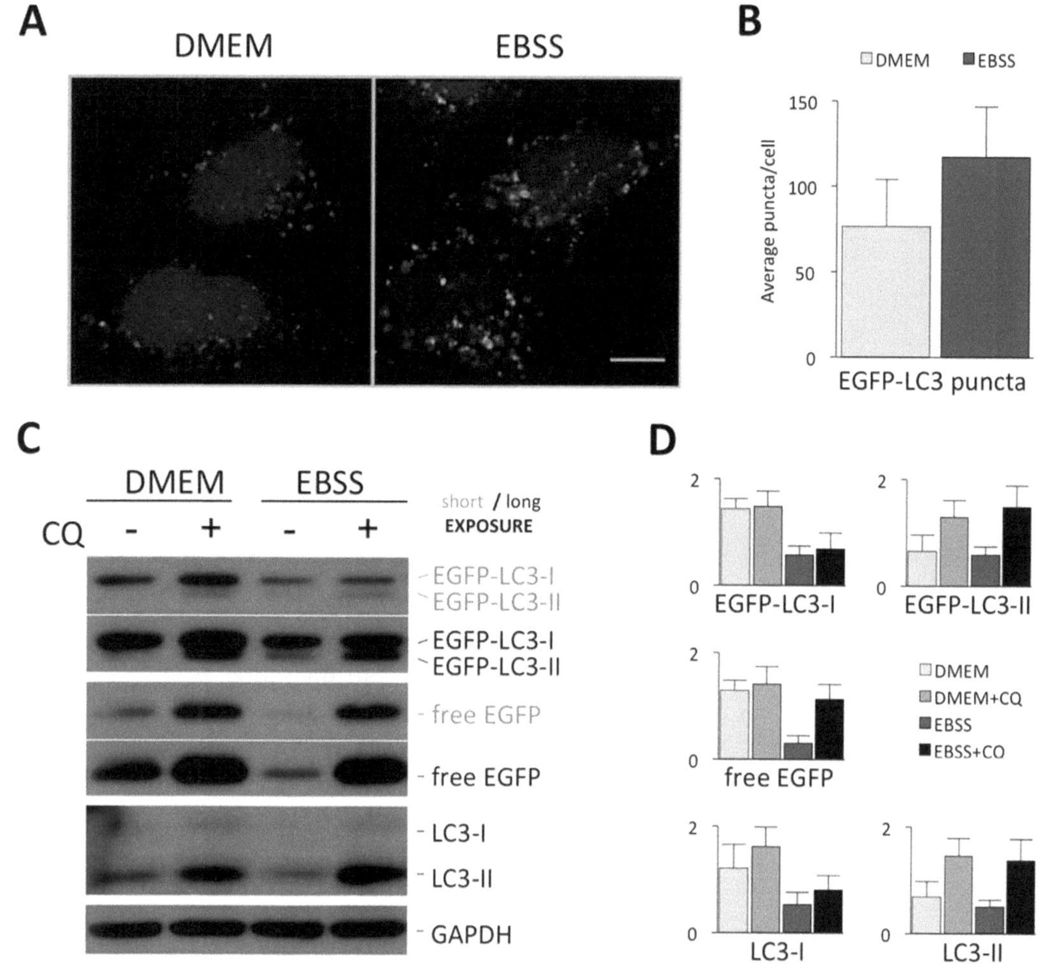

Fig. 3 HeLa cells stably expressing EGFP-LC3 and transfected with control siRNAs were incubated in complete (DMEM) or starvation (EBSS) medium for 2 h (**a**, **b**) or for 4 h in the absence or presence of 5 μM chloroquine (CQ) (**c**, **d**). Starvation causes the translocation of diffuse EGFP-LC3 fluorescence from the nucleus to the cytoplasm (**a**), an increase in the number of puncta per cell (**b**), a decrease in the amount of EGFP-LC3-I (45 kDa) and endogenous LC3-I (18 kDa) and the faster degradation of the free EGFP (27 kDa) generated by the cleavage of EGFP-LC3 (**c**, **d**). The block of the degradation caused by the presence of chloroquine shows the accumulation of the free EGFP fragment and the EGFP-LC3-II (43 kDa) and LC3-II (16 kDa) (**c**, **d**). The graphs show the mean values and standard deviations of the quantification of the puncta per cell in more than 300 cells for each condition from seven independent experiments (**b**) and of the densitometry of the protein bands observed in western blots of nine independent experiments, showing the comparison of the amounts of protein in arbitrary units (**d**). Scale bar: 10 μm

3.4 Western Blot Analysis

1. The day before the experiment, seed approximately 200,000 transfected cells in six-well plates. Cell confluence should be 60–70% at the time of the experiment (*see* **Note 3**).
2. The day of the experiment, aspirate off the medium, wash the cells twice with 2 ml of prewarmed PBS and once again with 1 ml of the prewarmed incubation medium for each experimental condition (DMEM or EBSS, without or with

chloroquine) to avoid any undesirable variation in the final composition and chloroquine concentration of the experimental incubation medium. Aspirate off the medium and add 2 ml of DMEM or EBSS medium, for incubation in rich nutrient or starvation conditions, respectively. To the samples in which autophagic degradation will be blocked, chloroquine has to be added to the medium at a final concentration of 5 μM from a prediluted stock solution (*see* **Notes 10** and **11**).

3. Incubate the cells at 37 °C and 5 % CO_2 for 4 h (*see* **Note 12**).
4. Put the plate on ice, discard the medium, and wash the cells with ice-cold PBS.
5. Add approximately 50 μl of RIPA lysis buffer with the protease inhibitor cocktail added prior to use.
6. Harvest the cells using a cell scraper and transfer them into a cooled microcentrifuge tube.
7. Leave the cells on ice during 30 min, mixing the solution each 10 min by pipetting or vortexing.
8. Centrifuge at 13,500 × *g* for 15 min at 4 °C and transfer the supernatant to a new microcentrifuge tube.
9. Take an aliquot of the sample and make a 1:4 dilution for measuring protein concentrations by the BCA protein assay. Follow the manufacturer's instructions to determine protein concentration.
10. Prepare the samples to load 7 μg of protein per well in sample loading buffer (*see* **Note 13**).
11. Boil the samples at 98 °C for 5 min. If the samples are not going to be loaded immediately, store them at −80 °C.
12. Place the previously prepared gels in the electrophoresis chamber and fill it with SDS running buffer. For preparing the gels, pour the freshly prepared resolving gel solution (APS and TEMED are added just prior pouring) in the cassette of cleaned glasses. Pour on top some milliliters of water or isopropanol to facilitate polymerization. Retire the water, pour the stacking solution, place the comb on top, and allow the gel to polymerize.
13. Load the samples and the protein marker in the polyacrylamide gel. Proceed with the electrophoresis at a constant voltage of 150 V at room temperature until the sample buffer is at the bottom of the gel.
14. Disassemble the gel cassette, discard the stacking gel, wash the gel in transfer buffer and assemble the transference sandwich (gel and methanol-activated PVDF membrane between filter papers and sponges) (*see* **Note 14**).
15. For protein transference, we routinely use a wet-type transfer system. Place the sandwich in the electrophoretic transfer cell

and fill it with transfer buffer. Transfer the proteins at 100 V constant current at 4 °C for 1 h (*see* **Note 15**).

16. Optionally, after transference the membrane can be stained with 0.1 % Ponceau S in 5 % acetic acid for 5 min to verify transference efficiency.
17. Block the membrane by incubating with 5 % skim milk in TBS-T on a shaker at room temperature for at least 1 h.
18. Incubate the membrane first with anti-LC3 antibody diluted 1:1000 in 5 % skim milk TBS-T on a shaker at 4 °C overnight (*see* **Note 16**).
19. Wash the membrane with TBS-T at room temperature five times for 10 min each.
20. Incubate the membrane with secondary antibody (goat anti-rabbit IgG-HRP) diluted 1:5000 in 2 % skim milk TBS-T on a shaker at room temperature for 1 h.
21. Wash the membrane with TBS-T at room temperature five times for 10 min each.
22. Mix the developing ECL solution, add it to the membrane and incubate for 1 min.
23. Use an X-ray film to capture the chemiluminescent signal on the membrane (*see* **Note 17**).
24. Repeat **steps 18–23** using anti-GFP antibody diluted 1:4000 and the goat anti-rabbit IgG-HRP antibody (*see* **Note 18**).
25. Repeat **steps 18–23** for detection of GAPDH as a loading control using anti-GAPDH antibody at 1:2000. In this case, use the secondary goat anti-mouse IgG-HRP antibody at 1:4000. Actin or tubulin can be also used as loading control proteins. Figure 3c shows the protein bands of the control sample of a representative experiment.
26. Scan the film and quantify bands signal intensity by densitometry using ImageJ software (*see* **Notes 19–21**). Figure 3d contains the graphs obtained from the protein bands densitometry of the control sample in independent experiments.

4 Notes

1. It is preferably to perform the procedure in a single population of cells that will be split at least 16–18 h before the planned experiment and no longer than 24 h. This is important to avoid heterogeneity in protein overexpression or depletion levels and even in confluence rates, which can alter basal autophagy and even LC3 expression. Concerning confluence, in general, cells should be maintained subconfluent but at a density that does not compromise transfection efficiency.

2. In certain cases, a large level of downregulation might be necessary to observe an effect. This can be particularly important for long-lived proteins. Longer periods of silencing can be achieved by clonal selection using expression of small hairpin RNA (shRNA); however, knockdown for prolonged time periods is not recommended for autophagy-related proteins [19]. We prefer to perform two consecutive siRNA transfections to assure the maximal depletion of the protein. However, it should be kept in mind that repetitive transfection is an additional stress to the cells. The decision to perform single or double transfections will depend on the protein of interest.
3. As stated above, confluence has an effect on basal autophagy levels. Thus, density of the cultures evaluated within and between experiments should be similar and confluence should be avoided.
4. This time period has been established for HeLa cells stably expressing EGFP-LC3. It might have to be modified if other cell lines are used as it depends on the autophagic activity of the cell [7]. In any case, it is not recommended to extend too much this incubation time when neither chloroquine nor other similar compound are added to block lysosomal degradation because it has been reported that LC3-puncta fluorescence decreases when cells are starved for longer periods [20].
5. Although the expression of the fluorescent marker allows live cell imaging, we describe here a protocol for fixing cells. Live cell imaging has the advantage that the response to the autophagy induction stimulus can be monitored along the time and that mobility of autophagosomes can also be traced. The major drawback is that the proper equipment is required to control temperature and CO_2 concentration in the environment, necessary to maintain the culture in suitable healthy conditions during the period of imaging. Another consideration is that GFP signal diminishes at an acidic pH. As a consequence, GFP fluorescence is quenched inside the autolysosomes [20]. This attenuation of fluorescence can be circumvented by fixation of the samples. Paraformaldehyde fixation maintains the sample at a neutral pH and thus, GFP fluorescence is retained. Fixed samples have the additional advantage that they can be stored and also used for immunodetection of other proteins and colocalization studies.
6. For immunofluorescence, it is important to keep in mind that certain detergents can lead to the appearance of artifactual GFP-LC3 puncta [21]. Similar cautions apply with regard to methanol, which can also be used for cell fixation and permeabilization, but might reduce GFP intensity [22].
7. To avoid misinterpretation of results that could arise from an uneven distribution of autophagosomes within the cell and the selection of random sections along the *z*-axis, we routinely

image and count the puncta present of the whole cell. For this, we capture the necessary slices to imaging the entire cell. For showing purposes, we also consider that the maximum intensity *z*-stack projection provides a more objective view of the cell.

8. Puncta can be automatically quantified by using specialized imaging software (ImageJ, Imaris, or CellProfiler). In that case, manual evaluation of the analysis is highly recommended to verify the quantification or, if necessary, adjust the parameters that define what is considered "puncta". Manual quantification might be more accurate, but similarly, uniform criteria must be applied regarding the definition of puncta. An advantage of automated analysis is that other parameters, such as area, can be monitored. In particular, the average percentage of the total area of GFP-LC3 puncta on a per cell basis can be a more appropriate index in the cases when individual puncta quantification is not possible due to autophagosome clustering.
9. GFP-LC3 overexpression results in most of the cells displaying some puncta regardless of the experimental condition. Therefore, the percentage of cells with puncta is not a good indicator of autophagic activity. To circumvent this problem, a threshold can be established to classify the population in cells with basal or induced autophagy. However, the definition of this cut-off value is rather subjective, and can be dependent on the expression levels of GFP-LC3. GFP-LC3 expression is more uniform in cell lines stably expressing the marker, but variability of autophagic activity is still significant. We think that a more appropriate index is the average number of puncta per cell. As a consequence of the variability of the autophagic activity within the same population, this parameter also considerably fluctuates across cells. Thus, for a good quantification of puncta, a large number of cells (around 100) from multiple sections should be documented for each condition in at least three independent experiments.
10. We have observed that this nonsaturated concentration of chloroquine is enough to allow the simultaneous accumulation of free GFP and LC3-II in HeLa cells stably expressing GFP-LC3. If this concentration has to be increased for other cell types, it is important to ensure that it is low enough to allow the visualization of the cleavage of the GFP-LC3, which requires a nonsaturating concentration of chloroquine as described previously [16]. It is believed that GFP is relatively resistant to lysosomal degradation and can be accumulated using low concentrations of lysosomal inhibitors. Higher concentrations (saturating concentrations) would totally inhibit lysosomal proteases hampering the GFP-LC3 cleavage, so free GFP fragments will not be generated and the GFP cleavage assay would not be applicable. Besides, high concentrations of

chloroquine could induce autophagy or could also affect other pathways independent of autophagy. This kind of side-effect has to be also taken into account if other compounds are used to raise the lysosomal pH.

11. When deciding the experimental incubation conditions, it should be kept in mind that acidity of lysosomes regulates GFP-LC3 cleavage and free GFP accumulation. This also accounts for the use of EBSS as a starvation medium to induce autophagy. EBSS lowers lysosomal pH and this provokes that, while free GFP fragments can still be detected when the cells are incubated in complete medium in the absence of lysosomal inhibitors, free GFP is further degraded and might be undetectable when EBSS is used without lysosomal inhibitors.

12. Note that we have extended the incubation period with respect to the protocol for fluorescence microscopy because accumulation of the lipidated GFP-LC3-II and LC3-II as well as free GFP will increase in a time-dependent manner due to the altered lysosomal pH and thus, the differences, in comparison to the samples in the absence of chloroquine, will be more evident. Longer incubations are not advisable because the expression of some proteins can change. One example is the autophagy substrate p62/SQSTM1 [23] but changes in other autophagy-related proteins, even in LC3, might also occur [24–26]. Moreover, a secondary autophagic response could be induced due to the accumulation of nondegraded autophagosomes if the incubation period is too long.

13. Using small amounts of protein is advisable so the chemiluminiscent signal is less saturated, and minor differences between the bands of accumulated proteins are easier visualized. In contrast, it is probably necessary to load higher amounts of protein to visualize the LC3-I band, which is difficult to detect even after long exposure periods in this cell type.

14. PVDF membranes are preferred rather than nitrocellulose membranes due to their better retention of the lipidated LC3-II.

15. Time of transfer should be short because of the low molecular weight of endogenous LC3.

16. We recommend incubating first with anti-LC3 antibody because LC3 signal is more easily lost during subsequent hybridizations than the GFP signal, due to the lesser sensitivity of the anti-LC3 antibody. This might also help in the detection of little differences in the fusion protein GFP-LC3-I and GFP-LC3-II amounts, which might be more difficult to perceive using the more sensitive anti-GFP antibody.

17. Capture the signal at different exposure times to cover all the range of signal saturation levels. Discard for densitometry those with too low or too strong signals as they might not be

in the linear range and would not represent the real differences between the samples. Short exposure times can lead to too faint bands in conditions of induced and nonblocked autophagy, while long exposure times can mask differences between bands in conditions where autophagy degradation has been blocked. Thus, for illustrating purposes, it is usually helpful to show both a short and a long exposure. Nevertheless, for a quantitative evaluation, the lanes from all the conditions have to be measured from a unique exposure time.

18. Stripping of the membrane is not necessary as the molecular weights of the proteins are different and this procedure could be aggressive and affect latter signals.
19. ImageJ software tool is useful for this purpose. As with puncta quantification, there are some practical aspects to be considered. The most important one is that the bands that are going to be subjected to densitometry must not be overexposed. This accounts, in particular, for the samples corresponding to chloroquine incubation, where the accumulation of GFP and the lipidated forms of overexpressed and endogenous LC3-II results in a strong signal. As stated above, for comparison between all the samples, the densitometry has to be done from a unique exposure; and although it might be difficult in practice, this exposure should be in the linear range for all the bands, avoiding saturation as much as possible. The data must be normalized to the loading control protein and then, the mean and the standard deviations of the independent experiments can be used to compare between samples and determine the autophagic flux in each experimental setting.
20. We would like to add some important considerations regarding data interpretation in the following notes. LC3-I levels decrease in response to autophagy induction due to the conversion to the LC3-II form and therefore, changes in the LC3-I amount in a given experimental setting compared to controls suggest that LC3-I conversion is affected in that condition. The comparison of LC3-II levels between samples in the presence and absence of lysosomal inhibitors better represents autophagic activity than the comparison of the ratio of LC3-II to LC3-I. Although the later has been employed as a measure of autophagic induction, we believe that it is not a trustworthy indicator because LC3-II is degraded within the autophagolysosomes and its levels can increase or decrease depending on the rate of conjugation and degradation if the later is not blocked. Besides, we have experienced that this ratio might change depending on the antibody used for immunodetection. Certain antibodies barely detect LC3-I although they give a strong signal for LC3-II and, inversely, LC3-I is sensitive to detection by other antibodies that might result in a

less intense LC3-II band [7]. Moreover, it should be kept in mind than LC3-I seems to be more fragile and poorly conserved in stored samples, especially when they have been repetitively frozen and thawed [15].

21. The amounts of the soluble and the lipidated forms of LC3 can also be analyzed when the protein is fused to GFP. Conversion of GFP-LC3 parallels that of endogenous LC3 [18]. Therefore, the evaluation of the fused protein, in addition to that of the endogenous one, might be helpful to interpret the autophagic activity within the cell. As in the case of endogenous protein, GFP-LC3-I amount decreases in response to autophagy induction and thus its comparison between growth and starvation conditions can be used as a reliable parameter. While the detection of the endogenous LC3-I have some drawbacks, GFP-LC3-I detection is clear due to the overexpression and the higher sensitivity of anti-GFP detection. Similarly to endogenous LC3-II, GFP-LC3-II band intensity can increase or decrease in response to starvation or other autophagy induction stimulus if the autophagic degradation is not blocked. But in reference to its accumulation in the presence of lysosomal inhibitors, it may be less evident for GFP-LC3-II, which tends to accumulate less than endogenous LC3-II in certain conditions [20]. This effect might arise from GFP-LC3-II cleavage within the lysosome when a nonsaturating concentration of lysosomal inhibitor is being used. Regarding the evaluation of free GFP, we believe that the comparison between its levels in the presence and absence of lysosomal inhibitors informs about the autophagic flux mainly in starvation conditions.

Acknowledgments

This work was supported by grants BFU2012-32536 and BFU2015-64440-P from the Spanish Ministerio de Ciencia e Innovación. SMB is recipient of a predoctoral fellowship from Universidad Autónoma de Madrid. The authors would like to acknowledge networking support by the Proteostasis COST Action (BM1307).

References

1. Shibutani ST, Yoshimori T (2014) A current perspective of autophagosome biogenesis. Cell Res 24(1):58–68
2. Schneider JL, Cuervo AM (2014) Autophagy and human disease: emerging themes. Curr Opin Genet Dev 26C:16–23
3. Choi AM, Ryter SW, Levine B (2013) Autophagy in human health and disease. N Engl J Med 368(19):1845–1846
4. Mizushima N, Yoshimori T, Ohsumi Y (2011) The role of Atg proteins in autophagosome formation. Annu Rev Cell Dev Biol 27:107–132

5. Itakura E, Mizushima N (2010) Characterization of autophagosome formation site by a hierarchical analysis of mammalian Atg proteins. Autophagy 6(6):764–776
6. Feng Y, He D, Yao Z, Klionsky DJ (2014) The machinery of macroautophagy. Cell Res 24(1):24–41
7. Kimura S, Fujita N, Noda T, Yoshimori T (2009) Monitoring autophagy in mammalian cultured cells through the dynamics of LC3. Methods Enzymol 452:1–12
8. Kabeya Y, Mizushima N, Ueno T, Yamamoto A, Kirisako T, Noda T, Kominami E, Ohsumi Y, Yoshimori T (2000) LC3, a mammalian homologue of yeast Apg8p, is localized in autophagosome membranes after processing. EMBO J 19(21):5720–5728
9. Tanida I, Tanida-Miyake E, Ueno T, Kominami E (2001) The human homolog of Saccharomyces cerevisiae Apg7p is a protein-activating enzyme for multiple substrates including human Apg12p, GATE-16, GABARAP, and MAP-LC3. J Biol Chem 276(3):1701–1706
10. Tanida I, Tanida-Miyake E, Komatsu M, Ueno T, Kominami E (2002) Human Apg3p/Aut1p homologue is an authentic E2 enzyme for multiple substrates, GATE-16, GABARAP, and MAP-LC3, and facilitates the conjugation of hApg12p to hApg5p. J Biol Chem 277(16):13739–13744
11. Mizushima N, Sugita H, Yoshimori T, Ohsumi Y (1998) A new protein conjugation system in human. The counterpart of the yeast Apg12p conjugation system essential for autophagy. J Biol Chem 273(51):33889–33892
12. Kirisako T, Ichimura Y, Okada H, Kabeya Y, Mizushima N, Yoshimori T, Ohsumi M, Takao T, Noda T, Ohsumi Y (2000) The reversible modification regulates the membrane-binding state of Apg8/Aut7 essential for autophagy and the cytoplasm to vacuole targeting pathway. J Cell Biol 151(2):263–276
13. Mizushima N, Yoshimori T, Levine B (2010) Methods in mammalian autophagy research. Cell 140(3):313–326
14. Kuma A, Matsui M, Mizushima N (2007) LC3, an autophagosome marker, can be incorporated into protein aggregates independent of autophagy: caution in the interpretation of LC3 localization. Autophagy 3(4):323–328
15. Klionsky DJ, Abdalla FC, Abeliovich H, Abraham RT, Acevedo-Arozena A, Adeli K, Agholme L, Agnello M, Agostinis P, Aguirre-Ghiso JA, Ahn HJ, Ait-Mohamed O, Ait-Si-Ali S, Akematsu T, Akira S, Al-Younes HM, Al-Zeer MA, Albert ML, Albin RL, Alegre-Abarrategui J, Aleo MF, Alirezaei M, Almasan A, Almonte-Becerril M, Amano A et al (2012) Guidelines for the use and interpretation of assays for monitoring autophagy. Autophagy 8(4):445–544
16. Ni HM, Bockus A, Wozniak AL, Jones K, Weinman S, Yin XM, Ding WX (2011) Dissecting the dynamic turnover of GFP-LC3 in the autolysosome. Autophagy 7(2):188–204
17. Mizushima N, Yamamoto A, Matsui M, Yoshimori T, Ohsumi Y (2004) In vivo analysis of autophagy in response to nutrient starvation using transgenic mice expressing a fluorescent autophagosome marker. Mol Biol Cell 15(3):1101–1111
18. Bampton ET, Goemans CG, Niranjan D, Mizushima N, Tolkovsky AM (2005) The dynamics of autophagy visualized in live cells: from autophagosome formation to fusion with endo/lysosomes. Autophagy 1(1):23–36
19. Staskiewicz L, Thorburn J, Morgan MJ, Thorburn A (2013) Inhibiting autophagy by shRNA knockdown: cautions and recommendations. Autophagy 9(10):1449–1450
20. Tanida I, Minematsu-Ikeguchi N, Ueno T, Kominami E (2005) Lysosomal turnover, but not a cellular level, of endogenous LC3 is a marker for autophagy. Autophagy 1(2):84–91
21. Ciechomska IA, Tolkovsky AM (2007) Non-autophagic GFP-LC3 puncta induced by saponin and other detergents. Autophagy 3(6):586–590
22. Patergnani S, Pinton P (2015) Mitophagy and mitochondrial balance. Methods Mol Biol 1241:181–194
23. Sahani MH, Itakura E, Mizushima N (2014) Expression of the autophagy substrate SQSTM1/p62 is restored during prolonged starvation depending on transcriptional upregulation and autophagy-derived amino acids. Autophagy 10(3):431–441
24. Settembre C, Di Malta C, Polito VA, Garcia Arencibia M, Vetrini F, Erdin S, Erdin SU, Huynh T, Medina D, Colella P, Sardiello M, Rubinsztein DC, Ballabio A (2011) TFEB links autophagy to lysosomal biogenesis. Science 332(6036):1429–1433
25. Bartholomew CR, Suzuki T, Du Z, Backues SK, Jin M, Lynch-Day MA, Umekawa M, Kamath A, Zhao M, Xie Z, Inoki K, Klionsky DJ (2012) Ume6 transcription factor is part of a signaling cascade that regulates autophagy. Proc Natl Acad Sci U S A 109(28):11206–11210
26. Jin M, Klionsky DJ (2014) Regulation of autophagy: modulation of the size and number of autophagosomes. FEBS Lett 588(15):2457–2463

Chapter 21

Analysis of Protein Oligomeric Species by Sucrose Gradients

Sandra Tenreiro, Diana Macedo, Zrinka Marijanovic, and Tiago Fleming Outeiro

Abstract

Protein misfolding, aggregation, and accumulation are a common hallmark in various neurodegenerative diseases. Invariably, the process of protein aggregation is associated with both a loss of the normal biological function of the protein and a gain of toxic function that ultimately leads to cell death. The precise origin of protein cytotoxicity is presently unclear but the predominant theory posits that smaller oligomeric species are more toxic than larger aggregated forms. While there is still no consensus on this subject, this is a central question that needs to be addressed in order to enable the design of novel and more effective therapeutic strategies. Accordingly, the development and utilization of approaches that allow the biochemical characterization of the formed oligomeric species in a given cellular or animal model will enable the correlation with cytotoxicity and other parameters of interest.

Here, we provide a detailed description of a low-cost protocol for the analysis of protein oligomeric species from both yeast and mammalian cell lines models, based on their separation according to sedimentation velocity using high-speed centrifugation in sucrose gradients. This approach is an adaptation of existing protocols that enabled us to overcome existing technical issues and obtain reliable results that are instrumental for the characterization of the types of protein aggregates formed by different proteins of interest in the context of neurodegenerative disorders.

Key words Protein misfolding, Yeast model, Mammalian cell lines, Proteinopathies, Oligomers, Protein aggregates, Sucrose gradients, Velocity sedimentation, High-speed centrifugation, Neurodegenerative diseases

1 Introduction

Proteinopathies are a group of diseases that share in common the misfolding, aggregation, and accumulation of specific proteins in different tissues, from peripheral nerves and organs, to the central nervous system. The aggregation process is thought to evolve through the formation of protein dimers, oligomers, protofibrils and, ultimately, amyloid fibrils organized in large aggregates [1]. Two consequences might derive from the aggregation process: a loss of the normal function of the protein involved, and a gain of

Rune Matthiesen (ed.), *Proteostasis: Methods and Protocols*, Methods in Molecular Biology, vol. 1449,
DOI 10.1007/978-1-4939-3756-1_21, © Springer Science+Business Media New York 2016

toxic function of the misfolded and aggregated forms. However, the precise nature of the toxic forms is still controversial, with recent data suggesting oligomeric forms might be more toxic than larger, insoluble, aggregated species. The clarification of these questions is central for the understanding of the molecular basis of many disorders and for the development of effective therapeutic strategies. Thus, intense efforts have been dedicated to the clarification of this question, through the use of in vitro approaches combined with the use of cellular and animal models. Accordingly, the establishment of a correlation between aggregation and cytotoxicity induced by a given misfolded protein in the various model systems is crucial.

Yeast models of proteinopathies, such as Parkinson's disease (PD), have provided important contributions to the current knowledge on the molecular pathways associated with the toxicity of alpha-synuclein (aSyn), the protein that misfolds and aggregates in PD [2]. In particular, aSyn was found to be toxic and to form cytoplasmic inclusions in yeast cells [3]. However, the precise biochemical nature of the aSyn species formed in yeast was unclear. Here, we demonstrate the adaptation of a technique that allows the biochemical analysis of the aggregation state of aSyn. This approach can also be used to study other proteins of interest and is applicable for the study of protein extracts obtained from different cells and tissues. This protocol is based on density gradient sedimentation, where oligomeric species sediment through the gradient in separate zones based on their sedimentation rate.

Briefly, yeast cell lysates are generated by spheroplasting, in order to preserve the proteins in their physiological state, and then a defined protein amount is applied at the top of a discontinuous sucrose gradient (5–30% (w/v)). The various protein species are separated by ultracentrifugation, and fractions are collected from the top of the gradient. The protein is then precipitated, washed, solubilized, and resolved in sodium dodecyl sulfate (SDS) polyacrylamide gel electrophoresis (SDS-PAGE), after which it is transferred electrophoretically to a nitrocellulose membrane. The protein of interest can then be visualized by immunoblotting with specific antibodies (Fig. 1).

Using this method we demonstrated that, in yeast cells, aSyn aggregation is promoted by coexpression with the polo-like kinase 2 PLK2 [4] and is inhibited by the chemical chaperone mannosylglycerate [5], by S129 phosphorylation [6], or by coexpression with Hsp31–34, the yeast DJ-1 orthologs [7].

2 Materials

All solutions are prepared using ultrapure water (prepared by purifying deionized water to attain a resistivity of 18 M Ω cm at 25 °C) and analytical grade reagents. Solutions are stored at room

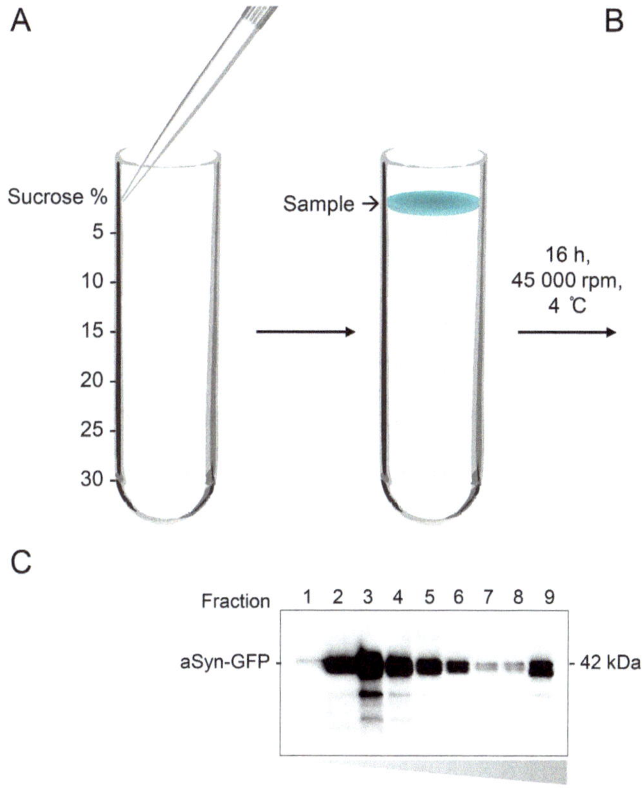

Fig. 1 Sucrose gradient preparation and immunoblotting. (**a**) The different sucrose solutions are carefully pipetted into the centrifuge tubes and the sample is loaded on top. (**b**) After centrifugation the components of the sample are separated based on their size and fractions 1–9 are collected. (**c**) The fractions, corresponding to different proteins sizes, are subjected to an SDS-PAGE and immunoblotting for the protein of interest

temperature unless indicated otherwise. Materials needed are: falcon and eppendorf centrifuges, ultracentrifuge, 30 and 95 °C incubator or bath, 2.5 mL syringes, 25G needles, 5 mL thinwall polypropylene centrifuge tubes, trichloroacetic acid (TCA), acetone, protease inhibitors.

2.1 Sucrose Gradient Solutions

1. Phosphate-buffered saline (PBS): 0.13 M NaCl, 2.7 M KCl, 12.5 M Na_2HPO_4, 1.76 M KH_2PO_4. Mix 80 g of NaCl, 2 g of KCl, 17.8 g of Na_2HPO_4, and 2.4 g of KH_2PO_4, in 800 mL of water. Adjust pH to 7.4. Complete the volume of the solution to 1 L with water. Before using, dilute 10 mL of PBS solution in 100 mL of water.
2. 80% (w/v) Sucrose: Weight 8 g of sucrose, adjust the volume to 100 mL of water and dissolve it well. Filter it using a 0.22 μm pore size filter (for aqueous solutions) and store at 4 °C to avoid contamination with microorganisms.

3. 1 M Tris: Weigh 121.1 g of Tris base and add water to a volume of 800 mL. Mix and adjust pH to 7.4 or 6.8 by adding concentrated HCl. Adjust the volume of the solution to 1 L with water and sterilize by autoclaving 15 min at 121 °C (*see* **Note 1**).
4. 1 M Magnesium chloride ($MgCl_2$): Dissolve 57.6 g of $MgCl_2$ in 400 mL of water. Adjust the volume to 500 mL with water, dispense into aliquots and sterilize by autoclaving 15 min at 121 °C (*see* **Note 2**).
5. 5000 U/mL Zymolyase 100T: Weight 1 g of zymolyase 100T and dissolve it in 20 mL of water or specific supplied buffer (*see* **Note 3**).
6. 5 M Sodium Chloride (NaCl): Dissolve 292.2 g of NaCl in 800 mL of water. Adjust the volume to 1 L with water, dispense into aliquots, and sterilize by autoclaving.
7. 350 mM Sodium dodecyl sulfate (SDS): Dissolve 100 g of SDS in 900 mL of water. Heat to 68 °C to assist dissolution. Adjust the pH to 7.2 by adding a few drops of concentrated HCl. Adjust the volume to 1 L with water and dispense into aliquots (*see* **Note 4**).
8. Protein sample buffer: 0.5 M Tris pH 6.8, 10% glycerol, 3.5 mM SDS, 6 mM bromophenol blue. Mix 25 mL of 1 M Tris pH 6.8, 5 mL of glycerol, 0.5 mL of 350 M SDS and 0.2 g of bromophenol blue. Complete the volume to 50 mL with water and aliquot. Store at −20 °C (*see* **Note 5**).
9. Spheroplasting solution: 20 mM Tris pH 7.4, 0.5 mM $MgCl_2$, 50 mM beta-mercaptoethanol, 1.2 M D-sorbitol, 50 U/mL zymolyase 100T. In a vial, add 5 mL of water, 200 μL of 1 M Tris pH 7.4, 5 μL of 1 M $MgCl_2$, 2.19 g of D-Sorbitol. Adjust the volume to 10 mL with water and mix. Store at −20 °C. Just before starting the experiment add 36 μL of beta-mercaptoethanol and 100 μL of zymolyase 100T at 5000 U/mL (*see* **Note 6**).
10. Lysis buffer: 20 mM Tris pH 7.4, 100 mM NaCl, 14 mM SDS, 0.2% Triton X-100. Prepare the lysis buffer by adding 5 mL of 1 M Tris pH 7.4, 5 mL of 5 M NaCl, 10 mL of 350 mM SDS and 0.5 mL of Triton X-100. Adjust the volume to 250 mL with water and mix very well. Aliquote and store at −20 °C (*see* **Note 7**).

3 Methods

All procedures are carried out at room temperature unless otherwise specified.

3.1 Preparation of Yeast Cells

1. Thaw the spheroplasting solution and lysis buffer, and keep them on ice.
2. Measure the OD_{600nm} of the yeast culture in a spectrophotometer.
3. Calculate the volume of culture corresponding to 5 OD_{600nm}/mL.
4. Place the determined volume in a vial and centrifuge at 800 × *g* for 4 min.
5. Discard the supernatant.
6. Resuspend cells in 1 mL of sterile water and transfer to a 1.5 mL eppendorf.
7. Pellet the cells by centrifuging at 800 × *g* for 3 min and remove the supernatant.
8. Resuspend cells in 1 mL of spheroplasting solution and incubate at 30 °C for 30 min (*see* **Note 8**).
9. Centrifuge at 800 × *g* for 5 min.
10. Remove the supernatant.
11. Add protease inhibitors to the lysis buffer, mix by vortexing. Add 1 tablet of cOmplete Mini, EDTA-free (Roche, Mannheim, Germany) to 10 mL of lysis buffer. It can be stored at −20 °C for 2 months.
12. Resuspend cells in 500 μL of lysis buffer for velocity with inhibitors and keep them 20 min on ice.
13. Break the cells by forcing the solution to pass through a 25G needle syringe 6 times. Use a different syringe and needle for each solution (*see* **Note 9**).
14. Proceed with protein quantification by the method of your preference.
15. Proceed to Subheading 3.3.

3.2 Preparation of Mammalian Cells

1. Grow cells as usual in six well plates.
2. When cells are 90% confluent place the plates on ice.
3. Remove the medium and gently wash cells three times with phosphate-buffered saline (PBS).
4. Add 100 μL of lysis buffer with inhibitors and keep them 20 min on ice.
5. Collect protein by scrapping the wells and pipette it to a 1.5 mL eppendorf.
6. Break the cells by forcing the solution to pass through a 25G needle syringe 6 times. Use a different syringe and needle for each solution (*see* **Note 9**).
7. Proceed with protein quantification by the method of your preference.
8. Proceed to Subheading 3.3.

3.3 Sucrose Gradients

1. While protein is being quantified prepare the following sucrose solutions:

#	Sucrose %	Sucrose 80 % (mL)	Lysis buffer (mL)
1	30	2.2	3.8
2	25	1.9	4.1
3	20	1.5	4.5
4	15	1.1	4.9
5	10	0.75	5.3
6	5	0.375	5.6

2. Approximately 15 min before the protein quantification protocol finishes, pipette 700 μL of each sucrose solution, from 1 to 6, in a 5 mL thinwall polypropylene centrifuge tube (from Beckman) (Fig. 1a). Do this gently and with the pipette tip close to the tube wall. You should be able to see an interface between sucrose solutions (*see* **Note 10**).
3. On top of each centrifuge tube, containing the gradient, place the desired total amount of protein: we use 1 mg (*see* **Note 11**).
4. Place the centrifuge tubes in the appropriate ultracentrifuge rotor tubes.
5. Carefully weight and calibrate the rotor tubes by weighting the opposing pairs and adjusting the weight with sucrose solution 6 (5 % sucrose in lysis buffer, in Subheading 3.3, **step 1**).
6. Carefully place the rotor tubes in the rotor and ultracentrifuge at 246,000 × *g*, 4 °C during 16 h in a swinging bucket rotor (SW-55Ti rotor, Beckman Instruments, Co., Palo Alto, CA), Beckman XL-90 S/N ultracentrifuge (*see* **Note 9**).
7. Immediately collect volumes of 500 μL from the gradients to 1.5 mL eppendorfs (Fig. 1b). Number them fraction 1–9 and keep them on ice (*see* **Note 12**).
8. Add 125 μL (40 M) of trichloroacetic acid (TCA) to each fraction and place for 4 h at 4 °C. TCA is corrosive and should be handled in the fume hood.
9. Centrifuge at 16,000 × *g* for 5 min.
10. Add 800 μL of acetone to each fraction, mix, and centrifuge at 16,000 × *g* for 5 min. Repeat this step two more times, keeping the samples on ice.
11. Place 1–2 min at 95 °C, but do not dry the samples excessively.
12. Add 20 μL of PSB and resuspend very well (*see* **Note 13**).

13. At this point the fractions can be stored at –20 °C until further used.
14. Place the samples for 10 min at 95 °C.
15. Run samples in an SDS-PAGE and do immunoblotting for the protein of your interest (Fig. 1c) (*see* **Notes 14** and **15**).

4 Notes

1. Having water at the bottom of the cylinder helps to dissolve Tris. Use a magnetic stir bar if necessary. Tris can be dissolved faster using warmed water to about 37 °C. Allow the solution to cool down to room temperature before making the final adjustments to the pH. After adjusting the volume of the solution to 1 L with water, dispense into aliquots, and sterilize by autoclaving 15 min at 121 °C. If the 1 M solution of Tris has a yellow color discard it and obtain better quality Tris. The pH of Tris solutions is temperature-dependent, and decreases approximately 0.03 pH units for each 1 °C increase in temperature. For example, a 0.05 M solution has pH values of 9.5, 8.9, and 8.6 at 5, 25, and 37 °C, respectively. Concentrated HCl (12 N) can be used at first to narrow the gap from the starting pH to the required pH. From then on it would be better to use a series of HCl (e.g., 6 and 1 N) to avoid a sudden drop in pH below the required pH.
2. $MgCl_2$ is extremely hygroscopic. Buy small bottles (e.g., 100 g) and do not store open bottles for long periods of time.
3. The solubility of Zymolyase 100T is very low, it may not be completely dissolved in buffers, you may use as suspension. It is stable for over 1 year at –20 °C or many years below –70 °C. However, about 70% of the lytic activity is lost when stored at 30 °C for 3 months.
4. Wear mask when weighing SDS. Wipe down the weighing area and balance after use, because the fine crystals of SDS disperse easily. There is no need to sterilize 10% SDS.
5. Care should be taken to add SDS solution last, since it makes bubbles. SDS precipitates at 4 °C. Thus, the solutions with SDS needs to be warmed prior to use, if needed vortex the solutions.
6. Work with beta-mercaptoethanol in the fume hood, since it is very toxic and has a pungent smell.
7. To obtain the correct volume of Triton X-100 pipette it slowly, because it is very tick.
8. This step could require optimization. Spheroplasting efficiency depends on yeast strain, growth stage, and culture conditions.

Spheroplasting efficiency could be evaluated by microscopy or by absorbance at OD_{600nm}. Under phase-contrast light microscopy, spheroplasts will display a round morphology while intact yeast cells will present an ovoid shape. For OD_{600nm} evaluation, dilute 10 μL of spheroplasting solution into 1 mL of water. Total spheroplasting conversion will result in a reduction of 5–10% of the initial OD_{600nm} [8].

9. The cell lysates should be maintained on ice during protein quantification and should be used in the sucrose gradient immediately without being frozen, to avoid the risk of alteration of the oligomeric forms of the protein of interest.
10. The more concentrated sucrose solution, solution 1 in Subheading 3.3, **step 1**, will stay in the bottom and the less concentrated, solution 6 in Subheading 3.3, **step 1**, on top. Tubes should be placed securely in a customized support to protect the gradients from disturbance during preparation. Add each one of the solutions, from 1 to 6, very slowly, placing the tip against the internal wall of the centrifuge tube. It is very important to avoid the formation of bubbles. Do not take more that 20–30 min between preparing the tubes and starting the centrifugation. From this step onwards carefully transport the gradient centrifuge tubes, to avoid mixing the sucrose solutions.
11. The volume of different samples should be adjusted to the same final volume with lysis buffer. Volume to be applied to the gradient should not be more than 300 μL.
12. Collect the fraction immediately to avoid mixing the sucrose fractions. Fraction 1 will be the less concentrated and fraction 9 the most concentrated. The heavier proteins or aggregates will be found at the bottom fractions and the lighter at the top fractions (Fig. 1).
13. The more concentrated fractions are harder to dissolve and can acquire a yellow coloration, due to the accumulation of higher proteins/aggregates and concomitant change in pH. We found useful to do cycles of adding protein sample buffer and vortexing, but care should be taken to keep the volume of sample within the capacity of the SDS-PAGE wells. Also do a spin-down before loading the samples into the wells.
14. When comparing treatments/conditions make sure that the differences obtained in the sucrose gradients are not due to different protein expression levels. For this, load the same concentration of total protein in an SDS-PAGE and do immunoblotting for the protein of interest and for a loading control (e.g. beta-actin, GAPDH). The sucrose gradient bands can be quantified by doing densitometry of the bands in each fraction and normalizing in comparison to the sum of all bands. The results are represented as percentage of your protein in each fraction.

15. For protocol optimization native molecular mass markers, of known size and sedimentation coefficients, to determine the molecular weights corresponding to each sucrose gradient fraction, might be useful [9]. Calibration proteins for gel filtration chromatography could be used as marker proteins. e.g. gel filtration calibration kit from GE Healthcare, Boehringer Mannheim, or Sigma. Briefly, a mixture containing sufficient quantity of each protein to be visualized by staining (usually 1–5 μg), is applied to a parallel gradient and it is processed exactly as the protein of interest. The sedimentation positions of the marker proteins are determined by SDS-PAGE and subsequent staining with Coomassie brilliant blue or silver.

Acknowledgments

This work was supported by Fundação para a Ciência e Tecnologia project PTDC/BIA-BCM/117975/2010, and fellowships SFRH/BPD/101646/2014 (ST) and SFRH/BD/73429/2010 (DM). TFO is supported by the DFG Center for Nanoscale Microscopy and Molecular Physiology of the Brain (CNMPB). The authors would like to acknowledge networking support by the Proteostasis COST Action (BM1307).

References

1. Soto C (2003) Unfolding the role of protein misfolding in neurodegenerative diseases. Nat Rev Neurosci 4(1):49–60. doi:10.1038/nrn1007
2. Tenreiro S, Munder MC, Alberti S, Outeiro TF (2013) Harnessing the power of yeast to unravel the molecular basis of neurodegeneration. J Neurochem 127(4):438–452. doi:10.1111/jnc.12271
3. Outeiro TF, Lindquist S (2003) Yeast cells provide insight into alpha-synuclein biology and pathobiology. Science 302(5651):1772–1775, 10.1126/science.1090439 302/5651/1772 [pii]
4. Basso E, Antas P, Marijanovic Z, Goncalves S, Tenreiro S, Outeiro TF (2013) PLK2 modulates alpha-synuclein aggregation in yeast and mammalian cells. Mol Neurobiol 48(3):854–862. doi:10.1007/s12035-013-8473-z
5. Faria C, Jorge CD, Borges N, Tenreiro S, Outeiro TF, Santos H (2013) Inhibition of formation of alpha-synuclein inclusions by mannosylglycerate in a yeast model of Parkinson's disease. Biochim Biophys Acta 1830(8):4065–4072. doi:10.1016/j.bbagen.2013.04.015
6. Tenreiro S, Reimao-Pinto MM, Antas P, Rino J, Wawrzycka D, Macedo D, Rosado-Ramos R, Amen T, Waiss M, Magalhaes F, Gomes A, Santos CN, Kaganovich D, Outeiro TF (2014) Phosphorylation modulates clearance of alpha-synuclein inclusions in a yeast model of Parkinson's disease. PLoS Genet 10(5):e1004302. doi:10.1371/journal.pgen.1004302
7. Miller-Fleming L, Antas P, Pais TF, Smalley JL, Giorgini F, Outeiro TF (2014) Yeast DJ-1 superfamily members are required for diauxic-shift reprogramming and cell survival in stationary phase. Proc Natl Acad Sci U S A 111(19):7012–7017. doi:10.1073/pnas.1319221111
8. Rothblatt J, Schekman R (1989) A hitchhiker's guide to analysis of the secretory pathway in yeast. Methods Cell Biol 32:3–36
9. Tanese N (1997) Small-scale density gradient sedimentation to separate and analyze multiprotein complexes. Methods 12(3):224–234. doi:10.1006/meth.1997.0475

Chapter 22

Analysis of Protein Oligomerization by Electrophoresis

Monica Cubillos-Rojas, Taiane Schneider, Susana Sánchez-Tena, Ramon Bartrons, Francesc Ventura, and Jose Luis Rosa

Abstract

A polypeptide chain can interact with other polypeptide chains and form stable and functional complexes called "oligomers." Frequently, biochemical analysis of these complexes is made difficult by their great size. Traditionally, size exclusion chromatography, immunoaffinity chromatography, or immunoprecipitation techniques have been used to isolate oligomers. Components of these oligomers are then further separated by sodium dodecyl sulfate polyacrylamide gel electrophoresis and identified by immunoblotting with specific antibodies. Although they are sensitive, these techniques are not easy to perform and reproduce. The use of Tris-acetate polyacrylamide gradient gel electrophoresis allows the simultaneous analysis of proteins in the mass range of 10–500 kDa. We have used this characteristic together with cross-linking reagents to analyze the oligomerization of endogenous proteins with a single electrophoretic gel. We demonstrate how the oligomerization of p53, the pyruvate kinase isoform M2, or the heat shock protein 27 can be studied with this system. We also show how this system is useful for studying the oligomerization of large proteins such as clathrin heavy chain or the tuberous sclerosis complex. Oligomerization analysis is dependent on the cross-linker used and its concentration. All of these features make this system a very helpful tool for the analysis of protein oligomerization.

Key words Oligomerization, Electrophoresis, Protein cross-linking, Gradient gel, Tris-acetate, PAGE

1 Introduction

In protein homeostasis (proteostasis), the balance between protein synthesis, protein folding and transport, posttranslational modifications, and degradation is fundamental for the biological activity of the protein. The assembly of proteins into functional oligomers may start during translation (for cytosolic homo-oligomers when protein subunits are synthesized from the same polyribosome) or after translation and folding (for homo- and hetero-oligomers) [1, 2]. Oligomerization from folded monomers regulated protein activity. The equilibrium between the monomeric and oligomeric states reflects the activity of the proteins. Some examples of proteins whose biological activities have been associated with their oligomerization state include pyruvate kinase isoform M2 (PKM2), tumor suppressor

Rune Matthiesen (ed.), *Proteostasis: Methods and Protocols*, Methods in Molecular Biology, vol. 1449,
DOI 10.1007/978-1-4939-3756-1_22,

p53, or G-protein-coupled receptors. Thus, whereas PKM2 activity is regulated by switching between an active tetramer and a less active dimer form that shifts the cellular metabolism accordingly during cell division [3, 4], the transcriptional activity of p53 is dependent on its tetrameric form [5–7], and the oligomerization of G-protein-coupled receptors regulates receptor trafficking and signaling [8].

Imbalances in proteostasis may produce disorders associated with the abnormal accumulation of protein aggregates, which are generated by the misfolding and oligomerization of specific proteins. Examples include the formation of intracellular inclusions containing aggregated α-synuclein in Parkinson's disease, huntingtin proteins in Huntington's disease, and extracellular β-amyloid plaques in Alzheimer's disease [1]. Deficiencies in protein oligomerization are also associated with diseases. For example, p53 mutations that impair its oligomerization have been associated with a rare hereditary cancer predisposition disorder called Li-Fraumeni syndrome [9, 10] or the low-activity-dimeric form of PKM2 in tumor progression [3, 4].

The assembly of proteins into oligomers results in large complexes difficult to detect and analyze. Traditionally, the biochemical analysis of protein oligomerization has used size exclusion chromatography, immunoaffinity chromatography, or immunoprecipitation techniques to isolate oligomers. Components of these oligomers are then further separated by sodium dodecyl sulfate polyacrylamide gel electrophoresis (SDS-PAGE), transferred to a nitrocellulose, or PVDF membrane and identified by immunoblotting with specific antibodies. Although they are sensitive, these techniques are not easy to perform and reproduce. Cross-linking reagents such as glutaraldehyde lead to covalent binding of interacting proteins. Several authors have used this approach to cross-link oligomers and further analyze them by SDS-PAGE and immunoblotting [7]. Frequently, large covalent complexes are formed that are difficult to analyze by SDS-PAGE. We have previously reported the use of Tris-acetate polyacrylamide gradient gel electrophoresis (Tris-acetate PAGE) to simultaneously analyze proteins in the mass range of 10–500 kDa [11, 12]. We have used this characteristic and, in combination with cross-linking reagents, we describe how to use them for the analysis of protein oligomerization. In this chapter, we show how this method can efficiently detect the oligomerization of p53, PKM2, and heat shock protein 27 (Hsp27) (Fig. 1). This system is useful for studying the oligomerization of large proteins such as clathrin heavy chain (CHC) or the tuberous sclerosis complex (TSC1/TSC2) (Fig. 2). Oligomerization analysis is dependent on the cross-linker used. We show how CHC can be analyzed better with disuccinimidyl suberate (DSS) cross-linker than with glutaraldehyde. Both cross-linkers react with the amino groups of amino acid residues. However, they have spacer arms of different lengths. Thus, whereas

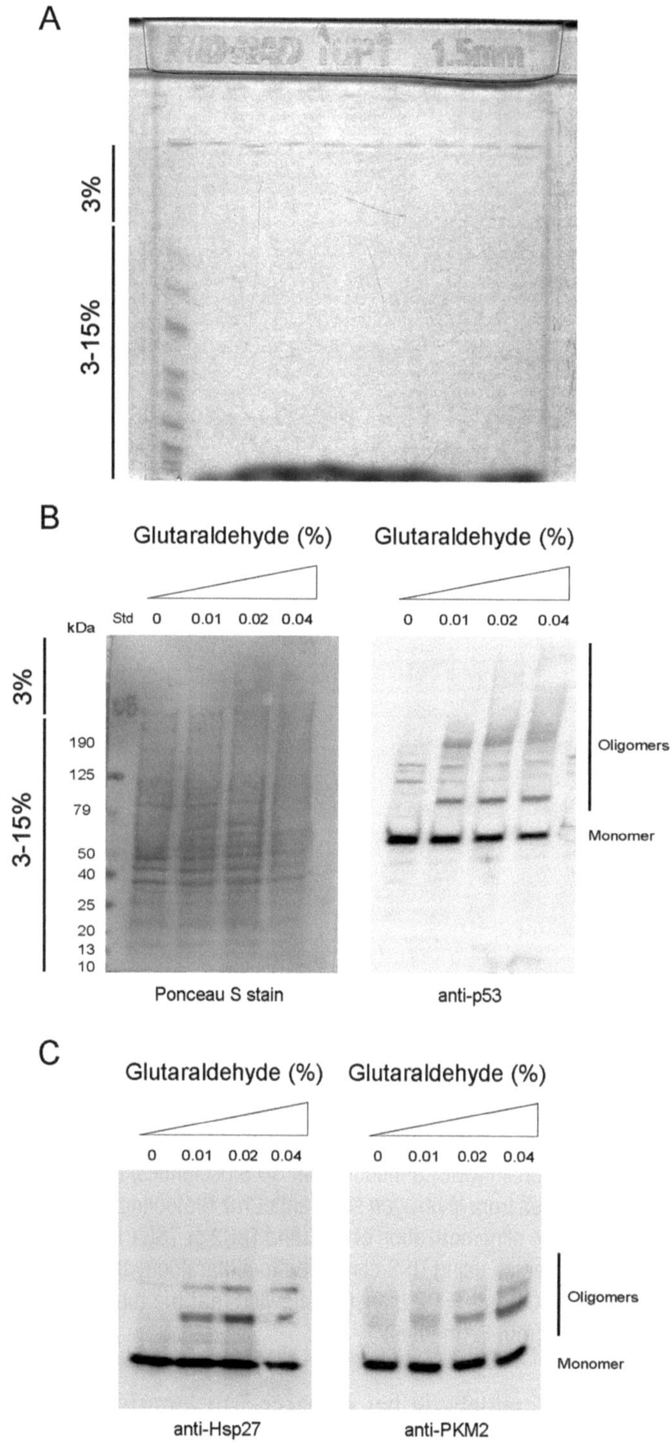

Fig. 1 Analysis of protein oligomerization using the Tris-acetate polyacrylamide gel. Lysates of U2OS cells were treated with glutaraldehyde at the indicated concentrations, run in a 3–15 % polyacrylamide gradient gel (panel **a**), transferred to a PVDF membrane, stained with Ponceau S (panel **b**, *left*), and analyzed by immunoblotting using specific antibodies against p53 (clone DO-7; Neomarkers) (panel **b**, *right*), Hsp27 (Santa Cruz Biotechnology), and PKM2 (Cell Signaling) to detect monomers and oligomers (panel **c**). Std, prestained protein standard (from Fermentas)

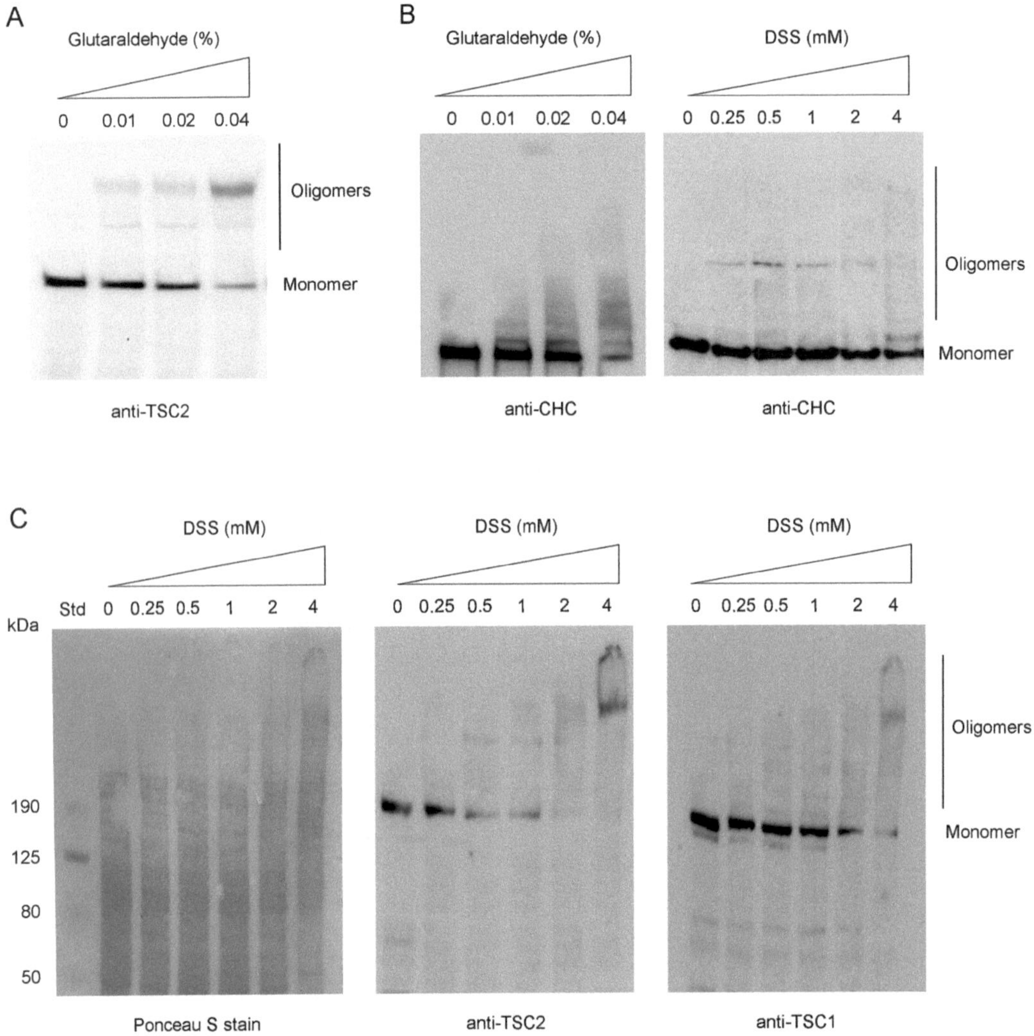

Fig. 2 The Tris-acetate polyacrylamide gel system is useful for studying the oligomerization of large proteins. Lysates of U2OS cells were treated with DSS or glutaraldehyde at the indicated concentrations, run in a 3–8 % polyacrylamide gradient gel, transferred to a PVDF membrane, and analyzed by immunoblotting with specific antibodies against clathrin heavy chain (CHC; antibody from BD Biosciences) (panel **b**) or the tuberous sclerosis complex (TSC1/TSC2; antibodies from Invitrogen and Santa Cruz Biotechnology, respectively) (panels **a** and **c**). Using DSS as a cross-linker, the oligomerization of CHC and TSC2 or TSC1 is also detected (panels **b** and **c**, respectively). CHC was better detected using DSS compared to glutaraldehyde (panel **b**). Similar oligomerization profiles were obtained for TSC1 and TSC2 (panel **c**) showing that the tuberous sclerosis complex forms. Std, prestained protein standard (from Fermentas)

glutaraldehyde has a spacer arm length of 5 Å, DSS has a spacer with a longer length of 11.4 Å [13]. These data seem to indicate that a steric effect, that is, the distance between the amino acid residues involved in the cross-linking, is the limiting factor. We have observed that the cross-linker concentration is also a critical parameter. The optimal detection and resolution of analyzed proteins were obtained with concentrations between 0.01 and 0.04 %

for glutaraldehyde and between 0.25 and 4 mM for DSS. At higher concentrations, large complexes are formed, and they cannot be resolved with this system.

2 Materials

All solutions were prepared using ultrapure water (Milli-Q quality) and analytical-grade reagents.

2.1 Cell Lysis

1. CHAPS buffer: 10 mM Tris–HCl, pH 7.5, 100 mM NaCl, 0.3% CHAPS, and protease and phosphatase inhibitors: 50 mM NaF, 1 mM sodium vanadate, 1 mM phenylmethylsulfonyl fluoride, 5 μg/mL leupeptin, 5 μg/mL aprotinin, 1 μg/mL pepstatin A, 50 mM β-glycerophosphate, 100 μg/mL benzamidine, and 1 μM E-64.
2. NP40 buffer: 50 mM Tris–HCl, pH 7.5, 150 mM NaCl, 0.5% NP40, and protease and phosphatase inhibitors indicated above.
3. HEPES buffer: 50 mM HEPES, pH 7.5, 50 mM NaCl, 1% Triton X-100, 1.5 mM $MgCl_2$, 1% glycerol, and protease and phosphatase inhibitors indicated above.

2.2 Electrophoresis and Immunoblotting

1. 15× Tris-acetate gel buffer: 3 M Tris base adjusted with acetic acid to pH 7.0, store at 4 °C.
2. 40% Acrylamide solution: 37.5:1 acrylamide/bisacrylamide mix, store at 4 °C.
3. 10% Ammonium persulfate (APS): solution in water, aliquot and store at −20 °C.
4. *N,N,N,N′*-tetramethyl-ethylenediamine (TEMED), store at 4 °C.
5. 1 M 1,4-dithiothreitol (DTT): solution in water, aliquot and store at −20 °C.
6. 1× Running buffer: dissolve 8.95 g of Tricine, 6.06 g of Tris base, 1 g of SDS, and 0.25 g of sodium bisulfite in 1 L of water. Mix and store at 4 °C. The pH of this solution is 8.24.
7. Gradient Maker (Hoefer Instruments, San Francisco, CA).
8. Stir bar and agitator.
9. MiniPROTEAN II electrophoresis system (Bio-Rad, Hercules, CA, USA).
10. Bicinchoninic acid (BCA) assay kit (Thermo Scientific, Rockford, Ill, USA).
11. 4× LDS sample buffer: 4 mL of 2.5 M Tris–HCl, pH 8.5, 0.8 g of LDS (Sigma, St Louis, MO, USA), 0.006 g of EDTA, 5 mL of 80% glycerol (Roche, Indianapolis, IN, USA),

0.75 mL of 1% Coomassie Brilliant Blue G solution (CBB) and 0.25 mL of 1% phenol red solution, store at room temperature.

12. PVDF membranes (Millipore, Bedford, MA, USA).
13. 20× Transfer buffer: dissolve 81.6 g of Bicine, 104.8 g of Bis–Tris and 6 g of EDTA in 1 L of water, store at 4 °C. For western analysis, dilute this buffer to 1× in water with 20% methanol and add 0.25 g of the antioxidant sodium bisulfite. The pH of this solution is 7.2.
14. Whatman chromatography papers (17 CHR, Whatman).
15. Mini Trans-Blot electrophoretic transfer cell (Bio-Rad).

2.3 Cross-Linking Reagents

1. 25% Glutaraldehyde solution, store at 4 °C. Prepare the stock solution immediately before use in PBS. Do not prepare stock solutions for storage.
2. Disuccinimidyl suberate (DSS), store at 4 °C. Prepare the stock solution immediately before use. DSS is soluble in DMSO at room temperature. Discard any unused cross-linker.

3 Methods

Carry out all procedures at room temperature unless otherwise specified.

3.1 Casting the Gel

1. We used the MiniPROTEAN II gel cassette to cast the gel.
2. Select and prepare the appropriate gradient solutions to separate the proteins of interest. The volume of reagents required for a mini 3–15% polyacrylamide gradient gel (dimensions: 8 cm×8 cm; spacers: 1.5 mm) is 7 mL for a 3% solution (0.47 mL Tris-acetate gel buffer (15×), 0.53 mL of 40% acrylamide solution, 6 mL of Milli-Q water, 8.75 μL of TEMED and 33.25 μL of 10% APS) and 4 mL for a 15% solution of (0.27 mL of Tris-acetate gel buffer (15×), 1.5 mL of 40% acrylamide solution, 2.23 mL of Milli-Q water, 5 μL of TEMED and 19 μL of 10% APS) (*see* **Note 1**).
3. We used the gradient Maker (Hoefer Instruments) to make the gradient. Add 4 mL of the lower concentration polyacrylamide solution (3%) into the non-outlet chamber (reservoir) and 4 mL of the higher concentration polyacrylamide solution (15%) into the outlet chamber equipped with a stir bar at the same time and open the valves. When the solutions of both chambers are nearing completion, add the remaining volume (3 mL) of the lower concentration solution into the non-outlet chamber, put in the comb and allow the gel to solidify.

3.2 Sample Preparation and Protein Loading

1. Lyse the samples in CHAPS or NP40 buffer for glutaraldehyde treatment or HEPES buffer for treatment with DSS.
2. Centrifuge the samples (e.g., 15,000 ×*g* for 15 min at 4 °C). Collect the supernatant and determine the protein concentration by the bicinchoninic acid assay or a similar method (*see* **Note 2**).
3. Add cross-linker to the protein sample at the indicated concentration and incubate for 30 min on ice with agitation (*see* **Note 3**).
4. Quench the reaction using sample 4× buffer without DTT to a final concentration 1×. After this step, you can use these samples or store them at −20 °C.
5. Add DTT to reach a final concentration of 100 mM when the protein samples are used in PAGE.
6. Heat the samples for 10 min at 100 °C and load them into the wells of the gel.

3.3 Electrophoresis Conditions

1. Run at $\Delta V = 130$ V ($I = 115$ mA). The run time estimated for a 3–15 % gradient gel is approximately 1 h and 15 min.

3.4 Transfer Conditions

1. Perform the transfer for 2 h ($I = 200$ mA) or overnight ($\Delta V = 20$ V) (*see* **Note 4**).

4 Notes

1. A 3–15 % gradient gel is recommended for the analysis of proteins of approximately 20–60 kDa. For the analysis of large proteins (>100 kDa), we recommend a 3–8 % gradient gel.
2. The use of a more concentrated lysate is most effective for the detection of oligomers. Generally, a concentration of 50–80 μg of lysate/lane is recommended.
3. Prepare the cross-linkers immediately before use. Do not prepare stock solutions for storage. Discard any unused cross-linker.
4. The semi-dry blotting system (Bio-Rad) is compatible with the gradient gel electrophoretic system (Tris-acetate PAGE). Using the standard protocol recommended by Bio-Rad, we obtain the optimal results for the analysis of protein oligomerization.

Acknowledgments

This study was supported by the Spanish Ministerio de Ciencia e Innovación Grant BFU2011-22498 and the Instituto de Salud Carlos III Grant RETIC, RD06/0020. T. Schneider was supported by a fellowship from the CAPES Foundation, Ministry of Education of Brazil. S. Sánchez-Tena was supported by a grant (PDJ 2013) from Agència de Gestió d'Ajuts Universitaris i de Recerca (AGAUR), Generalitat de Catalunya, Spain. The authors would like to acknowledge networking support by the Proteostasis COST Action (BM1307).

References

1. Kim YE, Hipp MS, Bracher A, Hayer-Hart M, Hart FU (2013) Molecular chaperone functions in protein folding and proteostasis. Annu Rev Biochem 82:323–355
2. Matthews JM, Sunde M (2012) Dimers, oligomers, everywhere. Adv Exp Med Biol 747:1–18
3. Yang W, Lu Z (2013) Regulation and function of pyruvate kinase M2 in cancer. Cancer Lett 339:153–158
4. Iqbal MA, Gupta V, Gopinath P, Mazurek S, Bamezai RN (2014) Pyruvate kinase M2 and cancer: an updated assessment. FEBS Lett 588:2685–2692
5. Chène P (2001) The role of tetramerization in p53 function. Oncogene 20:2611–2617
6. Kawaguchi T, Kato S, Otsuka K, Watanabe G, Kumabe T, Tominaga T, Yoshimoto T, Ishioka C (2005) The relationship among p53 oligomer formation, structure and transcriptional activity using a comprehensive missense mutation library. Oncogene 24:6976–6981
7. Itahana Y, Ke H, Zhang Y (2009) p53 Oligomerization is essential for its C-terminal lysine acetylation. J Biol Chem 284:5158–5164
8. Ng HK, Chow BK (2015) Oligomerization of family B GPCRs: exploration in inter-family oligomer formation. Front Endocrinol (Lausanne) 6:10. doi:10.3389/fendo.2015.00010
9. Davison TS, Yin P, Nie E, Kay C, Arrowsmith CH (1998) Characterization of the oligomerization defects of two p53 mutants found in families with Li-Fraumeni and Li-Fraumeni-like syndrome. Oncogene 17:651–656
10. DiGiammarino EL, Lee AS, Cadwell C, Zhang W, Bothner B, Ribeiro RC, Zambetti G, Kriwacki RW (2002) A novel mechanism of tumorigenesis involving pH-dependent destabilization of a mutant p53 tetramer. Nat Struct Biol 9:12–16
11. Cubillos-Rojas M, Amair-Pinedo F, Tato I, Bartrons R, Ventura F, Rosa JL (2010) Simultaneous electrophoretic analysis of proteins of very high and low molecular mass using Tris-acetate polyacrylamide gels. Electrophoresis 31:1318–1321
12. Cubillos-Rojas M, Amair-Pinedo F, Tato I, Bartrons R, Ventura F, Rosa JL (2012) Tris-acetate polyacrylamide gradient gels for the simultaneous electrophoretic analysis of proteins of very high and low molecular mass. Methods Mol Biol 869:205–213
13. Krohn NM, Yanagisawa S, Grasser KD (2002) Specificity of the stimulatory interaction between chromosomal HMGB proteins and the transcription factor Dof2 and its negative regulation by protein kinase CK2-mediated phosphorylation. J Biol Chem 277:32438–32444

Chapter 23

Blot-MS of Carbonylated Proteins: A Tool to Identify Oxidized Proteins

Rita Ferreira, Pedro Domingues, Francisco Amado, and Rui Vitorino

Abstract

The efficiency of proteostasis regulation declines during aging and the failure of protein homeostasis is common in age-related diseases. Protein oxidation is a major contributor to the loss of proteome homeostasis, also called "proteostasis," precluding protein misfolding and aggregation. So, the identification of the molecular pathways impaired by protein oxidation will increase the understanding of proteostasis and the pathophysiological conditions related to the loss of proteostasis. Sample derivatization with dinitrophenyl hydrazine and western blot immunoassay detection of carbonylated proteins (commonly known as Oxyblot™) coupled to mass spectrometry (blot-MS) is an attractive methodological approach to identify proteins that are more prone to carbonylation, a typical oxidative modification of amino acid residues. The integration of blot-MS data of carbonylated proteins with bioinformatics tools allows the identification of the biological processes more affected by protein oxidation and that, eventually, result in the loss of proteostasis.

In this chapter, we describe a blot-MS methodology to identify the proteins more prone to oxidation in biological samples, as cell and tissue extracts, and biofluids. Analysis of mitochondria isolated from cardiac tissue is provided as an example. Bioinformatic strategy to deal with data retrieved from blot—MS experiments are proposed for the identification of relevant biological processes modulated by oxidative stress stimuli.

Key words Oxidative posttranslational modifications, Immunodetection, Mitochondria, Proteomics, Mass spectrometry, Two-dimensional gel electrophoresis

1 Introduction

Protein oxidation is an important contributor to the progressive loss of cell function and, eventually, to cell death. As major sources of reactive oxygen species (ROS), mitochondria themselves, and particularly oxidative phosphorylation complexes, are especially susceptible to oxidative damage [1, 2]. Among the nonenzymatic modifications of specific amino acid residues promoted by ROS is the introduction of aldehyde or ketone functional groups, usually known as carbonylation [3]. Oxidation of amino acid residues impairs protein homeostasis through alterations in protein structure and function, leading to protein misfolding and aggregation, and to reduced proteolytic removal of oxidized proteins [4–7].

Rune Matthiesen (ed.), *Proteostasis: Methods and Protocols*, Methods in Molecular Biology, vol. 1449,
DOI 10.1007/978-1-4939-3756-1_23, © Springer Science+Business Media New York 2016

Protease-resistant protein aggregates are considered highly toxic and can mediate cell death [8, 9]. Consequently, protein carbonylation is likely to induce protein misfolding and aggregation, and the identification of the proteins more prone to this oxidative modification will increase the understanding of the molecular basis underlying the disruption of proteome homeostasis, also termed "proteostasis." Proteostasis is dynamically maintained in the cell by coordinating protein synthesis, folding, and degradation for which contribute several proteolytic pathways as the ubiquitin–proteasome system and the lysosome-dependent autophagy (reviewed by [10]). The loss of proteostasis is among the common features of pathophysiological stresses generated by distinct diseases as cardiovascular and neurodegenerative diseases [11].

The methods usually used in the determination of protein carbonylation rely on the reaction of carbonyl groups with 2,4-dinitrophenylhydrazine (DNPH) to form protein-bound 2,4-dinitrophenylhydrazones. Using anti-dinitrophenyl antibodies for the western blot immunoassay detection of carbonylated proteins allows not only to quantify carbonylated proteins using immunochemically approaches as ELISA or slot-blot but also to identify proteins more prone to carbonylation using with one-dimensional (1D) or two-dimensional (2D) sodium dodecyl sulfate (SDS) gel electrophoresis followed by western blot immunoassay and tandem mass spectrometry (MS/MS) (1D- or 2D-blot-MS/MS) (Fig. 1) [12–15]. Since chemical derivatization with DNPH might alter the electrophoretic migration pattern of proteins, 2D separation of cell or tissue extracts prior to derivatization would seemingly provide the best approach for the identification of the modified proteins [16]. Moreover, 2D-blot offers the advantage of giving a fingerprint of proteome susceptibility to oxidation. Since some proteins are modified at a greater extent than others, to obtain an accurate picture of oxidation-sensitive proteins within the overall cell or tissue proteome it is advised to simultaneously analyze the total protein pattern [16]. Herein, we describe a methodology to identify carbonylated proteins that combines immunodetection with MS identification of oxidation-sensitive proteins, giving as example the analysis of protein carbonylation in mitochondria isolated from cardiac muscle.

2 Reagents and Solutions

2.1 Mitochondria Isolation

1. Homogenization buffer: 250 mM sucrose, 0.5 mM EGTA, 10 mM HEPES–KOH (pH 7.4), and 0.1% defatted BSA (Catalog. No A6003, Sigma).
2. Isolation buffer: homogenization buffer containing protease subtilopeptidase A type VIII (Catalog No. P5380, Sigma; 1 mg/g tissue) (prepare at the day of the experiments) (*see* **Note 1**).
3. Wash buffer: 250 mM sucrose, 10 mM HEPES–KOH (pH 7.4).

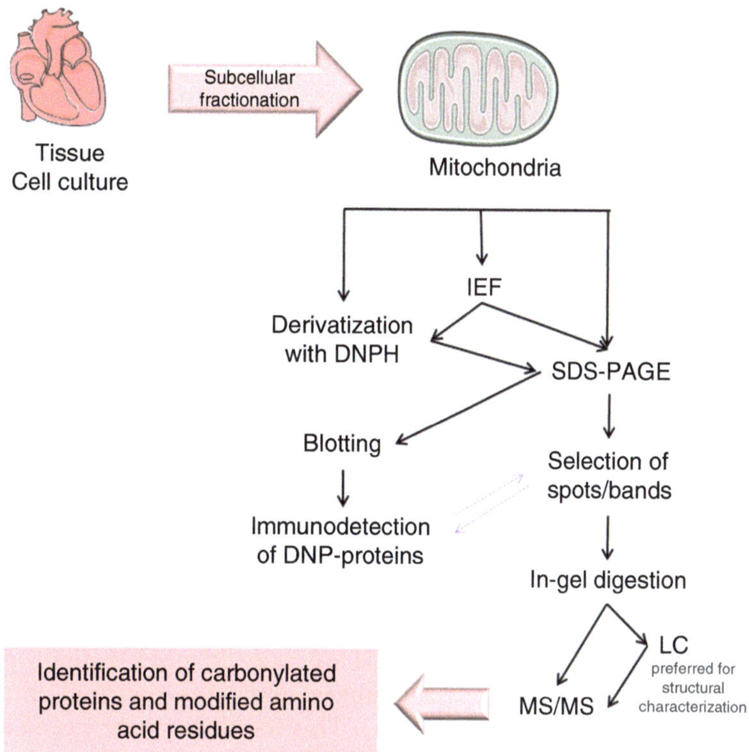

Fig. 1 Different procedures used for the identification of oxidized proteins detected by blot coupled to mass spectrometry and characterization of oxidative PTMs in mitochondria isolated from heart

2.2 Determination of Protein Concentration

1. Reagent A: DC Protein assay kit (Bio Rad Laboratories, Berkeley, CA).
2. Reagent B: DC Protein assay kit (Bio Rad Laboratories, Berkeley, CA).
3. Bovine serum albumin (BSA) solution: dissolve 1.5 mg of BSA in 1 ml of water.

2.3 Sodium Dodecyl Sulfate Electrophoresis

1. Running Buffer (10×): 0.25 M Tris, 1.92 M Glycine, 1% SDS pH 8.6 (*see* **Note 2**).
2. Buffer of the running gel (Buffer RG): 1.5 mM Tris pH 8.8.
3. SDS: 10% (w/v).
4. APS solution: 10% (w/v) (prepare fresh or prepare aliquots preserved at −20 °C).
5. TEMED: Tetramethylethylenediamine.
6. Acrylamide: 40% (BioRad, Hercules, CA).
7. Bis: bisacrylamide 2% (BioRad, Hercules, CA).

2.4 Isoelectric Focusing Electrophoresis and Two-Dimensional Gel Electrophoresis

1. IPGphor system (GE Healthcare, Freiburg, Germany).
2. Isoelectric focusing electrophoresis (IEF) rehydration buffer: 8 M Urea, 2 M Thiourea, 1% Chaps, 12.9 mM DTT, 0.1% Pharmalyte 3–10 NL (GE Healthcare, Freiburg, Germany), 0.01% Bromophenol Blue (dissolved in deionized water made fresh before use) (*see* **Note 3**).
3. Equilibration buffer: 0.05 M Tris-glycine buffer (pH 8.8), 4 M Urea, 35% Glycerol, 2% SDS (*see* **Note 4**).
4. Sealing solution: agarose 0.5% in running buffer (1×), 0.01% Bromophenol Blue (*see* **Note 5**).
5. Unstained and prestained molecular weight marker (Fermentas, St. Leon-Rot, Germany).
6. IPG strip 3–10 NL, 7 cm (GE Healthcare, Freiburg, Germany).
7. Drystrip cover fluid (GE Healthcare, Freiburg, Germany).
8. Gel stain: BlueSafe (Nzytech, Portugal).

2.5 Sample Preparation for IEF and Carbonylated Protein Detection

1. SDS Solution: 12% sodium dodecyl sulfate (w/v).
2. DNPH Solution: 20 mM 2,4-dinitrophenyl hydrazine (Aldrich, Milwaukee, WI) was prepared in 10% trifluoroacetic acid (TFA). This solution can be stored at room temperature in a dark flask for at least 3 months.
3. Stop equilibration buffer: 2% (w/v) SDS, 6 M urea, 30% glycerol, 0.05 M Tris–HCl pH 8.8, 20 mg/mL DTT.

2.6 Immunochemical Detection

1. ChemiDoc System (BioRad, Hercules, CA).
2. GelDoc XR+ (BioRad, Hercules, CA).
3. PdQuest (BioRad, Hercules, CA).
4. Transfer Buffer: 25 mM Tris, 192 mM glycine, pH 8.3 and 20% methanol.
5. Wash blot (TBST): 100 mM Tris, 1.5 mM NaCl, pH 8.0 and 0.5% Tween 20.
6. Blocking Solution: 5% (w/v) dry nonfat milk dissolved in TBST made fresh before use.
7. Primary Antibody Solution: Anti-dinitrophenyl hydrazone (anti-DNP) antibody (Chemicon International, Temecula, CA) diluted in Blocking Solution (1:1000).
8. Secondary Antibody Solution: Anti-rabbit antibody conjugated to horseradish peroxidase (Sigma-Aldrich, St Louis, MO) diluted in TBST (1:3000) directly before use.
9. Detection Solution: Enhanced chemiluminescence ECL (GE Healthcare).
10. Signal Developing/Acquisition: Chemidoc System or alternatively use X-ray films (Kodak Biomax light Film, Sigma);

Developer solution (Sigma; prepare according to manufacturer's instructions); Fixer solution (prepare according to manufacturer's instructions) (Sigma-Aldrich, St Louis, MO).

2.7 Tryptic Digestion and LC–MS

1. nano-HPLC: Ultimate 3000 (Dionex Corporation, Sunnyvale, CA).
2. Chromeleon (v.6.8, Dionex Corporation, Sunnyvale, CA).
3. Probot: (Dionex Corporation, Sunnyvale, CA).
4. TS2Mascot (v1.0 MatrixScience, UK).
5. Calmix 5 (Applied Biosystems).
6. Wash solution: 0.1 M ammonium bicarbonate (NH_4HCO_3) (Sigma-Aldrich Co., Karlsruhe, Germany).
7. ACN: acetonitrile.
8. FA: 10% Formic acid.
9. Reducing reagent: 10 mM DTT in 0.1 M ammonium bicarbonate.
10. Alkylation reagent: 50 mM iodoacetamide in 0.1 M ammonium bicarbonate.
11. Trypsin (Promega, Madison, WI).
12. Matrix: 3 mg/mL of α-Cyano-4-hydroxycinnamic acid, 0.1% TFA, 15 fmol Glu-fib (Sigma-Aldrich Co., Karlsruhe, Germany).
13. Eluant A: 95% H_2O, 5% ACN, 0.05% TFA.
14. Eluent B: 90% ACN, 10% H_2O, 0.045% TFA.
15. Solubilizing solution: 50% ACN, 50% H_2O, 0.1% TFA.

3 Methods

The methodology described in this chapter is suitable for detection of carbonylated proteins present in, e.g., mitochondria isolated from heart tissue. The annotated spots in 2DE maps were subjected to trypsin digestion followed by nano-HPLC MALDI-TOF/TOF analysis. The data analysis was made through bioinformatic tools freely available. Figure 2 overviews the main steps and time-course of blot-MS analysis for the characterization of oxidized proteins.

3.1 Mitochondria Enrichment

1. Wash and mince the excised hearts in an ice-cold homogenization buffer (*see* **Note 6**).
2. Resuspend the minced heart in isolation buffer and homogenized with tightly fitted Potter–Elvehjen glass homogenizer and a teflon pestle.
3. Incubate the suspension for 1 min (4 °C), re-homogenize and centrifuge at 14,500 × *g* during 10 min at 4 °C.

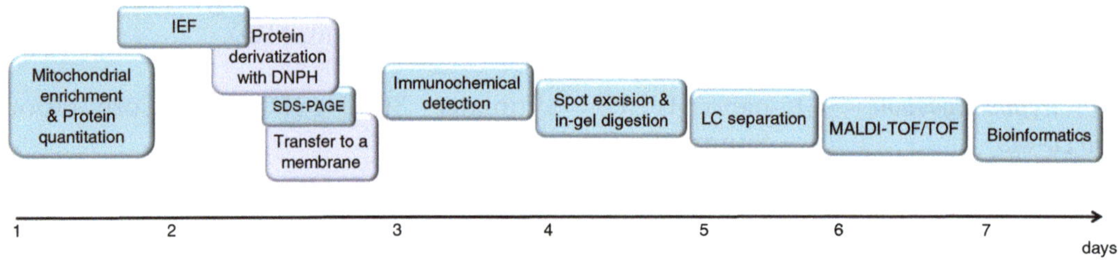

Fig. 2 Time-course of the main experimental steps for blot-MS of carbonylated proteins

4. Decant the supernatant fluid and gently resuspend the pellet in isolation buffer.
5. The suspension is centrifuged at 750 × *g* for 10 min and the resulting supernatant is centrifuged at 12,000 × *g* for 10 min.
6. The pellet is resuspended in isolation buffer and centrifuged again at 12,000 × *g* for 10 min.
7. The final pellet, containing the mitochondrial fraction, is gently resuspended in the washing buffer.
8. All procedures should be performed at 4 °C.
9. Determine the protein concentrations in the mitochondria extracts by using RC DC reagent kit using the manufacturer instructions.

3.2 IEF of Mitochondria Extracts

1. Take an aliquot of approx. 70–100 μg of the protein for blot analysis of protein carbonyls.
2. Add IEF rehydration buffer to the sample in order to obtain a volume of 120 μL (buffer should exceed 70 % of total volume) and incubate at room temperature on a vortex for 10 min.
3. IEF is performed with IPGphor system using 7 cm, pH 3–10 NL immobilized pH gradient (IPG) strips.
4. Transfer 120 μL of the samples into the anode side (+) of the IEF holder carefully by using a micropipette (avoid bubbles).
5. Remove the protective layer of the strip with the aid of a forceps in the anode side.
6. Put the strip slowly in the strip holder (certify that the gel is facing down in order to stain in contact with the sample) starting in the anode side. Be careful with the air bubbles under the strip.
7. Cover the strip with the drystrip cover fluid.
8. Put the strip holder in the isoelectric focusing system.
9. Turn on the apparatus and select the desired program.
10. Initiate the program by indicating the number of strips: e.g., 12 h of rehydration at 50 mW, 20 °C; 1 h 30 m at 150 V-linear, 1 h at 500 V-linear, 1 h at 1000 V-linear, 2 h at 5000 V-step-n-hold (*see* **Note 7**).

11. After the isoelectric focusing remove the strips and dry them slightly with filter paper. Then place the strips in a plastic tube. At this step the strips can be kept at −80 °C until the second dimension.

3.3 Strip Preparation for Protein Carbonyls Detection

1. Following the IEF, derivatize the IPG strips for protein carbonyls detect carbonyl, otherwise skip for Subheading 3.4.
2. Incubate the IPG strip in 5 mL of SDS 12% for 20 min at room temperature.
3. Discard the solution and add 5 mL of DNPH solution to the IPG strip and incubate for 30 min in the dark at room temperature with gentle agitation.
4. Discard the solution and add 5 mL of stop equilibration buffer. Incubate for 15–30 min under gentle agitation at room temperature.
5. Discard the solution and prepare the strips for two-dimensional gel electrophoresis.

3.4 Two-Dimensional Gel Electrophoresis

1. Prepare gel assembly for gel casting following the manufacturer instructions with 1 mm spacers.
2. Cast the gel with the appropriate acrylamide concentration prepared in accordance with the table below in 1-mm cassettes for 7 cm gels (*see* **Note 8**).

	Running gel		
Volumes in mL	**10%**	**12.5%**	**15%**
Acrylamide (40%)	2.43	3.04	3.65
Bis-acrylamide (2%)	1.34	1.68	2.01
Buffer RG	2.50	2.50	2.50
H_2O	2.18	1.24	0.29
SDS	0.10	0.10	0.10
APS[a]	0.05	0.05	0.05
TEMED[a]	0.005	0.005	0.005
Final volume	10.0	10.0	10.0

[a]Add just prior to pouring in the casting frames

3. Boil the 0.5% of agarose solution.
4. Add 100 mg of fresh DTT per 10 mL of equilibrium buffer.
5. Add 5 mL of equilibrium buffer to each strip with the gel side facing up and agitate for 15 min at room temperature.
6. While waiting for equilibration, prepare 1× running buffer by diluting 100 mL of 10× buffer with 900 mL of deionized water.

7. Dip the IPG strips into 1× running buffer to remove excess equilibration buffer.
8. Place the IPG strips with gel side facing up into gel assembly and gently push to put in contact with the running gel.
9. Add warm agarose solution into the gel assembly, avoiding bubbles, wait for 5 min for agarose to solidify and place the gels in tank filled with running buffer.
10. Load unstained molecular weight marker into standard well adjacent to the IPG strip for stained gels or prestained molecular weight markers for western blotting.
11. Run the gels at 200 V for approx. 45 min at room temperature, until the dye front (bromophenol blue) exits the gel into the lower tank.
12. Open the gel assembly and remove the gels.
13. For the blotting of carbonylated proteins follow to the next section, otherwise stain the gel with BlueSafe™ for 15 min until the spots will be visible.

3.5 Immunochemical Detection

1. Gels containing DNPH derivatized proteins are transferred to nitrocellulose membranes for immunochemical detection of protein carbonyls.
2. Soak the gels, nitrocellulose membrane, and filter papers in cold transfer buffer for 5 min.
3. Prepare the transfer setup in the following order (from positive to the negative pole): first place one soaked filter paper on the transfer unit platform, followed by nitrocellulose membrane, gel, and one more filter paper.
4. Remove bubbles that are trapped between the nitrocellulose membrane, gels, and filter paper. Once the sandwich is ready roll the glass rod once again to ensure no bubbles are trapped in the transfer sandwich.
5. Close the lid of the transfer unit and activate power supply. Transfers are performed at 200 mA for 2 h at room temperature (voltage, transfer time, and methanol concentration in the transfer buffer should be optimized according to proteins MW).
6. Once the transfer is completed, the transfer unit is disconnected and the unit is carefully disassembled, and the nitrocellulose membrane is taken out. Since prestained molecular weight markers are used, orientation of protein MW and pI separation on the nitrocellulose membrane is easy to follow. The gel and filter papers can then be discarded.
7. The nitrocellulose membrane is then incubated in 50 mL of blocking buffer for 1 h at room temperature (or overnight at 4 °C) on a rocking platform.

8. Then, the membrane is incubated in anti-DNPH antibody solution for 1 h at room temperature (or overnight at 4 °C) on a rocking platform.
9. The primary antibody is then removed and the membrane washed three times for 10 min each with TBST.
10. The membrane is then incubated with secondary antibody solution (Anti-Rabbit HRP-conjugated) on a rocking platform for 1 h at room temperature.
11. The secondary antibody is discarded and the membrane is washed three times for 5 min each with wash blot.
12. After final wash the blot is developed using Enhanced chemiluminescence ECL (prepared before use according to manufacturer's instructions) for 1 min at room temperature.
13. Signal might be acquired in a ChemiDoc System; alternatively, in a dark room, expose the membrane to a X-ray film in an exposition cassette for 5 min (the exposition time should be optimized) and then develop the film by immerse it in the developer solution until visualize the spots reactive to DNP, wash the film with water and immerse it on fixer solution for 5 min.

3.6 Image Acquisition and Analysis

This procedure is outside the scope of this chapter. However, the main steps consist in:

1. Scan the 2DE gel images using GelDoc XR$^+$ (*see* **Note 9**).
2. After production and digitalization of the 2D gel of interest, import to the image analysis program, e.g., PdQuest for spot detection and matching.
3. Ensure that images are cropped to match each other exactly in size and define faint and saturated spots.
4. Manually inspect images to determine if all visible spots are marked and proceed for spot matching.
5. After spot matching, normalize spot volume intensity ratios for each spot, and export to excel the spot list with respective normalized volume.
6. Calculate the volume ratio for spot that matched between blot and Coomassie Gel stain (*see* **Note 10**).
7. Determine which spots indicated as being significantly changed in expression are present to be further analyzed by MS.

3.7 Protein Digestion

1. Matched spots between blot and Coomassie Gel stained gel are cut from the gel with an excisor and transferred to an empty eppendorf.
2. Wash the spots with 50 μL of Wash solution and incubate for 15 min.

3. Add 50 μL of ACN and incubate for 15 min.
4. Remove Supernatant and repeat washing (**steps 2–4**).
5. Remove supernatant.
6. Add 50 μL of ACN and incubate for 10 min (*see* **Note 11**).
7. Remove the ACN and dry the spots in the Speed Vac.
8. For disulfide bonds, reduction and alkylation add 50 μL of DTT to dried gels and incubate at 60 °C for 1 h (*see* **Note 12**).
9. Remove supernatant, add 50 μL of iodoacetamide and incubate at room temperature for 30 min in the dark.
10. Wash the bands/spots with 50 μL of wash solution and incubate for 15 min.
11. Add 50 μL of ACN and incubate for 15 min.
12. Remove supernatant and repeat washing (**steps 10** and **11**).
13. Remove supernatant. Add 50 μL of ACN and incubate for 10 min.
14. Remove the ACN and dry the spots in the Speed Vac.
15. For protein in-gel digestion with trypsin, add 25 μL of trypsin and incubate at 37 °C for 60 min (*see* **Note 13**).
16. After 1 h add wash solution until all the spots are immersed and reincubate overnight at 37 °C.
17. After approx. 18 h, perform the acid extraction of the digested peptides.
18. Collect the supernatant to an empty eppendorf.
19. Add 25 μL of FA to the spots and incubate for 30 min.
20. Collect the supernatant to the same eppendorf as before.
21. Add 25 μL of FA and 25 μL of ACN and incubate for 30 min.
22. Collect the supernatant to the same eppendorf as before and repeat **step 21**.
23. Collect the supernatant to the same eppendorf as before and dry the collected solutions in the Speed Vac.

3.8 LC Separation and Mass Spectrometry

1. Separation and analysis of tryptic peptides is performed in a NanoHPLC-UltiMate 3000 connected to Probot.
2. Define the separation program in Chromeleon.
3. Prepare the tryptic digests (*see* Subheading 3.4) and fill them in respective autosampler racks.
4. Prepare eluent A and eluent B.
5. Define Chromeleon separation program the following points: 7 °C for autosampler temperature, 25 °C for column-oven temperature, 214 nm for UV detector, 30 μL for loading pump flow rate, 0.3 μL micro pump flow rate, 1_2 for inject valve position, 5

syringe wash cycles, 12 min for wait time for signal start acquisition, 95 % of eluent A for initial conditions.

6. Connect the HPLC column to the fraction collector μ-tee using a fused capillary: 30 μm I.D.
7. Prepare the Probot™ using μCarrier software and fill the syringe with MALDI matrix solution.
8. Purge the line until bubbles are absent.
9. Dispense matrix to certify that the line is completely full and ready to mix with sample.
10. Using the application editor, define the matrix addition flow rate; and collection time per fraction and number of spots and wait for signal defined on Chromeleon program to start the fraction collection.

3.9 Peptide Separation

1. Prepare the sample by solubilizing it in 4 μL of solubilizing solution, sonicate for 30 s and spin down in a microcentrifuge.
2. Dilute the sample by adding 20 μL of eluent A, sonicate and spin down in a microcentrifuge at maximum speed of 3 min, and place it in the autosampler (*see* **Note 14**).
3. Set the initial program parameters: Column oven 25 °C, LoadingPump.Flow = 30 μL/min, MicroPump.Flow = 0.3 μL/min, and MicroPump.%B = 5.0 %.
4. Start the acquisition/fraction collection following the gradient as defined.

Time (min)	*B* %
0	5
3	5 (valve shift)
35	45
40	80
45	5
50	5 (valve shift)
55	5

3.10 MALDI Spectra Acquisition

1. Prepare mass spectrometer for measurement according to manufacturer's instructions, and calibrate the instrument using the Calmix 5.
2. Acquire the spectra in the MS positive reflector mode using 1200 laser shots in the *m/z* range between 700 and 4500 Da. MS/MS analysis of automatically selected precursors is performed at collision energy of 2 kV with air as collision gas at a pressure of 2×10^{-7} Torr. MS spectra are internally calibrated

using Glu-Fib (*m/z* 1570.68). Up to 16 of the most intense ion signals per spot position with S/N above 50 are selected as precursors for MS/MS.

3.11 Protein Identification and PTM Analysis

Identify the proteins doing a search against an appropriate database (e.g., Swiss-Prot protein database) using the Mascot algorithm:

1. Open the TS2Mascot and select the spotset run of interest.
2. Click in save peak list to create the .mgf file.
3. Select the location to save the .mgf file.
4. Open Mascot web page: http://www.matrixscience.com/.
5. Browse and select .mgf file for search.
6. Select the database that you want to search against the dataset.
7. Select the enzyme used for digestion (e.g., trypsin).
8. Select peptide tolerance and MS/MS tolerance at appropriate values (e.g., precursor tolerance 25 ppm and fragment tolerance 0.3 Da).
9. Select the following modifications as variable: oxidation (C, H, W, D, K, N, P, F, Y, R, M), dioxidation (W, M, F, Y, C, P, R, K, N), arginine oxidation to glutamic semialdehyde and lysine oxidation to aminoadipic semialdehyde (*see* **Note 15**) (Table 1).
10. Select fixed modifications according to the experimental set up.
11. Select data format and instrument used for acquiring spectra.
12. Select automatic decoy database to determine the false discovery rate (FDR) (*see* **Note 16**).
13. Start the search.

3.12 Data Analysis Using the Bioinformatic Tools: ClueGO and CluePedia

ClueGo and CluePedia are two Cytoscape plug-ins utilized in order to give biological meaning to interaction networks. ClueGO integrates Gene Ontology (GO) terms as well as KEGG/BioCarta pathways and creates a functionally organized GO/pathway term network [17]. CluePedia extends ClueGO functionality by giving the possibility to enrich those networks with known and experimental data [18]. In this chapter, the two plug-ins were combined to explore the biological processes associated to proteins annotated in the blot (*see* **Note 17**).

1. Open Cytoscape Subheading 3.1.
2. Import glycoproteins network (File → Import → Network → File, e.g., excel file).
3. Run ClueGo plug-in (Apps → ClueGO v2.1.2 + CluePedia v1.1.2).
4. After open ClueGo panel it is necessary to define a number of variables: analysis type, cluster list(s), organism, identifiers type,

Table 1
List of some oxidative modifications and mass shifts

Modification	Description	Monoisotopic mass shift (Da)
Arg → GluSA	Arginine oxidation to glutamic semialdehyde	-43.0534
Pro → Pyrrolidinone	Proline oxidation to pyrrolydinone	-30.0105
Pro → Pyrrolidone	Proline oxidation to pyrrolidone	-27.9949
His → Asn	Histidine oxidation to asparagine	-23.0159
His → Asp	Histidine oxidation to aspartic acid	-22.0319
Cys → oxoalanine	Cysteine oxidation to oxoalanine	-17.9928
Lys → Allysine	Lysine oxidation to aminoadipic semialdehyde	-1.0316
Deamidation (Arg, Asn, Gln)	Deamidation	0.984
Trp → Kynurenin	Tryptophan oxidation to kynurenin	3.9949
Pro → Pyro-Glu	Proline oxidation to pyroglutamic acid	13.9792
Trp → Oxolactone	Tryptophan oxidation to oxolactone	13.9792
Arg, Gln, Glu, Leu, Ile, Lys, Val	Carbonylated amino acid	13.9793
Lys → Aminoadipic acid	Lysine oxidation to α-aminoadipic acid	14.9632
Amino (Tyr)	Tyrosine oxidation to 2-aminotyrosine	15.0108
Oxidation	Oxygen addition and hydroxylation	15.9949
Trp → Hydroxykynurenin	Tryptophan oxidation to hydroxykynurenin	19.9898
Quinone (Tyr)	Quinone	29.9741
Dioxidation	Oxygen addition and hydroxylation	31.9898
Carbamylation (Lys, Arg, Cys, Met)	Isocyanate reaction with amino groups	43.0058
Carboxy (Lys, Asp, Glu)	Carboxylation	43.9898
Trioxidation (Cys)	Cysteine oxidation to cysteic acid	47.9847

ontology, statistical test, PV correction, advanced statistical options, network specificity, and advanced settings (*see* **Note 18**).

(a) Select the analysis type (single gene set (cluster) or comparison of clusters).

(b) Select the organism (*see* **Note 19**).

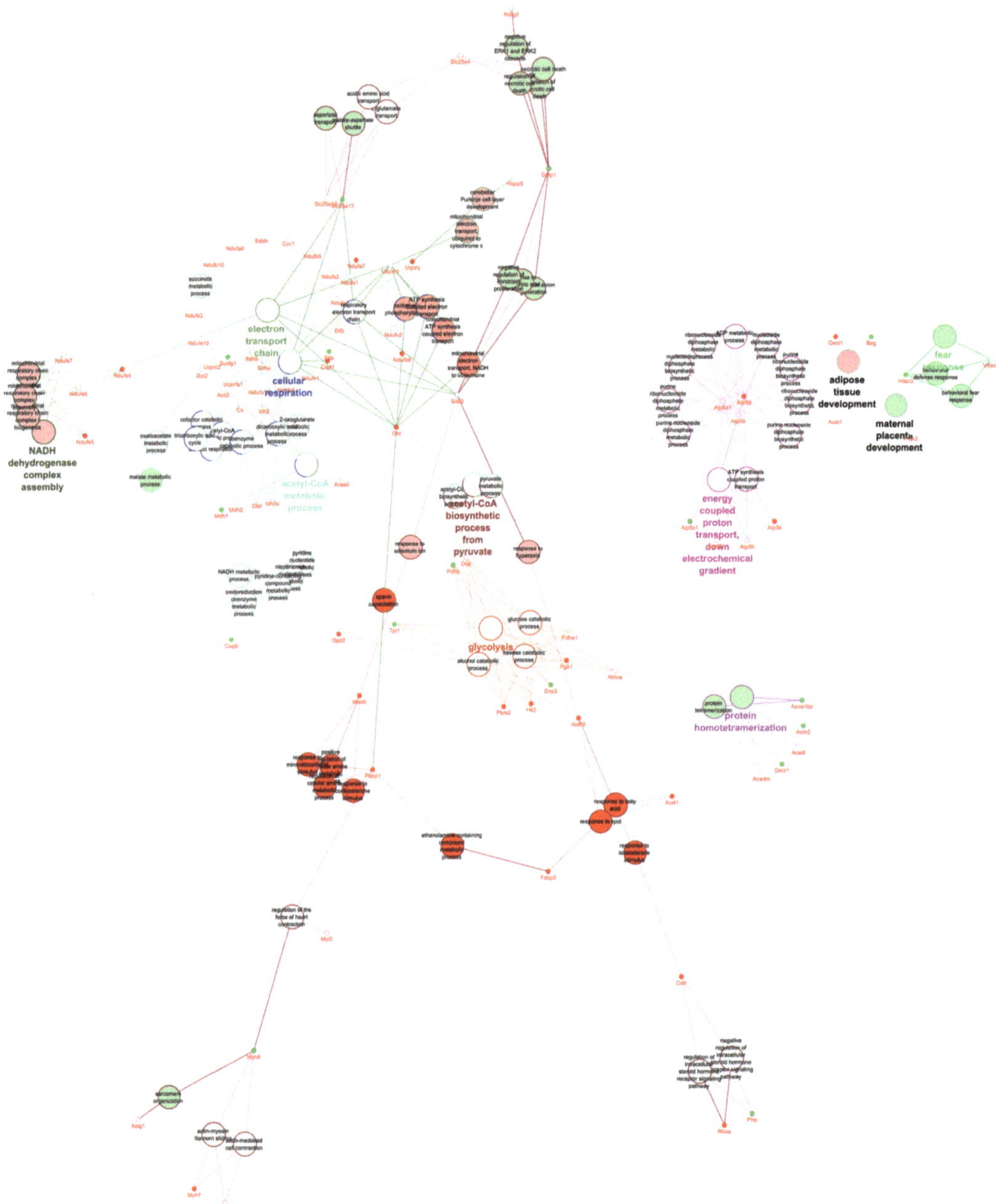

Fig. 3 ClueGO and CluePedia analyses of the carbonylated proteins annotated in the blot from heart mitochondria of sedentary mouse (*red* nodes) and active mouse (*green* nodes) using biological processes as the selected ontology

(c) Select the identifiers type (e.g., AffymetrixID, AccessionID, and SymbolID).

(d) Select the ontology or ontologies and for the GO based files select the evidence (*see* **Note 20**).

(e) Select the statistical test (*see* **Note 21**).

(f) Select *p* value (PV) correction (Bonferroni, Bonferroni step-down, or Benjamini-Hochberg).

(g) Select network specificity (Global, Medium, or Detailed network) (*see* **Note 22**).

(h) Start the analysis.

In our example (Fig. 3), the analysis mode used two clusters created for the proteins identified by blot-MS from mitochondria samples: cluster 1 (red) represents the oxidized proteins associated to a sedentary lifestyle, cluster 2 (green) symbolizes the ones captured associated to the active lifestyle (evaluated in mice). Regarding the ClueGO settings, the "Biological Process" ontology (updated), the right-sided hypergeometric statistical test (enrichment), and Bonferroni step down PV correction were selected before start the analysis.

5. Open the CluePedia Panel and Update to Visualize the Proteins Associated to Each Cluster

The network created can also be enriched trough new interaction files (subnetworks). For that, select the genes/proteins from the subnetwork and click "Enrich" (right mouse click on the subnetwork or in CluePedia Enrichment panel). Keep standard settings, set the number of genes and color before click in the "Start" button. For example, if the number of genes set were 5, and the color blue, a top 5 new enriched genes based on the subnetwork selected will be added to the network as blue nodes.

4 Notes

1. For the isolation of mitochondria from "hard" tissues it is recommended a pretreatment with a protease to promote breakdown of the cellular structure. The enzyme must be chosen according to the sample used. Trypsin was used for the isolation of mitochondria from skeletal and cardiac muscle [19, 20]. Nevertheless, the majority of the studies use Nagarse (Sigma) for skeletal muscle [21] and Subtilisin A type VIII (Sigma P5380) for cardiac muscle [22].
2. The pH of this solution is approximately 8.6 but there is no need to adjust it. This solution is usually preserved concentrated and should be diluted in the ratio of 1:10 before its use.
3. If necessary, warm the solution at 30 °C (maximum) to dissolve completely the urea and then complete the volume with distilled water. DTT and ampholytes should always be added just before use.
4. If necessary, warm the solution at 30 °C (maximum) to dissolve completely the urea and then complete the volume with distilled water. Divide in aliquots and store at –20 °C.

5. Boil the agarose until everything is solubilized. Divide in aliquots and store at room temperature.
6. Mitochondria should be isolated from fresh tissue to ensure a high yield and high purity of isolated mitochondria, which is difficult to achieve with frozen tissue.
7. The program should be adjusted in accordance to the sample characteristics.
8. IPG strip and agarose seal replace the stacking gel.
9. Adjust the capture time to avoid saturated pixels during the acquisition.
10. For more accurate quantitation of differential expression, especially in the case of subtle differences in expression level, the entire process should be repeated in triplicate and the results analyzed for statistical significance.
11. After the addition of ACN the bands/spots should be dehydrated and white.
12. The following steps of reduction and alkylation can be skipped. First step is the reduction of disulfide bonds by DTT and formation of thiols (-SH). The second step consists on blocking of thiols by alkylation with iodoacetamide to prevent reoxidation of proteins by formation of S-carboxyamidomethylcysteine (CAM; adduct: CH2 CONH2). The specific mass of the aminoacid cysteine is thereby increased from 103.01 to 160.03 Da.
13. Other enzymes (such as Lys-C, Arg-C) can be used as an alternative to trypsin. Keep in mind that the optimal conditions of each enzyme may vary in respect to solubilization buffer, incubation temperature and time of incubation.
14. Be careful and avoid bubbles while sample is being transferred to sample vial.
15. Mascot has the limitation of allowing only nine modifications per search.
16. False discovery rate associated with the similarity score of a peptide spectrum match (PSM) is estimated by searching the experimental data against a randomized peptide or protein sequence database; a decoy database. A decoy search can be performed automatically in Mascot by choosing the *Decoy checkbox* on the search form. Nevertheless, other decoy databases can be created to estimate FDR. Following decoy analysis, inspect the MS/MS annotation in spectrum to certify the correct PTM.
17. The use of software that analyzes complex networks, as Cytoscape, is relevant to give biological meaning to the data obtained in order to find potential biomarkers and targets for disease and health surveillance. However, it is necessary to

understand the principles of the software and know how to use it to achieve a specific purpose.

18. The ClueGo uses ontology sources: GO, Kyoto Encyclopedia of Genes and Genomes (KEGG) and BioCarta, which exist in order to obtain biological information in a meaningful way. GO describes gene products in terms of their associated biological processes, cellular components, and molecular functions [23]. Between the terms there is a hierarchical relationship (parent-child). KEGG is a database of biological systems that integrates genomic, chemical, and systemic functional information [24]. BioCarta provides useful pathway information [25]. If the primary aim is a complete view on the studied process, several ontology sources should be consulted in order to integrate their complementary information. For that reason, ClueGo was created and represents an open-source Java tool that extracts the nonredundant biological information for large clusters of genes, using GO, KEGG, and BioCarta [17].
19. At the moment, ClueGo supports *Arabidopsis thaliana*, *Bos taurus*, *Caenorhabditis elegans*, *Danio rerio*, *Dictyostelium discoideum*, *Drosophila melanogaster*, *Escherichia coli*, *Galus galus*, *Homo sapiens*, *Magnaporthe grisea*, *Mus musculus*, *Oryza sativa*, *Rattus norvegicus*, and *Saccharomyces cerevisiae*.
20. Do not forget to update GO categories in the "show ontology update" options.
21. The standard test is hypergeometric test two-sided.
22. The "Global network" provides a general biological information whereas the "Detailed network" shows more specific and informative ontologies underlying particular aspects of the studied gene product [17].

Acknowledgments

This work was supported by Portuguese Foundation for Science and Technology (FCT), European Union, QREN, FEDER, and COMPETE for funding the QOPNA research unit (project PEst-C/QUI/UI0062/2013; FCOMP-01-0124-FEDER-037296), iBiMED, (UID/BIM/04501/2013), the Cardiovascular R&D Unit (UID/IC/00051/2013) the research projects (PTDC/DES/114122/2009; FCOMP-01-0124-FEDER-014077; EXPL/DTP-DES/1010/2013; FCOMP-01-0124-FEDER-041115), and RNEM (REDE/1504/REM/2005 that concerns the Portuguese Mass Spectrometry Network). The authors would like to acknowledge the support of COST action BM1307.

References

1. Padrao AI, Ferreira RM, Vitorino R, Alves RM, Neuparth MJ, Duarte JA et al (2011) OXPHOS susceptibility to oxidative modifications: the role of heart mitochondrial subcellular location. Biochim Biophys Acta 1807(9):1106–1113
2. Murray J, Oquendo CE, Willis JH, Marusich MF, Capaldi RA (2008) Monitoring oxidative and nitrative modification of cellular proteins; a paradigm for identifying key disease related markers of oxidative stress. Adv Drug Deliv Rev 60(13–14):1497–1503
3. Stadtman ER, Berlett BS (1997) Reactive oxygen-mediated protein oxidation in aging and disease. Chem Res Toxicol 10(5):485–494
4. Starke PE, Oliver CN, Stadtman ER (1987) Modification of hepatic proteins in rats exposed to high oxygen concentration. FASEB J 1(1):36–39
5. Starke-Reed PE, Oliver CN (1989) Protein oxidation and proteolysis during aging and oxidative stress. Arch Biochem Biophys 275(2):559–567
6. Dasgupta A, Zheng J, Perrone-Bizzozero NI, Bizzozero OA (2013) Increased carbonylation, protein aggregation and apoptosis in the spinal cord of mice with experimental autoimmune encephalomyelitis. ASN Neuro 5(1):e00111
7. Dalle-Donne I, Rossi R, Giustarini D, Gagliano N, Lusini L, Milzani A et al (2001) Actin carbonylation: from a simple marker of protein oxidation to relevant signs of severe functional impairment. Free Radic Biol Med 31(9):1075–1083
8. Nystrom T (2005) Role of oxidative carbonylation in protein quality control and senescence. EMBO J 24(7):1311–1317
9. Maisonneuve E, Fraysse L, Lignon S, Capron L, Dukan S (2008) Carbonylated proteins are detectable only in a degradation-resistant aggregate state in Escherichia coli. J Bacteriol 190(20):6609–6614
10. Hartl FU, Bracher A, Hayer-Hartl M (2011) Molecular chaperones in protein folding and proteostasis. Nature 475(7356):324–332
11. Hipp MS, Park SH, Hartl FU (2014) Proteostasis impairment in protein-misfolding and -aggregation diseases. Trends Cell Biol 24(9):506–514
12. Dalle-Donne I, Carini M, Orioli M, Vistoli G, Regazzoni L, Colombo G et al (2009) Protein carbonylation: 2,4-dinitrophenylhydrazine reacts with both aldehydes/ketones and sulfenic acids. Free Radic Biol Med 46(10):1411–1419
13. Padrao AI, Ferreira R, Vitorino R, Alves RM, Figueiredo P, Duarte JA et al (2012) Effect of lifestyle on age-related mitochondrial protein oxidation in mice cardiac muscle. Eur J Appl Physiol 112(4):1467–1474
14. Alves RM, Vitorino R, Figueiredo P, Duarte JA, Ferreira R, Amado F (2010) Lifelong physical activity modulation of the skeletal muscle mitochondrial proteome in mice. J Gerontol A Biol Sci Med Sci 65(8):832–842
15. Conrad CC, Choi J, Malakowsky CA, Talent JM, Dai R, Marshall P et al (2001) Identification of protein carbonyls after two-dimensional electrophoresis. Proteomics 1(7):829–834
16. Talent JM, Kong Y, Gracy RW (1998) A double stain for total and oxidized proteins from two-dimensional fingerprints. Anal Biochem 263(1):31–38
17. Bindea G, Mlecnik B, Hackl H, Charoentong P, Tosolini M, Kirilovsky A et al (2009) ClueGO: a Cytoscape plug-in to decipher functionally grouped gene ontology and pathway annotation networks. Bioinformatics 25(8):1091–1093
18. Bindea G, Galon J, Mlecnik B (2013) CluePedia Cytoscape plugin: pathway insights using integrated experimental and in silico data. Bioinformatics 29(5):661–663
19. Ferreira R, Vitorino R, Alves RM, Appell HJ, Powers SK, Duarte JA et al (2010) Subsarcolemmal and intermyofibrillar mitochondria proteome differences disclose functional specializations in skeletal muscle. Proteomics 10(17):3142–3154
20. Padrao AI, Carvalho T, Vitorino R, Alves RM, Caseiro A, Duarte JA et al (2012) Impaired protein quality control system underlies mitochondrial dysfunction in skeletal muscle of streptozotocin-induced diabetic rats. Biochim Biophys Acta 1822(8):1189–1197
21. Magalhães J, Ascensão A, Soares JMC, Ferreira R, Neuparth MJ, Marques F et al (2005) Acute and severe hypobaric hypoxia increases oxidative stress and impairs mitochondrial function in mouse skeletal muscle. J Appl Physiol 99(4):1247–1253
22. Ascensão A, Magalhães J, Soares JMC, Ferreira R, Neuparth MJ, Marques F et al (2005) Moderate endurance training prevents

doxorubicin-induced in vivo mitochondriopathy and reduces the development of cardiac apoptosis. Am J Physiol Heart Circ Physiol 289(2):H722–H731
23. Ashburner M, Ball CA, Blake JA, Botstein D, Butler H, Cherry JM et al (2000) Gene ontology: tool for the unification of biology. The Gene Ontology Consortium. Nat Genet 25(1):25–29
24. Kanehisa M, Goto S, Kawashima S, Nakaya A (2002) The KEGG databases at GenomeNet. Nucleic Acids Res 30(1):42–46
25. LLC B. BioCarta 2000. http://www.biocarta.com.

Chapter 24

Quantitation of Protein Translation Rate In Vivo with Bioorthogonal Click-Chemistry

Borja Belda-Palazón, Alejandro Ferrando, and Rosa Farràs

Abstract

The development of novel bioorthogonal reactives that can be used to tag biomolecules in vivo has revolutionized the studies of cellular and molecular biology. Among those novel reactive substances, amino acid analogs can be used to label nascent proteins, thus opening new avenues for measuring protein translation rates in vivo with a limited manipulation of the sample. Here, we describe the use of Click-chemistry to tag and separate newly synthesized proteins in mammalian cells that can be used, coupled with western analysis, to estimate the translation rate of any protein of interest.

Key words Click-chemistry, Bioorthogonal amino acid, Translation rate, Affinity purification, Western blot

1 Introduction

The comprehension of the mechanisms that control the translation process is fundamental for the global understanding of the gene expression in every organism. Proteins are not equally translated and the differences in translation efficiencies have been shown to be involved in the phenotypic divergence of species [1]. Moreover, the changing of the rate at which the ribosome translates an mRNA can alter the behavior of the newly synthesized protein with consequences on protein homeostasis. Therefore, the analysis of protein translation rates over diverse cell growth conditions, such as inhibitor or drug treatments are critical aspects to be studied in cell biology [2]. Early studies on protein turnover relied on isotopic labeling of amino acids, typically labeled with [^{35}S]-Methionine and pulse-chase experiments after blocking protein synthesis. Novel techniques based on mass-spectrometry proteomics allow the determination of protein turnover of large number of proteins after pulse labeling with amino acids that incorporate stable isotopes as it has been shown for human cells [3]. In spite of the power of these global approaches, individual studies of protein

Rune Matthiesen (ed.), *Proteostasis: Methods and Protocols*, Methods in Molecular Biology, vol. 1449,
DOI 10.1007/978-1-4939-3756-1_24, © Springer Science+Business Media New York 2016

turnover at relative low cost may be required in many cases. The recent development of "click"-chemistry with bioorthogonal chemical reactives which serve to selectively tag biomolecules such as proteins, DNA, RNA, lipids, glycans, etc., have provided unvaluable tools to carry out these type of studies [4–6].

Here, we focus on the use of the bioorthogonal noncanonical amino acid L-Azidohomoalanine (AHA), which is a methionine analog, for the metabolic labeling of the newly synthesized proteins in human cells [7]. The newly synthesized AHA-containing proteins can click-react with modified biotin-alkyne groups forming stable bioconjugates. Next, the biotinylated de novo synthesized proteins are affinity purified by using a Streptavidin-Agarose chromatography for their subsequent western blot analysis. Eventually, different AHA time-incubation periods are needed to observe and quantify the AHA incorporation to determine the translation rate of the protein of interest.

In conclusion, this method can be used to specifically compare the translation rate for the proteins of interest among different experimental conditions by analyzing the incorporation rate of the bioorthogonal noncanonical amino acid AHA into newly synthesized proteins.

2 Materials

Unless otherwise indicated prepare all solutions using sterile ultrapure 18 MΩ water and store all the reagents at room temperature.

2.1 Cell Culture Media

1. 6 Wells cell culture plates: growth area 9.5 cm^2 per well (*see* **Note 1**).
2. DMEM Mixture F-12 (DMEM/F12), supplemented with 10% fetal bovine serum (FBS), 10 U/mL penicillin, and 10 μg/mL streptomycin. To prepare DMEM/F12 *see* **Note 2**.
3. DMEM Methionine free (DMEM-Met), supplied with 10% FBS, 10 U/mL penicillin, and 10 μg/mL streptomycin. To prepare DMEM-Met *see* **Note 2**.
4. 50 mM L-Azidohomoalanine (AHA) stock solution (Click-iT® AHA, Life Technologies): add 387 μL of DMSO to 5 mg of Click-iT® AHA (MW = 258.16) and mix well. Store at −20 °C.
5. Phosphate-Buffered Saline (PBS). Store at 4 °C.
6. Sterile cell scraper.

2.2 Protein Extraction and Quantification

1. 1× Protein Extraction Buffer (PEB): 50 mM Tris–HCl pH 8.0, 150 mM NaCl, 0.1% sodium dodecyl sulfate (SDS), 1% IGEPAL® CA-630 (Sigma-Aldrich), 0.5% sodium deoxycholate (DOC), 2 μg/mL leupeptine, 2 μg/mL aprotinin, and 1 mM phenylmethylsulfonyl fluoride (PMSF). To prepare 1 mL of 1× PEB, mix 50 μL of 1 M Tris–HCl pH 8.0 (*see* **Note 3a**),

30 μL of 5 M NaCl (*see* **Note 3d**), 10 μL of 10% SDS (*see* **Note 3b**), 10 μL of 100% IGEPAL® CA-630, 50 μL of 10% DOC (*see* **Note 3e**), 1 μL of 2 mg/mL leupeptine, 1 μL of 2 mg/mL aprotinin, and 1 μL of 1 M PMSF. Bring the volume to 1 mL with water and store at 4 °C.

2. Bradford Reagent (Sigma-Aldrich). Store at 4 °C.
3. 2 mg/mL Quick Start Bovine Serum Albumin (BSA) Standard (Bio-Rad). Store at −20 °C.
4. Phosphate-Buffered Saline (PBS). Store at 4 °C.
5. Refrigerated centrifuge 5430 R (Eppendorf) and its fixed-angle rotors: F-35-6-30 for 15 or 50 mL tubes and FA-45-30-11 for 1.5 or 2 mL tubes.
6. 96 Wells microtiter plate.
7. Microplate reader.
8. Microplate centrifugator.

2.3 Click-It Reaction

1. 4 mM Biotin-PEG4-alkyne (Sigma-Aldrich) stock solution: add 2.732 mL of DMSO to 5 mg Biotin-PEG4-alkyne (MW = 457.58) and mix well. Store at −20 °C.
2. Click-iT® Protein Reaction Buffer Kit (Life Technologies), includes contents of 2× concentrate Click-iT® reaction buffer (Component A), 40 mM Copper (II) sulfate ($CuSO_4$; Component B), Click-iT® reaction buffer additive 1 (Component C), Click-iT® reaction buffer additive 2 (Component D). Fully dissolve the Component C and Component D in 500 μL and 540 μL of distilled deionized (DDI) water, respectively. Store all the reagents at −20 °C.
3. Sterile ultrapure 18 MΩ water.
4. End over end rotator.

2.4 Purification of Biotinylated Proteins

1. PD MiniTrap G-25 (GE Healthcare).
2. 1× Equilibration Buffer (EB): 1% IGEPAL® CA-630 (Sigma-Aldrich) and 0.1% SDS in PBS. To prepare 200 mL of 1× EB, mix 2 mL of 100% IGEPAL® CA-630, 2 mL of 10% SDS (*see* **Note 3b**), and 196 mL of PBS.
3. Centrifugable and autoclavable 1 mL columns Mobicol classic and filters of 10 μm pore size (MoBiTec GmbH).
4. Streptavidin-Agarose (Sigma-Aldrich) stored a 4 °C
5. 1× Washing Buffer (WB): 1% IGEPAL® CA-630 in PBS. To prepare 200 mL of 1× WB, mix 2 mL of 100% IGEPAL® CA-630 and 198 mL of PBS.
6. 5× Laemmli Buffer (LB): 0.25 M Tris–HCl pH 6.8, 10% SDS, 0.5 M dithiothreitol (DTT), 30% (w/v) sucrose, 0.5% (w/v)

bromophenol blue (BPB). To prepare 1 mL of 5× LB, mix 250 μL of 1 M Tris–HCl pH 6.8 (*see* **Note 3a**), 300 mg of sucrose, 5 mg of BPB, 76 mg of DTT (MW = 154.25), and 100 mg of SDS. Bring the volume to 1 mL with water and dissolve well. Store at −20 °C.

7. Uncooled Benchtop Centrifuge.
8. Refrigerated centrifuge 5430 R (Eppendorf) and its fixed-angle rotors: F-35-6-30 for 15 or 50 mL tubes and FA-45-30-11 for 1.5 mL tubes.
9. 1.5, 2 and 15 mL polypropylene tubes and tube adapters.
10. Thermoblock for 1.5 mL tubes.

2.5 SDS Polyacrylamide Gel Electrophoresis (SDS-PAGE)

1. Mini-Protean® Tetra handcast systems (Bio-Rad) (*see* **Note 4**).
2. Resolving gel: 14 % acrylamide–bis-acrylamide (Sigma-Aldrich) (*see* **Notes 5** and **6**), 375 mM Tris–HCl pH 8.8, 0.1 % SDS, 0.2 % *N,N,N,N′*-tetramethyl-ethylenediamine (TEMED) and 0.08 % ammonium persulfate (APS). For 10 mL mix 3.5 mL of 40 % acrylamide–bis-acrylamide, 3.75 mL of 1 M Tris–HCl pH 8.8 (*see* **Note 3a**), 100 μL of 10 % SDS (*see* **Note 3b**), 2.55 mL of water, 20 μL of 100 % TEMED (*see* **Note 6**), and 80 μL of 10 % APS (*see* **Notes 3c** and **4**).
3. Stacking gel: 4 % acrylamide–bis-acrylamide (Sigma-Aldrich) (*see* **Note 6**), 125 mM Tris–HCl pH 6.8, 0.1 % SDS, 0.8 % TEMED, and 0.1 % APS. For 5 mL mix 0.5 mL of 40 % acrylamide–bis-acrylamide, 625 μL of 1 M Tris–HCl pH 6.8 (*see* **Note 3a**), 50 μL of 10 % SDS (*see* **Note 3b**), 3.785 mL of water, 40 μL of 100 % TEMED (*see* **Note 6**), and 50 μL of 10 % APS (*see* **Notes 3c** and **4**).
4. 1× Running buffer (RB): 25 mM Tris, 192 mM glycine, 0.1 % SDS, pH 8.3. To prepare a 10× RB stock solution weigh 144.12 g of glycine (MW = 75.07 kDa) and dissolve it in 600 mL of water. Adjust the pH to 8.3 with Tris base, add 100 mL of 10 % SDS (*see* **Note 3b**) and bring the volume to 1 L. Prepare working solution 1× RB by mixing 100 mL of 10× RB and 900 mL of water.
5. Molecular weight marker, Precision Plus Protein™ Prestained Standards (Bio-Rad).

2.6 Protein Transfer from SDS-PAGE Gel to Membrane

1. Polyvinylidene fluoride (PVDF) membrane (*see* **Note 7**).
2. Mini Trans-Blot® cell system (Bio-Rad).
3. 1× Transfer Buffer (TB): 6 g/L Tris base, 3.1 g/L boric acid.
4. Extra Thick Blot Paper (7.5 × 10 cm; Bio-Rad).
5. Methanol.
6. Magnetic stirrer and magnetic stir bar.

2.7 Protein Immunodetection by Western Analysis

1. 1× Tris-Buffered Saline (TBS): 20 mM Tris–HCl pH 7.6, 150 mM NaCl. To prepare 1 L of 10× TBS stock solution, mix 200 mL of 1 M Tris–HCl pH 7.6 (*see* **Note 3a**), 300 mL of 5 M NaCl (*see* **Note 3d**), and 500 mL of water. Prepare working solution 1× TBS by mixing 100 mL of 10× TBS and 900 mL of water.
2. 1× Blocking Buffer (BB): 0.05 % Tween 20 and 5 % nonfat dry milk in 1× TBS. To prepare 1 L of 1× BB, weigh 50 g of nonfat dry milk and dissolve it in 500 mL of 1× TBS. Add 500 μL of 100 % Tween 20 and dissolve well. Bring the volume to 1 L with 1× TBS.
3. Primary antibodies against the proteins of interest. In the case example, mouse anti-GAPDH antibodies (Chemicon International). Store at −20 °C.
4. Secondary antibodies conjugated to enzyme, such as horseadish peroxidase (HRP), against the Fc domain of the primary antibodies. In the case example, anti-Mouse IgG (Fab specific)-Peroxidase antibody produced in goat (Sigma-Aldrich). Store at −20 °C.
5. Chemiluminescence detection reagents: Amersham™ ECL Western Blotting analysis system (GE Healthcare). Use SuperSignal West Femto Chemiluminescent Substrate (Thermo Scientific) if more sensitivity is required. Store the reagents at 4 °C.
6. Platform Rocker.
7. Saran Wrap plastic.
8. X-ray films.
9. Cassette.
10. X-ray film processor.

2.8 Estimation of Protein Synthesis Rate

1. Image scanner.
2. ImageJ software (free download at http://imagej.net/Downloads).

3 Methods

Carry out all procedures at room temperature unless otherwise specified. Figure 1 summarizes the methods used in this protocol.

3.1 Amount of Starting Biological Material

The translation rate of one protein of interest is analyzed in HeLa cells for this case example. Six AHA time-incubation periods are used in order to calculate the translation rate. Therefore in this example the experiment is composed of six samples/cultures grown in a six wells culture plate, with a growth area of 9.5 cm^2 per well.

1. Grow HeLa cells until reaching the subconfluent state

↓

2. Replace the growth medium with methionine-free medium and incubate for 1 hour

↓

3. Add AHA to the culture media and incubate the cells further for the different time-incubation periods

↓

4. Collect the cells and perform protein extraction and quantification

↓

5. Use the same amount of total protein for the click-it reactions

↓

6. Affinity purify the biotinylated *de novo* synthesized proteins with streptavidin agarose

↓

7. Separate the proteins by SDS-PAGE and transfer them onto membrane

↓

8. Immunodetect the protein of interest by western blot analysis and quantify by densitometry the intensity of the immunoreactive bands

↓

9. Plot the relative band intensity against the different time-incubation periods to calculate the protein translation rate

Fig. 1 Summary of the methods used for quantitation of protein translation rate in vivo with bioorthogonal Click-chemistry

1. Seed 3×10^5 exponentially growing HeLa cells in 2 mL of DMEM/F12 in each well. After 24 h, replace medium with fresh DMEM/F12 and incubate at 37 °C and 5% CO_2 for 24 h until reaching the subconfluent state (1.5×10^6 cells/cm^2). Each sample/culture of subconfluent HeLa cell monolayers contributes sufficient tissue to extract at least 100 μg of total protein to prepare the click-it reactions.
2. Once at the subconfluent state, wash cells twice with 1 mL of PBS.
3. Remove PBS and add 1 mL of DMEM-Met prewarmed at 37 °C. Incubate the cultures for 1 h at 37 °C.

4. Add 1 μL of 50 mM AHA (50 μM final concentration) to each growth culture media and incubate for different time periods. In this case example the time-incubation periods are 0, 1, 2, 4, 6, and 8 h (*see* **Note 8**).
5. After each time-incubation period, wash three times with 1 mL of cold PBS.
6. After the last wash, add 1 mL of cold PBS once again and harvest the cells with a cell scraper.
7. Collect the cell suspensions in 15 mL tubes and put them on ice.
8. Centrifuge the samples at 7000 × *g* for 5 min at 4 °C and discard supernatant. Store the samples at −80 °C before use.

3.2 Total Protein Extraction and Quantification

1. Add 60 μL of 1× PEB to the cell pellets and mix by using the vortex. Incubate the lysates for 30 min on ice vortexing every 5 min.
2. Centrifuge the lysates at 12,000 × *g* for 15 min at 4 °C, and collect individually the supernatants, which are the total protein extracts, into new 1.5 mL tubes and keep them on ice.

   The quantification of the total proteins in each sample (Bradford assay) is depicted in the next steps. First, a calibration curve representing the absorbance at 595 nm (A_{595nm}) versus the concentration of protein (mg/mL) is needed. The calibration curve is made by measuring the A_{595nm} of different BSA concentrations.
3. Use a BSA stock solution (2 mg/mL) to make different dilutions by mixing with cold PBS. Maintain the serial dilutions on ice:
   (a) Prepare 1.4 mg/mL BSA by mixing 70 μL of 2 mg/mL BSA with 30 μL of PBS. Vortex.
   (b) Prepare 1 mg/mL BSA by mixing 71.4 μL of 1.4 mg/mL BSA with 28.4 μL of PBS. Vortex.
   (c) Prepare 0.50 mg/mL BSA by mixing 50 μL of 1 mg/mL BSA with 50 μL of PBS. Vortex.
   (d) Prepare 0.25 mg/mL BSA by mixing 50 μL of 0.50 mg/mL BSA with 50 μL of PBS. Vortex.
   (e) Prepare solution 0 mg/mL BSA with 100 μL of PBS.
4. Dilute the protein extract five times by mixing 3 μL of each protein extract with 12 μL of cold PBS. Vortex and keep on ice.
5. Mix 5 μL of each BSA or protein extract dilutions with 200 μL of Bradford reagent. Do it in duplicate.
6. Pippete the mixes separately in a 96-well microtiter plate and measure the A_{595nm} with the microplate reader.

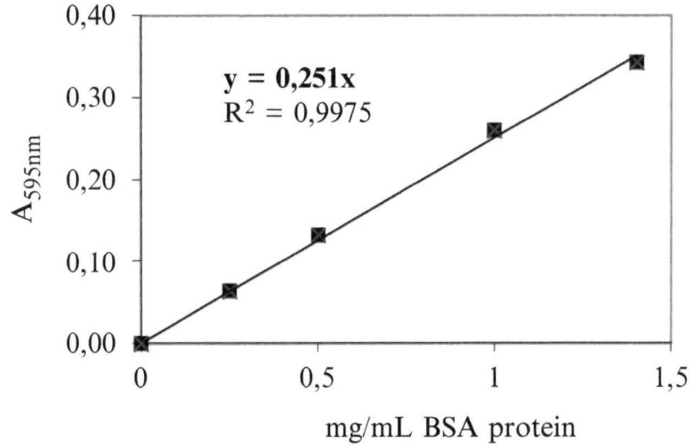

Fig. 2 Standard calibration curve for protein quantification from data obtained with the Bradford assay. The resulting linear regression equation is shown in bold. The *y* means the A_{595nm} while the *x* refers to the protein concentration

7. Represent the obtained standard calibration curve. Figure 2 represents the calibration curve and the linear regression equation obtained for this case example.
8. After measuring the A_{595nm} for every sample, calculate the protein concentration in the extracts by using the linear regression equation obtained from the standard calibration curve. Table 1 shows the protein concentration of the extracts for this case example.

Table 1

Protein concentration of the different samples/time-incubation periods with AHA are determined by using the linear regression equation obtained with the calibration curve

Samples/time-incubation periods with AHA (hours)	A_{595nm}	mg/mL of protein in the extract
0	0.103	2.06
1	**0.102**	**2.03**
2	0.103	2.05
4	0.137	2.73
6	0.115	2.29
8	0.132	2.63

The limiting concentration sample is highlighted in bold (*see* **Note 9**)

3.3 Performing the Click-It Reactions

The same amount of total protein has to be used to prepare the click-it reactions for every time-incubation sample (*see* **Note 9**).

1. For each click-it reaction mix 10 μL of 4 mM biotin-PEG4-alkyne with 90 μL of Click-iT® reaction buffer (Component A) in a 1.5 mL tube.
2. Add up to 200 μg of total protein extract labeled with AHA in a maximum volume of 50 μL (*see* **Note 10**).
3. Bring the volume to 160 μL with ultrapure 18 MΩ water and vortex for 5 s.
4. Add 10 μL of 40 mM $CuSO_4$ (Component C) and vortex for 5 s.
5. Add 10 μL of Click-iT® reaction buffer additive 1 (Component C) and vortex for 5 s. Wait for 2–3 min, but not longer than 5 min before **step 7**.
6. Add 20 μL of Click-iT® reaction buffer additive 2 (Component D) and vortex for 5 s. This mixture turns color into bright orange.
7. Incubate the reactions for 20 min in the end-over-end rotator.

3.4 Purification of Biotinylated "De Novo" Synthesized Proteins

1. To clean up the excess of free biotin-PEG4-alkyne, use the PD pipette MiniTrap G-25 (GE Healthcare) columns. Remove the top and bottom caps and transfer the columns into 15 mL tubes (*see* **Note 11**).
2. To equilibrate the columns add 2 mL of 1× EB per column and allow the EB to enter the packed bed completely. Discard the flow-through and repeat this step twice.
3. Centrifuge the columns at 1000 × *g* for 2 min.
4. Place the columns into new 15 mL tubes and add separately the samples (0.2 mL) into the columns in the middle of the packed bed.
5. Elute and collect the samples by centrifugation at 1000 × *g* for 2 min (*see* **Note 12**).
6. To purify the biotinylated proteins from the total extract protein background, use a column of Streptavidin-Agarose (Sigma-Aldrich). First, place the 10 μm pore filter into the 1 mL Mobicol classic column until well adjusted. Place the columns into 2 mL tubes.
7. Add 50 μL of Streptavidin-Agarose slurry per column and wash twice with 400 μL of PBS centrifuging at 1800 × *g* for 2 min. Discard the flow-through and place the columns into new 2 mL tubes.
8. Put a bottom cap into the column, add the samples to each Streptavidin-Agarose column and incubate the columns in the end-over-end rotator for 1 h.

9. Remove the bottom cap, put the columns into new 1.5 mL tubes, centrifuge at 1800 × *g* for 2 min and collect the unbound sample to streptavidin protein extract. Store it at −20 °C as a control to check binding to the agarose matrix if necessary.
10. Place the columns into new 2 mL tubes, wash the Streptavidin-Agarose beads with 400 μL of 1× WB and centrifuge at 1800 × *g* for 2 min. Repeat this step three times.
11. Place the columns into new 1.5 mL tubes and add 50 μL of 2× LB. Incubate at 95 °C for 10 min in the Thermoblock (*see* **Note 13**).
12. Elute and collect completely the biotinylated proteins by centrifugation at 1800 × *g* for 2 min. Store the samples at −20 °C until use.

3.5 Separation of the Purified Biotinylated Proteins in a SDS-PAGE

1. Place the previously polymerized 10–14 % SDS-PAGE gel and the Mini Cell Buffer Dam (Bio-Rad) facing each other into the Mini-Protean® Tetra handcast systems (Bio-Rad). Fill completely the inner space between plates with 1× RB. Check the wells to be entirely filled with 1× RB. Fill the outside space until the line marked for 2 gels in the buffer tank.
2. Pippete 25 μL of each sample (the order of the pipetting in this case example is 0, 1, 2, 4, 6, and 8 h of time-incubation periods with AHA) into the well. Pipette in the right or in the left corner side of the gel, the prestained MW markers (*see* **Note 14**).
3. Run the gel electrophoresis until the dye has reached the bottom of the gel.
4. Disassemble the plates and place the gel in a cuvette with 1× TB.

3.6 Protein Transfer from SDS-PAGE Gel to Membrane

1. Activate the PVDF membrane by rinsing in methanol for 1 min and next place the membrane in a cuvette with 1× TB.
2. Prepare the transfer in wet conditions with 1× TB by using the Mini Trans-Blot® cell system (Bio-Rad) and two pieces of Extra Thick Blot Paper (Bio-Rad).
3. Fill totally the tank with 1× TB and place the magnetic stir bar inside.
4. Place the system on a magnetic stirrer during the transfer.
5. Transfer the proteins from the SDS-PAGE gel to the membrane at 2–3 V/cm overnight.

3.7 Protein Immunodetection by Western Analysis

1. Disassemble the transfer system and place the PVDF membrane in a cuvette with 1× TBS.
2. Discard the 1× TBS and incubate to block the membrane with abundant 1× BB for 1 h in the rocking platform (*see* **Note 15**).

3. Discard the 1× BB and incubate the membrane with the primary antibody dissolved in 1× BB for 1 h in the rocking platform (*see* **Note 15**).
4. Discard the solution, and wash with abundant 1× BB by incubating for 10 min in the rocking platform. Repeat this step three times.
5. Discard the 1× BB and incubate the membrane with the secondary antibody dissolved in 1× BB for 1 h in the rocking platform (*see* **Note 15**).
6. Discard the solution, and wash with abundant 1× BB by incubating for 10 min in the rocking platform. Repeat this step three times.
7. Prepare the chemiluminescence detection mix. To prepare this, mix one volume of solution I with one volume of solution II (*see* **Note 16**).
8. Place face up the wet membrane on a Saran Wrap piece and pipette the chemiluminescence detection mix on the membrane. Cover the membrane with another piece of Saran Wrap to spread completely the detection reagent mix on the membrane and incubate for 1–5 min.
9. Unwrap the membrane and discard the excess of detection mix contacting the corner of the membrane.
10. Place face down the membrane on a new Saran wrap piece (*see* **Note 17**) and cover with another one.
11. Place face up the wrapped membrane into the cassette.
12. In the darkness, expose the X-ray film with the membrane (*see* **Note 18**).
13. Develop the X-ray film by using the X-ray film processor. The Fig. 3a shows the autoradiography results obtained in this case example.

3.8 Estimation of Protein Synthesis Rate

1. Scan the autoradiography with the image scanner.
2. Quantify by densitometry the intensity of the bands by using the ImageJ software.
3. Calculate the relative band intensity as the ratio with respect to the reference value (*see* **Note 19**).
4. Plot the relative band intensity against the time-incubation periods.
5. The slope of the curve reflects the translation rate of the protein of interest. Figure 3b shows the case example of GAPDH protein analysis (*see* **Note 18**).

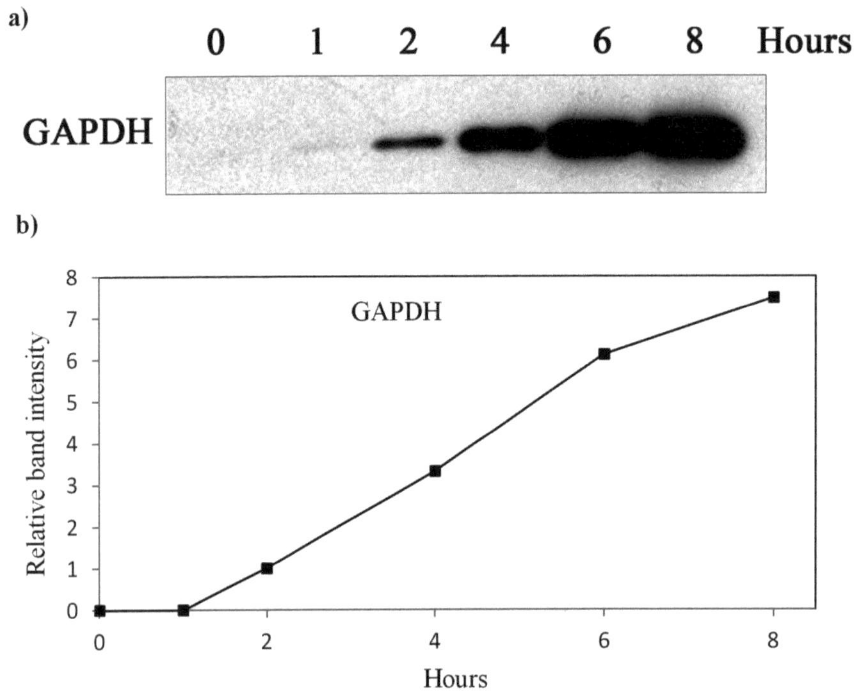

Fig. 3 Estimation of GAPDH protein synthesis rate can be calculated from the slope of the plot of the relative band intensity against the time-incubation periods. The immunodetected bands used are shown in (**a**). The relative values, depicted in (**b**), were obtained using as the reference value the scanned signal of the first immunodetectable band of *lane 2* (*see* **Note 18**)

4 Notes

1. In the case example described in this chapter, six AHA time-incubation periods are used. In the case of adherent cultures of HeLa cells, the growth area (9.5 cm^2/sample) provides suitable quantity of protein to perform the click-it assays.
2. To prepare the DMEM cell culture solutions, mix 50 mL of 100% FBS (Gibco®, Life Technologies) and 450 mL of the specific DMEM. Then add 5 mL of 10,000 U/mL Penicillin-Streptomycin (Gibco®, Life Technologies). Store at 4 °C.
3. It is useful to prepare the following stock solutions in order to facilitate the elaboration of all the buffers and reagents depicted in this chapter. Unless otherwise indicated all the solutions should be autoclaved for 20 min at 120 °C and 1 atm of pressure and stored at room temperature:
   (a) 1 M Tris–HCl pH 6.8, 7.6, 8.0, or 8.8 (500 mL): weigh 60.57 g of Tris base (MW = 121.14) and mix with 250 mL of water. Adjust the pH to 6.8, 7.6, 8.0, or 8.8 with HCl and bring the volume to 500 mL with water.

(b) 10% SDS (500 mL): weigh 50 g of SDS and dissolve it in water up to 500 mL.

(c) 10% APS (10 mL): weigh 1 g of APS and dissolve it in water up to 10 mL. Do not autoclave. Store at −20 °C.

(d) 5 M NaCl (500 mL): weigh 146.10 g of NaCl (MW = 58.44) and dissolve it in water up to 500 mL.

(e) 10% DOC (100 mL): weigh 10 g of DOC and dissolve it in water up to 100 mL. Do not autoclave.

4. In this case example 1.0 mm integrated spacer plates were used to pour 10 mL of resolving gel and 5 mL of stacking gel mixes to prepare two SDS-PAGEs gels. Scale the volumes to perform more than two SDS-PAGEs gels or to use different spacer plates. Add the APS just before pipetting the liquid mixes inside the plates in order to start polymerization of the gels. Immediately pour the resolving gel mix (4.70 mL per gel) and cover with a layer of 2-butanol until polymerization is completed. Next, eliminate the 2-butanol by decantation and add the stacking gel mix (freshly prepared) up to the upper limit of the plates. Immediately place the comb to set the wells.

5. The correct percentage of acrylamide–bis-acrylamide depends on the molecular weight of the proteins of interest.

6. The acrylamide–bis-acrylamide and the TEMED solutions should be stored at 4 °C and −20 °C respectively.

7. For each gel to transfer, prepare a PVDF membrane section of 7 × 9 cm.

8. The time-incubation periods with AHA depend on the protein turnover under study. This should be checked empirically in order to optimize the technique. As an approximation to address protein turnover under standard growth conditions in HeLa cells, the tool PepTracker (http://www.peptracker.com/epd/search/) can be helpful.

9. The maximum quantity of protein used is restricted to the sample with less protein concentration. This limiting amount of total protein has to be equal for every sample. In the case example, the sample of 1 h of time-incubation period with AHA (2.03 mg/mL) is the limiting one.

10. In this case example, the maximum quantity of protein of the limiting sample, which has to be used for the click-it reactions, is 50 μL × 2.03 μg/μL = 101.5 μg of total protein. Next, calculate the volume for 101.5 μg of total protein to add of each sample/time-incubation period for the click-it reactions. For example, for the sample of 4 h of time-incubation period with AHA (2.73 mg/mL), add 37.18 μL of protein extract.

11. The use of 15 mL tube centrifuge adapters is recommended.

12. The sample will still have bright orange color.

13. The elution of the biotinylated proteins will start in this step. Take care to avoid losing the eluted proteins.
14. The prestained MW marker helps visualization of the SDS-PAGE progression and the transfer onto PVDF membrane.
15. The blocking or the antibody-incubation conditions depends on the antibody used. Check them empirically before use. In this case example, the PVDF membrane was blocked for 1 h with 1× BB. Then the membrane was first incubated for 1 h with the anti-GAPDH primary antibody diluted 1:5000 in 1× BB and, after washing, with the anti-mouse IgG-HRP secondary antibody diluted 1:50,000 in BB for 1 h.
16. 500 μL of the detection reagent mix is enough to cover a 7×9 cm PVDF membrane.
17. Try to avoid the appearance of bubbles between the plastic and the membrane.
18. The exposure time depends on the protein of interest or the antibody used. Optimize the exposure time empirically.
19. The reference value is the intensity data of the first detectable scanned band of the time-incubation periods. The slope of the curve reflects the protein translation rate and can be used to compare translation rates of any protein of interest under different conditions such growth conditions, pharmacological treatments, genetic alterations, etc.

Acknowledgment

The authors would like to acknowledge networking support by the Proteostasis COST Action (BM1307).

References

1. Man O, Pilpel Y (2007) Differential translation efficiency of orthologous genes is involved in phenotypic divergence of yeast species. Nat Genet 39(3):415–421. doi:10.1038/ng1967
2. Nissley DA, O'Brien EP (2014) Timing is everything: unifying codon translation rates and nascent proteome behavior. J Am Chem Soc 136(52):17892–17898. doi:10.1021/ja510082j
3. Boisvert F-M, Ahmad Y, Gierliński M, Charrière F, Lamont D, Scott M, Barton G, Lamond AI (2012) A quantitative spatial proteomics analysis of proteome turnover in human cells. Mol Cell Proteomics 11(3). doi: 10.1074/mcp.M111.011429
4. Best MD (2009) Click chemistry and bioorthogonal reactions: unprecedented selectivity in the labeling of biological molecules. Biochemistry 48(28):6571–6584. doi:10.1021/bi9007726
5. Breinbauer R, Köhn M (2003) Azide-alkyne coupling: a powerful reaction for bioconjugate chemistry. Chembiochem 4(11):1147–1149. doi:10.1002/cbic.200300705
6. Prescher JA, Bertozzi CR (2005) Chemistry in living systems. Nat Chem Biol 1(1):13–21. doi:10.1038/nchembio0605-13
7. Dieterich DC, Link AJ, Graumann J, Tirrell DA, Schuman EM (2006) Selective identification of newly synthesized proteins in mammalian cells using bioorthogonal noncanonical amino acid tagging (BONCAT). Proc Natl Acad Sci 103(25):9482–9487. doi:10.1073/pnas.0601637103

Chapter 25

A Simple Protocol for High Efficiency Protein Isolation After RNA Isolation from Mouse Thyroid and Other Very Small Tissue Samples

Panos G. Ziros, Dionysios V. Chartoumpekis, and Gerasimos P. Sykiotis

Abstract

As a dedicated hormone-secreting organ, the thyroid gland possesses a complement of proteostatic systems, including antioxidant, unfolded protein, and autophagic responses. The vast majority of animal investigations of thyroid physiology and, more recently, proteostasis, have utilized as model the rat, rather than the mouse. This is due to the very small size of the thyroid gland in the latter, with a total weight of ~2 mg (~1 mg per thyroid lobe). However, this strategy has limited the utilization of genetic approaches, such as taking advantage of the various transgenic and knockout mouse models. Here, we describe a simple and highly efficient protocol for the simultaneous isolation of mRNA, micro-RNA and 150–200 μg of protein from as little as 1 mg of mouse thyroid tissue, the average weight of one of the two thyroid lobes, thus preserving the other lobe for immunohistochemical or other analyses. While our workflow is similar to other protocols published in the literature and/or proposed by commercial reagent providers, we have introduced a key modification that addresses efficiently the most challenging step of the protein isolation process: the solubilization of the protein pellet after RNA extraction and protein precipitation. We demonstrate the feasibility of our approach and its utility for downstream analyses (including Western blotting) that facilitate the comparative study of proteostatic pathways in the mouse thyroid. We have also successfully applied this protocol on samples from mouse liver, brown and white adipose tissue, as well as from rodent cell lines.

Key words Simultaneous, Isolation, RNA and protein, Micro-RNA, Guanidinium, TRIzol, QIAzol, TRI reagent, Thyroid, Proteostasis

1 Introduction

The guanidinium thiocyanate–phenol–chloroform (GTPC) extraction method, also called "TRIzol" extraction, introduced by Chomczynski and Sacchi in 1987 [1], is widely used in molecular biology to isolate both nucleic acids and proteins. This method is based on the different solubilities of RNA, DNA, and protein molecules in water and organic solvents such as phenol and chloroform. It also exploits the ability of the chaotropic agent guanidinium thiocyanate to denature all proteins, including those that degrade

Rune Matthiesen (ed.), *Proteostasis: Methods and Protocols*, Methods in Molecular Biology, vol. 1449,
DOI 10.1007/978-1-4939-3756-1_25,

nucleic acids (DNAses and RNAses). When a cell or tissue lysate that has been incubated with these reagents is subjected to centrifugation, the solution separates into a lower organic phase that contains DNA and proteins, and an upper aqueous phase that contains RNA. The RNA is recovered from the aqueous phase with isopropanol precipitation (after transfer to a different tube); the DNA and proteins can also be recovered from the organic phase following a different procedure (discussed in detail below).

Homemade and commercially available GTPC reagents (such as TRIzol, QIAzol, TRI Reagent, etc.) are widely used in the literature for the extraction of RNA. The simultaneous isolation of protein from the same sample is advantageous for several reasons: (1) it saves time; (2) it permits the reliable assessment of and correlation between coordinated changes in gene and protein expression levels; and (3) it is especially critical when the quantity of the starting biological sample is limited (such as for small, precious, or rare samples). Nevertheless, compared to the huge use of GTPC to isolate RNA in the literature, there are relatively very few reports with simultaneous protein isolation. This is due to the fact that when the standard recommended protocol for protein isolation is used, it is very difficult to dissolve the protein pellet in the final step, leading to experimental failure. Indeed, there are some reports in the literature that have tried to overcome this problem by using different approaches. For example, one approach replaced protein precipitation with dialysis of phenol–ethanol supernatants against a 100× volume of 1 % SDS; this was repeated three times, and the extracted protein was subsequently concentrated using commercially available columns [2]; this is a laborious and complicated technique. Another approach is to attempt the solubilization of TRIzol-extracted proteins not with the standard 1 % SDS solution but with alternative solutions. Different solutions have been reported to improve solubilization efficiency, such as 9.5 M Urea and 2% CHAPS ([3-[(3-cholamidopropyl)-dimethylammonio]propanesulfonate]) (UREA-CHAPS) [3]; diethylamine [4]; or Urea-SDS solubilization and sonication [5]. Another proposed method is the precipitation of proteins from the phenol–ethanol phase by an ethanol–bromochloropropane–water solution, followed by solubilization of the protein pellet with 4 % SDS and heating at 50 °C [6].

Here, we demonstrate that a simple modification of the standard protocol for protein isolation from the phenol–ethanol supernatant after DNA precipitation facilitates the easy and complete solubilization of the protein pellet in the final step without the need for additional treatments or special reagents. We have found that modification of the steps where the protein pellet is washed with 0.3 M Guanidine hydrochloride in 95 % ethanol in order to remove the remaining phenol from the protein has a dramatic impact on the solubility of the protein in the final step. Specifically, we first completely dissolve the protein pellet in 7 M guanidine hydrochloride solution. Proteins can be efficiently precipitated from a guanidine hydrochloride solution by addition of ethanol [7]; we thus add at least 9 volumes of 100% ethanol. We believe that this modification improves significantly the removal of phenol from the protein, which in turn increases dramatically protein solubilization in the final step.

We demonstrate the feasibility and simplicity of our approach and its utility for downstream analyses that facilitate the comparative study of proteostatic pathways in the mouse thyroid (~1 mg of starting tissue material). Western blotting analyses show that the protocol can isolate detectable amounts of proteins that span a broad range of molecular weights and are present in varying abundance in the starting material. We have also successfully applied this protocol on samples from mouse liver, brown and white adipose tissue, as well as from rodent cell lines.

2 Materials

2.1 Required Chemicals, Commercial Reagents, Kits, and Equipment

1. Guanidine hydrochloride (Sigma).
2. 1-Bromo-3-chloropropane (1-BCP, Sigma).
3. Isopropyl alcohol.
4. 100% ethanol (absolute ethanol).
5. Urea.
6. 20% SDS sodium dodecyl sulfate solution (Applichem).
7. 1 M Tris-Buffer pH 8.0 (Applichem).
8. TRIzol reagent (Life Technologies).
9. RNeasy Mini Kit (QIAGEN).
10. Tissue Grinder homogenizer.
11. Benchtop centrifuge.
12. Swing rotor centrifuge.
13. Vortex device.
14. Water bath or hot plate and magnetic stirrer.
15. 0.22 μm syringe filters.
16. Ultrapure water.
17. 100× Protease inhibitor cocktail (any commercial source).

2.2 Solutions to be Prepared Before Starting the Procedure

1. 7 M Guanidine hydrochloride solution (GndCl) solutions: Weigh 66.87 g GndCl in a plastic or glass beaker, add ultrapure water to a volume of 90 ml, and dissolve by stirring (*see* **Note 1**). Once dissolved, bring the volume to 100 ml with water and filter the solution through a 0.22 μm syringe filter.
2. Protein solubilization solution: 8 M Urea, 40 mM Tris, pH 8, 1% SDS. Because urea solutions are unstable, always make

fresh small quantities. To make 2 ml of this solution, place 961 mg Urea in a 2 ml tube, add water to about 1.8 ml and dissolve by vortexing, then add 80 μl of 1 M Tris pH 8, and 100 μl of 20% SDS solution. Finally, bring the volume to 2 ml with water. Just before solubilizing the protein pellet, add to the solution 20 μl of the 100× Protease inhibitor cocktail.

3 Methods

3.1 RNA Isolation

1. In fume hood, add 1100 μl TRIzol reagent to a 15 ml Falcon tube (*see* **Notes 2–4**). Place the thyroid tissue inside the tube and homogenize immediately until the sample is homogeneous (requires about 30 s using the QIAGEN TissueRuptor with disposable probes at maximum speed; *see* **Note 5**).
2. Incubate for 5 min at room temperature, and then centrifuge briefly the samples for 30 s on a bench top centrifuge at maximum speed.
3. Transfer the homogenized samples to 1.5 ml tubes and store them at −70 °C for least 30 min.
4. Move the samples directly from −70 °C to a water bath of 60 °C for 5 min, and then place them on ice for another 5 min (*see* **Note 6**).
5. Add 100 μl of 1-BCP (*see* **Note 7**) to the tubes containing the samples and shake vigorously by hand or vortexing for 30 s.
6. Leave the samples on the bench top at room temperature for 5 min.
7. Centrifuge for 15 min at 12,000×*g* at 4 °C. After centrifugation the mixture separates into a lower phenol–chloroform phase of red color, an interphase, and a colorless upper aqueous phase. The upper aqueous phase that contains RNA comprises ~40–50% of the total volume.
8. Transfer the upper aqueous phase (400–450 μl) to a new collection tube (*see* **Note 8**). Add 1.5 volume of 100% ethanol and mix thoroughly by pipetting. Save the tube containing the interphase and the organic phenol/1-BCP interphase for the protein isolation procedure (*see* **Note 9**).
9. Transfer up to 700 μl of the sample, including any precipitate that may have formed, to an RNeasy spin column placed in a 2 ml collection tube. Close the lid gently and centrifuge for 15 s at ≥8000×*g* (≥10,000 rpm). Discard the flow-through. Repeat using the same column and the remainder of the sample. Discard the flow-through.
10. Add 700 μl Buffer RW1 to the RNeasy spin column. Close the lid gently and centrifuge for 15 s at ≥8000×*g* (≥10,000 rpm) to wash the spin column membrane. Discard the flow-through.

11. Add 500 μl Buffer RPE to the RNeasy spin column. Close the lid gently and centrifuge for 15 s at ≥8000 × *g* (≥10,000 rpm) to wash the spin column membrane. Discard the flow-through.
12. Add 500 μl Buffer RPE to the RNeasy spin column. Close the lid gently and centrifuge for 2 min at ≥8000 × *g* (≥10,000 rpm) to wash the spin column membrane.
13. After centrifugation, carefully remove the RNeasy spin column from the collection tube so that the column does not contact the flow-through. Place the RNeasy Mini spin column into a new 2 ml collection tube. Centrifuge at full speed for 2 min to dry the RNeasy Mini spin column membrane.
14. Place the RNeasy spin column in a new 1.5 ml collection tube. Add 30–50 μl RNase-free water directly to the spin column membrane. Close the lid gently and centrifuge for 1 min at 10,000 × *g* to elute the RNA (*see* **Note 10**).
15. The RNA can be stored and used in downstream applications for mRNA and microRNA analyses as usual.

3.2 Protein Isolation

1. Proteins are isolated from the interphase and organic phenol/1-BCP interphase that was saved in **step 9** of the RNA isolation procedure.
2. Centrifuge the samples briefly (1 min at 10,000 × *g*) and remove any remaining aqueous phase overlying the interphase.
3. Add 0.3 ml of 100 % ethanol per 1 ml TRIzol Reagent used for the initial homogenization to precipitate the DNA. Cap the tube and invert the sample several times to mix. Incubate samples for 5 min at room temperature.
4. Centrifuge the samples for 5 min at 2000 × *g* at 4 °C to precipitate the DNA into a pellet (*see* **Note 11**).
5. Taking care not to dislodge the DNA pellet, transfer the phenol–ethanol supernatant to a 15 ml tube. Add 1.5 volume of isopropanol to the phenol–ethanol supernatant. Incubate the samples for 10–20 min at room temperature until a visible precipitate is formed (*see* **Note 12**). This is the precipitated protein.
6. Centrifuge the samples for 10 min at 4000 × *g* at 4 °C in a swing rotor centrifuge to pellet the protein. Remove and discard the supernatant (*see* **Note 13**).
7. Resuspend the protein pellet in 100 μl 7 M GndCl solution (*see* **Note 14**).
8. Transfer the protein solution to 2 ml tubes. Add 1900 μl 100 % ethanol and vortex for 10 s (*see* **Note 15**). Incubate samples for at least 30 min at −70 °C; a visible protein precipitate will be formed (*see* **Note 16**).

9. Centrifuge the samples for 5 min at full speed at 4 °C to pellet the protein. Remove and discard the supernatant.
10. Resuspend the protein pellet in 100 μl 7 M GndCl solution (*optional*, *see* **Note 17**). Add 1900 μl 100% ethanol and vortex for 10 s. Incubate the samples for at least 30 min at −70 °C; a visible protein precipitate is formed.
11. Centrifuge the samples at full speed for 5 min at 4 °C to pellet the protein. Remove and discard the supernatant.
12. Add 2 ml 100% ethanol and vortex three times for 10 s each time over a period of 10 min at room temperature.
13. Centrifuge the samples for 10 min at full speed at 4 °C to pellet the protein. Discard the ethanol supernatant and air dry the protein pellet for about 10 min (*see* **Note 18**).
14. Dissolve the protein pellet in 100–200 μl of 8 M Urea, 40 mM Tris pH 8, 1% SDS, 1× protease inhibitors solution (*see* **Note 19**).
15. The protein can be stored and used in downstream applications as usual. We routinely test visually our extracts on a protein gel (Fig. 1), and use them for Western immunoblotting (Fig. 2) (*see* **Notes 20–23**).

4 Notes

1. Warming up the GndCl solution to 40 °C will help to dissolve faster the guanidine hydrochloride. Use a water bath or a hot plate and a magnetic stirrer.
2. TRIzol Reagent contains phenol (toxic and corrosive) and guanidine isothiocyanate (an irritant), and may be a health hazard if not handled properly. The manual and protocol of TRIzol Reagent recommend to always work with TRIzol Reagent in a fume hood, and to always wear a lab coat, gloves, and safety glasses.
3. During the disruption and homogenization of the starting material, we always lose about 5–10% of the initial TRIzol volume. That is why we use 10% more than the recommended volume (i.e., 1100 μl instead of 1000 μl).
4. The volume of TRIzol reagent depends on the quantity and the type of starting material. For thyroid tissue, which in mice weighs about 0.5–1 mg per lobe, 1 ml is more than enough. For liver, we use 50 mg of tissue when the tissue is fresh or stored at −70 °C, and no more than 10 mg when it is preserved in RNAlater RNA Stabilization Reagent. For adipose tissue, we use 100–200 mg of fresh tissue, or about 50 mg of tissue preserved in RNAlater. Tissues stored in RNAlater lose most of their water content, and therefore shrink and weigh less than the corresponding amount of fresh tissue.

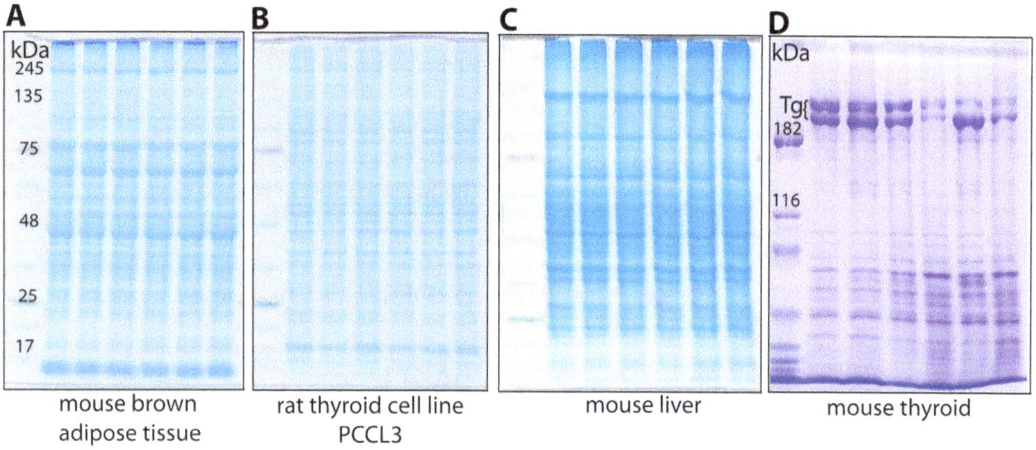

Fig. 1 SDS-PAGE gel analysis of proteins extracted by our protocol from thyroid, liver, and brown adipose tissue, as well as from the rat thyroid cell line PCCL3 grown in 12-well culture dishes. Various genotypes and/or treatment conditions are used (not indicated). Proteins from liver, brown adipose tissue, and PCCL3 cells were separated on a 10 % Bis–Tris gel (MOPS buffer) that was then stained with Blue Silver G-250 [8]. Proteins from thyroid tissue were separated on a 4–10 % gradient Tris-Glycine gel that was then stained with Coomassie Brilliant Blue R-250

5. After storage in RNAlater, tissues become harder than fresh or thawed tissues. This must be taken into account during the disruption and homogenization of these tissues, optimizing the duration and magnitude of the disruption and homogenization method.
6. We have found that performing a freeze-thaw treatment of the samples increases RNA yields, possibly due to the complete dissociation of the nucleoprotein complex formed after lysis with TRIzol. In particular, for cells in culture dishes whereas a viscous material is formed after addition of TRIzol, this freeze-thaw cycle eliminates the need to homogenize the samples with the TissueRuptor.
7. Chloroform is commonly used instead of 1-BCP. Since chloroform is a neurotoxicant, an endocrine disruptor, and possibly also a carcinogen, substitution with the less hazardous and less volatile 1-BCP is advantageous, resulting in safer working conditions; see: http://www.subsport.eu/case-stories/071-en?lang=en. 1-BCP can be used as a fully functional substitute for chloroform without any changes to the experimental protocol [9]. Moreover, Ambion supports that use of 1-BCP results in better phase separation, and thus better purification of the RNA, DNA, and protein fractions; see: https://www.lifetechnologies.com/ch/en/home/references/ambion-tech-support/rna-isolation/tech-notes/rna--dna--and-protein-from-a-single-sample.html.

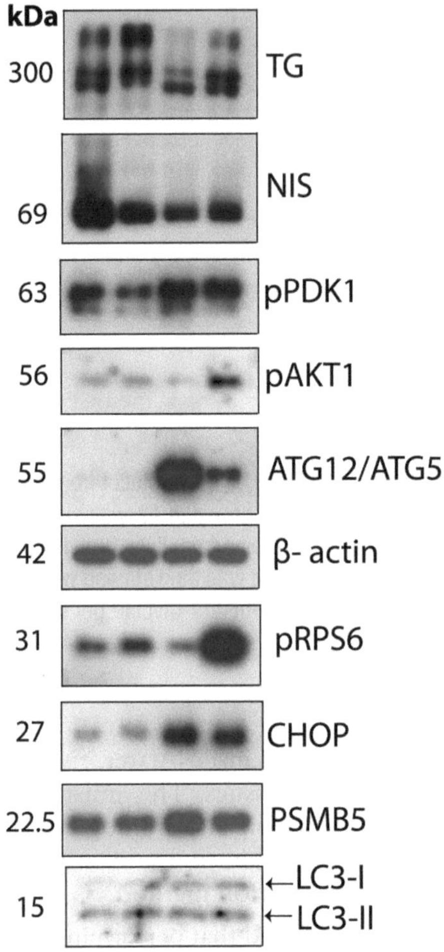

Fig. 2 Western blot analysis of thyroid-specific or proteostasis-related proteins extracted from mouse thyroid tissues by our protocol. Various genotypes and/or treatment conditions are used (not indicated). Proteins of a wide range of molecular weights are successfully detected, including site-specific phosphorylations. Five to 10 μg of extracted proteins from mouse thyroid tissues were separated under reducing conditions on 7.5 %, 10 % or 12 % Bis–Tris gels and immunoblotted with the indicated antibodies: anti-thyroglobulin anti-TG (A0251, DAKO); anti-sodium-iodide symporter antibody anti-NIS (a kind gift by Prof. Nancy Carrasco, [10]). The following antibodies were all from Cell Signaling Technology: anti-autophagy-related 12 (anti-ATG12, #2011); anti-C/EBP-homologous protein (anti-CHOP, #2895); anti-Phospho-S6 Ribosomal Protein (anti-pS6RP, #8207); anti-LC3I/II (#12741); anti-Phospho-PDK1(Ser241) (#3438); anti-Phospho-Akt (Ser473) (#4060). The anti-beta-actin (ab6276) and anti-Proteasome Subunit Beta type-5 (anti-PSMB5, ab3330) antibodies were from Abcam

8. At this step, in order to ensure the quality of isolated RNA, it is critical to carefully take the upper aqueous phase without disrupting the interphase. For this reason, we always leave behind 50–100 μl of aqueous phase. For thyroid tissue, where the starting material is very limited, and the expected RNA quantity is accordingly small, we do a second extraction: we add 200 μl of RNAse-free water to the remaining sample; vortex; centrifuge; take 200 μl from the aqueous phase and combine with the first one.

9. Store the samples dedicated for protein isolation at 4 °C, if you plan to make the extraction on the same day. Otherwise, place them at −20 °C or at −70 °C for long-term storage.
10. We routinely quantify the RNA on a NanoDrop spectrophotometer. From each mouse thyroid tissue we obtain about 1 μg of total RNA.
11. When the starting material is limited, as is the case with thyroid tissue, most of the time the DNA pellet will not be visible. For this reason, we always leave behind a small portion of phenol–ethanol supernatant (about 50 μl) to ensure that we do not carry over DNA into the protein samples. Moreover, if isolation of DNA is not needed, then it is better in **step 4** to centrifuge the samples at full speed for 2 min rather than at 2000 × *g* for 5 min. Carryover of DNA into the protein solution will lead to a solution with some very viscous and sticky parts in the final step. If this is observed, then sonicate the samples briefly (5–10 s) to shear the DNA.
12. If the protein concentration is high, the precipitate will form within 1–2 min. But if the starting tissue material is too little (<1 mg of tissue), or if protein is isolated from small numbers of cells (e.g., cells grown in 24-well culture plates), then place the samples at −20 °C or −70 °C for 1 h to facilitate protein precipitation.
13. After removal of the supernatant, let the tubes drain in an upside down position on a clean piece of absorbent paper for 5 min to completely remove the phenol-isopropanol supernatant.
14. In this step, protein solubilization usually takes 10–20 min. Especially if the protein pellet is big, the volume of GndCl should be increased such that the protein pellet is resuspended in 150–200 μl. Leave the samples with the GndCl solution at room temperature for 10 min, and then solubilize the pellet by pipetting. Protein degradation under these conditions is not a concern, because GndCl is a potent denaturant, and therefore proteins in the GndCl solution are fully protected from degradation.
15. The minimal volume of ethanol that needs to be added is nine times the volume of the GndCl solution. Thus, if one wants to keep working with 2 ml tubes for practical reasons, then the maximum volume of GndCl solution in which the pellet can be resuspended is 200 μl.
16. Precipitation of proteins will take place also at room temperature or at −4 °C, but we prefer to place the samples at −70 °C in order to ensure the quantitative (i.e., maximal) precipitation of the proteins from GndCl solution, especially for samples with low protein concentration.
17. Perform these extra steps (10, 11) only when the protein pellet has a red tint (indicative of residual phenol), or when it is big (suggesting it may not have been thoroughly washed). Because the

protein pellet is completely dissolved in the GndCl solution (**step 7**), the first precipitation (**steps 5** and **6**) and the ethanol washes (**steps 8** and **12**) result in complete removal of the phenol.

18. After the final ethanol wash and centrifugation, the protein pellet detaches from the tube quite easily. Therefore, in order to not lose the pellet, drain off carefully the ethanol supernatant immediately after centrifugation, and then place the tube upside down on absorbent paper for 10 min to drain out the remaining ethanol. Then place the tubes in normal position with the lids open to evaporate the traces of ethanol remaining inside the protein. Depending upon the size of the pellet, this usually needs about 5–10 min. It is very important not to allow the protein pellet to dry completely. Stop the drying when the tube no longer smells of ethanol, and the protein pellet still has a white milky appearance. Waiting too long will result in a protein pellet that has dried completely, has become transparent, and is very difficult to dissolve.
19. The protein pellet is easily solubilized in this buffer. The pellet can also be dissolved in any other buffer that contains at least 1 % SDS.
20. We have applied this protocol to isolate proteins from thyroid, liver, brown and white fat adipose tissue, as well as from cell lines. Simultaneous isolation of both RNA and protein from a single biological sample permits the reliable assessment of coordinated changes in gene and protein expression levels.
21. We measure the protein concentration using the Thermo Scientific Pierce BCA Protein Assay Kit that is compatible with the components of this buffer. For each thyroid tissue we recover about 150–200 μg of protein. We run no more than 5–10 μg of protein per lane for SDS-PAGE and Western immunoblotting, and we use high sensitive ECL reagents like the Amersham ECL Prime Western Blotting Detection Reagent or the Advansta WesternBright Quantum kit.
22. Thyroglobulin, the main protein of the thyroid tissue, accounts for approximately half of the protein content of the thyroid gland. For this reason, when Western blots are performed with thyroid samples coming from different experimental conditions, equal protein loading is not always consistent with the results obtained from the loading controls such as beta-actin, tubulin, etc. This discrepancy is observed when the experimental conditions lead to a significant change in the protein abundance of thyroglobulin.
23. When the total number of samples to be processed does not exceed 10–20, both the RNA and the protein protocols can be completed in a single day. We usually store the phenol/1-BCP supernatant at −20 °C, and we perform the protein extraction the following day. The protein isolation procedure is very

flexible regarding time lines and is highly amenable to interruption of the protocol; it can be stopped at any step, storing the samples at −20 °C or −70 °C to continue on the same day or a subsequent day.

Acknowledgments

This work was supported by Swiss National Science Foundation Project 31003A_153062 and by Swiss Society for Endocrinology and Diabetology 2014 Young Independent Investigator Award, both to GPS. The authors would like to acknowledge networking support by the Proteostasis COST Action (BM1307).

References

1. Chomczynski P, Sacchi N (1987) Single-step method of RNA isolation by acid guanidinium thiocyanate-phenol-chloroform extraction. Anal Biochem 162(1):156–159. doi:10.1006/abio.1987.9999
2. Hummon AB, Lim SR, Difilippantonio MJ, Ried T (2007) Isolation and solubilization of proteins after TRIzol extraction of RNA and DNA from patient material following prolonged storage. Biotechniques 42(4):467–470, 472
3. Man TK, Li Y, Dang TA, Shen J, Perlaky L, Lau CC (2006) Optimising the use of TRIzol-extracted proteins in surface enhanced laser desorption/ionization (SELDI) analysis. Proc Natl Acad Sci U S A 4:3. doi:10.1186/1477-5956-4-3
4. Nolan RL, Teller JK (2006) Diethylamine extraction of proteins and peptides isolated with a mono-phasic solution of phenol and guanidine isothiocyanate. J Biochem Biophys Methods 68(2):127–131. doi:10.1016/j.jbbm.2006.04.002
5. Simoes AE, Pereira DM, Amaral JD, Nunes AF, Gomes SE, Rodrigues PM, Lo AC, D'Hooge R, Steer CJ, Thibodeau SN, Borralho PM, Rodrigues CM (2013) Efficient recovery of proteins from multiple source samples after TRIzol((R)) or TRIzol((R))LS RNA extraction and long-term storage. BMC Genomics 14:181. doi:10.1186/1471-2164-14-181
6. Chey S, Claus C, Liebert UG (2011) Improved method for simultaneous isolation of proteins and nucleic acids. Anal Biochem 411(1):164–166. doi:10.1016/j.ab.2010.11.020
7. Pepinsky RB (1991) Selective precipitation of proteins from guanidine hydrochloride-containing solutions with ethanol. Anal Biochem 195(1):177–181
8. Candiano G, Bruschi M, Musante L, Santucci L, Ghiggeri GM, Carnemolla B, Orecchia P, Zardi L, Righetti PG (2004) Blue silver: a very sensitive colloidal Coomassie G-250 staining for proteome analysis. Electrophoresis 25(9):1327–1333. doi:10.1002/elps.200305844
9. Chomczynski P, Mackey K (1995) Substitution of chloroform by Bromo-chloropropane in the single-step method of RNA isolation. Anal Biochem 225:163–164
10. Levy O, Dai G, Riedel C, Ginter CS, Paul EM, Lebowitz AN, Carrasco N (1997) Characterization of the thyroid Na+/I- symporter with an anti-COOH terminus antibody. Proc Natl Acad Sci U S A 94(11):5568–5573

Chapter 26

Monitoring Target Engagement of Deubiquitylating Enzymes Using Activity Probes: Past, Present, and Future

Jeanine Harrigan and Xavier Jacq

Abstract

Deubiquitylating enzymes or DUBs are a class of enzymes that selectively remove the polypeptide posttranslational modification ubiquitin from a number of substrates. Approximately 100 DUBs exist in human cells and are involved in key regulatory cellular processes, which drive many disease states, making them attractive therapeutic targets. Several aspects of DUB biology have been studied through genetic knock-out or knock-down, genomic, or proteomic studies. However, investigation of enzyme activation and regulation requires additional tools to monitor cellular and physiological dynamics. A comparison between genetic ablation and dominant-negative target validation with pharmacological inhibition often leads to striking discrepancies. Activity probes have been used to profile classes of enzymes, including DUBs, and allow functional and dynamic properties to be assigned to individual proteins. The ability to directly monitor DUB activity within a native biological system is essential for understanding the physiological and pathological role of individual DUBs. We will discuss the evolution of DUB activity probes, from in vitro assay development to their use in monitoring DUB activity in cells and in animal tissues, as well as recent progress and prospects for assessing DUB inhibition in vivo.

Key words ABP, Activity-based probe, Acyloxymethyl ketone, DUB, Deubiquitylating enzyme, Hemagglutinin, JAMM, JAB1/Mov34/Mpr1 Pad1 N-terminal+protease, MJD, MACHADO-Josephin domain proteas, MP, Mpr1/Pad1 N-terminal+, OTU, Ovarian tumor protease, PA, Propargy, SENP, Sentrin-specific protease, Ubl, Ubiquitin-like protein, UCH, Ubiquitin C-terminal hydrolase, USP, Ubiquitin-specific protease, VMS, Vinyl methyl sulfone, VME, Vinyl methyl ester

1 Ubiquitin–Proteasome System

Protein homeostasis is essential for most cellular processes. The ubiquitin–proteasome system is responsible for much of the regulated proteolysis in the cell, as well as many other regulatory processes such as transcriptional regulation, DNA damage, quality control, trafficking, inflammation, and autophagy. Ubiquitin is a small 76-amino acid protein that can be reversibly attached to protein substrates. Several ubiquitin-like proteins (Ubls) have also been identified including ISG15, NEDD8, and SUMO, which share a characteristic three-dimensional fold with ubiquitin but are

Rune Matthiesen (ed.), *Proteostasis: Methods and Protocols*, Methods in Molecular Biology, vol. 1449,
DOI 10.1007/978-1-4939-3756-1_26, © Springer Science+Business Media New York 2016

otherwise distinct. The ubiquitin–proteasome system has multiple essential biological roles, and thus its function and dysfunction, are important factors in various human diseases, including cancer, infection, inflammation, and neurodegeneration [1–4].

Ubiquitylation of substrate proteins first involves an ATP-dependant activation of the ubiquitin polypeptide by the activating enzyme E1. Activation involves covalent linkage between the carboxy terminus of ubiquitin and a cysteine residue present on the E1, forming a thioester bond. The activated ubiquitin is then transferred to an E2 ubiquitin-conjugating enzyme forming a thioester linkage. In the final step, an E3 ligase transfers the ubiquitin from the E2 to the substrate protein. The majority of E3 ligases are classified as RING finger E3s and act by bringing the substrate and E2 enzyme in close proximity. The RING finger E3s directly transfer ubiquitin from the E2 to the substrate. The HECT domain E3s act by forming an intermediate thioester linkage with ubiquitin before transfer to the substrate (reviewed in [5]). More recently, a third class of E3 ligases with an intermediate mechanism of action has been identified. The RING-in-between-RING (RBR) E3s are an unusual family of ubiquitin E3-ligases composed of a dozen proteins. Their activities are autoinhibited, causing a requirement for activation by protein–protein interactions or posttranslational modifications. They catalyze ubiquitin conjugation by a concerted RING/HECT-like mechanism in which the RING1 domain facilitates E2-discharge to directly form a thioester intermediate with a cysteine in RING2. This short-lived, HECT-like intermediate then modifies the target [6, 7].

Following monoubiquitylation of a substrate, the process can either stop, forming monoadducts of ubiquitin, or be repeated forming an elongated chain of ubiquitin residues. Polyubiquitin chains can be formed using the N-terminus (linear) or any of the seven internal lysine residues found in ubiquitin, and these various chain topologies lead to different functional outcomes. Of the most well studied linkages, K63-linked polyubiquitin chains are often involved in nonproteolytic signal transduction while K48-linked chains generally target substrates for proteasomal degradation. A number of additional linkages such as Met1, K6, K11, K27, K29, and K33 have been identified and their nondegradative cellular signaling roles are still subject to a number of investigations. The complexity of ubiquitin chain signaling is further enhanced by the existence of mixed-lineage chains [8, 9].

Proteins destined for degradation via the ubiquitin–proteasome system include proteins that are damaged, improperly folded, or that have short half-lives [10]. Proteins that have been appropriately polyubiquitylated are recognized and degraded by the 26S macromolecular proteasome complex [11]. The 26S complex consists of a 20S catalytic core particle that is capped at both ends by 19S regulatory particles. The 19S regulatory particle can be further subdivided into lid and base components. Following recruitment to the proteasome, polyubiquitylated proteins undergo deubiquitylation and unfolding. The removal of ubiquitin is accomplished by deubiquitylating enzymes (DUBs) associated with the 19S lid. Ubiquitin polypeptides that are removed from substrate proteins can be directly recycled by the cell. The 19S base component plays a key role in the unfolding of the substrate protein and delivery of the deubiquitylated, unfolded protein into the 20S catalytic core particle. The 20S consists of four layers of ring-like structures [12]. The outer rings are composed of seven α subunits with the inner rings composed of seven β subunits. The β1 subunits exhibit caspase-like activity, the β2 subunits trypsin-like activity, and β5 subunits chymotrypsin-like activity, collectively degrading proteins into short oligopeptides as well as recycling amino acids [5, 13].

2 DUBs

Ubiquitin is covalently linked to many cellular proteins and regulates their activity, stability, localization, or interactions. Ubiquitylation is a reversible process carried out by the opposing activities of ubiquitin ligases and DUBs. The human genome encodes approximately 100 DUBs [14–16]. Of the five families of DUBs, four (UCH, USP, MJD, and OTU) belong to the cysteine peptidase class, while one (JAMM) belongs to the metallopeptidase class. As DUBs have been shown to play critical roles in many pathological processes, particularly cancer, infectious disease, and neurodegeneration, they have begun to attract significant attention from the pharmaceutical industry [17–20]. Unlike most posttranslational modifications, ubiquitin is able to form polymeric chains [21]: the ubiquitin linkage in the chain as well as the length of the chain will impact on the fate of the protein modified by the polymer of ubiquitin [9, 22].

Pharmacological modulation of DUBs using a multitude of approaches in the last decade has seen limited success to date; however, recent progress is beginning to identify DUB inhibitors with the potential for drug development [23–27]. A number of conceptual and technological obstacles need to be overcome in order to progress genuine DUB therapies. A major challenge in characterization of DUB inhibitors is the development of high throughput assays monitoring “on-target” inhibition in cells and in vivo. Monitoring DUB target engagement by small molecule inhibitors in vivo has a number of implications. Firstly, as a biomarker readout of inhibition and for understanding the physiological implication of inhibiting a class of enzymes for which there is usually no known unique ubiquitylated substrate. Secondly, for assessment of the selectivity of compounds as well as understanding the mechanism of action of the inhibition, including duration, reversibility, and pharmacodynamic parameters.

3 Activity Probes

Activity-based probes (ABPs) rely on the design of chemical warheads which selectively react with the active site of an enzyme. ABPs are usually composed of a reactive electrophile, to covalently modify an active-site residue, and a reporter group to allow detection of the labeled enzyme [28], *see* Fig. 1a, b. Activity probes have been designed for a number of enzyme classes such as serine hydrolases [29], metalloproteases [30, 31], proteasomes [32], and oxidoreductases [33]. Epitope-tagged ubiquitin and ubiquitin-like derivatives have been utilized in a variety of assays to identify or monitor active DUBs in biological samples [34, 35] (Fig. 1c). Ubiquitin ABPs have been instrumental in the identification of a number of new DUBs [36] including a novel class of DUBs: OTUs [37]. Unlike other proteolytic enzymes, for optimal recognition, DUBs require not only an electrophilic trap but also a very large portion of ubiquitin or chains of ubiquitin for binding and recognition in the enzyme active site: truncated portions of ubiquitin are usually not sufficient to trap DUBs. In addition, the isopeptide nature of the covalent linkage of ubiquitin to the target protein

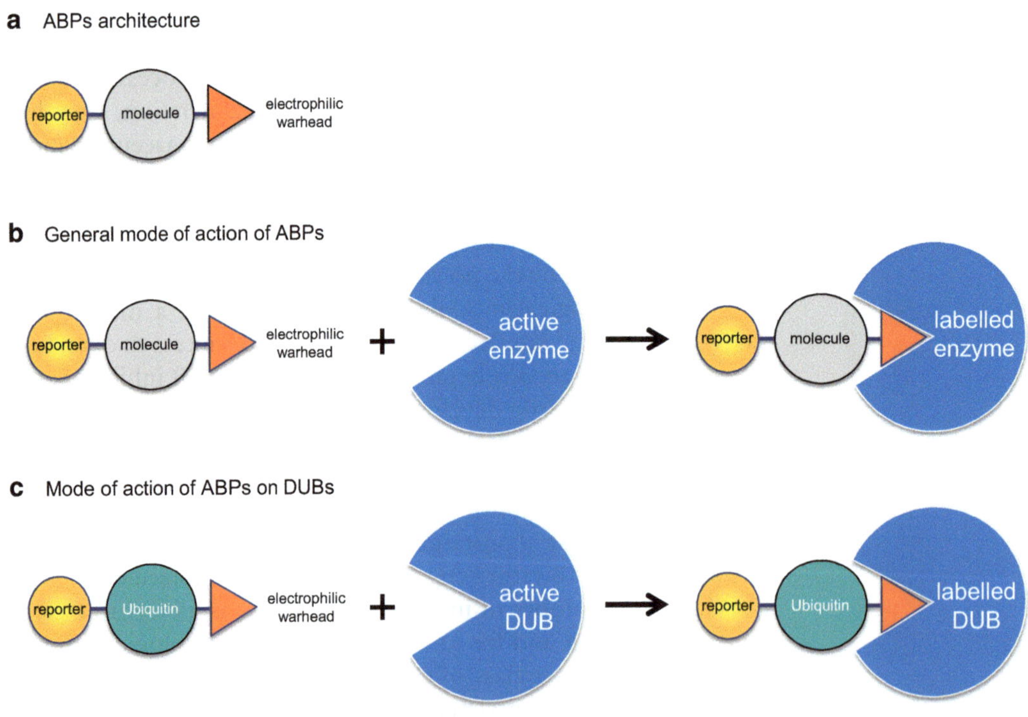

Fig. 1 (**a**) General structure of an ABP consisting of a reporter (tag), specific molecule (protein), and warhead. (**b**) General mechanism of action of ABPs. Catalytically competent enzymes react with the electrophilic warhead resulting in a covalently labeled protein. (**c**) Mechanism of action for labeling DUBs by ubiquitin ABPs

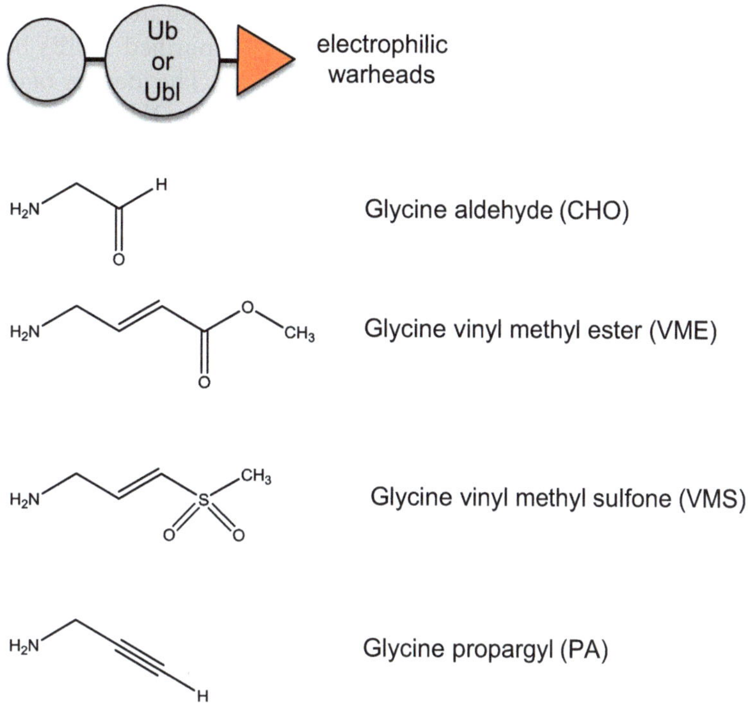

Fig. 2 Common warheads used for ubiquitin ABPs

imposes a restricted number of choices of electrophilic warheads. Monitoring the activity of endogenous enzymes such as DUBs in their native, full-length status as well as under all possible naturally occurring posttranslational modifications or interference/allosteric regulation from binding partners is a major advantage of ABPs. The irreversible covalent nature of ABPs toward their enzyme targets has a number of advantages when compared to many other analytical technologies that rely on weak, naturally transient and difficult to capture interactions between an enzyme and its substrate. Various warheads (Fig. 2) have been employed including alkyl halides (chloroethyl, bromoethyl, bromopropyl), Michael acceptors (vinyl methyl ester (VME), vinyl methyl sulfone (VMS), vinyl phenyl sulfone, vinyl cyanide) and more recently propargyl (PA) [36, 38, 39].

4 Activity Probes for Monitoring DUB Activity in Cells

The first attempt at generating activity probes to label DUBs on their catalytic site thiol group was described by Hidde Ploegh and colleagues [35]. Using a trypsin catalyzed transpeptidation to modify ubiquitin at its carboxy terminus with a vinyl sulfone group, they were able to demonstrate that ubiquitin vinyl sulfone labeled not only recombinant purified DUBs but also a number of yeast DUBs in a crude lysate. The identity of each labeled band was verified using individual yeast DUB mutant strains. The initial version of the ubiquitin vinyl sulfone probe was labeled with iodine^{125}and allowed for detection of a number of DUBs in mouse tissues as well as in mouse cell lysates. In the same study, Borodovsky et al. described the use of unlabeled ubiquitin vinyl sulfone to detect a specific DUB by monitoring a shift in the apparent molecular weight in SDS-PAGE followed by immunoblotting: USP7 was labeled efficiently in mammalian cell lysates. Finally, the authors were also able to identify USP14 as a novel DUB associated with the proteasome thanks to the use of ubiquitin vinyl sulfone in fractionation and immune-purification assays.

In a second generation of activity probes, the thiol-reactive group was added to ubiquitin using an intein-based chemical ligation method [36]. The reactivity of the DUBs depends on the type of electrophilic warhead fused to ubiquitin. The second generation of probes were additionally used for the identification of bound DUBs by affinity purification/mass spectrometry [34]. More recently, ABPs using a fluorescent reporter tag have been generated to replace the initial tags (e.g., HA) to allow replacement of the immunoblot procedure with fluorescent imaging [39–41].

While the historical production of ubiquitin ABPs was based on a trypsin catalyzed transpeptidation to modify ubiquitin at its carboxy terminus with a vinyl sulfone group or based on the addition of the electrophilic warhead via intein-based chemical ligation methods, recent approaches have moved toward the full-chemical synthesis of ubiquitin ABPs [41, 42]. This latest improvement has the added advantage of allowing the incorporation of modified amino acid residues at any position in the ABPs, whether natural or not.

In addition their major role in monitoring or identifying active DUBs in biological samples, ubiquitin-based probes are useful tools for structural analysis of DUBs. A number of cocrystals of DUBs with ubiquitin have been solved [34, 39] and in some cases, the structure of the apo-DUB was not achieved in the absence of the modified ubiquitin ABP [43]. ABPs are sometimes the only option available for cocrystallizing DUBs with ubiquitin substrates or ubiquitin chains.

5 Activity Probes for Monitoring DUB Activity in Tissues, Viruses, or Parasites

A limited number of studies have demonstrated the utility of activity probes for monitoring DUB activity in normal or diseased animal tissues. In earlier DUB activity probe publications, mouse tissues were examined and significant differences in the profile of active DUBs in tissues was observed [35]. More recently, in a very detailed study, Altun et al. investigated the activity of DUBs in models of aging and dietary restriction [44]. Dramatic differences in the levels of active DUBs in cell lines derived from various tissues as well as in primary tissues has been observed [45]. For a small number of DUBs, the activity as monitored by activity probe binding can be correlated with the malignant status of the cell line or tissue, suggesting a possible therapeutic window for DUB inhibitors [45]. Given the sensitivity and significance of such techniques, we can expect an increase in the number of studies taking full advantage of ubiquitin ABPs to monitor differential DUB activity in pathological versus normal conditions in the near future.

ABPs have been used to identify and monitor the activity of bacterial, viral, or parasitic DUBs including Herpes viridae, Chlamydia trachomatis, Toxoplasma gondii, and Plasmodium falciparum: ABPs are invaluable tools to identify functionally active DUBs in complex in sometimes relatively poor or difficult to annotate organisms [46–50]. While viruses or bacteria do not encode a full complement of ubiquitin proteasome enzyme systems, they express DUBs to evade the detection of their proteins by the immune system or otherwise enhance virulence [51]. In addition, since DUBs are essential for viral proliferation, viral DUBs have been considered as possible therapeutic strategies for the treatment of certain viral infections such as SARS or MERS [52].

6 Chemical Proteomics-Activity Probes for Characterizing DUB Inhibitors

Mass spectrometry has emerged as an important tool for characterizing the various forms of ubiquitin. Initial global characterization of the ubiquitin-modified proteome has been made possible in proteomic studies taking advantage of a monoclonal antibody that recognizes (di-Gly)-containing isopeptides following trypsin digestion of complex cellular lysates [53, 54]. In the Ubiquitin-AQUA approach, synthetic isotopically labeled internal standard peptides are used to quantify branched peptides and the branched -GG signature peptides generated by trypsin digestion of ubiquitin signals [55]. Proteomic studies looking at DUB interaction partners have also generated a great deal of information about their substrates, regulation, and function [56]. Additional studies have evaluated the functional role of DUBs using RNAi libraries [57, 58] or GFP-DUB fusions [59, 60], and have linked DUBs to specific cellular pathways. While such studies are very informative and have generated a wealth of data on the biological roles of DUBs, they provide only limited information regarding the dynamic activity profile of DUBs, and are not able to distinguish the catalytic state (active versus inactive) of DUBs. As the cellular activity of DUBs can be controlled by multiple factors including protein interactions [61], stoichiometric changes to the structure of the protein [62, 63], and posttranslational modifications [64, 65], the advantage of activity probes is their specific reactivity with catalytically active DUBs.

One of the benefits of ABPs for the characterization of DUB inhibitors is the ability to monitor compound selectivity. A chemical activity-based proteomic approach using HA-tagged ubiquitin labeled with electrophilic warheads (HA-UbBr2 or HA-Ub-VME) was undertaken to characterize the selectivity of two USP7 inhibitors either in immunoblots or by quantitative mass spectrometry following treatment of cells or cell lysates with compounds [66]. An independent study using another USP7 inhibitor displaying selectivity in a panel of biochemical DUB assays, was also subjected to cellular selectivity profiling using HA-Ub-VMS followed by immunoblotting [24]. In a more targeted approach, an active-site ubiquitin probe (HA-Ub-VMS) has been used to demonstrate that USP14/UCHL5 inhibition by a small molecule (b-AP15) inhibits the 19S proteasome in a reconstituted biochemical assay. A similar probe approach was also used to demonstrate that b-AP15 is not a general inhibitor of DUBs in a cell lysate probed with an anti-HA antibody detecting the conjugated ubiquitin species [67]. While the studies mentioned above are paving the way for elucidating DUB selectivity profiles in a cellular context, coverage of the "DUBome" is still limited. Technological improvements are still required to increase sensitivity and accurately monitor DUBs in a given cell or tissue experiment.

While DUB proteomic studies using activity probes have mainly been used for monitoring the selectivity of first generation DUB inhibitors, the potential for ubiquitin ABPs is much broader. Indeed, it is possible to determine the dynamic nature of DUB inhibitors by using ABPs to monitor the reversibility or the duration of DUB inhibition. Furthermore, most of the work so far on DUBs using ABPs has been restricted to cellular studies. Recent progress in developing DUB inhibitors with in vivo preclinical potential is currently driving the tools for pharmacodynamic as well as mode-of-action understanding of DUB inhibitors in vivo. Activity probes based on selective inhibitors of peptidases have already been developed such as probes targeting proteasomes [68], cathepsins [69] or caspases [70] and are proving their usefulness for in vivo imaging studies as well as for diagnostic purposes [71].

7 Activity Probes for Ubiquitin-Like Deconjugating Enzymes

The utility of ubiquitin activity probes to identify and characterize DUBs in a number of conditions is not limited to ubiquitin. Indeed, probes for enzymes that remove Ubls have been generated. The exquisite selectivity of DUBs for their cognate substrates suggested that specific probes are also required for Ubl peptidases. An initial approach based on the synthesis of peptide vinyl sulfones harboring various portions of the ubiquitin-like carboxy terminus has suggested that truncated Ubls are able to bind Ubl-specific proteases in a manner similar to the ubiquitin-based vinyl sulfone polypeptides [72]. Ubl-based probes for Nedd8, SUMO-1, ISG15, GATE-16, MAP1-LC3, GABARAP, and Apg8L have been successfully synthesized [73–75].

An alternative to classical activity probes containing a full ubiquitin or ubiquitin-like polypeptide is based on the use of small molecule inhibitors to label the catalytic site of desumoylating enzymes (sentrin-specific proteases, SENPs). A peptide acyloxymethyl ketone (AOMK) containing a large aromatic O-acyl group are selective covalent inhibitors of SENPs and can be modified using fluorescent labels to detect SENPs activity in biological samples [76]. A similar approach has been described using a different family of proteins: glycine fluoromethylketones, which serve as probes to selectively target SENPs [77]. A more conventional derivatization of the carboxy-terminal end of Ubls with electrophilic warheads has also been pursued and a general derivatization procedure to produce any Ubl domain chemically activated at its C-terminus by formation of a thiol ester. Reaction of the thiol with a nucleophile produces the desired derivatives taking advantage of the intein fusion technology [78]. There is no technical challenges preventing the development of fully synthetic Ubl ABPs and indeed, a number of such reagents are already commercially available from various sources.

As the mechanism for the removal of Ubls by specific enzymes has not yet been fully characterized, ABPs will certainly play a key role in the elucidation of such understanding. Similarly, the biological or mechanistic functions of a number of DUBs or SENPs remains poorly understood, and existing ABPs or novel more selective ABPs can serve as tools for extending our knowledge.

8 Activity Probes Using Ubiquitin Chains or Modified Ubiquitin

In parallel with the development of monoubiquitin ABPs, a number of groups have also achieved total (semi)-synthesis of di-ubiquitin [42, 79–81] or even tetra-ubiquitin chains [82, 83]. However, incorporation of electrophilic warheads into polyubiquitin chains remains problematic. An intermediate approach to the generation of polyubiquitin ABPs was elaborated on the basis of the synthesis of branched-peptides incorporating an isopeptide-linked ubiquitin and an electrophilic warhead [84]. In addition, the synthesis and characterization of K48- or K63-linked di-ubiquitin probes bearing dehydroalanine as a warhead near the isopeptide bond has been described [85]. Finally, ABPs engineered for di-ubiquitin chains incorporating the 8 known ubiquitin linkages have been successful and now allow DUB ubiquitin-linkage specificity in a cellular context to be addressed[86–88]. Structural studies of DUBs with di-ubiquitin have demonstrated that in addition to the peptide flanking the ubiquitylated residues, more extensive interactions between DUBs and the proximal ubiquitin in the chain also contribute to the recognition by DUBs. Probing DUB selectivity with the latest generation of probes not only generates a distinct pattern from that obtained using mono-ubiquitin ABPs, but also suggests that the promiscuity of some DUBs for their substrates is probably much less pronounced than initially anticipated.

In the last couple of years, posttranslational modifications of ubiquitin, especially phosphorylation of ubiquitin at specific residues (e.g., Ser57 and Ser65) have been shown to play important roles in a number of cellular processes [89, 90]. Ubiquitin ABPs bearing the phosphorylated variants of ubiquitin have been generated and used to probe the selectivity of the modifications for conjugating and deconjugating enzymes. E1 and E2 enzymes are usually able to tolerate phosphorylated ubiquitin, however, a number of DUBs have difficulty recognizing the modified substrates [91, 92]. Studies evaluating additional posttranslational modifications of ubiquitin such as methylation, acetylation, hydroxylation, or other phosphorylation will certainly be unraveled in the near future: the corresponding ABPs will again serve as useful tools to understand the mechanistic and physiological role of novel variants of ubiquitin.

9 Activity Probes to Measure Target Engagement

A key issue facing researchers involved in deciphering the roles of DUBs in a cellular context is the lack of understanding of the most direct or relevant substrate of specific DUBs in a given cellular pathway. Some DUBs have very well characterized substrates (e.g., USP1 or USP7) [15] that are clearly linked to the function of the DUBs, however, the known substrate specificity is still relatively poor or partial at best for most DUBs. In certain cases, it is quite clear that unique substrates do not exist: e.g., USP14 or UCHL5 are DUBs that indiscriminately recognize any ubiquitylated substrates which is targeted to the proteasome [93]. Ubiquitin ABPs can play a critical role as tools to monitor the dynamics of the activation or inhibition of DUBs under specific physiological or pharmacological pathway alterations. The problem is especially acute for the monitoring of DUB activity upon inhibition with specific inhibitors: the pharmaceutical development of DUB inhibitors requires a good understanding of the pharmacokinetic modulation of the target upon treatment with compounds. The development of ABPs for proteomic evaluation of target engagement is currently being investigated by a number of groups. In addition, higher throughput ABP-based strategies are also under development for the determination of DUB target engagement in cellular contexts as well as in tissues or eventually for clinical sample evaluation (Fig. 3).

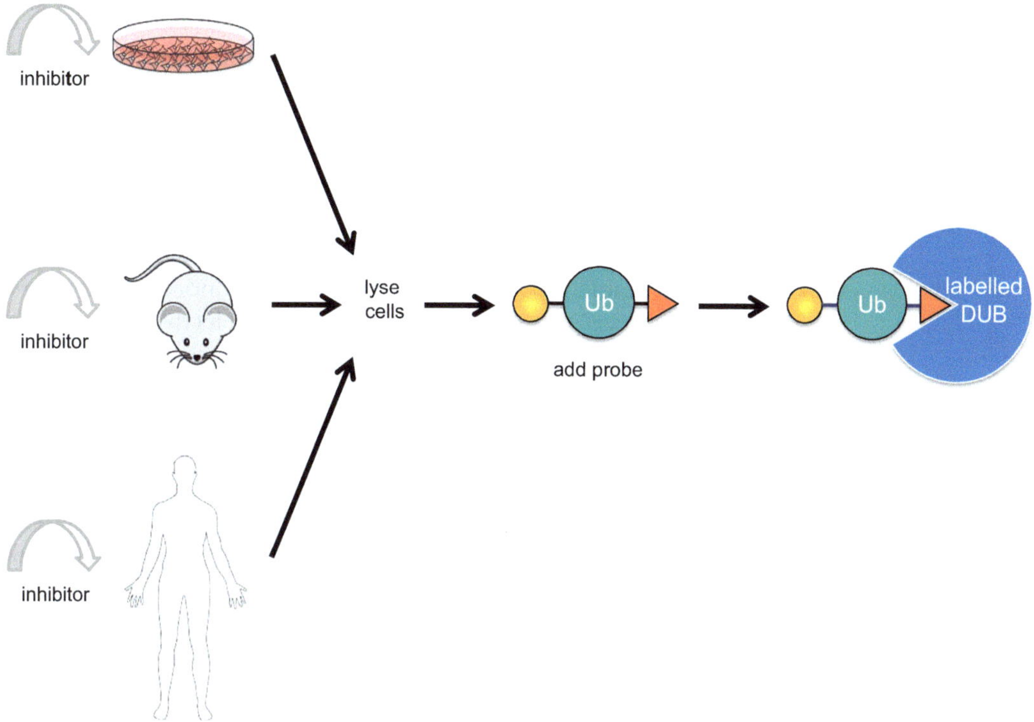

Fig. 3 High-throughput assay design to monitor DUB target engagement using ubiquitin ABPs in cells, animals, or patients tissues: (1) treatment of cells, animals, or patients with DUB inhibitor; (2) generation of protein lysates; (3) incubation of lysates with ubiquitin ABPs; (4) visualization of DUB activity or inhibition

10 What Is Next for Chemical Probes Targeting DUBs?

The development of activity probes for DUBs has lagged behind the development of probes for more classical proteases. Indeed, the complexity of the recognition site of DUBs, which requires the binding of full-length ubiquitin in the catalytic site as well as the challenges in the characterization of potent and selective DUB inhibitors, has hindered production of ABPs for DUBs. However, following on from the ground-breaking evolution of cell-permeable and in vivo-compatible activity-based imaging probes developed for other proteases such as caspases or cathepsins [69, 70], the next generation of probes for DUBs will certainly be agents that enable direct visualization and quantification of DUB activity in vivo. Such noninvasive agents have great potential for early diagnosis as well as pharmacodynamic evaluation of DUB inhibition in preclinical as well as clinical settings. One attractive avenue to explore for the development of selective DUB activity probes is based on the design of copper-catalyzed click-labeled DUB inhibitors with quenchable or nonfluorescent labels [94]. Click-labeled ABPs allows for selective labeling, visualization, and enrichment of active enzymes in a complex proteome. Another approach will likely be based on the generation of noninvasive substrate probes that do not bind covalently to the enzyme. The advantage of this approach is based on the theoretically higher signal that can be generated, in contrast to covalent activity probes which are limited by the stoichiometric labeling of the enzyme (the signal being proportional to the amount of enzyme in various tissues). So far a very limited number of reporter substrates are available, none being cell permeable, or suitable for in vivo applications. Again, noninvasive permeable substrates will likely be derived from selective inhibitors of individual DUBs or knowledge around selectively ubiquitylated sites on DUB substrates. Probably one of the most promising avenues for developing cell- and tissue-permeable selective ubiquitin ABPs for DUBs will rely on the modification of selective small molecule inhibitors of DUBs. Similar approaches have already achieved some preliminary success for other enzymes of the UPS such as E1 enzymes [95] and proteasome probes [96]. The limiting step in developing such probes for DUBs is currently a lack of potent, specific and selective DUB inhibitors available, however, the community is successfully designing novel generations of selective DUB inhibitors.

While a number of ABPs have been successfully designed for monitoring the activity of cysteine peptidase DUBs, there is still a gap in the development of ABPs for DUBs of the metalloenzyme class (MPN+/JAMMs). A number of approaches are currently being investigated for the design of ABPs for metallo-DUBs and will certainly aid the characterization of inhibitors for that class of enzymes which is showing great promise as therapeutic targets [97–99].

11 Summary

Protein ubiquitylation is critical for the control of protein half-life, localization, and function. Deregulation of this process is a causative factor of many diseases. The development of ABPs has allowed for major advancement in the identification and characterization of cysteine DUBs. Significant progress has been made in terms of probe design and preparation. For example, five papers have been published in the past 3 years describing di-ubiquitin ABPs, underscoring the importance of these tools for DUB research. The JAMM family remains difficult to target using ABPs due to the catalytic mechanism which does not involve a covalent DUB-substrate intermediate. Hopefully new approaches and novel probe designs will yield better tools to investigate this class of metalloproteases. ABPs will ultimately shed light on the function and relevance of DUBs involved in various chain-specific ubiquitin signaling, and will continue to advance our knowledge of DUB regulation and function in a cellular context. Furthermore, ABPs will aid the development and characterization of DUB inhibitors, allowing the monitoring of target engagement as well as selectivity in vivo. Finally, while ubiquitin ABPs have not yet been as broadly used as one might expect to monitor DUBs in developmental or pathological evaluations, they can provide a unique dynamic assessment of the activity of DUBs, and will undoubtedly become a more familiar option for many researchers.

Acknowledgements

The authors would like to acknowledge networking support by the Proteostasis COST Action (BM1307).

References

1. Bedford L et al (2011) Ubiquitin-like protein conjugation and the ubiquitin-proteasome system as drug targets. Nat Rev Drug Discov 10(1):29–46
2. Goldberg M et al (2003) MDC1 is required for the intra-S-phase DNA damage checkpoint. Nature 421(6926):952–956
3. Ravid T, Hochstrasser M (2008) Diversity of degradation signals in the ubiquitin-proteasome system. Nat Rev Mol Cell Biol 9(9):679–690
4. Schmidt M, Finley D (2014) Regulation of proteasome activity in health and disease. Biochim Biophys Acta 1843(1):13–25
5. Johnson DE (2015) The ubiquitin-proteasome system: opportunities for therapeutic intervention in solid tumors. Endocr Relat Cancer 22(1):T1–T17
6. Wenzel DM, Klevit RE (2012) Following Ariadne's thread: a new perspective on RBR ubiquitin ligases. BMC Biol 10:24
7. Smit JJ, Sixma TK (2014) RBR E3-ligases at work. EMBO Rep 15(2):142–154
8. Rahighi S, Dikic I (2012) Selectivity of the ubiquitin-binding modules. FEBS Lett 586(17):2705–2710
9. Komander D, Rape M (2012) The ubiquitin code. Annu Rev Biochem 81:203–229
10. Ciechanover A (2005) Proteolysis: from the lysosome to ubiquitin and the proteasome. Nat Rev Mol Cell Biol 6(1):79–87
11. Gallastegui N, Groll M (2010) The 26S proteasome: assembly and function of a destructive machine. Trends Biochem Sci 35(11):634–642
12. Groll M et al (1997) Structure of 20S proteasome from yeast at 2.4 A resolution. Nature 386(6624):463–471
13. Suraweera A et al (2012) Failure of amino acid homeostasis causes cell death following proteasome inhibition. Mol Cell 48(2):242–253
14. Komander D, Clague MJ, Urbe S (2009) Breaking the chains: structure and function of the deubiquitinases. Nat Rev Mol Cell Biol 10(8):550–563
15. Clague MJ et al (2013) Deubiquitylases from genes to organism. Physiol Rev 93(3):1289–1315
16. Eletr ZM, Wilkinson KD (2014) Regulation of proteolysis by human deubiquitinating enzymes. Biochim Biophys Acta 1843(1):114–128
17. Mattern MR, Wu J, Nicholson B (2012) Ubiquitin-based anticancer therapy: carpet bombing with proteasome inhibitors vs surgical strikes with E1, E2, E3, or DUB inhibitors. Biochim Biophys Acta 1823(11):2014–2021
18. Sippl W, Collura V, Colland F (2011) Ubiquitin-specific proteases as cancer drug targets. Future Oncol 7(5):619–632
19. Lill JR, Wertz IE (2014) Toward understanding ubiquitin-modifying enzymes: from pharmacological targeting to proteomics. Trends Pharmacol Sci 35(4):187–207
20. Jacq X et al (2013) Deubiquitylating enzymes and DNA damage response pathways. Cell Biochem Biophys 67(1):25–43
21. Ye Y et al (2012) Ubiquitin chain conformation regulates recognition and activity of interacting proteins. Nature 492(7428):266–270
22. Heideker J, Wertz IE (2015) DUBs, the regulation of cell identity and disease. Biochem J 467(1):191
23. Colland F et al (2009) Small-molecule inhibitor of USP7/HAUSP ubiquitin protease stabilizes and activates p53 in cells. Mol Cancer Ther 8(8):2286–2295

24. Reverdy C et al (2012) Discovery of specific inhibitors of human USP7/HAUSP deubiquitinating enzyme. Chem Biol 19(4):467–477
25. Chauhan D et al (2012) A small molecule inhibitor of ubiquitin-specific protease-7 induces apoptosis in multiple myeloma cells and overcomes bortezomib resistance. Cancer Cell 22(3):345–358
26. Tian Z et al (2014) A novel small molecule inhibitor of deubiquitylating enzyme USP14 and UCHL5 induces apoptosis in multiple myeloma and overcomes bortezomib resistance. Blood 123(5):706–716
27. Peterson LF et al (2015) Targeting deubiquitinase activity with a novel small molecule inhibitor as therapy for B-cell malignancies. Blood 125(23):3588–3597
28. Cravatt BF, Wright AT, Kozarich JW (2008) Activity-based protein profiling: from enzyme chemistry to proteomic chemistry. Annu Rev Biochem 77:383–414
29. Liu Y, Patricelli MP, Cravatt BF (1999) Activity-based protein profiling: the serine hydrolases. Proc Natl Acad Sci U S A 96(26):14694–14699
30. Chan EW et al (2004) Developing photoactive affinity probes for proteomic profiling: hydroxamate-based probes for metalloproteases. J Am Chem Soc 126(44):14435–14446
31. Saghatelian A et al (2004) Activity-based probes for the proteomic profiling of metalloproteases. Proc Natl Acad Sci U S A 101(27):10000–10005
32. Berkers CR et al (2005) Activity probe for in vivo profiling of the specificity of proteasome inhibitor bortezomib. Nat Methods 2(5):357–362
33. Adam GC, Sorensen EJ, Cravatt BF (2002) Proteomic profiling of mechanistically distinct enzyme classes using a common chemotype. Nat Biotechnol 20(8):805–809
34. Galardy P, Ploegh HL, Ovaa H (2005) Mechanism-based proteomics tools based on ubiquitin and ubiquitin-like proteins: crystallography, activity profiling, and protease identification. Methods Enzymol 399:120–131
35. Borodovsky A et al (2001) A novel active site-directed probe specific for deubiquitylating enzymes reveals proteasome association of USP14. EMBO J 20(18):5187–5196
36. Borodovsky A et al (2002) Chemistry-based functional proteomics reveals novel members of the deubiquitinating enzyme family. Chem Biol 9(10):1149–1159
37. Balakirev MY et al (2003) Otubains: a new family of cysteine proteases in the ubiquitin pathway. EMBO Rep 4(5):517–522
38. Hemelaar J et al (2004) Chemistry-based functional proteomics: mechanism-based activity-profiling tools for ubiquitin and ubiquitin-like specific proteases. J Proteome Res 3(2):268–276
39. Ekkebus R et al (2013) On terminal alkynes that can react with active-site cysteine nucleophiles in proteases. J Am Chem Soc 135(8):2867–2870
40. McGouran JF et al (2012) Fluorescence-based active site probes for profiling deubiquitinating enzymes. Org Biomol Chem 10(17):3379–3383
41. de Jong A et al (2012) Ubiquitin-based probes prepared by total synthesis to profile the activity of deubiquitinating enzymes. Chembiochem 13(15):2251–2258
42. El Oualid F et al (2010) Chemical synthesis of ubiquitin, ubiquitin-based probes, and diubiquitin. Angew Chem Int Ed Engl 49(52):10149–10153
43. Renatus M et al (2006) Structural basis of ubiquitin recognition by the deubiquitinating protease USP2. Structure 14(8):1293–1302
44. Altun M et al (2010) Muscle wasting in aged, sarcopenic rats is associated with enhanced activity of the ubiquitin proteasome pathway. J Biol Chem 285(51):39597–39608
45. Ovaa H et al (2004) Activity-based ubiquitin-specific protease (USP) profiling of virus-infected and malignant human cells. Proc Natl Acad Sci U S A 101(8):2253–2258
46. Kattenhorn LM et al (2005) A deubiquitinating enzyme encoded by HSV-1 belongs to a family of cysteine proteases that is conserved across the family Herpesviridae. Mol Cell 19(4):547–557
47. Schlieker C et al (2007) Structure of a herpesvirus-encoded cysteine protease reveals a unique class of deubiquitinating enzymes. Mol Cell 25(5):677–687
48. Misaghi S et al (2006) Chlamydia trachomatis-derived deubiquitinating enzymes in mammalian cells during infection. Mol Microbiol 61(1):142–150
49. Frickel EM et al (2007) Apicomplexan UCHL3 retains dual specificity for ubiquitin and Nedd8 throughout evolution. Cell Microbiol 9(6):1601–1610
50. Artavanis-Tsakonas K et al (2006) Identification by functional proteomics of a deubiquitinating/deNeddylating enzyme in Plasmodium falciparum. Mol Microbiol 61(5):1187–1195
51. Le Negrate G et al (2008) ChlaDub1 of Chlamydia trachomatis suppresses NF-kappaB activation and inhibits IkappaBalpha ubiquitination and degradation. Cell Microbiol 10(9):1879–1892

52. Baez-Santos YM, St John SE, Mesecar AD (2015) The SARS-coronavirus papain-like protease: structure, function and inhibition by designed antiviral compounds. Antiviral Res 115:21–38
53. Kim W et al (2011) Systematic and quantitative assessment of the ubiquitin-modified proteome. Mol Cell 44(2):325–340
54. Phu L et al (2011) Improved quantitative mass spectrometry methods for characterizing complex ubiquitin signals. Mol Cell Proteomics 10(5):M110 003756
55. Bennett EJ et al (2010) Dynamics of cullin-RING ubiquitin ligase network revealed by systematic quantitative proteomics. Cell 143(6):951–965
56. Sowa ME et al (2009) Defining the human deubiquitinating enzyme interaction landscape. Cell 138(2):389–403
57. Dirac AM et al (2005) Functional annotation of deubiquitinating enzymes using RNA interference. Methods Enzymol 398:554–567
58. Buus R et al (2009) Deubiquitinase activities required for hepatocyte growth factor-induced scattering of epithelial cells. Curr Biol 19(17):1463–1466
59. Urbe S et al (2012) Systematic survey of deubiquitinase localization identifies USP21 as a regulator of centrosome- and microtubule-associated functions. Mol Biol Cell 23(6):1095–1103
60. Nishi R et al (2014) Systematic characterization of deubiquitylating enzymes for roles in maintaining genome integrity. Nat Cell Biol 16(10):1016–1026, 1–8
61. Faesen AC et al (2011) Mechanism of USP7/HAUSP activation by its C-terminal ubiquitin-like domain and allosteric regulation by GMP-synthetase. Mol Cell 44(1):147–159
62. Lander GC et al (2012) Complete subunit architecture of the proteasome regulatory particle. Nature 482(7384):186–191
63. Wiener R et al (2013) E2 ubiquitin-conjugating enzymes regulate the deubiquitinating activity of OTUB1. Nat Struct Mol Biol 20(9):1033–1039
64. Cotto-Rios XM, Jones MJ, Huang TT (2011) Insights into phosphorylation-dependent mechanisms regulating USP1 protein stability during the cell cycle. Cell Cycle 10(23):4009–4016
65. Huang OW et al (2012) Phosphorylation-dependent activity of the deubiquitinase DUBA. Nat Struct Mol Biol 19(2):171–175
66. Altun M et al (2011) Activity-based chemical proteomics accelerates inhibitor development for deubiquitylating enzymes. Chem Biol 18(11):1401–1412
67. D'Arcy P et al (2011) Inhibition of proteasome deubiquitinating activity as a new cancer therapy. Nat Med 17(12):1636–1640
68. Li H et al (2014) Assessing subunit dependency of the plasmodium proteasome using small molecule inhibitors and active site probes. ACS Chem Biol 9(8):1869–1876
69. Edgington LE et al (2009) Noninvasive optical imaging of apoptosis by caspase-targeted activity-based probes. Nat Med 15(8):967–973
70. Blum G et al (2007) Noninvasive optical imaging of cysteine protease activity using fluorescently quenched activity-based probes. Nat Chem Biol 3(10):668–677
71. Sanman LE, Bogyo M (2014) Activity-based profiling of proteases. Annu Rev Biochem 83:249–273
72. Borodovsky A et al (2005) Small-molecule inhibitors and probes for ubiquitin- and ubiquitin-like-specific proteases. Chembiochem 6(2):287–291
73. Ovaa H, Galardy PJ, Ploegh HL (2005) Mechanism-based proteomics tools based on ubiquitin and ubiquitin-like proteins: synthesis of active site-directed probes. Methods Enzymol 399:468–478
74. Hemelaar J et al (2003) A single protease, Apg4B, is specific for the autophagy-related ubiquitin-like proteins GATE-16, MAP1-LC3, GABARAP, and Apg8L. J Biol Chem 278(51):51841–51850
75. Hemelaar J et al (2004) Specific and covalent targeting of conjugating and deconjugating enzymes of ubiquitin-like proteins. Mol Cell Biol 24(1):84–95
76. Albrow VE et al (2011) Development of small molecule inhibitors and probes of human SUMO deconjugating proteases. Chem Biol 18(6):722–732
77. Dobrota C et al (2012) Glycine fluoromethylketones as SENP-specific activity based probes. Chembiochem 13(1):80–84
78. Wilkinson KD, Gan-Erdene T, Kolli N (2005) Derivitization of the C-terminus of ubiquitin and ubiquitin-like proteins using intein chemistry: methods and uses. Methods Enzymol 399:37–51
79. Virdee S et al (2010) Engineered diubiquitin synthesis reveals Lys29-isopeptide specificity of an OTU deubiquitinase. Nat Chem Biol 6(10):750–757
80. Kumar KS et al (2010) Total chemical synthesis of di-ubiquitin chains. Angew Chem Int Ed Engl 49(48):9126–9131
81. Yang R et al (2010) Synthesis of K48-linked diubiquitin using dual native chemical ligation

at lysine. Chem Commun (Camb) 46(38):7199–7201

82. Kumar KS et al (2011) Total chemical synthesis of a 304 amino acid K48-linked tetraubiquitin protein. Angew Chem Int Ed Engl 50(27):6137–6141

83. Bavikar SN et al (2012) Chemical synthesis of ubiquitinated peptides with varying lengths and types of ubiquitin chains to explore the activity of deubiquitinases. Angew Chem Int Ed Engl 51(3):758–763

84. Iphofer A et al (2012) Profiling ubiquitin linkage specificities of deubiquitinating enzymes with branched ubiquitin isopeptide probes. Chembiochem 13(10):1416–1420

85. Haj-Yahya N et al (2014) Dehydroalanine-based diubiquitin activity probes. Org Lett 16(2):540–543

86. Mulder MP et al (2014) A native chemical ligation handle that enables the synthesis of advanced activity-based probes: diubiquitin as a case study. Chembiochem 15(7):946–949

87. McGouran JF et al (2013) Deubiquitinating enzyme specificity for ubiquitin chain topology profiled by di-ubiquitin activity probes. Chem Biol 20(12):1447–1455

88. Li G et al (2014) Activity-based diubiquitin probes for elucidating the linkage specificity of deubiquitinating enzymes. Chem Commun (Camb) 50(2):216–218

89. Koyano F et al (2014) Ubiquitin is phosphorylated by PINK1 to activate parkin. Nature 510(7503):162–166

90. Kane LA et al (2014) PINK1 phosphorylates ubiquitin to activate Parkin E3 ubiquitin ligase activity. J Cell Biol 205(2):143–153

91. Wauer T et al (2015) Ubiquitin Ser65 phosphorylation affects ubiquitin structure, chain assembly and hydrolysis. EMBO J 34(3):307–325

92. Bondalapati S et al (2015) Chemical synthesis of phosphorylated ubiquitin and diubiquitin exposes positional sensitivities of E1-E2 enzymes and deubiquitinases. Chemistry 21(20):7360–7364

93. Lander GC, Martin A, Nogales E (2013) The proteasome under the microscope: the regulatory particle in focus. Curr Opin Struct Biol 23(2):243–251

94. Martell J, Weerapana E (2014) Applications of copper-catalyzed click chemistry in activity-based protein profiling. Molecules 19(2):1378–1393

95. An H, Statsyuk AV (2013) Development of activity-based probes for ubiquitin and ubiquitin-like protein signaling pathways. J Am Chem Soc 135(45):16948–16962

96. Carmony KC, Kim KB (2013) Activity-based imaging probes of the proteasome. Cell Biochem Biophys 67(1):91–101

97. Clague MJ, Urbe S (2006) Endocytosis: the DUB version. Trends Cell Biol 16(11):551–559

98. Ambroggio XI, Rees DC, Deshaies RJ (2004) JAMM: a metalloprotease-like zinc site in the proteasome and signalosome. PLoS Biol 2(1):E2

99. Hannss R, Dubiel W (2011) COP9 signalosome function in the DDR. FEBS Lett 585(18):2845–2852

Chapter 27

Activity Based Profiling of Deubiquitylating Enzymes and Inhibitors in Animal Tissues

Lauren McLellan, Cassie Forder, Aaron Cranston, Jeanine Harrigan, and Xavier Jacq

Abstract

The attachment of ubiquitin or ubiquitin-like modifiers to proteins is an important signal for the regulation of a variety of biological processes including the targeting of substrates for degradation, receptor internalization, regulation of gene expression, and DNA repair. Posttranslational modification of proteins by ubiquitin controls many cellular processes, and aberrant ubiquitylation can contribute to cancer, immunopathologies, and neurodegeneration. Thus, deubiquitylating enzymes (DUBs) that remove ubiquitin from proteins have become attractive therapeutic targets. Monitoring the activity of DUBs in cells or in tissues is critical for understanding the biological function of DUBs in particular pathways and is essential for determining the physiological specificity and potency of small-molecule DUB inhibitors. Here, we describe a method for the homogenization of animal tissues and incubation of tissue lysates with ubiquitin-based activity probes to monitor DUB activity in mouse tissues and target engagement following treatment of animals with small-molecule DUB inhibitors.

Key words Activity-based probe, Deubiquitylating enzyme, Target engagement, Tissue, Ubiquitin

1 Introduction

The activity of DUBs can be monitored by activity probes consisting of ubiquitin (with or without an epitope tag) containing a C-terminal warhead that introduces a reactive group in the position of glycine 76 [1]. Various warheads have been employed including alkyl halides (chloroethyl, bromoethyl, bromopropyl), Michael acceptors (vinyl methyl ester (VME), vinyl methyl sulfone, vinyl phenyl sulfone, vinyl nitrile) and propargyl [2]. The design and synthesis of active site-directed probes that target DUBs has been instrumental in the identification of isopeptidase activity from crude cell lysates, as well as the discovery of novel DUBs [3–5]. Here, we describe the use of ubiquitin-based activity probes to monitor DUB inhibition in animal tissues. Tissues are first homogenized in a buffer

Rune Matthiesen (ed.), *Proteostasis: Methods and Protocols*, Methods in Molecular Biology, vol. 1449, DOI 10.1007/978-1-4939-3756-1_27, © Springer Science+Business Media New York 2016

compatible with the preservation of DUB activity. Subsequently, lysates are incubated with an ubiquitin-based activity probe. Proteins are separated by SDS-PAGE and DUB activity is monitored by immunoblotting. The covalent, irreversible bond formed between the warhead of the activity probe and the catalytic cysteine of the DUB results in a slower migrating DUB-ubiquitin complex. Here, we characterize the activity of several DUBs in various mouse tissues, and demonstrate DUB inhibition following the treatment of mice with small-molecule inhibitors.

2 Materials

Prepare all solutions using ultrapure water and analytical grade reagents. Prepare and store all reagents at room temperature, unless otherwise indicated. Diligently follow all waste disposal regulations and local health and safety rules when disposing of waste materials.

2.1 Tissue Homogenization Components

1. Homogenization buffer: 50 mM Tris–HCl, pH 7.5, 150 mM NaCl, 0.1 % IGEPAL® CA-630, 0.5 % CHAPS, 5 mM $MgCl_2$, 5 mM β-mercaptoethanol, 10 % glycerol, protease inhibitors (Roche cOmplete tablets, mini EDTA-free, #04693159001), phosphatase inhibitors (Roche PhosSTOP, #04906837001).
2. 0.5, 1.5 and 2 mL microcentrifuge tubes.
3. Forceps.
4. Scalpels.
5. Balance.
6. Tissue homogenizer (Retsch Mixer Mill MM400) and metal discs (Retsch, #22.455.0006C).
7. Refrigerated microcentrifuge.

2.2 Protein Quantitation and Activity Probe Assay Components

1. Coomassie Plus (Bradford) protein assay reagent (Perbio, #23200).
2. Bovine serum albumin (BSA) standard (Thermo Scientific, #23209).
3. Activity probe assay buffer: 50 mM Tris–HCl, pH 7.5, 150 mM NaCl, 0.1 % IGEPAL® CA-630, 0.5 % CHAPS, 5 mM $MgCl_2$, 5 mM β-mercaptoethanol.
4. HA-Ubiquitin-VME activity probe (UbiQ, catalog #UbiQ-035).

2.3 Immunoblotting Components

1. 4–12 % Novex NuPage, 10 well, 1.5 mm gels (Invitrogen, #12020166).
2. 20× MES running buffer (Invitrogen, #10515383): use at 1×.

3. 5× SDS loading buffer: 156.25 mM Tris–HCl pH 6.8, 5% SDS, 0.025% Bromophenol blue, 25% glycerol, 12.5% β-mercaptoethanol.
4. Nitrocellulose membranes (GE Healthcare, #10600041).
5. 10× Protein Running Buffer (PRB): 250 mM Tris-base and 1.92 M Glycine.
6. 1× Transfer buffer: 20% Ethanol and 2× PRB stored at 4 °C.
7. Ponceau S (Sigma P7170).
8. 10× Tris buffered saline (TBS): 150 mM Tris–HCl, 45 mM Tris-base and 1.5 M NaCl.
9. TBS containing Tween-20 (TBS-T): 1× TBS and 0.05% Tween-20.
10. Blocking solution: 5% dried skimmed milk (Marvel) in 1× TBS. Store at 4 °C.
11. Primary antibodies: USP11 (Bethyl, A301-613A, rabbit, 1:1000); USP14 (Cell Signaling, 11931S, rabbit, 1:2000); UCHL3 (Santa Cruz, sc-100340, mouse, 1:1000); UCHL1 (Abnova, PAB12509, rabbit, 1:10,000); USP4 (Bethyl, A300-830A, rabbit, 1:1000); USP7 (Abcam, AB4080, rabbit, 1:1000); USP5 (Bethyl, A301-542A, rabbit, 1:1000).
12. Secondary antibodies: anti-rabbit (Thermo, #31460, 1:10,000); anti-mouse (Thermo, #31430, 1:10,000).
13. Enhanced chemiluminescence (ECL; GE Healthcare, #RPN2109).

3 Methods

3.1 Animal Welfare and Procedures

Carry out all procedures involving the use of animals to the highest level of welfare and in line with best veterinary and modern husbandry practices. Ethical approval for the use of animals in research should be sought in advance and conducted in accordance with local, national and/or internationally recognized guidelines. At all times, consideration should be given to the humane principles of the 3Rs (namely, *replacement, reduction, refinement*; http://www.understandinganimalresearch.org.uk/how/three-rs/) and animals should not be reused (see also ref. [6]).

Purchase animals of a high health status from accredited suppliers and allow to acclimate for at least 5–7 days before using in procedures. House animals in individually ventilated cages (IVCs, Techniplast) in social groups at approved stocking densities for the cage size (e.g., 100 cm^2 floor area per 30 g mouse), provide with environmental enrichment (e.g., nesting material and chew blocks), and supply with sterilized food and water ad libitum. Tightly control the environmental conditions (e.g., temperature: 19–23 °C; humidity: 55 % ± 10 %; 12 h light/dark cycle). Identify animals by the mildest and most appropriate method in line with the study duration. Dosing and sampling volumes should be in line with the size of the species [7].

1. Source mice (typically female, 6–8 weeks of age) from Charles River Laboratories, UK.
2. After a period of acclimation, place mice into experimental groups and identify by tail marking with an indelible marker.
3. Prior to dosing, draw up the test compound into a 1 mL syringe and dip the end of the gavage needle in a sucrose solution (10 g/mL (w/v)) to aid acceptance of the gavage needle, assist in lubricating the esophagus and reduce stress in gavaged mice [8].
4. Scruff the mice and dose with the test compound by oral gavage (*per os* (*p.o.*)). Administer a single bolus dose (10 μL/g body weight) using a stainless steel 20 gauge oral gavage needle (Interfocus, Fine Science Tools, UK) (Fig. 1).
5. Typically, use two naïve mice for each dose level and for each time-point; mice were not reused in experiments (Figs. 2 and 3).
6. Following dosing, observe mice closely and frequently for the onset of any adverse clinical signs using the Mouse Grimace Scale to assess pain [9].
7. Immediately prior to the designated time-point, terminally anesthetize mice with Euthatal (Merial Animal Health Limited, UK) diluted 1:1 with sterile water and administer *via* intraperitoneal injection (6 μL/g body weight) sufficient to ensure nonrecovery. Check mice for an appropriate depth of anesthesia using the pedal withdrawal reflex before confirming death using cervical dislocation.
8. Obtain samples postmortem by dissection, snap-freeze in liquid nitrogen or freeze rapidly on dry ice in labeled bijou vials and store at −80 °C pending analysis (*see* **Note 1**).

3.2 Tissue Homogenization

1. Label homogenization tubes (2 mL microcentrifuge tube) and weigh them using a balance.
2. Remove tissue samples from −80 °C freezer and thaw on ice.
3. Using clean forceps and a new scalpel for each sample, carefully slice a piece of tissue from the main sample. Depending on the size of the sample, aim for 30–50 mg of tissue final, although as little as 10 mg can be used. Clean forceps with 70 % ethanol between samples.
4. Weigh the microcentrifuge tube containing the tissue and calculate tissue weight.

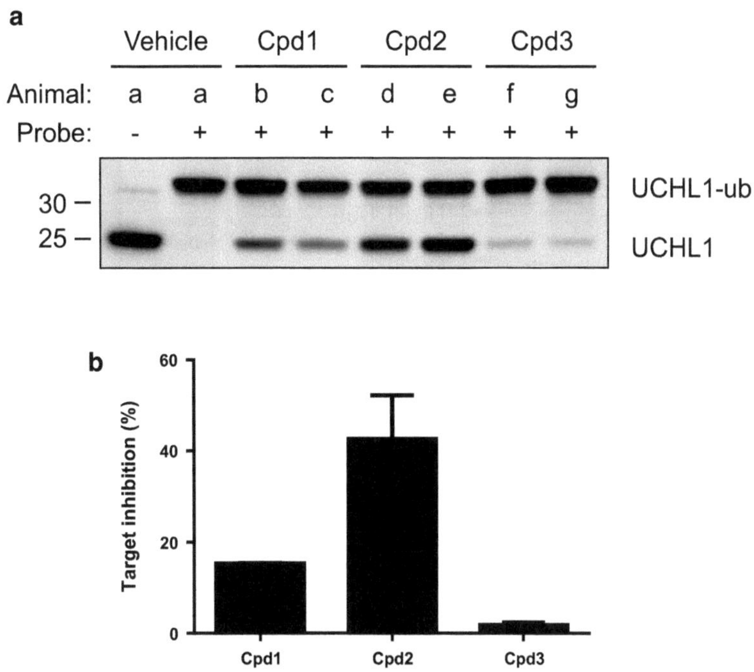

Fig. 1 In vivo target engagement in mouse surrogate tissues following treatment of animals with small-molecule DUB inhibitors. (a) Mice were dosed orally with vehicle or various small-molecule UCHL1 inhibitors (two independent animals each) with 100 mg/kg, *p.o.* Ovaries were harvested 2 h after dosing. Mouse ovary lysates were incubated in the absence or presence of an ubiquitin-based activity probe as indicated, separated by SDS-PAGE and Western blots were performed using an anti-UCHL1 antibody. (**b**) Quantitation of the results shown in (**a**). Error bars represent the SD from two independent biological replica

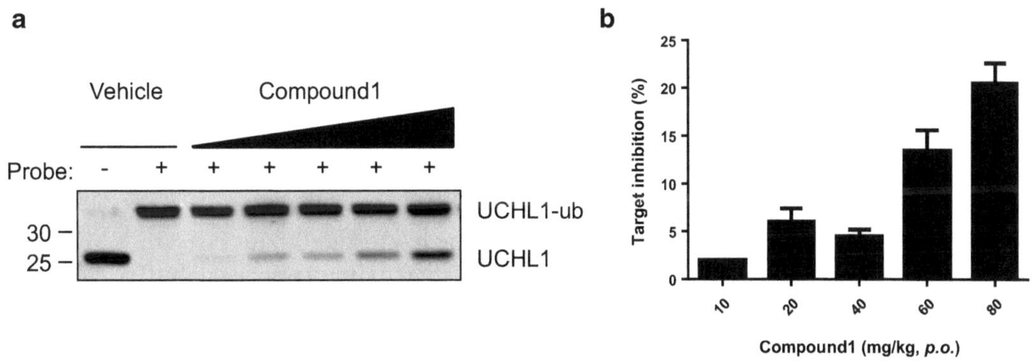

Fig. 2 Dose-dependent target engagement in mouse surrogate tissues following treatment of animals with a small-molecule UCHL1 inhibitor. (**a**) Mice were dosed with vehicle or increasing concentrations (10, 20, 40, 60, or 80 mg/kg, *p.o.*) of UCHL1 inhibitor. Ovaries were harvested 2 h after dosing. Mouse ovary lysates were incubated in the absence or presence of an ubiquitin-based activity probe as indicated, separated by SDS-PAGE and Western blots were performed using an anti-UCHL1 antibody. (**b**) Quantitation of results from (**a**). Error bars represent the SD from two independent biological replica

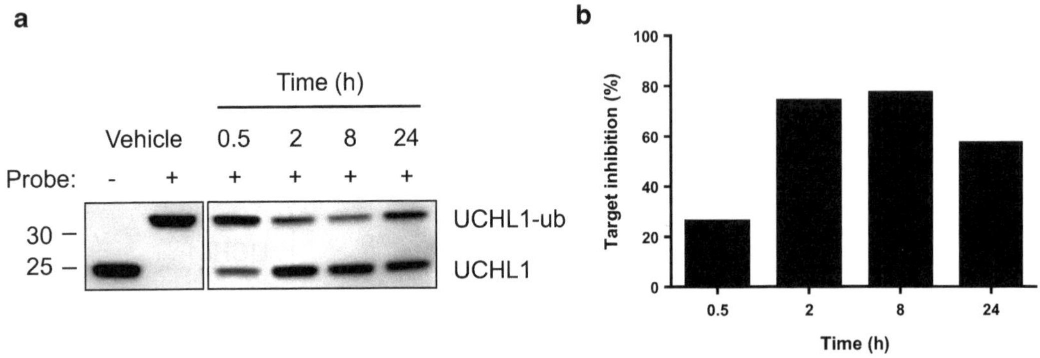

Fig. 3 Duration of target engagement in mouse surrogate tissues following treatment of animals with a small-molecule UCHL1 inhibitor. (**a**) Mice were dosed with vehicle or UCHL1 inhibitor (80 mg/kg, *p.o.*). Ovaries were harvested at various time points after dosing as indicated. Mouse ovary lysates were incubated in the absence or presence of an ubiquitin-based activity probe, separated by SDS-PAGE and Western blots were performed using an anti-UCHL1 antibody. (**b**) Quantitation of results shown in (**a**)

5. Add 1–3 volumes of homogenization buffer to the tissue (depending on tissue type). For example, if the tissue weighs 30 mg, add 30–90 μL homogenization buffer.
6. Homogenize tissues for 45 s (1 cycle), frequency 25, in the tissue homogenizer using 2–3 small discs per microcentrifuge tube.
7. Transfer the supernatant to a new 1.5 mL microcentrifuge tube being careful to avoid any solid material or beads.
8. Centrifuge lysates at 13,000 rpm (16,200 × *g*) in a microcentrifuge for 15 min at 4 °C. Transfer the supernatant to a new microcentrifuge tube and aliquot supernatant into 0.5 mL microcentrifuge tubes. Snap freeze in liquid nitrogen and store at −80 °C.
9. Retrieve discs with forceps or a magnet and clean/sterilize these for repeat usage.

3.3 Quantitation of Lysates and Incubation with Activity Probe

1. Quantitate protein lysates with the Coomassie (Bradford) assay, a BSA standard curve, and using the homogenization buffer as a negative control.
2. Combine 20 μg of tissue lysate and 0.5 μg HA-Ub-VME activity probe (*see* **Note 2**) in activity probe assay buffer (*see* **Note 3**) to a final reaction volume of 20 μL. Incubate at room temperature for 60 min (*see* **Note 4**).
3. Stop the reaction with 10 μL of 5× SDS loading buffer.
4. Heat at 95 °C for 5 min.
5. Store samples at −20 °C or proceed to SDS-PAGE and Western blotting steps.

3.4 SDS-PAGE and Western Blotting

1. Load samples on SDS-PAGE gel, and when the dye front reaches the end of the gel, turn off the power supply (*see* **Note 5**). Separate the gel plates and remove the top of the gel containing the wells.
2. Rinse the gel and nitrocellulose membrane carefully with 1× transfer buffer.
3. Transfer the gel for 2 h at 300 mA.
4. Rinse the membrane with water.
5. If desired, discard the water and stain the membrane with Ponceau S with gentle agitation until bands appear. Remove the Ponceau S and rinse with water until the desired level of signal is obtained.
6. Remove the Ponceau S by incubating the membrane in TBS-T.
7. Incubate the membrane with blocking solution for at least 30 min.
8. Incubate the membrane with the primary antibody (in blocking buffer or in accordance with the antibody datasheet) for 1 h at room temp or overnight at 4 °C (*see* **Note 6**).
9. Wash 3 × 8 min with TBS-T with gentle agitation.
10. Incubate the membrane with the secondary antibody (1:10,000 in blocking buffer) for 1 h.
11. Wash 3 × 8 min with TBS-T with gentle agitation.
12. Develop using ECL according to the manufacturer's instructions.
13. Capture the image using an imaging system (GE Healthcare ImageQuant LAS 4000) or X-ray film.
14. Analyze DUB activity and inhibition using ImageQuant TL software (GE Healthcare). Quantify the bands corresponding to DUB and DUB-Ub in each lane and calculate maximal DUB activity using the vehicle treated samples [DUB-Ub/(DUB-Ub + DUB) × 100]. In compound treated samples, the ratio of DUB to DUB-Ub is proportional to inhibition.

4 Notes

1. Determine the expression level of the DUB of interest in various tissues (Fig. 4a). Additionally, the catalytic competency of the DUB, and the subsequent reactivity with the activity probe may differ depending on the tissue examined (Fig. 4b) [3].
2. The concentration of tissue lysate can be adjusted based on the expression level of the individual DUB as well as the lysate: probe ratio [5]. Importantly, the reactivity of a DUB with the ubiquitin-based activity probe depends on the specific warhead

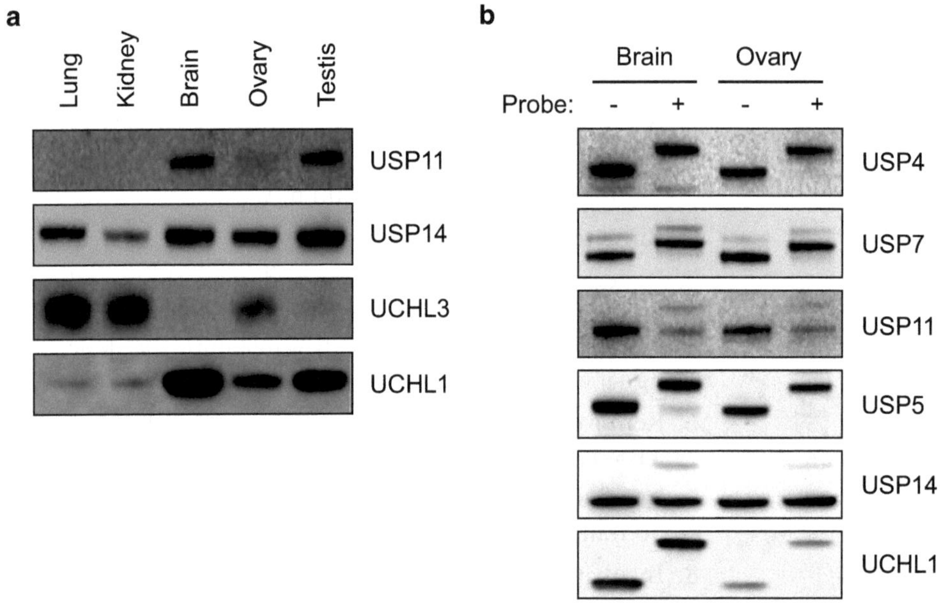

Fig. 4 Expression and activity of DUBs in mouse tissues. (**a**) Lysates from various mouse tissues were separated by SDS-PAGE and Western blots were performed using antibodies against the following DUBs: USP11, USP14, UCHL3, and UCHL1. (**b**) Lysates from mouse brain and ovary were incubated in the absence or presence of an ubiquitin-based activity probe as indicated. Subsequently, lysates were separated by SDS-PAGE and Western blots were performed using antibodies against the following DUBs: USP4, USP7, USP11, USP5, USP14, and UCHL1

employed and should be optimized for each individual DUB. It has been shown previously that ubiquitin-VME works well for the majority of DUBs, but there are specific cases where changing the warhead dramatically improves the reactivity and the DUB-ubiquitin complex. There are several commercially available ubiquitin-based activity probes with different warheads (e.g., bromoethyl, chloroethyl, vinyl sulfone, and propargyl) [2, 5].

3. The standard homogenization and activity probe assay buffers work well for many DUBs. However, due to the intracellular localization of some DUBs, the homogenization buffer may not be sufficient to solubilize the protein of interest. Increasing the detergent or salt concentration, or including enzymes such as micrococcal nuclease, DNAse I, or Benzonase® to solubilize chromatin associated proteins may be required [10, 11]. In addition, if the reactivity of the DUB with the activity probe is not optimal using the standard activity probe assay buffer, altering the pH, salt, detergent, or reducing agent may improve the activity of the DUB [12].

4. The incubation time and temperature can be optimized for individual DUBs. Some DUBs react quickly with the ubiquitin-based activity probe and may only require short incubation

times (5–15 min), while others may require longer reaction times or higher temperatures in order to see a desirable shift with the activity probe [5].

5. For large DUBs, running the gel longer may be necessary in order to see a separation between the native DUB and DUB-ubiquitin complex. It may also be worth considering the use of a different gel percentage (3–8 %) or gel type (Tris-acetate) [13].
6. Check the cross-reactivity of the antibody in various species (e.g., mouse, rat, and human). This is usually indicated on the antibody datasheet or can be determined based on the sequence alignment of the protein in different species and the immunogen used to generate the antibody [14].

Acknowledgment

The authors would like to acknowledge networking support by the Proteostasis COST Action (BM1307).

References

1. Ovaa H, Galardy PJ, Ploegh HL (2005) Mechanism-based proteomics tools based on ubiquitin and ubiquitin-like proteins: synthesis of active site-directed probes. Methods Enzymol 399:468–478
2. Ekkebus R et al (2013) On terminal alkynes that can react with active-site cysteine nucleophiles in proteases. J Am Chem Soc 135(8): 2867–2870
3. Ovaa H et al (2004) Activity-based ubiquitin-specific protease (USP) profiling of virus-infected and malignant human cells. Proc Natl Acad Sci U S A 101(8):2253–2258
4. Galardy P, Ploegh HL, Ovaa H (2005) Mechanism-based proteomics tools based on ubiquitin and ubiquitin-like proteins: crystallography, activity profiling, and protease identification. Methods Enzymol 399:120–131
5. Altun M et al (2011) Activity-based chemical proteomics accelerates inhibitor development for deubiquitylating enzymes. Chem Biol 18(11):1401–1412
6. Workman P et al (2010) Guidelines for the welfare and use of animals in cancer research. Br J Cancer 102(11):1555–1577
7. Diehl KH et al (2001) A good practice guide to the administration of substances and removal of blood, including routes and volumes. J Appl Toxicol 21(1):15–23
8. Hoggatt AF et al (2010) A spoonful of sugar helps the medicine go down: a novel technique to improve oral gavage in mice. J Am Assoc Lab Anim Sci 49(3):329–334
9. Langford DJ et al (2010) Coding of facial expressions of pain in the laboratory mouse. Nat Methods 7(6):447–449
10. Dignam JD, Lebovitz RM, Roeder RG (1983) Accurate transcription initiation by RNA polymerase II in a soluble extract from isolated mammalian nuclei. Nucleic Acids Res 11(5): 1475–1489
11. Mendez J, Stillman B (2000) Chromatin association of human origin recognition complex, cdc6, and minichromosome maintenance proteins during the cell cycle: assembly of prereplication complexes in late mitosis. Mol Cell Biol 20(22):8602–8612
12. Altun M et al (2015) The human otubain2-ubiquitin structure provides insights into the cleavage specificity of poly-ubiquitin-linkages. PLoS One 10(1):e0115344
13. Maniatis T, Fritsch EF, Sambrook J (1982) Molecular cloning: a laboratory manual. Cold Spring Harbor Laboratory, Cold Spring Harbor, NY, x, 545 p
14. Harlow E, Lane D (2006) Immunoblotting: preparing cell lysates. CSH Protoc 2006(3). pii: pdb.prot4299

Chapter 28

High-Throughput siRNA Screening Applied to the Ubiquitin–Proteasome System

Esben G. Poulsen, Sofie V. Nielsen, Elin J. Pietras, Jens V. Johansen, Cornelia Steinhauer, and Rasmus Hartmann-Petersen

Abstract

The ubiquitin–proteasome system is the major pathway for intracellular protein degradation in eukaryotic cells. Due to the large number of genes dedicated to the ubiquitin–proteasome system, mapping degradation pathways for short lived proteins is a daunting task, in particular in mammalian cells that are not genetically tractable as, for instance, a yeast model system. Here, we describe a method relying on high-throughput cellular imaging of cells transfected with a targeted siRNA library to screen for components involved in degradation of a protein of interest. This method is a rapid and cost-effective tool which is also highly applicable for other studies on gene function.

Key words Ubiquitin, Proteasome, siRNA, Screening, Degradation

1 Introduction

The ubiquitin–proteasome system (UPS) is responsible for the majority of intracellular protein degradation in eukaryotic cells [1]. In this pathway, the specificity is ensured by conjugating substrates to a chain of ubiquitin moieties. This polyubiquitylation process requires a cascade of three types of enzymes: E1, E2, and E3. First, ubiquitin is bound and activated by the E1 ubiquitin-activating enzyme. Then the activated ubiquitin moiety is transferred to the E2 ubiquitin-conjugating enzyme which finally, via the E3 ubiquitin-protein ligase, transfers ubiquitin to the target protein. Once ubiquitylated, the target protein gains affinity for the 26S proteasome and is rapidly degraded [1]. However, ubiquitylation is a reversible process and deubiquitylating enzymes (DUBs) may deubiquitylate the target protein back into its unmodified form [2] and thus ultimately rescue a protein from degradation. The human genome encodes two ubiquitin-specific E1s, more than 30 E2s [3], more than 600 E3s [4] and approximately 100 DUBs [2]. Accordingly, the specificity in protein degradation lies primarily

Rune Matthiesen (ed.), *Proteostasis: Methods and Protocols*, Methods in Molecular Biology, vol. 1449,
DOI 10.1007/978-1-4939-3756-1_28, © Springer Science+Business Media New York 2016

with the substrate-binding E3s, however, both E2s and DUBs also contribute to specificity. Hence, linking a substrate of the UPS to the relevant E2(s), E3(s), and DUB(s) is a major challenge, which is further complicated by a significant amount of redundancy and cross-talk between the individual degradation pathways. For instance, multiple E3s, including Mdm2, Hrd1/Synoviolin, and CHIP, have been linked to ubiquitylation of p53 [5]. For these reasons, comprehensive high-throughput approaches are often required to identify components involved in the degradation of a specific protein, particularly in mammalian cells that are not tractable for systematic genetic analyses.

Here, we describe a method relying on high-throughput cellular imaging of cells transfected with a targeted siRNA library to screen for components involved in the degradation of a protein of interest (Fig. 1). Briefly, a reporter cell line expressing the protein of interest is reverse transfected by seeding in 384-well plates where each well contains transfection agent and a unique siRNA targeting a component of the UPS. After a defined time, the cells are fixed and processed for high-throughput fluorescence microscopy. Finally, high-content image analysis is used to determine cell-based fluorescence intensities from acquired images, allowing the user to identify targets that, upon knockdown, lead to an increased level of the protein of interest.

This is a powerful tool for studying gene function which has previously been successfully applied to mapping protein degradation pathways [6–8]. In comparison with other methods, such as those utilizing the CRISPR/Cas9 system [9], this technique is simpler, but like most high-throughput technologies prone to yield false positives and false negatives. Careful statistical analyses and subsequent verification of screening hits by orthogonal approaches, such as pulse-chase or cycloheximide shut-off analyses, is therefore essential.

2 Materials

2.1 Cell Culture

1. Any mammalian cell line that displays adherent growth in a monolayer and is easily transfected, such as HeLa or U2OS.
2. The cells are propagated using standard culture media such as Dulbecco's Modified Eagle's Medium (DMEM) supplemented with 10% fetal calf serum, 200 μM glutamine, 90 U/mL penicillin, and 70 U/mL streptomycin sulfate.
3. 0.25% Trypsin in 10 mM Na-citrate pH 7.8, 100 mM NaCl.
4. Phosphate buffered saline (PBS).
5. Hemocytometer.

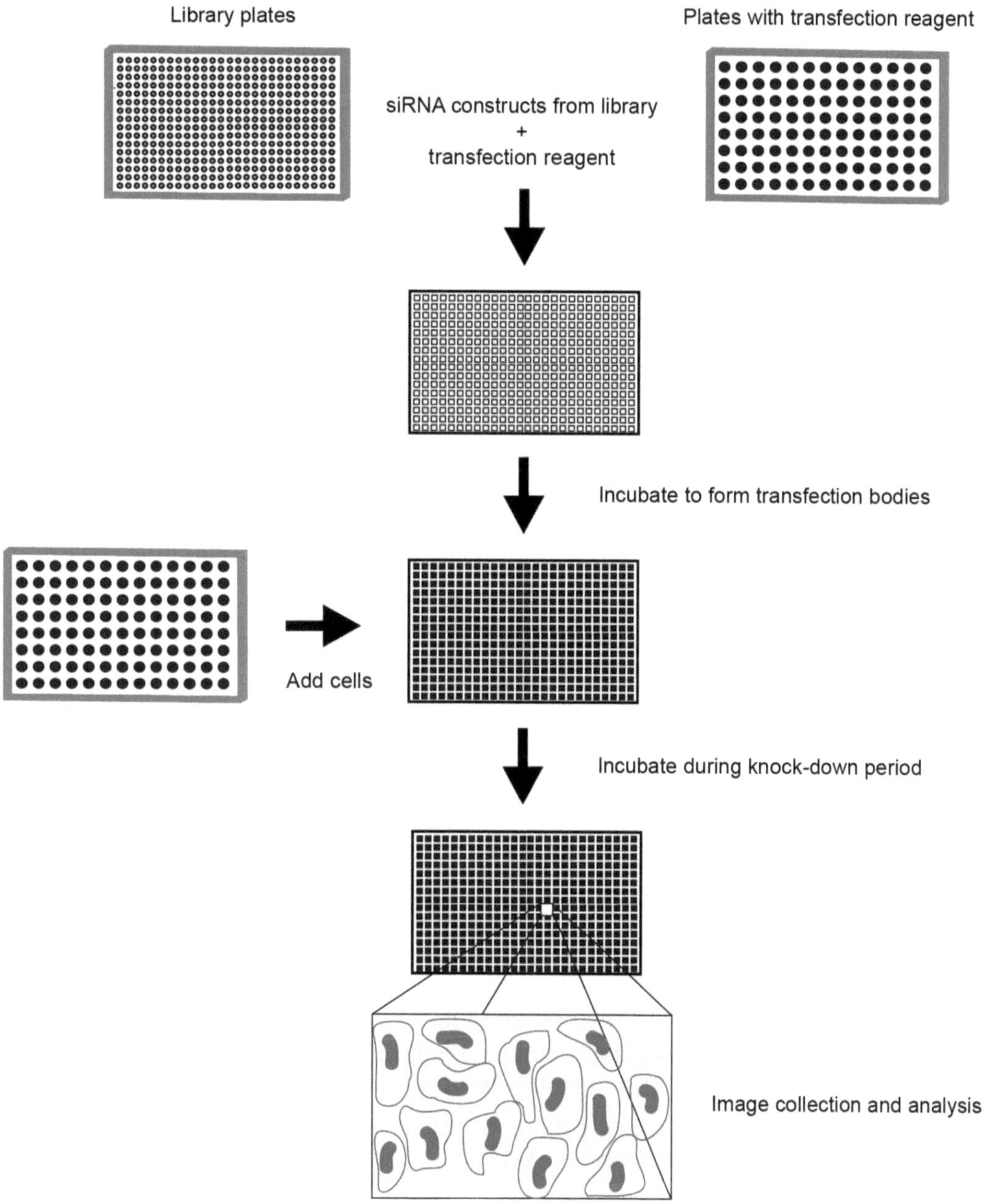

Fig. 1 *Workflow of the siRNA screening method.* The figure depicts a schematic overview of the described siRNA screening protocol

2.2 Chemical Inhibitors

1. 50 mM MG132 stock in DMSO (use at 25 μM final concentration) or 25 mM Bortezomib stock in DMSO (use at 25 μM final concentration).
2. 10 mg/mL cycloheximide stock in DMSO (use at 10 μg/mL final concentration).

2.3 siRNA Transfection

1. Any lipid-based siRNA transfection agent suitable for the cell line in question, such as Lipofectamine RNAiMAX (Invitrogen), MISSION siRNA transfection reagent (Sigma-Aldrich), DharmaFECT (GE Healthcare), or HiPerFect (Qiagen). In general we use RNAiMAX.
2. Serum reduced medium, such as OptiMEM (Invitrogen), UltraMEM (Lonza), or TransfectaGRO (Corning).

2.4 siRNA

1. An siRNA library, such as ON-target plus library (GE Healthcare), Silencer select (Invitrogen), or MISSION siRNA (Sigma-Aldrich) (*see* **Note 1**). Our siRNA library is based on ON-target plus product.
2. Positive and negative control siRNAs (*see* **Note 2**).
3. siRNA resuspension buffer: 10 mM Tris–HCl pH 8.0, 20 mM NaCl, 1 mM EDTA in RNase-free water.

2.5 Nuclear Counter Stain

1. A nuclear stain, such as 10 μM final concentration of DAPI or Hoechst 33342 in PBS (*see* **Note 3**).

2.6 Plates

1. Assay Plate 384-well, flat bottom, tissue culture treated, black with clear bottom, sterile, polystyrene (Corning).
2. 96- and 384-well plates for short and long term storage of reagents.
3. Microplate seals.
4. Plate centrifuge.
5. Multichannel pipette.

2.7 Immunofluorescence

1. Cell fixation reagent, such as 4 % formaldehyde in PBS or Gurrs Histological Fixative (Sigma-Aldrich).
2. Permeabilization buffer: 0.2 % Triton X-100 in PBS.
3. Blocking solution: 5 mg/mL Bovine Serum Albumin (BSA) in PBS containing 20 mM glycine, and 0.2 % Triton X-100.
4. Antibody dilution buffer: 5 mg/mL BSA in PBS containing 0.2 % Triton X-100.

2.8 Liquid Handling

1. Microlab STARlet Liquid Handling Workstation utilizing a 384 Probe Head (Hamilton Robotics).
2. Venus Software.

2.9 Automated Fluorescence Microscopy

1. GE Healthcare INCell1000.
2. INCell1000 imaging software.
3. 10× Objective.
4. 360–420 nm filter for Hoechst or DAPI visualization.
5. Second filter for visualization of protein of interest.

2.10 Software

1. In Cell Analyzer 1000 Workstation software.

2.11 Statistical Analysis

1. The statistical software R 3.0.2 and R Studio.
2. Redundant siRNA Activity (RSA) analysis algorithm [10].

3 Methods

3.1 Cell Culture

1. For maintenance, keep cells in DMEM containing 10% fetal calf serum supplemented with glutamine, and penicillin/streptomycin in a humidified atmosphere containing 5% CO_2 at 37 °C.
2. When passaging the cells, split the culture 1:10 using trypsin.

3.2 MG132 or Bortezomib Treatment

1. Dilute the proteasome inhibitor in serum-free DMEM to a final working concentration of 25 μM Bortezomib or MG132.
2. Both proteasome inhibitors have a tendency to precipitate when first added to DMEM. We recommend warming the inhibitor-containing medium to 37 °C followed by thorough mixing prior to use.
3. Remove the growth medium from cells and add the proteasome inhibitor-DMEM solution prepared in **step 1**.

3.3 Cycloheximide Treatment

1. Dilute the cycloheximide stock in serum-free DMEM to obtain a final concentration of 10 μg/mL.
2. Aspire growth medium from cells and add the cycloheximide-DMEM solution.

3.4 Library Dilution and Storage

1. The layout of the master plates is designed in collaboration with the retailer during purchasing (*see* **Note 4**).
2. The 96-well master plates containing lyophilized siRNA constructs should be handled under strict RNase-free conditions.
3. The siRNA constructs are first resuspended in siRNA resuspension buffer.
4. On the liquid handling robot, the siRNA is then transferred to 384-well library plates containing a siRNA resuspension buffer of a volume yielding the desired final library plate concentration (*see* **Note 5**).
5. Seal the plates and store at −20 °C (*see* **Note 6**).

3.5 Standard Screening Protocol for 384-Well Plates

1. Prior to commencing a full library screen, we advise that the substrate is thoroughly characterized (*see* **Note** 7) (Fig. 2) and a smaller pilot screening is attempted (*see* **Note 8**) where cell density (*see* **Note 9**), experimental range and the controls are tested (Fig. 3). For volumes relevant for 6-well and 96-well plates, consult Table 1 and *see* **Note 10**.

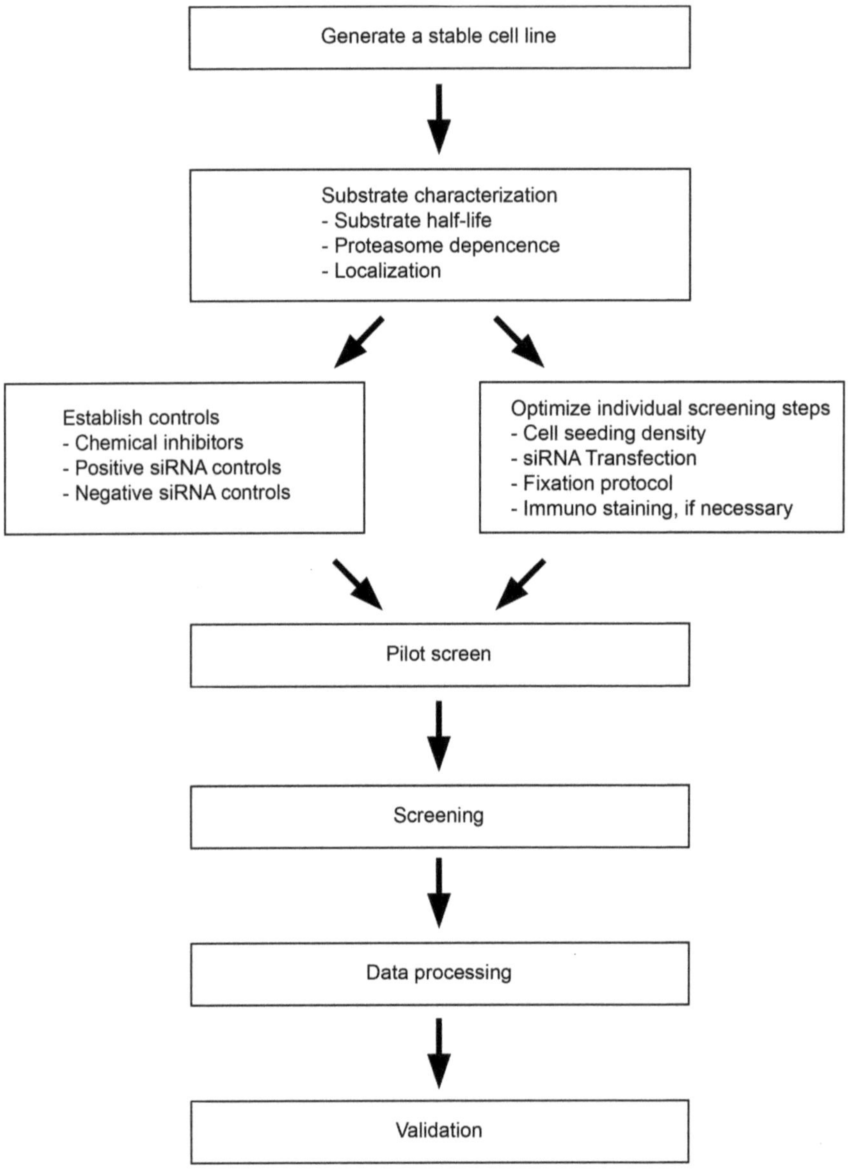

Fig. 2 *Prescreening work flow.* The figure depicts a schematic overview of the prescreening and actual siRNA screening protocol

2. On day 1, begin by preparing: 110 μL RNAi transfection agent dissolved in 24.89 mL reduced serum medium.
3. siRNA library from −20 °C freezer (in library plates) is thawed and centrifuged at 800 × *g* prior to using.
4. Using the liquid handling system, transfer 4 μL of tenfold concentrated siRNA into imaging plates.
5. Add 9 μL of the transfection reagent in reduced serum medium.
6. Incubate for 20 min at room temperature.

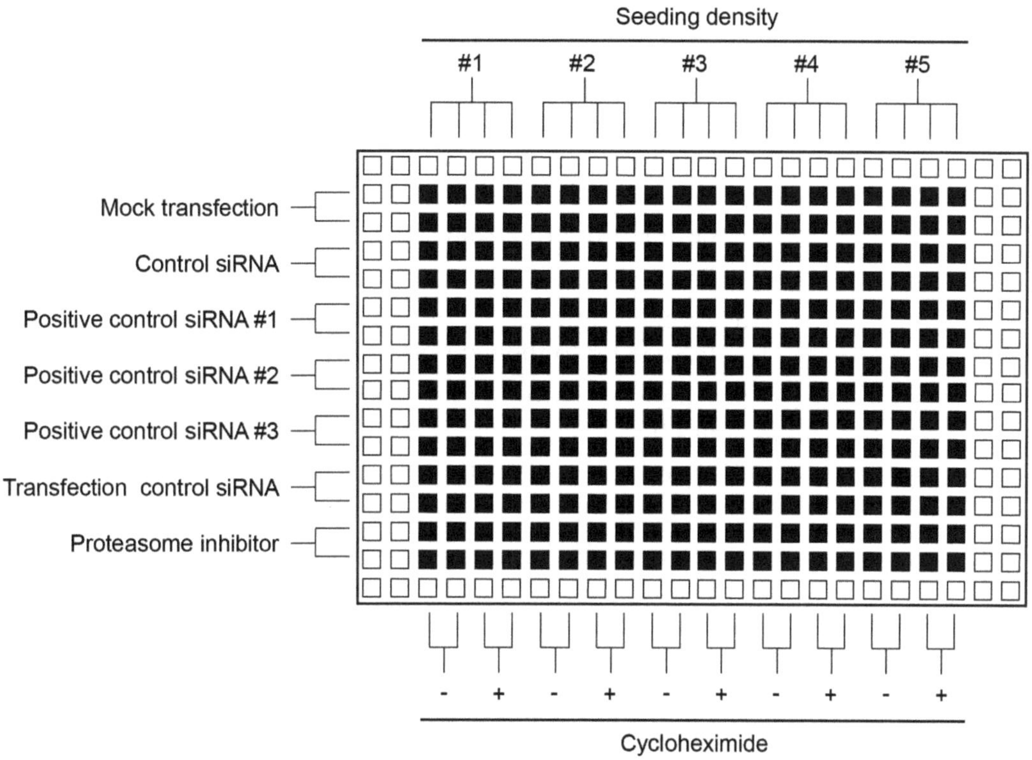

Fig. 3 *An example of a pilot screening plate.* The layout of a pilot screening plate could be as shown. This particular plate includes transfection controls and proteasome inhibitors (*vertical*), cycloheximide (*horizontal*), and testing of five different seeding densities (*horizontal*). Note that to avoid edge effects, the wells along the edge are not included

Table 1

Amounts required for siRNA transfection in various plate formats

Reagent	6-Well	96-Well	384-Well
500 nM siRNA in media (μL)	500	16	4
Transfection agent in media (μL)	812.5	36	9
Cell suspension (μL)	3375	108	27

7. While incubating siRNA/transfection reagent, split cells, count, and dilute to desired density.
8. Add 27 μL cell suspension.
9. Spin down plates at 800 × *g* and check the cell distribution by light microscopy.
10. Incubate the cells overnight under standard conditions.

11. On day 2, evaluate cell morphology and adhesion by light microscopy; make a note of any divergences from the pilot screens.
12. Incubate the cells for another 24 or 48 h under standard conditions.
13. On day 3 or day 4, add any required pretreatments, e.g., cycloheximide or proteasome inhibitors.
14. Proceed to stain and fix cells.
15. Prepare a tray with Gurr histological fixative or equivalent, to be used for fixation (*see* **Note 11**).
16. Remove growth medium from the cells and wash them twice in PBS.
17. Fix cells by adding 40 μL fixative to each well.
18. Incubate for 10–20 min at room temperature.
19. After fixation, wash cells in PBS and proceed with immunofluorescence staining if required, otherwise proceed with a nuclear staining.
20. For immunofluorescence (*see* **Note 12**), all steps are performed at room temperature. Permeabilize fixed cells by incubating in 0.2 % Triton X-100 in PBS for 5 min.
21. Block with PBS containing 5 mg/mL BSA, 20 mM glycine, and 0.2 % Triton X-100 for 30 min.
22. Dilute the primary antibody in PBS containing 5 mg/mL BSA and 0.2 % Triton X-100, and incubate cells with the antibody solution for 2 h.
23. Wash the cells with PBS to remove excess primary antibody.
24. Dilute the fluorophore-conjugated secondary antibody in PBS containing 5 mg/mL BSA and 0.2 % Triton X-100, and incubate cells with the antibody solution for 1 h.
25. After adding the secondary antibody, the experimental plate must be shielded from light to avoid bleaching of the fluorophore.
26. Wash the cells with PBS to remove excess secondary antibody.
27. For nuclear staining, prepare a tray with 50 μM Hoechst 33342 nuclear stain in culture media (*see* **Note 12**).
28. On the liquid handling system, add 40 μL of the Hoechst 33342 solution from **step 27** to each well.
29. Incubate for 20 min at room temperature.
30. Wash wells twice with PBS.
31. If not continuing directly with imaging, store the plates in the dark at 4 °C.

3.6 Liquid Handling

1. Automation of described transfection and immunostaining protocols (Subheading 3.5) within Venus software at the Microlab STARlet Liquid Handling Workstation (*see* **Note 13**).

3.7 Image Acquisition

1. Using the GE Healthcare INCell1000 automated microscope, capture four fields/well with 10× objective (*see* **Note 14**).
2. Image Hoechst staining using the 360 nm excitation filter and the 420 nm emission filter.
3. Image protein of interest using a primary antibody that can be visualized with a secondary antibody coupled to a fluorophore.

3.8 Image Analysis

1. Using INCell Analyzer 1000 Workstation (*see* **Note 15**) individual cells are segmented and counted by finding a nuclear-like shape of a minimum size of 45 μm^2 in the nuclear signal using the top hat segmentation option in the INCell Analyzer software. This algorithm emphasizes and extracts objects of a prespecified size.
2. Intensity of the cellular stain is then measured in the cellular stain channel by expanding each nucleus outwards to a minimum size of 80 μm^2 to represent the cytoplasm using multi-scale top hat segmentation in the INCell Analyzer software.
3. The average intensity of the nuclear compartment plus the cytoplasmic compartment is measured in the cellular channel and used as a read-out for further analysis (*see* **Note 16**).
4. An intensity threshold was set for each plate, in which ~1 % of the cells in the negative control wells were scored as positive (*see* **Note 17**). The percentage of positive cells in each well was then used for statistical analysis. For example of segmentation *see* Fig. 4.

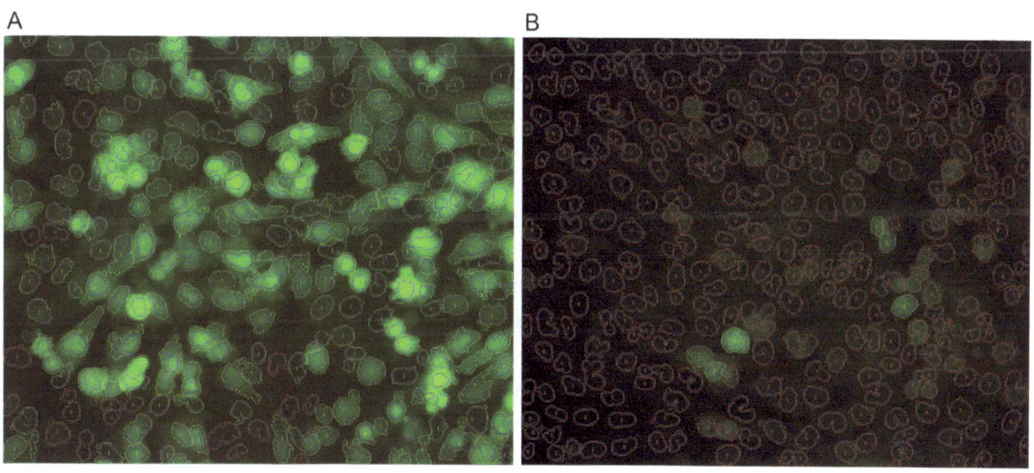

Fig. 4 *Example of segmentation of cells.* Segmentation example of field with 92 % positive cells (**a**) and 2 % positive cells (**b**) from INCell Analyzer 1000 workstation software. Images acquired during screening for members of the ubiquitin-fusion degradation pathway [7]

3.9 Statistical Analysis

Statistical analysis (Fig. 5) is performed based on an R pipeline developed in house (*see* **Note 18**).

1. Data Triage (*see* **Note 19**).
   (a) Observations during screening process.
   (b) Plate visualization by means of heatmaps for each plate, boxplot of all samples per plate and boxplot of controls vs. samples.
2. Normalization (*see* **Note 20**).
   (a) by median of negative control, positive control, and all samples.
3. Calculation of Quality Metrics (*see* **Note 21**) using Strictly Standardized Mean Difference (SSMD) [11].
4. Hit Identification (*see* **Note 22**) using RSA [10].

4 Notes

1. When faced with the task of performing siRNA screens, it is important to consider if a genome-wide approach is necessary, or if perhaps a library targeted toward a more narrow selection of specific genes of interest is more appropriate. Obviously, the advantage of a genome-wide approach is that the roles of all transcripts are evaluated. However, all genes with known or predicted functions in the UPS will, at most, comprise about 1000 different genes. Thus, focusing instead on such a targeted selection of genes, will significantly reduce the complexity and costs of the task. As typical with screening efforts, one often aims at obtaining a balanced number of hits. A high number of hits can leave you stranded in the validation phase for years, while screening a too narrow selection can leave you without any hits to pursue.

   For a successful siRNA screen it is obviously essential that the target genes are efficiently knocked-down. To achieve this it can be an advantage to pool 2–3 different siRNAs toward each target gene in the same well. However, pooled siRNAs also increase the risk of false positive signals since individual siRNAs with strong off-target effect can lead to an overestimation of the gene specific scoring. Vice versa, weak signals on multiple siRNAs per gene would be scored as false negatives. Thus, we prefer to use three different siRNAs directed against each target gene in separate wells, as this lowers the risk of both false positive and negative hits and ultimately allows for a much stronger statistical analysis on the obtained data.
2. Negative control siRNAs can be either control siRNAs designed not to target any transcripts from the genome, or siRNAs specific for exogenous transcripts such as luciferase or GFP. It is

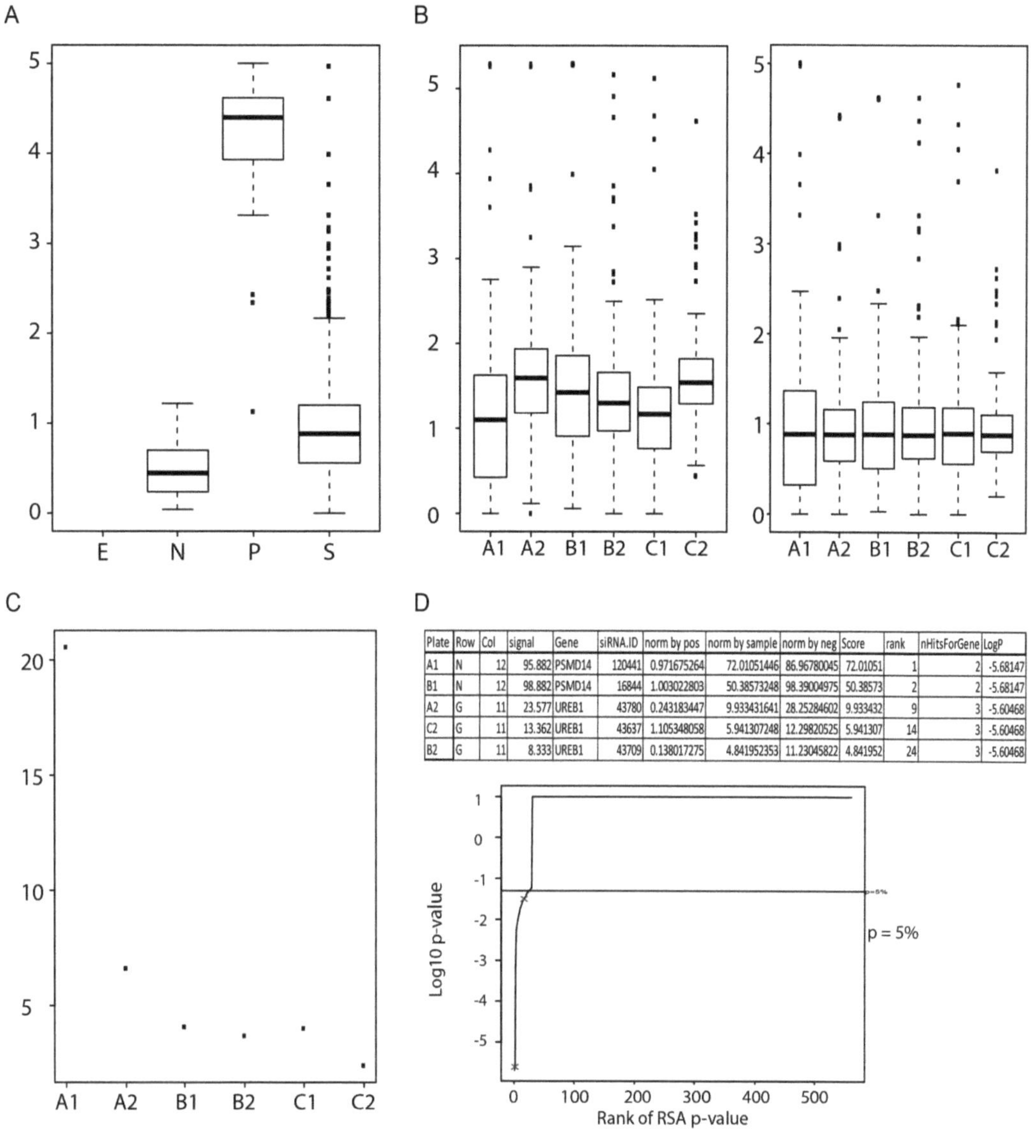

Plate	Row	Col	signal	Gene	siRNA.ID	norm by pos	norm by sample	norm by neg	Score	rank	nHitsForGene	LogP
A1	N	12	95.882	PSMD14	120441	0.971675264	72.01051446	86.96780045	72.01051	1	2	-5.68147
B1	N	12	98.882	PSMD14	16844	1.003022803	50.38573248	98.39004975	50.38573	2	2	-5.68147
A2	G	11	23.577	UREB1	43780	0.243183447	9.933431641	28.25284602	9.933432	9	3	-5.60468
C2	G	11	13.362	UREB1	43637	1.105348058	5.941307248	12.29820525	5.941307	14	3	-5.60468
B2	G	11	8.333	UREB1	43709	0.138017275	4.841952353	11.23045822	4.841952	24	3	-5.60468

Fig. 5 *Statistical Analysis.* The four different steps in the statistical analysis pipeline exemplified by data produced during screening for members of the ubiquitin-fusion degradation pathway [7]. (**a**) Boxplot representing all signals by sample type, i.e., empty wells (E), negative control (N), positive control (P), and samples from the screening library (S). (**b**) Boxplots representing all signals plate-wise. *Left*: raw data; right: normalized to the median of all samples on the plate. (**c**) SSMD values for individual screening plates. (**d**) Hit identification by RSA analysis. *Top*: top head of the results table ranked by gene specific *p*-values; *bottom*: diagram of ranked genes vs. their log10 *p*-values with significant hits scoring by a *p*-value below 0.05

necessary to consider the complexity of the assay when choosing positive controls. Ideally, each step of the protocol should have an internal positive control. A list of controls includes, but is not limited to, a control for any pretreatment of the cells prior to the siRNA transfection, a control for the siRNA transfection itself, and a control for any staining steps prior to data acquisition.

When working with the UPS, positive control siRNAs to assess the effect on the substrate could target certain proteasome subunits or other genes known to be required for degradation of the protein of interest. As a rule of thumb, knocking down the positive control transcript should give a stabilizing effect closely related to the expected positive hits, as this will provide insights on the range of signal to be expected in the screen. As an example, if one is looking to identify the E3 responsible for the degradation of a known substrate, and one already knows that a specific chaperone is involved in the shuttling of the substrate to the proteasome, then this chaperone would be the optimal choice for a positive control, despite the fact that a knockdown of a proteasomal subunits might yield a higher positive signal than a knockdown of the chaperone.

Other controls could include wells loaded with proteasome inhibitor or simply wells without any siRNA in the transfection mix. When choosing siRNA for the controls, it is recommended to use siRNA of the same brand and make as the siRNA in the library, and this also applies to any later additions to the library.

3. The nuclear stain must not overlap with the emission and excitation patterns of the fluorophore used for labeling the substrate in the screen. An online tool, such as the "Fluorescence SpectraViewer," provided by Invitrogen on their web page, may be of use when comparing spectra and choosing the correct nuclear stain.
4. High quality siRNA libraries can be purchased from several research supply companies, and are often delivered for convenience as aliquots in so-called master plates of 96- or 384-well format.

   We recommend distributing the siRNAs from these master plates between so-called library plates containing enough siRNA for five screens. For the actual screens the siRNAs are then collected from the library plates and distributed to screening plates.

   When receiving such a library in master plates, it may be lyophilized and require resuspension. We recommend siRNA to be diluted in siRNA resuspension buffer, or alternatively a buffer supplied from the siRNA manufacturer. The siRNA should also be diluted to a suitable concentration for high-throughput systems, and be aliquoted and distributed in multiple 384-well library plates, preferably using a liquid handling robot. When performing these initial dilutions, it is important to be attentive to the future experimental setups one wishes to have available. A good library layout will make future work much easier.

   The primary consideration for assembling a siRNA library against UPS components is if the entire UPS is to be screened every time the library is used, or if it would be sensible to divide the major UPS components onto separate plates. For instance, it might be practical to have the ability to perform a smaller screen where one is only looking at E2s, E3s, or proteasomal interaction partners. Such a division of the library does not hinder the use of the full library for a complete UPS screen, as long as it is not too finely subdivided.

To be able to account for plate-to-plate variation, each plate should contain a number of positive and negative control siRNAs in replicates. Also, leaving some wells empty will, if necessary, allow for spiking of the library with other siRNAs at a later point, but can otherwise function as mock controls. In addition, wells along the edge of the plate should be avoided due to potential edge effects, such as temperature and evaporation gradients. The library plate setup is mirrored on the experimental plate, and any experiments including siRNA from an edge well in the library plate will be performed in wells on the edge of the experimental plate.

5. When diluting the siRNA, it should be diluted to a concentration that allows for exact pipetting in an automated setup. This means that the concentration cannot be too high, as this would require too small a volume be pipetted during screening. Similarly, the siRNA must be of a sufficiently high concentration to not significantly dilute the media in the wells with siRNA buffer, and to not exceed the volume of the wells. A tenfold concentrated siRNA stock is a good starting point. Identifying the ideal working concentration for the type of siRNA in the library will then determine the stock concentration of the library plates. Example: for a 50 nM working concentration in the well, one would store the siRNA in 500 nM aliquots on the library plates.
6. The number of freeze/thaw cycles should also be considered, along with the dead volume of the pipetting system. It is recommended to have a standardized limit on freeze/thaw cycles, so that all plates are volume optimized for, e.g., five experiments, resulting in minimal waste of siRNA, and a consistent quality of knockdown.
7. The initial characterization of the substrate should ideally be performed using fluorescence microscopy as the primary readout since this will be the case for the final screen. Western blotting can be used in parallel. Also, if there is a need for an immunostaining, it is necessary to test this on the robotics setup, both in terms of the robot performing the stain, and in the quality of the collected data.

   The ideal screening substrate is undetectable in untreated cells due to the rapid degradation, and appears upon treatment with proteasome inhibitors, as it accumulates when degradation is blocked. Treating cells with proteasome inhibitor is toxic to the cells, and thus, the stabilization should be achieved within 6–10 h in order to prevent cell death. Note that the toxicity of proteasome inhibition varies from cell line to cell line. A positive siRNA control should copy the effect of proteasome inhibition after a 48 or 72 h knockdown period.

8. Prior to the full scale screening experiment, one should run at least one pilot screen to adjust the experimental setup to the high-throughput system and to ensure that the controls behave as expected. An example of such a pilot screening setup (Fig. 3) could include siRNA controls, cycloheximide and proteasome inhibitor treatments, and different seeding densities.
9. Attaining a uniform and reproducible cell density and distribution is of high priority during screens. There must be little to no variation between runs, and ideally there should also be little to no variation between wells. Cell counting should be as standardized as possible, but also the homogenization of the cells prior to seeding is important. Since some cell lines will tend to lump together after dislodgement with trypsin. To better obtain single cells, either leave the cells in trypsin a little longer, or add EDTA to the trypsin solution.

   Another important parameter affecting the final cell density is the cell loss during fixation, washing, and staining. Slow, steady pipetting and a minimum of movement of the plates prior to fixation are essential. To address the problem of varying cell adhesion in wells due to heat gradients across the wells and plates, it may be advisable to leave the plates at room temperature for 1 h after the cells are added to the wells, and prior to placing them in the CO_2 incubator [12].
10. Depending on the half-life of the substrate, the efficiency of the siRNA, and the turnover of the proteins being targeted, it may be necessary to adjust the knockdown period for the screen. Since the removal of some transcripts will have a detrimental effect over time, a longer experiment might actually lower the number of positive scoring cells in these wells, as the cells would cease to divide, die, or simply detach. This would be apparent in the cell count, why we recommend that this is determined (i.e., by nuclear staining with Hoechst or similar). We find that testing 24, 48, and 72 h of knockdown will often provide a workable window.
11. Ideally, volatile organic solvents, such as methanol, should be avoided, due to the difficulty in pipetting these in a precise manner on the robot and the risk of completely drying the wells due to evaporation. It is important to note that evaporation from small wells is rapid and is a general concern in all steps.
12. Since the segmentation algorithm used in the data analysis should be based on a nuclear staining, the fluorophore chosen

for the substrate should be matched so that there is a minimal overlap in excitation and emission energies between the substrate and the nuclear counter staining.

In general, the staining protocol should be as minimalistic as possible, in order to minimize potential wash-off of cells.

13. If available, automated liquid handling is clearly the fastest, most reproducible, and reliable way to pipet high-throughput screens. When choosing a robotic liquid handler for HT cell-based screens, one should however consider the following:
    - It should provide enough deck space to be able to handle several plates at the same time.
    - Preferably it should be equipped with a 384 or at least 96 probe head to pipet one plate at a time.
    - The pipetting height should be flexible to be able to find the optimal height for taking off as much medium as possible while at the same time the cells stay untouched.
    - It should be able to deal with different liquid classes. When pipetting small amounts of liquid it makes a big difference if the liquid is slightly viscous or volatile.
    - The pipetting speed should be adjustable. Especially, when working with adherent cells it is absolutely crucial to exchange medium carefully in order not to wash off the cells.
    - In the process of upscaling and automating lab-scale transfections, treatments and/or immunostainings, it is important to simplify the protocols as much as possible in order to minimize pipetting steps and thereby expensive consumables. Unfortunately, basic liquid handling robots do not provide cooling, shaking and plates cannot be tilted in order to remove maximum amount of liquid. Thus, we suggest reverse instead of forward transfections for high-content screening, i.e., pipetting cells on top of transfection reagents in order to avoid additional mixing steps. In most of the cases, reverse transfections turn out to be more effective as well.
    - Furthermore, every step has to take a certain amount of dead volume into account as the robot will not be able to take off total volumes.
    - A potential concern is that most liquid handling stations will not be setup in sterile environment, i.e., cells might be prone to infections during transfection process. Therefore we recommend using antibiotics during transfection despite common recommendations in standard transfection protocols. Usually, the transfection efficiencies are not affected and we never experienced any infections upon open environment transfections.

14. In order for the microscopy images to capture a true representation of the knockdown effect by each siRNA construct, a sufficient amount of cells has to be imaged in each well and ideally a similar amount of cells in all the wells. In general, the fluorescent signal corresponding to at least 1000 cells gives enough data for the result to statistically represent the biology of the cells in each well. In addition, spread out the fields evenly in the well to capture a good representation of the well, and make sure the signal is within the dynamic range of the instrument and does not reach saturation when setting up exposure length.
15. Several software options for image analysis of large scale image projects such as a siRNA screen have been developed. Most automated microscopes have accompanying proprietary software developed by the microscope company in question. These programs are developed to sort and handle large amounts of images and data and to provide the data generated from image analysis in an orderly and easily readable format. An alternative, free, open source software specifically designed to analyze large amounts of image data is CellProfiler, developed at the Broad Institute [13].
16. Image analysis software can be set to generate a large amount of measures and ratios of signals from the background, nuclear, and cytosolic compartment. As an alternative read-out, the ratio between the intensity in the whole cellular compartment and the local background intensity can be used. The integrated intensity of the cellular channel can also be used, which is the average intensity of the whole cell in the cellular stain channel multiplied by the cellular area. Finally, make sure that cells are of a proper confluency, not only for the biology of the cells, but also for the analysis software to be able to properly find and segment the individual cells.
17. The threshold should be set separately for each plate to account for differences in staining, variations in imaging and plate-to-plate variation. Make sure that the threshold gives the largest possible window to detect positively scoring constructs. By setting a threshold that is too high, the window might be too small. However, a threshold set too low increases the risk for false positives.
18. Small cell culture volumes, off-target effects, biological cross reactions but also unspecific staining makes HT siRNA screening prone to errors and variations which have to be taken into account during setup, screening, and data evaluation. As such, the choice of proper positive and negative siRNA-controls, preferably some additional chemical controls and relevant

antibody controls is absolutely crucial. In addition, comprehensive statistical analysis should be applied to the raw data. Here, we suggest a statistical pipeline, which has been developed in-house and is based on consideration by Birmingham et al. [14].

19. The first step in data evaluation should be a careful observation during and after the screening process. Missed wells, unequal volumes but also more or less exposed positions in the incubator over time could be sources of errors or variations over the plates. This can afterwards be further visualized by heatmaps illustrating potential edge effects, general shifts in intensities over the plates or variations between the controls. Additional boxplots comparing all the signals plate-wise or between samples and controls can further help identifying potential outliers.

20. To be able to account for unspecific variations within individual plates, we suggest placing all relevant controls on each of the screening plates. Data can then be plate-wise normalized to the median of either the controls or the overall samples on one plate. In general, we prefer the normalization to the median over the mean, as the latter is too dependent on either the performance of the controls or the number of hits per plate.

    The pipeline will calculate and visualize all three variants of normalization, to the median of the (a) negative control, (b) positive control, and (c) all samples, as the best normalization method will always be dependent on the number of replicates for the controls and the number of samples scoring. However so far, for standard 384-well screens with four replicates per control, such as the described UPS screen, the best normalization has always been to the median of all samples per plate.

21. There are several described ways to account for screen quality. Our preferred alternative to, e.g., Z- or *Z'*-factor [15], SSMD, has a solid statistical basis and was specifically developed for siRNA screens [11]. It is the ratio between the difference of the means and the standard deviations of two populations, e.g., the positive and negative controls.

$$SSMD = (\mu_P - \mu_N) / \surd\left(\delta_P^2 + \delta_N^2\right)t$$

    As a rule of thumb, for screens with upregulated signals acceptable SSMD should be above 1 for moderate, above 2 for strong and above 3 for very strong controls. Correspondingly, it should be below −1 for moderate, −2 for strong and −3 for very strong controls in screens looking at down-regulated signals.

22. Simple *Z*- or *Z'*-score ranking is still one of the most common ways to identify hits in RNAi screens. In deconvoluted screens,

i.e., screens with individual constructs per well, this leaves you with a long list of ranked siRNAs but no combined and statistically sound scoring of genes. Thus, we chose the RSA algorithm [10] for our deconvoluted screens because its combined scoring of all siRNA constructs per gene gives robust gene-based results.

As RNA interference is prone to off-target effects due to the complexity of knockdown in biological systems, single high scores are often less probable to represent true hits than low or medium scoring of several siRNA constructs against the same gene. The RSA algorithm accounts for this by first ranking all the constructs according to their normalized signal, then from these ranks calculating a single hypergeometric derived *p*-value per gene and finally ranking the genes according to their *p*-values. Thus, the list of hits provides a ranking of genes and corresponding *p*-values which can then be used for further validation.

RSA *p*-values from replicate screens can be combined using, e.g., Fisher's Method.

Acknowledgmevnts

The authors apologize to those whose work was not cited due to space constraints. S.V.N., E.G.P., and R.H.P. are supported financially by grants from the Lundbeck Foundation and the Danish Council for Independent Research (Natural Sciences). The authors would like to acknowledge networking support by the Proteostasis COST Action (BM1307).

References

1. Finley D (2009) Recognition and processing of ubiquitin-protein conjugates by the proteasome. Annu Rev Biochem 78:477–513
2. Reyes-Turcu FE, Ventii KH, Wilkinson KD (2009) Regulation and cellular roles of ubiquitin-specific deubiquitinating enzymes. Annu Rev Biochem 78:363–397
3. van Wijk SJ, Timmers HT (2010) The family of ubiquitin-conjugating enzymes (E2s): deciding between life and death of proteins. FASEB J 24:981–993
4. Deshaies RJ, Joazeiro CA (2009) RING domain E3 ubiquitin ligases. Annu Rev Biochem 78:399–434
5. Brooks CL, Gu W (2011) p53 regulation by ubiquitin. FEBS Lett 585:2803–2809
6. Christianson JC, Olzmann JA, Shaler TA, Sowa ME, Bennett EJ, Richter CM, Tyler RE, Greenblatt EJ, Harper JW, Kopito RR (2012) Defining human ERAD networks through an integrative mapping strategy. Nat Cell Biol 14:93–105
7. Poulsen EG, Steinhauer C, Lees M, Lauridsen AM, Ellgaard L, Hartmann-Petersen R (2012) HUWE1 and TRIP12 collaborate in degradation of ubiquitin-fusion proteins and misframed ubiquitin. PLoS One 7:e50548
8. Pegoraro G, Voss TC, Martin SE, Tuzmen P, Guha R, Misteli T (2012) Identification of Mammalian protein quality control factors by high-throughput cellular imaging. PLoS One 7:e31684
9. Barrangou R (2014) RNA events. Cas9 targeting and the CRISPR revolution. Science 344:707–708
10. Konig R, Chiang CY, Tu BP, Yan SF, DeJesus PD, Romero A, Bergauer T, Orth A, Krueger U, Zhou Y, Chanda SK (2007) A probability-

based approach for the analysis of large-scale RNAi screens. Nat Methods 4:847–849

11. Zhang XD (2007) A pair of new statistical parameters for quality control in RNA interference high-throughput screening assays. Genomics 89:552–561
12. Lundholt BK, Scudder KM, Pagliaro L (2003) A simple technique for reducing edge effect in cell-based assays. J Biomol Screen 8:566–570
13. Carpenter AE, Jones TR, Lamprecht MR, Clarke C, Kang IH, Friman O, Guertin DA, Chang JH, Lindquist RA, Moffat J, Golland P, Sabatini DM (2006) Cell Profiler: image analysis software for identifying and quantifying cell phenotypes. Genome Biol 7:R100
14. Birmingham A, Selfors LM, Forster T, Wrobel D, Kennedy CJ, Shanks E, Santoyo-Lopez J, Dunican DJ, Long A, Kelleher D, Smith Q, Beijersbergen RL, Ghazal P, Shamu CE (2009) Statistical methods for analysis of high-throughput RNA interference screens. Nat Methods 6:569–575
15. Zhang JH, Chung TD, Oldenburg KR (1999) A simple statistical parameter for use in evaluation and validation of high throughput screening assays. J Biomol Screen 4:67–73

Chapter 29

High-Throughput Yeast-Based Reporter Assay to Identify Compounds with Anti-inflammatory Potential

G. Garcia, C. Nunes do Santos, and R. Menezes

Abstract

The association between altered proteostasis and inflammatory responses has been increasingly recognized, therefore the identification and characterization of novel compounds with anti-inflammatory potential will certainly have a great impact in the therapeutics of protein-misfolding diseases such as degenerative disorders. Although cell-based screens are powerful approaches to identify potential therapeutic compounds, establishing robust inflammation models amenable to high-throughput screening remains a challenge. To bridge this gap, we have exploited the use of yeasts as a platform to identify lead compounds with anti-inflammatory properties. The yeast cell model described here relies on the high-degree homology between mammalian and yeast Ca^{2+}/calcineurin pathways converging into the activation of NFAT and Crz1 orthologous proteins, respectively. It consists of a recombinant yeast strain encoding the *lacZ* gene under the control of Crz1-recongition elements to facilitate the identification of compounds interfering with Crz1 activation through the easy monitoring of β-galactosidase activity. Here, we describe in detail a protocol optimized for high-throughput screening of compounds with potential anti-inflammatory activity as well as a protocol to validate the positive hits using an alternative β-galactosidase substrate.

Key words Calcineurin, Cell-based assays, Crz1, Drug screening, Inflammation, NFAT, Yeast

1 Introduction

Acute inflammation is an essential and transient biological process settled by the immune system against harmful stimuli such as pathogens, damaged, or apoptotic cells [1]. Its main functions are the removal of injury agents, the cleaning of necrotic and apoptotic cells, and the promotion of tissue repair. On the other hand, chronic inflammation is a process that persists indefinitely after the initial trigger generating a self-perpetuating event, which in turn leads to the progressive damage of the inflammation site and, in some cases, of the neighbor tissues [2]. Consequently, uncontrolled chronic inflammatory responses have a central role in the etiology and progression of several human pathologies [3] including protein-misfolding diseases such as neurodegenerative disorders [4] and type 2 diabetes [5]. In this

Rune Matthiesen (ed.), *Proteostasis: Methods and Protocols*, Methods in Molecular Biology, vol. 1449,
DOI 10.1007/978-1-4939-3756-1_29,

perspective, numerous studies have been focused on the identification of compounds with anti-inflammatory activity to be exploited in immuno-controlled diets or as immunosuppressant therapeutics [6–8] aiming at the restoration of protein homeostasis (proteostasis).

Several molecular players are involved in inflammatory processes [9], among them stands out the Nuclear Factor of Activated T-cells (NFAT) family of transcription factors involved in upregulation of proinflammatory genes such as TNFα [10, 11]. NFAT activation is tightly controlled by Ca^{2+}-dependent signaling pathways. In its phosphorylated state, NFAT localizes to the cytoplasm, where it remains inactive. The increase of cytosolic Ca^{2+} levels triggered by injury agents lead to Ca^{2+} binding to the Ca^{2+}-high affinity protein calmodulin (CaM), which activates the phosphatase activity of calcineurin (CaN) [10]. CaN then dephosphorylates cytosolic NFAT proteins exposing the nuclear localization signal and targeting proteins to the nucleus, where they bind their cognate recognition sequences in the promoter region of a wide array of proinflammatory genes [12, 13] (Fig. 1a). FK506 (Tacrolimus) is an immunosuppressant drug commonly used to prevent the rejection of organ transplants. It blocks CaN activation, consequently repressing NFAT activation [14, 15].

Saccharomyces cerevisiae is the most worthwhile model organism [16], being considered a robust primary drug-screening platform to filter for compounds with cytoprotective activity for further validation in more complex models. In many aspects, *S. cerevisiae*

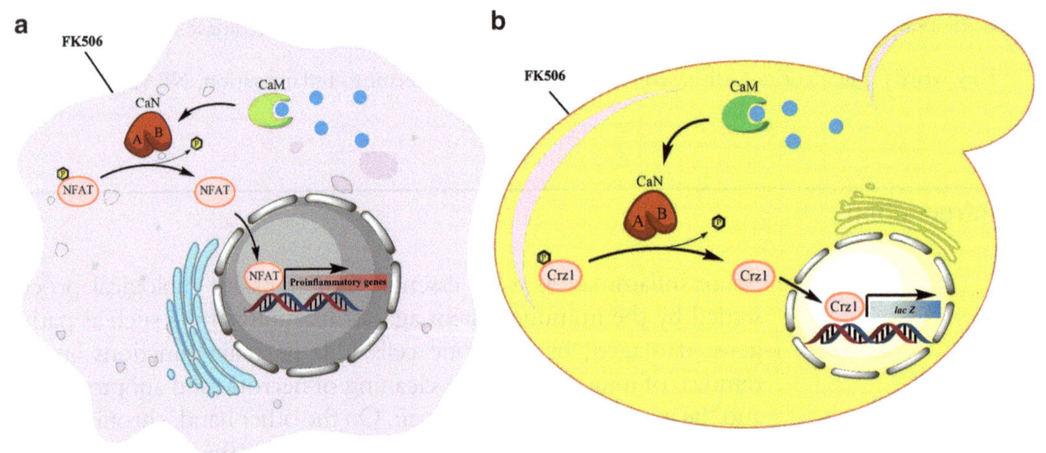

Fig. 1 Schematic representation of Ca^{2+}-dependent signaling pathways modulating NFAT and Crz1 activation in mammals and yeasts, respectively. (**a**) The increase of cytosolic Ca^{2+} levels lead to Ca^{2+} binding to calmodulin (CaM), which in turn activates calcineurin (CaN). NFAT is then dephosphorylated by CaN and translocate into the nucleus, where they are recruited to the promoter region of proinflammatory genes mediating their transcriptional activation. The immunosuppressant FK506 blocks CaN activity impairing NFAT activation. (**b**) The molecular mechanisms underlying activation of Crz1, the yeast orthologue of NFAT, are highly conserved among eukaryotes. *Saccharomyces cerevisiae* was genetically modified to express *lacZ* under the control of Crz1-recognition elements thereby providing a powerful tool for the bioprospection of compounds with the potential to inhibit Crz1 (NFAT) activation

recapitulates the molecular events associated to the induction of Ca^{2+}-signaling cascades leading to the activation of the NFAT yeast orthologue Crz1 [17]. Similarly to NFAT, Crz1 activation is controlled by Ca^{2+} cytosolic levels. Phosphorylated Crz1 is kept in an inactive form by compartmentalization in the cytosol and slight fluctuations in cytosolic Ca^{2+} levels trigger CaM/CaN activation in a similar manner to that of the mammalian orthologues [18]. Upon dephosphorylation by yeast CaN, Crz1 translocates into the nucleus stimulating calcineurin-dependent response element (CDRE)-driven gene expression [19, 20] (Fig. 1b). Interesting enough, FK506 also blocks yeast CaN and Crz1 activation [21] further reinforcing the high-degree evolutionary conservation of NFAT/Crz1 regulatory processes. These features support the use of *S. cerevisiae* as "in vivo test-tubes" for the screening of compounds modulating Crz1 (NFAT) activation [22]. Additionally, the easy laboratory handling of yeasts cells allied to the inexpensive culture conditions make *S. cerevisiae* the microorganism of choice for high-throughput studies.

2 Materials

Prepare growth media and stock solutions using double-distilled water (ddH_2O) (unless otherwise specified). When indicated, autoclave media and solutions at 121 °C during 25 min using a (slow exhaust) liquid cycle (*see* **Note 1**) and store at room temperature (exceptions are specified). Prepare solutions using analytical grade reagents and carefully follow waste disposal regulations when discarding toxic waste materials.

2.1 *Saccharomyces cerevisiae* Strains

1. A list of *S. cerevisiae* strains required for this protocol as well as their respective genotypes and original references are given in Table 1.
2. BY4742 is the parental wild-type strain.
3. BY4742_*CDRE-lacZ* is a recombinant strain derived from the BY4742 parental strain. It encodes an integrated copy of the bacterial *lacZ* gene under the control of a promoter containing four *in tandem* repeats of CDRE *cis*-elements. This reporter construction allows inferring the degree of Crz1 activation by measuring the activity of β-galactosidase encoded by *lacZ*.
4. BY4742_*CDRE-lacZ_crz1* is isogenic to BY4742_*CDRE-lacZ*. The *CRZ1* ORF was disrupted in this strain providing a negative control of the system.
5. BY4742_*CDRE-lacZ_cnb1* is also isogenic to BY4742_*CDRE-lacZ*. The *CNB1* ORF, encoding the regulatory subunit of CaN, was disrupted in this strain providing a second negative control of the system.

Table 1

***Saccharomyces cerevisiae* strains required to carry out this protocol**

Strain	Genotype	Source or reference
BY4742	*MATα his3 leu2 lys2 ura3*	EUROSCARF[a]
BY4742_*CDRE-lacZ*	*MATα his3 leu2 lys2 ura3 aur1::AUR1-C-4xCDRE-lacZ*	[26]
BY4742_*CDRE-lacZ*_ *crz1*	*MATα his3 leu2 lys2 ura3 YNL027W::HIS3MX4 aur1::AUR1-C-4xCDRE-lacZ*	[26]
BY4742_*CDRE-lacZ*_ *cnb1*	*MATα his3 leu2 lys2 ura3 YKL190W::kanMX4 aur1::AUR1-C-4xCDRE-lacZ*	[26]

[a]EUROpean Saccharomyces Cerevisiae ARchive for Functional analysis

2.2 Growth and Storage Media

1. YPD (Yeast extract, Peptone, Dextrose–glucose): 1% (w/v) yeast extract, 2% (w/v) bactopeptone and 2% (w/v) glucose. Add about 500 mL of ddH_2O and a magnetic stir into a 2 L-flask. Weight 10 g yeast extract, 20 g bactopeptone and 20 g glucose and transfer to the flask. Stir on a stir plate for approximately 10 min until all the components are dissolved into the solution. Transfer to a 1 L-graduated cylinder, add ddH_2O to 1 L and pour back into the 2 L-flask. Autoclave and store as indicated. Agar-YPD medium is prepared in a similar way, by adding 2% (w/v) agar (20 g) to the solution (*see* **Note 2**). After the medium has cooled to about 70 °C, pour *circa* 15 mL directly from the flask into sterile petri dishes in a sterile hood and let them cool to room temperature. Place plates back into plastic bags and close firmly. The plates can be stored at 4 °C for months.
2. $MnCl_2$ stock solution: 300 mM $MnCl_2$. Prepare a 100× stock solution by dissolving 378 mg $MnCl_2$ in 10 mL ddH_2O, sterile filter using a 0.2 μm membrane and store at room temperature.
3. SC (Synthetic Complete Supplement Mixture): 0.79 g/L CSM (Complete Supplement Mixture–QBiogene®, USA), 6.7 g/L YNB (Yeast Nitrogen Base–Difco®, USA), and 2% (w/v) glucose. Add about 500 mL of ddH_2O and a magnetic stir into a 2 L-flask. Weight 0.79 g CSM, 6.7 g YNB, and 20 g glucose, transfer to the flask and proceed as in the previous step. Prepare SC-inducing medium by supplementing SC with 3 mM $MnCl_2$ (*see* **Notes 3** and **4**). Agar-SC medium is prepared as described above, by adding 2% (w/v) agar (20 g) to the solution. Prepare agar-SC-inducing medium by adding 3 mM $MnCl_2$ just before pouring medium into the petri dishes.
4. FK506 monohydrate immunosuppressant (Tacrolimus) (Sigma-Aldrich®–Poole, Dorset, UK) stock solution: 1 mg/mL FK506. Prepare a 1000× stock solution by dissolving 1 mg FK506 in 1 mL Dimethyl sulfoxide (DMSO–Sigma–Aldrich®–Poole, Dorset, UK), aliquot and store at −20 °C (*see* **Note 4**).

2.3 Reagents for Monitoring β-Galactosidade Activity Using Ortho-nitrophenyl-β-galactoside

1. Y-PER Yeast Protein Extraction Reagent (Thermo Fisher Scientific Inc.–Life Technologies, USA).
2. LacZ buffer: 8.5 g/L Na_2HPO_4, 5.5 g/L $NaH_2PO_4{\cdot}H_2O$, 0.75 g/L KCl, 0.246 g/L $MgSO_4{\cdot}7H_2O$. Add about 500 mL of ddH_2O and a magnetic stir into a 2 L-flask. Weight 8.5 g Na_2HPO_4, 5.5 g NaH_2PO_4, 0.75 g KCl, 0.246 g $MgSO_4$ and transfer to the flask. Stir on a stir plate until the components are dissolved into the solution. Transfer to a 1 L-graduated cylinder, add ddH_2O to 1 L and pour back into the 2 L-flask. Autoclave and store as indicated.
3. Ortho-Nitrophenyl-β-galactoside (ONPG) (Sigma–Aldrich®–Poole, Dorset, UK).
4. LacZ/ONPG buffer: 2 mg ONPG/mL LacZ. Prepare according the number of conditions to be tested (*see* **Note 5**).

2.4 Reagents for Monitoring β-Galactosidade Activity in Solid Medium

1. Phosphate-buffered saline (PBS): 8 g/L NaCl, 0.2 g/L KCl, 1.44 g/L Na_2HPO_4, 0.24 g/L KH_2PO_4. Add 500 mL of ddH_2O and a magnetic stir into a 2 L-flask. Weight 8 g NaCl, 0.2 g KCl, 1.44 g Na_2HPO_4, 0.24 g KH_2PO_4 and transfer to the flask. Stir on a stir plate until the components are dissolved into the solution. Transfer to a 1 L-graduated cylinder, add ddH_2O to 1 L and pour back into the 2 L-flask. Autoclave and store as indicated.
2. Ultrafine agarose.
3. Sodium Dodecyl Sulfate (SDS) stock solution: 20% (w/v) SDS. Add 80 mL of ddH_2O into a 200 mL-flask. Weight 20 g SDS and transfer to the flask. Mix gently to avoid bubbles, transfer to a 200 mL-graduated cylinder, add ddH_2O to 100 mL and pour back into the 200 mL-flask.
4. 5-Bromo-4-chloro-3-indolyl-β-D-galactopyranoside (X-Gal) (ImmunoSource®–Ruiterslaan 29, Zoersel, Belgium).
5. *N,N*-Dimethylformamide (DMF), purity ≥99.8%.
6. X-Gal overlay solution: 0.5% (w/v) agarose, 50% (v/v) LacZ buffer, 0.2% (w/v) Sodium Dodecyl Sulfate (SDS), 2 mg/mL X-Gal. This solution should be prepared immediately before use. Prepare 20 mL X-Gal overlay solution per each set of two plates as follows: (1) weight 100 mg ultrafine agarose, transfer to a 10 mL-flask containing 10 mL ddH_2O and heat in the microwave for 30 s; (2) add 10 mL LacZ buffer; (3) add 200 μL SDS 20%; and (4) add 40 mg X-Gal previously dissolved in 500 μL DMF (*see* **Notes 5** and **6**). Scale up the quantities of each reagent according to number of plates to be processed.

2.5 Equipment Required

1. Sterile hood.
2. Orbital incubator set for 30 °C and 200 rpm.
3. Microcentrifuge.

4. Sterile 96-well plates.
5. Microplate spectrophotomer with temperature set.
6. Centrifuge with swing basket rotor.

3 Methods

Handle yeast cells in a sterile hood and carry out the incubation of yeast cultures at 30 °C. All yeast liquid cultures should be incubated with orbital agitation at 200 rpm. If it is the first time working with BY4742, BY4742_*CDRE-lacZ*, BY4742_*CDRE-lacZ_crz1*, and BY4742_*CDRE-lacZ_cnb1* yeast strains proceed as following: (1) streak cells from each strain onto separate agar-YPD plates; (2) after obtaining isolated colonies (48–72 h) grow one single colony of each strain in 1 mL YPD for 16 h; (3) make glycerol stocks by adding 1 mL 60% (v/v) sterile glycerol to 1 mL YPD stationary phase growth cultures; and (4) freeze at −80 °C. Yeast strains can be maintained almost indefinitely at −80 °C in glycerol stocks. Begin the experiment by growing yeast strains from glycerol stocks. Streak cells onto agar-YPD plates and incubate for 48–72 h. Restreak single colonies onto fresh agar-YPD plates and incubate for further 48 h. If properly stored at 4 °C these cells can be used for about a month. After this period, streak cells onto agar-YPD plates to refresh cultures. Regularly renew yeast cultures from the glycerol stocks. Continue with the preparation of yeast cultures and cells treatments and subsequently with the protocols for the identification of potential anti-inflammatory compounds. A schematic representation of the whole process is given in Fig. 2.

3.1 Yeast Cultures and Cell Treatments

1. For each yeast strain, pick one colony, inoculate into 5 mL SC liquid medium and incubate for 16 h.
2. Make a 1/10 culture dilution in fresh SC medium and incubate at the same conditions for further 8–10 h.
3. Read the optical density of cultures at 600 nm (OD_{600}) (*see* **Note 7**). Dilute cultures into fresh SC medium as to have a final $OD_{600} = 1$ after 16 h, using the equation

$$\mathrm{ODi} \times Vi = (\mathrm{ODf} / \left(2^{(t/\mathrm{gt})}\right) \times V\mathrm{f},$$

Where ODi = initial optical density of the culture, *V*i = initial volume of culture, ODf = final optical density of the culture, *t* = time (16 h), gt = generation time of the strain (90 min in SC medium), *V*f = final volume of culture (established by the number of conditions to be tested).

4. Monitor OD_{600} of the cultures and dilute to $OD_{600} = 0.1$ in SC medium. Aliquot 300 μL of cultures in microtubes (prepare as many microtubes as the number of conditions to be tested,

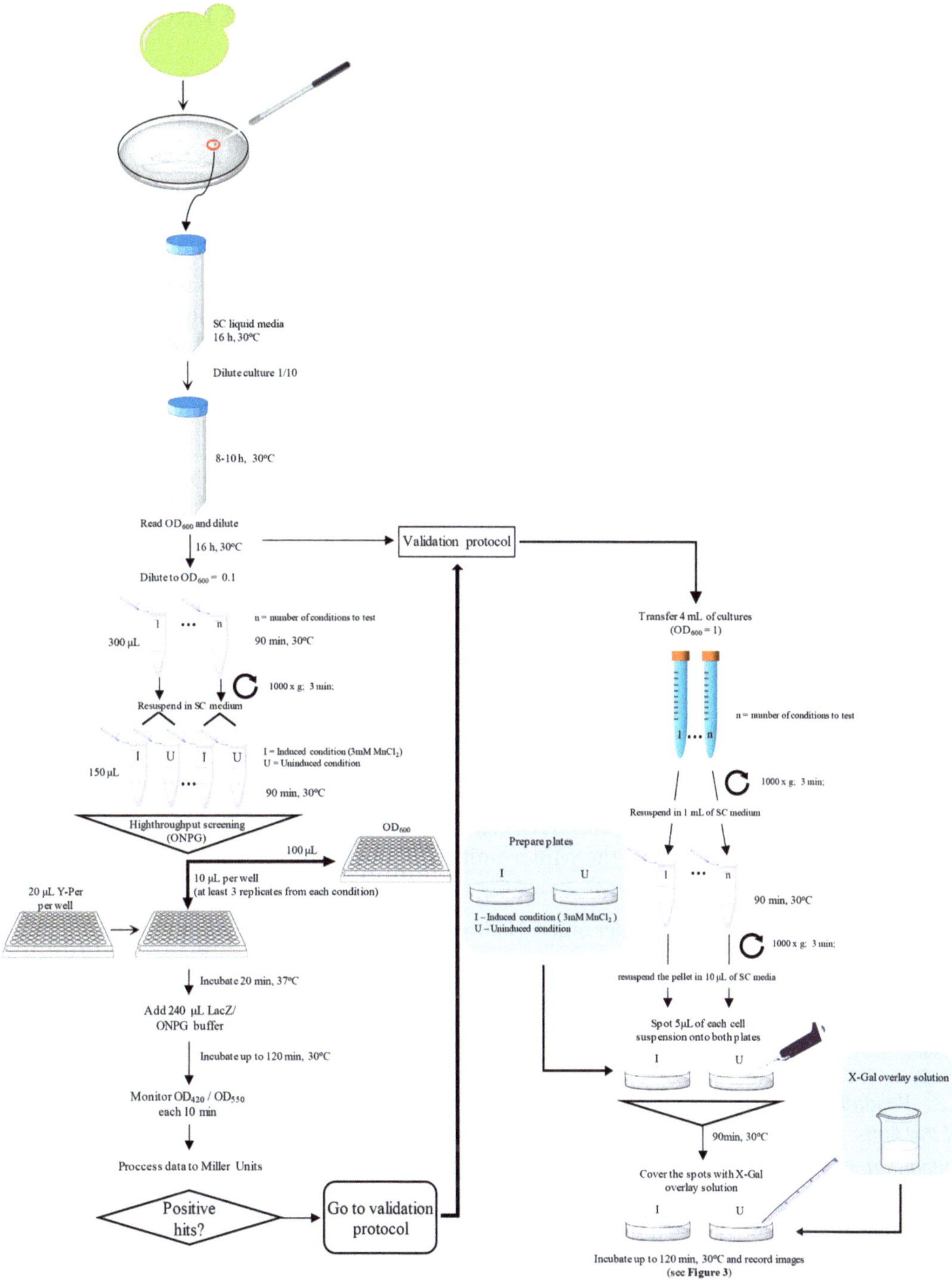

Fig. 2 Schematic representation of high-throughput and validation protocols. For detailed information see the text and notes

including the biological replicates and the positive and negative controls).

5. Add the compounds to be tested to each microtube individually using concentrated stock solutions. Leave on sample untreated (the negative control) and add FK506 to a final concentration of 10 μg/mL (positive control). Adjust the final volume of cultures with the appropriated solvents.
6. Incubate cultures for 90 min at 30 °C.
7. After thoroughly homogenizing the cells, transfer 150 μL of each sample into a new microtube containing the CaN/Crz1 inducer $MnCl_2$ to a final concentration of 3 mM.
8. Incubate cells for further 90 min at 30 °C.

3.2 High-Throughput Screening of Compounds with Potential Anti-inflammatory Activity: Monitoring β-Galactosidase Activity Using Ortho-nitrophenyl-β-galactoside

1. Thoroughly homogenize cell suspensions and save 100 μL of each sample in a microplate to monitor OD_{600}.
2. Transfer 10 μL of each sample into a 96-well plate containing 20 μL Y-PER Reagent per well. At least 5 replicates of each sample should be performed.
3. Incubate 20 min at 37 °C with no agitation.
4. Add 240 μL of LacZ/ONPG buffer to each well, incubate the 96-well plate at 30 °C in a microplate spectrophotometer and monitor the development of yellow color by measuring OD_{420}/OD_{550} every 10 min for up to 120 min.
5. Choose the OD_{420} read that best differentiate the β-galactosidase activities between the positive and negative controls and calculate the Miller units [23] according to the equation

$$\text{Miller unit} = 1000 \times \left(\text{OD}_{420} - 1.75 \times \text{OD}_{550}\right) / \left(t \times V \times \text{OD}_{600}\right),$$

where t = reaction time in minutes, V = volume of culture assayed in mL (used for lysis).

3.3 Validation of the Anti-inflammatory Potential of Candidate Compounds: Monitoring β-Galactosidase Activity Using 5-Bromo-4-chloro-3-indolyl-β-D-galactopyranoside

1. Grow yeast cultures as described in Subheading 3.1 (to **step 3**).
2. Transfer 4 mL of cell suspension at $OD_{600} = 1$ (approx. 4×10^7 cells) (*see* **Note 7**) of each culture into 15 mL Falcon tubes (prepare as many tubes as the number of conditions to be tested, including the positive and negative controls).
3. Centrifuge cell suspensions at $1000 \times g$ for 3 min, discard the supernatant and resuspend cells in 1 mL of SC media.
4. Add the compounds to be tested to each tube individually using concentrated stock solutions. Leave one sample untreated (the negative control) and add FK506 to a final concentration of 10 μg/mL (positive control). When required, adjust the final volume of cultures with the appropriated solvents.
5. Incubate cultures for 90 min.

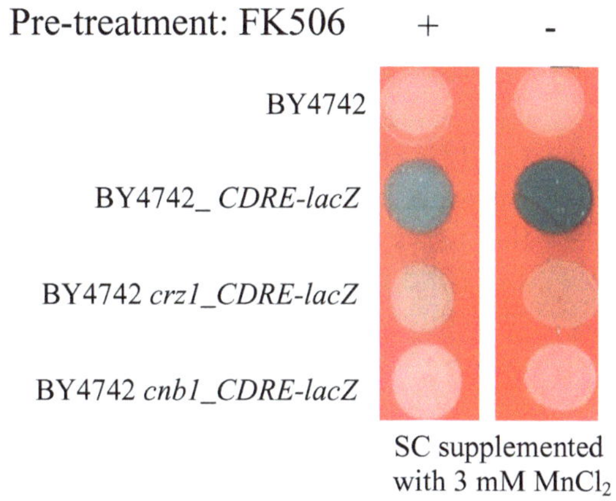

Fig. 3 Illustration of a typical result obtained in the validation phase of candidate compounds with anti-inflammatory potential: monitoring β-galactosidase activity using 5-bromo-4 chloro-3-indolyl-β-D-galactopyranosidase (X-Gal). BY4741 parental strain and wild type, *crz1* and *cnb1* (lacking the regulatory subunit of CaN) strains expressing the CDRE–*lacZ* reporter construct were pretreated with FK506 to a final concentration of 10 μg/mL for 90 min at 30 °C. Cells were centrifuged, the supernatant was discarded and after resuspension in SC fresh medium cells were spotted onto agar-SC-inducing media containing 3 mM $MnCl_2$. Images were recorded after 120 min incubation at 30 °C. The results show that (1) the BY4741 parental strain does not have endogenous β-galactosidade activity; (2) the wild-type strain encoding the CDRE–*lacZ* reporter construct exhibits a strong β-galactosidade activity, reflecting Crz1 activation, that is reduced in cells pre-exposed to the immunosuppressant FK506; and (3) the absence of both Crz1 and Cnb1 abrogated induction of the reporter gene, demonstrating the Ca^{2+}-responsiveness of the system and its dependence on the intact function of Crz1 and Cnb1

6. Centrifuge cell suspensions at 1000 × *g* for 3 min, discard the supernatant (*see* **Note 8**) and resuspend cells in 10 μL SC medium (*see* **Note 9**).
7. Spot 5 μL of each sample onto agar-SC and agar-SC-inducing media and let the spots dry with open lids for 5 min in the sterile hood before transferring the plates to the incubator chamber.
8. Incubate plates for 90 min at 30 °C.
9. Using a pipette, carefully transfer 10 mL of freshly prepared X-Gal overlay solution onto the surface of the plates without disturbing cell spots (*see* **Note 10**).
10. Let solidify for a few minutes and incubate plates at 30 °C for up to 120 min recording images each 30 min. Illustrative results are provided in Fig. 3.

4 Notes

1. It is critical to strictly follow autoclave conditions otherwise media will caramelize and boil over at the end of the autoclave cycle.

2. In the preparation of agar-media, agar will not go into solution before media is autoclaved.
3. Crz1 is under the control of Ca^{2+}-signaling pathways, therefore $CaCl_2$ is the most commonly used inducer of Ca^{2+}-signaling pathways [24]. However, $CaCl_2$ precipitation strongly affects the assays to monitor β-galactosidase activity, both in liquid and solid media, since it interferes with OD_{600} and OD_{550} of cell cultures and lysates as well as with the interpretation of data from the hydrolysis of X-Gal. The current strategy to circumvent this problem relies on the chelation of Ca^{2+} ions with EDTA, however it does not completely block $CaCl_2$ precipitation. After a screening of various conditions described to induce Crz1 activation [24–26], it was verified that $MnCl_2$ at a final concentration of 3 mM is potent inducer of the system without interfere with the readout of assays.
4. It is strongly suggested to prepare 100–1000× concentrated stock solutions of $MnCl_2$, FK506, and testing compounds, when possible, to minimize volume alterations/corrections and solvent toxicity.
5. ONPG hardly dissolves into LacZ buffer. It is recommend preparing the solution in advance and incubating it at 42 °C with occasional vortexing to speed up ONPG solubilization.
6. Preparation of X-Gal overlay solution is one of the most critical steps when monitoring β-galactosidase activity with X-Gal. Before adding X-Gal/DMSO to the overlay solution it is suggested to monitor the temperature of the solution. If it is close to the melting point of agarose, it will be necessary to heat the solution briefly since it is not recommended to heat the solution after adding X-Gal/DMSO. If it is still too hot, let it cool for a while at room temperature. These procedures will avoid unspecific hydrolysis of X-Gal substrate.
7. Only consider OD_{600} measurements ≤ 1 to ensure linearity. If OD_{600} of the sample is greater than 1, dilute the sample 1:10 with ddH_2O and read it again. $OD_{600} = 1$ is equivalent to approx. $1–3 \times 10^7$ cells/mL, depending on the spectrophotometer used.
8. At this step it is indispensable completely discarding the supernatant taking into consideration that cells will be resuspended in a small volume of medium. Any drop of remaining medium will interfere with the total volume of cell suspension and consequently the number of cells spotted onto the solid medium leading to misinterpretation of the results, which are performed in a comparative basis. It is recommended the use of a micropipette to discard the supernatant. If there is any uncertainty that the supernatant is not completely removed,

centrifuge again and use a 200 μL-micropipette to remove the remaining medium.

9. If testing a colored compound, resuspend the cell pellets in 1 mL PBS and centrifuge under the same conditions. Repeat the wash procedure three times (or as many times as required to remove any pigment) and finally resuspend cells in 10 μL SC medium. It is essential to remove all traces of color in cell spots otherwise they will interfere with data interpretation.
10. Cover the entire surface of the plate with overlay solution avoiding disturbing cell spots. Do not move the plates until the overlay solution solidifies.

Acknowledgments

This protocol was adapted and implemented in the ambit of the BacHBerry project, which is cofunded by the European Commission in the 7th Framework Programme (Project No. FP7-613793). Acknowledgments also to BacHBerry, for RM and GG fellowships, and to Fundação para Ciência e Tecnologia (FCT), for the individual fellowship of C.N. Santos (IF/01097/2013). The authors would like to acknowledge networking support by the Proteostasis COST Action (BM1307). iNOVA4Health - UID/Multi/04462/2013, a program financially supported by Fundação para a Ciência e Tecnologia/Ministério da Educação e Ciência, through national funds and co-funded by FEDER under the PT2020 Partnership Agreement is acknowledged.

References

1. Ferrero-Miliani L, Nielsen OH, Andersen PS, Girardin SE (2007) Chronic inflammation: importance of NOD2 and NALP3 in interleukin-1beta generation. Clin Exp Immunol 147:227–235
2. Janeway CA Jr, Travers P, Walport M et al (2001) Basic concepts in immunology. In: Immunobiology: the immune system in health and disease, 5th edn. Garland Science, New York, pp 1–34
3. Kumar V, Abbas AK, Aster JC (2015) Cellular responses to stress and toxic insults: adaptation, injury and death. In: Robbins and Cotran pathologic basis of disease, 9th edn. Elsevier Health Sciences, pp 43–76
4. Amor S, Puentes F, Baker D, Van Der Valk P (2010) Inflammation in neurodegenerative diseases. Immunology 129:154–169
5. Imai Y, Dobrian AD, Morris MA, Nadler JL (2013) Islet inflammation: a unifying target for diabetes treatment? Trends Endocrinol Metab 24:351–360
6. More SV, Kumar H, Kim IS, Song S-Y, Choi D-K (2013) Cellular and molecular mediators of neuroinflammation in the pathogenesis of Parkinson's disease. Mediators Inflamm. doi:10.1155/2013/952375
7. Vauzour D (2012) Dietary polyphenols as modulators of brain functions: biological actions and molecular mechanisms underpinning their beneficial effects. Oxid Med Cell Longev 2012:914273
8. Cabrera-Benitez NE, Pérez-Roth E, Casula M, Ramos-Nuez A, Ríos-Luci C, Rodríguez-Gallego C, Sologuren I, Jakubkiene V, Slutsky AS, Padrón JM, Villar J (2012) Anti-inflammatory activity of a novel family of aryl ureas compounds in an endotoxin-induced airway epithelial cell injury model. PLoS One 7:e48468

9. Libby P (2007) Inflammatory mechanisms: the molecular basis of inflammation and disease. Nutr Rev 65(Suppl 3):S140–S146. doi:10.1111/j.1753-4887.2007.tb00352.x
10. Serfling E, Berberich-Siebelt F, Chuvpilo S, Jankevics E, Klein-Hessling S, Twardzik T, Avots A (2000) The role of NF-AT transcription factors in T cell activation and differentiation. Biochim Biophys Acta 1498:1–18
11. Fric J, Zelante T, Wong AYW, Mertes A, Yu HB, Ricciardi-Castagnoli P (2012) NFAT control of innate immunity. Blood 120:1380–1389
12. Rao A, Luo C, Hogan PG (1997) Transcription factors of the NFAT family: regulation and function. Annu Rev Immunol 15:707–747
13. Kiani A, Rao A, Aramburu J (2000) Manipulating immune responses with immunosuppressive agents that target NFAT. Immunity 12:359–372
14. Matsuda S, Shibasaki F, Takehana K, Mori H, Nishida E, Koyasu S (2000) Two distinct action mechanisms of immunophilin-ligand complexes for the blockade of T-cell activation. EMBO Rep 1:428–434
15. Vafadari R, Hesselink DA, Cadogan MM, Weimar W, Baan CC (2012) Inhibitory effect of Tacrolimus on p38 mitogen-activated protein kinase signaling in kidney transplant recipients measured by whole-blood phosphospecific flow cytometry. Transplant J 93:1245–1251
16. Botstein D, Fink GR (2011) Yeast: an experimental organism for 21st century biology. Genetics 189:695–704
17. Thewes S (2014) Calcineurin-Crz1 signaling in lower eukaryotes. Eukaryot Cell 13: 694–705
18. Premack BA, Gardner P (1992) Signal transduction by T-cell receptors: mobilization of Ca and regulation of Ca-dependent effector molecules. Am J Physiol 263:C1119–C1140
19. Stathopoulos-Gerontides A, Guo JJ, Cyert MS (1999) Yeast calcineurin regulates nuclear localization of the Crz1p transcription factor through dephosphorylation. Genes Dev 13: 798–803
20. Matheos DP, Kingsbury TJ, Ahsan US, Cunningham KW (1997) Tcn1p/Crz1p, a calcineurin-dependent transcription factor that differentially regulates gene expression in Saccharomyces cerevisiae. Genes Dev 11:3445–3458
21. Garrett-Engele P, Moilanen B, Cyert MS (1995) Calcineurin, the Ca2+/calmodulin-dependent protein phosphatase, is essential in yeast mutants with cell integrity defects and in mutants that lack a functional vacuolar H(+)-ATPase. Mol Cell Biol 15:4103–4114
22. Prescott TA, Arino J, Kite GC, Simmonds MS (2012) Inhibition of human calcineurin and yeast calcineurin-dependent gene expression by Jasminum humile leaf and root extracts. J Ethnopharmacol 140:293–297
23. Miller JH (1972) Experiments in molecular genetics. Cold Spring Harb. Lab. Press, Cold Spring Harb, NY
24. Cyert MS, Thorner J (1992) Regulatory subunit (CNB1 gene product) of yeast Ca2+/calmodulin-dependent phosphoprotein phosphatases is required for adaptation to pheromone. Mol Cell Biol 12:3460–3469
25. Ferreira RT, Silva AR, Pimentel C, Batista-Nascimento L, Rodrigues-Pousada C, Menezes RA (2012) Arsenic stress elicits cytosolic Ca(2+) bursts and Crz1 activation in Saccharomyces cerevisiae. Microbiology 158:2293–2302
26. Araki Y, Wu H, Kitagaki H, Akao T, Takagi H, Shimoi H (2009) Ethanol stress stimulates the Ca2+-mediated calcineurin/Crz1 pathway in Saccharomyces cerevisiae. J Biosci Bioeng 107:1–6

Chapter 30

Using AlphaScreen® to Identify Small-Molecule Inhibitors Targeting a Conserved Host–Pathogen Interaction

Sisley Austin, Saïd Taouji, Eric Chevet, Harald Wodrich, and Fabienne Rayne

Abstract

AlphaScreen® is a technology particularly suitable for bi-molecular inhibitor screening assays, e.g. using protein–protein interactions with purified recombinant proteins. Each binding partner of the bi-molecular interaction is coupled either to donor or to acceptor beads. The technology is based on the quantifiable transfer of oxygen singlets from donor to acceptor microbeads brought together by a specific interaction between the partners. We identified the conserved interaction between WW domains of cellular ubiquitin ligases of the Nedd4 family and a short peptide motif (PPxY) present in several structural and non-structural viral proteins as a potential drug target. Using an AlphaScreen assay recapitulating the interaction between Nedd4.2 and the PPxY motif of the adenoviral capsid protein VI, we screened a library of small molecules and identified specific inhibitors of this interaction.

Key words AlphaScreen®, High-throughput, Protein-Protein interactions, Small-molecule inhibitors, [LP]PxY motif, Adenovirus, Ubiquitin ligase Nedd4, Antiviral, 384-Well plate

1 Introduction

Viruses infect host cells by several mechanisms but as obligate parasites in order to promote viral replication they must recruit and/or avoid certain cell signaling pathways (e.g. cell death or cell survival pathways). For this reason one can assume that some host–pathogen interactions are not restricted to a single virus family but could be rather common to several virus families. One such interaction, which we recently described, occurs between cellular ubiquitin ligases of the Nedd4 family and virus-encoded short peptide motifs of the [LP]PxY type (where *x* can be any amino acid). [LP]PxY motifs bind to WW domains, which are part of the N-terminal substrate recognition region used by Nedd4 ligases. [LP]PxY motifs are found in several structural and non-structural viral proteins [1, 2]. Most described functions for host–pathogen Nedd4 ⇔ [LP]PxY interactions contribute to egress of enveloped viruses by redirecting

Rune Matthiesen (ed.), *Proteostasis: Methods and Protocols*, Methods in Molecular Biology, vol. 1449,
DOI 10.1007/978-1-4939-3756-1_30, © Springer Science+Business Media New York 2016

parts of the vesicular sorting machinery to viral exit sites [3, 4]. More recently a role for the Nedd4⇔[LP]PxY interactions was also shown during entry of the non-enveloped adenovirus suggesting a broad range of functions for this interaction in the life cycle of many viruses [5]. Thus targeting this common viral–host interaction using small inhibitory compounds could potentially contribute to the development of antivirals with broad applicability.

Here, we describe the detailed protocol for a miniaturized screening assay to identify small compound inhibitors able to block Nedd4⇔[LP]PxY interactions. The prototypic system is based on the interaction between the adenoviral capsid protein VI, which encodes for a conserved PPSY motif and the cellular ubiquitin ligase Nedd4. The adenoviral capsid protein VI is involved in membrane lysis and escape of the entering particle from the endosomal compartment to access microtubule-dependent trafficking toward the nucleus and activation of viral transcription. Adenoviruses encoding a protein VI mutant (termed mutant M1) with the PPSY motif mutated into PGAA fail to bind Nedd4 resulting in strongly reduced infectivity, an altered intracellular targeting and lack efficient gene delivery [5, 6]. For the miniaturized screening system, we chose to use the AlphaScreen® technology. Amplified luminescence proximity homogeneous assay (AlphaScreen®) is a bead-based non-radioactive and homogeneous detection technology, which allows assessing the interaction between biological partners. For the assay, the cellular GST-tagged protein Nedd4 and the viral poly-histidine-tagged protein VI (His-VI) were purified and coupled to glutathione donor beads and nickel acceptor beads, respectively (*see* **Note 1**). After excitation at 680 nm, the donor beads which contain a photo-sensitizer, phtalocyanine, emit an excited form of O_2 (singlet oxygen), which has a lifetime compatible with an approximate diffusion range of 200 nm in solution. In case of a protein–protein interaction between the two partners, donor and acceptor beads are brought into <200 nm proximity and the transfer of energy from the donor to the acceptor beads containing thioxene derivatives leads to a cascade of chemiluminescence, resulting in the emission of light at 520–620 nm (Fig. 1). This energy transfer is proportional to the binding strength and can be quantified. As negative control of this interaction assay, we used the viral poly-histidine-tagged protein VI-M1 mutant, which does not bind to the Nedd4 protein.

The ideal inhibitory compound for host–pathogen interactions (e.g. to block the Nedd4⇔[LP]PxY interaction) must exhibit high bioavailability and a low toxic profile. To increase the number of potential inhibitory compounds that adhere to these criteria, we decided to screen the Prestwick chemical library composed of 1280 small compounds. They are 100% approved drugs (FDA, EMA, and other agencies), which potentially could facilitate their use for antiviral applications.

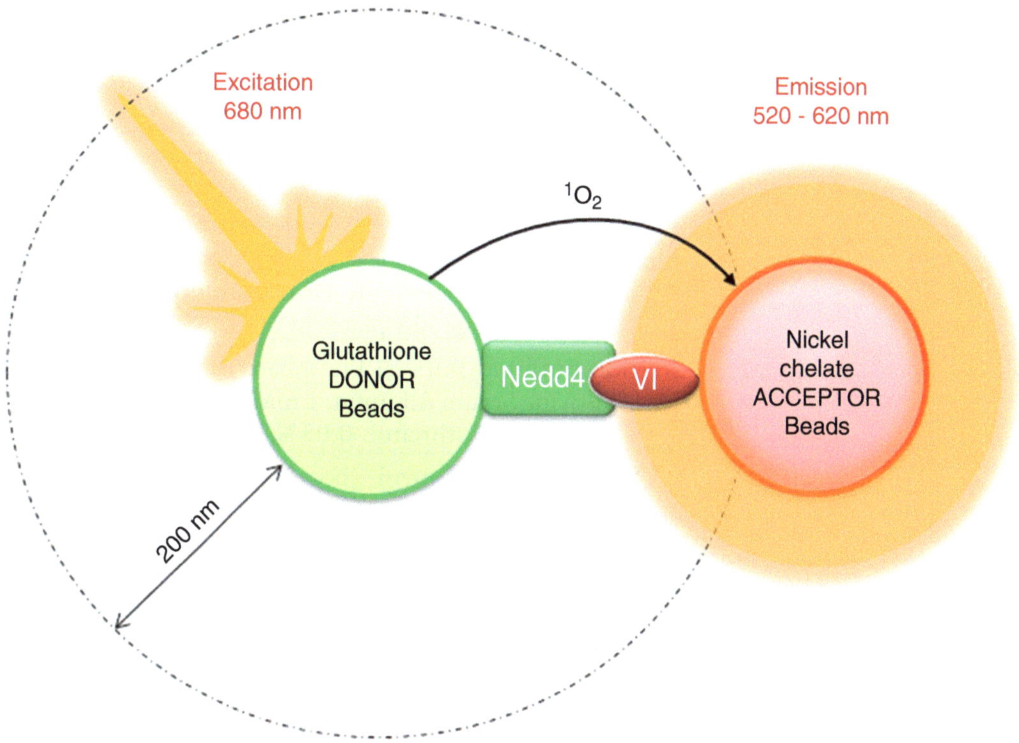

Fig. 1 Principle of AlphaScreen®-based protein-protein interaction assay. Glutathione donor beads are coated with recombinant bacterially purified GST-tagged Nedd4 protein and nickel chelate acceptor beads are coated with recombinant, bacterially purified viral poly-histidine-tagged protein VI. Following interaction between the 2 proteins through the Nedd4⇔PPxY interaction and excitation at 680 nm, the proximity of the 2 beads permits the transfer of singlet oxygen from the donor to the acceptor beads resulting in quantifiable light emission

In this chapter, we describe a detailed hands-on protocol to perform an AlphaScreen®-based high-throughput inhibitor screening using a chemical library to identify inhibitors of a conserved viral–host interaction. Although this assay is specific for Nedd4⇔[LP]PxY interaction, the step-by-step description with interaction assay, stoichiometry of the proteins, sensitivity, and the calculation of assay performance, can be easily adapted to other kinds of bi-molecular interaction screens and should be considered as general guidelines.

2 Materials

2.1 Platform

The platform (www.bmyscreen.com) comprises:

1. An integrated and automated pipeline: JANUS® Mini liquid handler with a MDT dispensing head placed in line with a plate stacker.
2. A JANUS® robot containing an 8 pins head and a gripper.
3. An Envision® Multidetection Plate Reader.

2.2 Environment

1. For optimal performance, the manufacturer (PerkinElmer) recommends to avoid temperature variations in the room containing the microplate reader machine.
2. Because the beads are sensitive to light, it is better to handle all reagents under low-light conditions.

2.3 Tools and Reagents

1. Prestwick Chemical Library® (www.prestwickchemical.com) contains 1280 small molecules, 100 % approved drugs (FDA, EMA, and other agencies), in DMSO solution supplied at 10 mM in 96-well plate (14 plates in total).
2. Interaction buffer (I-Buffer): Phosphate Buffer Saline 1× pH 7.4, 0.1 % Bovine Serum Albumin, 1 mM Phenylmethylsulfonyl fluoride, 2 mM Dithiothreitol, 0.05 % Tween20.
3. DMSO solution as control (because it is the carrier for compounds).
4. Epitope tagged proteins of interest: His-VI-wt, His-VI-M1 mutant as negative control of interaction (*see* **Note 2**) and GST-Nedd4 (*see* **Note 3**).
5. A robotic liquid handler, such as JANUS® Mini liquid handler for the transfer of compounds between 96- and 384-well plates.
6. Pipette tips 200 μL Robot rack compatible and pipette tips 20 μL Robot rack compatible.
7. 96-Well plates.
8. 384-Well white opaque plates (*see* **Note 4**).
9. Plastic 384-well plate covers.
10. Adhesive plate seals.
11. Nickel Chelate AlphaLISA® Acceptor 45 Beads (PerkinElmer) (*see* **Note 5**).
12. AlphaScreen® Glutathione Donor beads (PerkinElmer) (*see* **Note 5**).
13. Vortex.
14. Low-speed centrifuge.
15. Automatic monochannel pipette (*see* **Note 6**).
16. Incubator at 25 °C.
17. Plate reader, such as Envision® Multidetection Plate Reader (*see* **Note 7**).

2.4 Storage

1. Purified tagged proteins should be aliquoted, snap-frozen, and stored at −80 °C for long-term storage in glycerol supplemented transport buffer (*see* **Note 3**). To avoid freeze–thaw cycles, aliquoted proteins were thawed the day of use and discarded (*see* **Notes 8** and **9**).

2. Prestwick Chemical Library® supplied in DMSO at 10 mM concentration and stored at −20 °C.
3. Beads should be stored at 4 °C protected from light.

3 Methods

3.1 Characterization of the AlphaScreen® Assay as Model for a Conserved Host–Pathogen Interaction

For practical reasons, the method section starts with a detailed description of the interaction assay (*see* Subheading 3.1.1). Prior to performing the screen one should determine the optimal stoichiometry (described in Subheading 3.1.2) of the binding partner, the assay sensitivity (described in Subheading 3.1.3) and the assay robustness (described in Subheading 3.1.4).

For each well of the 384-well opaque plate, the final volume for the reaction is 25 μL.

3.1.1 Interaction Assay

For this assay, each condition is performed in duplicate.

1. Using the automatic monochannel pipette or the robotic liquid handler distribute 10 μL of I-Buffer in eight empty control wells and 5 μL of I-Buffer in 4 wells of a 384-well opaque plate.
2. Distribute 5 μL/well of purified histidine-tagged protein VI-wt protein (His-VI-wt) at 6.25 ng/μL (resulting in 31.25 ng/well, optimal quantity determined, *see* Subheading 3.1.2) in 2 wells containing 10 μL of I-Buffer prepared at **step 1** (to determine the background signal, Fig. 2 *Control 2*) (*see* **Note 10**).
3. Distribute 5 μL/well of purified His-VI-wt protein at 6.25 ng/μL in 2 wells containing 5 μL of I-Buffer prepared at **step 1** (*see* **Note 11**).
4. Distribute 5 μL/well of purified histidine-tagged protein VI-M1 mutant protein (His-VI-M1) at 6.25 ng/μL in 2 wells containing 10 μL of I-Buffer prepared at **step 1** (to determine the background signal, Fig. 2 *Control 3*).
5. Distribute 5 μL/well of purified His-VI-M1 protein at 6.25 ng/μL in 2 wells containing 5 μL of I-Buffer prepared at **step 1**.
6. Distribute 5 μL/well of purified GST-tagged Nedd4 protein (GST-Nedd4) at 20 ng/μL (resulting 125 ng/well, optimal quantity determined, *see* Subheading 3.1.2) in 2 wells containing 10 μL of I-Buffer prepared at **step 1** (to determine the background signal, Fig. 2 *Control 4*).
7. Distribute 5 μL/well of purified GST-Nedd4 protein at 20 ng/μL in 2 wells prepared at **step 3** (constitutes the duplicate positive control of the interaction, Fig. 2 *Control 5*).

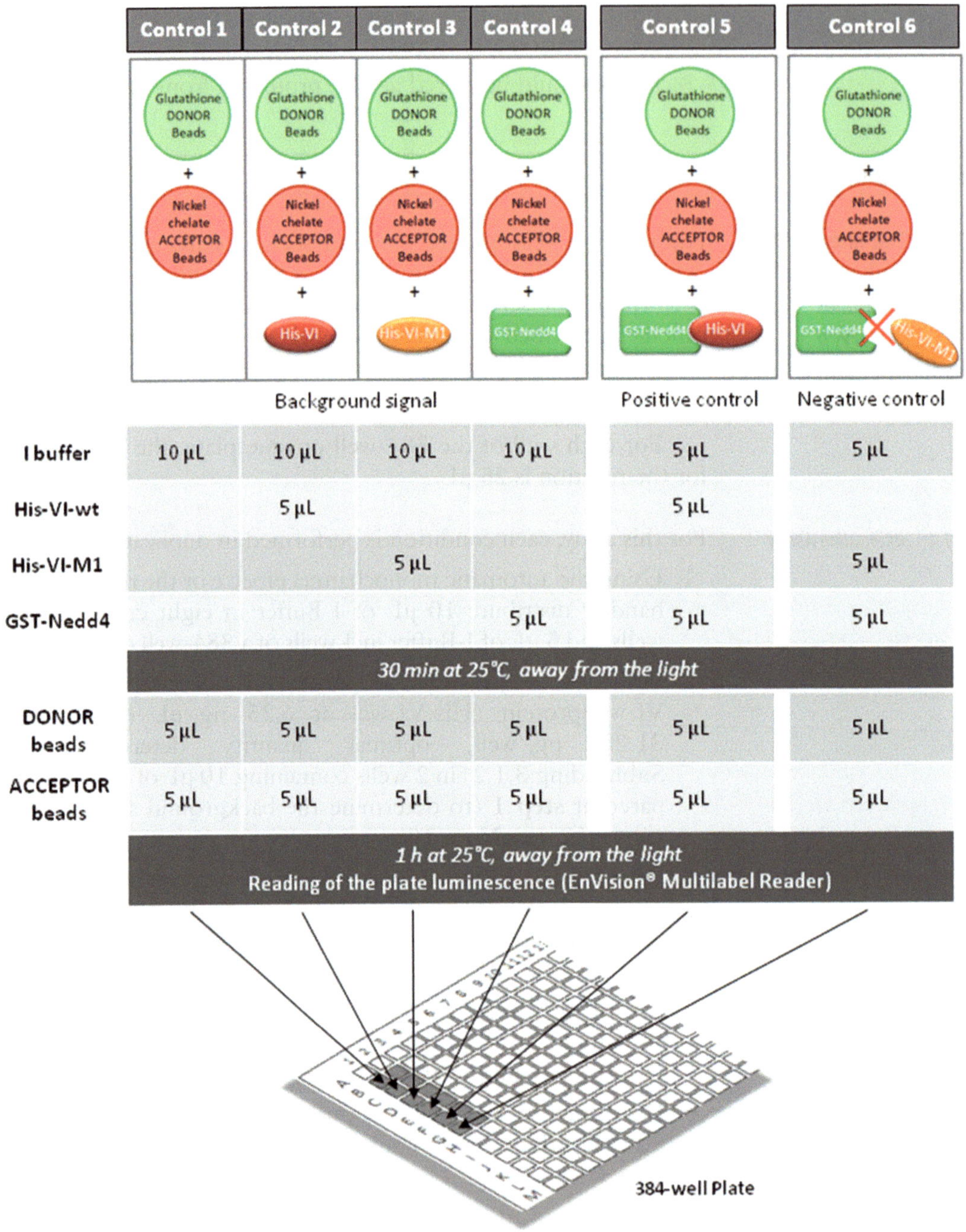

Fig. 2 Interaction assays. Schematic distribution of reagents added into different 384-well plates to determine the assay controls

8. Distribute 5 μL/well of purified GST-Nedd4 protein at 20 ng/μL in 2 wells prepared at **step 5** (constitutes the duplicate negative control of the interaction, Fig. 2 *Control 6*) (*see* **Note 12**).
9. Cover the plate with a plastic 384-well plate cover.
10. Incubate the plate in the dark in the incubator at 25 °C (*see* **Note 13**) during 30 min.
11. Add 5 μL/well of 0.1 μg/μL of nickel chelate AlphaLISA® acceptor beads solution in all wells prepared at **steps** 7 and **8** and in 2 wells containing 10 μL of I-Buffer prepared at **step 1** (to determine the background signal, Fig. 2 *Control 1*).
12. Add 5 μL/well of 0.1 μg/μL of AlphaScreen® glutathione donor beads solution in all wells prepared at **steps** 7, **8**, and **10**.
13. Cover the plate with adhesive seals.
14. Transfer all drops to the bottom of the wells by gently tapping the edges of the plate with your hand (*see* **Note 14**).
15. Incubate the plate in the dark in the incubator at 25 °C during 1 h.
16. Read the luminescence of the plate at the EnVision® Multilabel Reader after 1, 4, and 24 h (*see* **Note 15**) of incubation to determinate the absolute signal intensity and the stability of the signal over time (*see* **Notes 16** and **17**) (Fig. 2).

3.1.2 Stoichiometry

For this assay, each condition is performed in duplicate.

We estimated the stoichiometry between both partners: His-VI-wt and GST-Nedd4 to find the optimal ratio between proteins to obtain an optimal interaction resulting in the best AlphaScreen® signal over noise ratio. This was achieved by defining the amount of one partner (GST-tagged Nedd4 protein) and increasing the amount of the second partner (His-protein VI-wt) at a fixed bead ratio (*see* **Note 18**).

1. Distribute 5 μL of I-Buffer in 10 wells of a 384-well opaque plate.
2. Add 5 μL/well of purified GST-Nedd4 protein at 200 ng/μL (10 pmol) in 10 wells of the 384-well containing 5 μL of I-Buffer.
3. Add 5 μL/well of a variable amount of purified His-VI-wt protein from 57 pmol to 3.5 pmol with a 1:2 serial dilution in I-Buffer; 2 wells with 57 pmol, 2 wells with 28.5 pmol, 2 wells with 14.25 pmol, 2 wells with 7.12 pmol, and 2 wells with 3.5 pmol in the 10 wells prepared at **step 2**.
4. Follow the protocol (*see* Subheading 3.1.1) from **step 9**.

3.1.3 Sensitivity

For this assay, each condition is performed in duplicate.

After estimating the optimal stoichiometry, we next estimate the sensitivity of interaction detection between His-VI-wt and

GST-Nedd4 to determine the minimal quantity of protein that is necessary to obtain a robust AlphaScreen® signal.

1. Distribute 5 μL of I-Buffer in 14 wells of a 384-well opaque plate.
2. Add 5 μL/well of a variable amount of purified GST-Nedd4 protein from 10 pmol to 78 fmol with a 1:2 serial dilution; 2 wells with 10 pmol, 2 wells with 2.5 pmol, 2 wells with 1.25 pmol, 2 wells with 0.62 pmol, 2 wells with 0.31 pmol, 2 wells with 0.15 pmol, and 2 wells with 78 fmol in the 14 wells of the 384-well containing I-Buffer.
3. Add 5 μL/well of a variable amount of purified His-VI-wt protein from 10 pmol to 78 fmol with a 1:2 serial dilution (maintaining the optimal stoichiometry between His-VI-wt and GST-Nedd4) in the 14 wells prepared at **step 2**.
4. Follow the protocol (*see* Subheading 3.1.1) from **step 9**.

3.1.4 Assay Robustness (Z Prime Value)

Assay robustness was confirmed through calculation of the *Z′* factor using 30 replicates. As described by Zhang et al. [7], the *Z′* factor is a simple statistical parameter to evaluate the quality of the assay using data points obtained with the controls (*see* **Note 19**).

1. Distribute 5 μL of I-Buffer in 60 wells of a 384-well opaque plate.
2. Distribute 5 μL/well of purified His-VI-wt protein at 6.25 ng/μL (resulting in 31.25 ng/well) in 30 wells of the 384-well opaque plate.
3. Distribute 5 μL/well of purified His-VI-M1 mutant protein at 6.25 ng/μL (in 31.25 ng/well) in 30 different wells of the 384-well opaque plate.
4. Distribute 5 μL/well of purified GST-Nedd4 protein at 20 ng/μL (resulting in 125 ng/well) in the 60 wells prepared at **steps 2** and **3**.
5. Follow the protocol (*see* Subheading 3.1.1) from **step 9**.
6. Calculate the *Z′* factor (Fig. 3) to determine the robustness of the assay.

$$Z' \text{ factor} = 1 - \frac{(3\text{SD of positive control} + 3\text{SD of negative control})}{|\text{mean of positive control} - \text{mean of negative control}|}$$

Fig. 3 Formula of the *Z′* factor; a statistical parameter for evaluation of AlphaScreen® assay

3.2 Preparation of Intermediate 96-Well and Final 384-Well Prestwick Library Plates

1. Prestwick library compounds are supplied at 10 mM in 96-well plate (14 plates in total). Each plate contains 80 compounds, column 1 and 12 are empty.
2. The robot of the platform was programmed to first perform a dilution 1:100 (100 μM) of each of the compounds into a new 96-well plate, called intermediate diluted Prestwick Library 96-well plates 1–14.
3. Then the robot disposes 2.5 μL of I-Buffer in each well of a 384-well opaque plate.
4. The robot adds 2.5 μL of each compound from the intermediate diluted Prestwick Library 96-well plates 1 in corresponding wells containing I-Buffer of the 384-well plate (resulting compounds concentration is 50 μM) (Fig. 4).
5. The robot repeats the transfer with the intermediate diluted Prestwick Library 96-well plates 2, 3, and 4 to constitute the final Prestwick library 384-well plate A (intermediate diluted Prestwick Library 96-well plates 5–8 correspond to Prestwick library 384-well plate B, etc.). Finally, each well contains 5 μL of one Prestwick library compound at 50 μM (*see* **Note 20**) (Fig. 5).

3.3 Screening of the Prestwick Library Using AlphaScreen® Assay

1. Using automatic monochannel pipette or the robotic liquid handler prepare control wells (follow protocol (*see* Subheading 3.1.1) from **step 1–8**) in the four empty column on either side of the wells containing 50 μM compounds.
2. Add 2.5 μL of DMSO + 2.5 μL of I-Buffer in two empty control wells of the final Prestwick library 384-well plate A.
3. Add 5 μL/well of purified His-VI-wt protein at 6.25 ng/μL (resulting in 31.25 ng /well) in each well containing 50 μM compounds or DMSO of the final Prestwick library 384-well plate A.

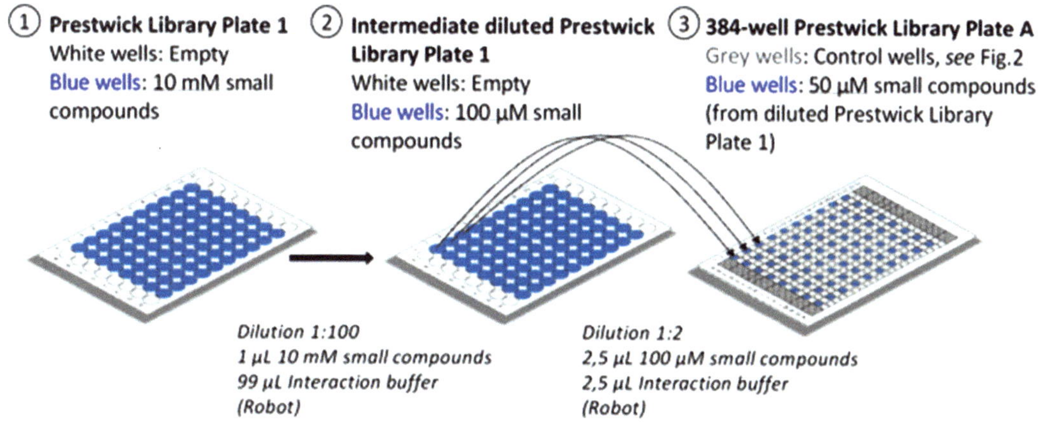

Fig. 4 Schematic representation of the preparation of the intermediate diluted Prestwick library plate 1 resulting in compounds at 100μM concentration starting from the original initial Prestwick library plate 1 at 10 mM. The 96-well intermediate diluted Prestwick chemical library plate 1 is transferred into one 384-well plate, called plate A, as indicated. The 2 flanking columns on each side serve as control wells

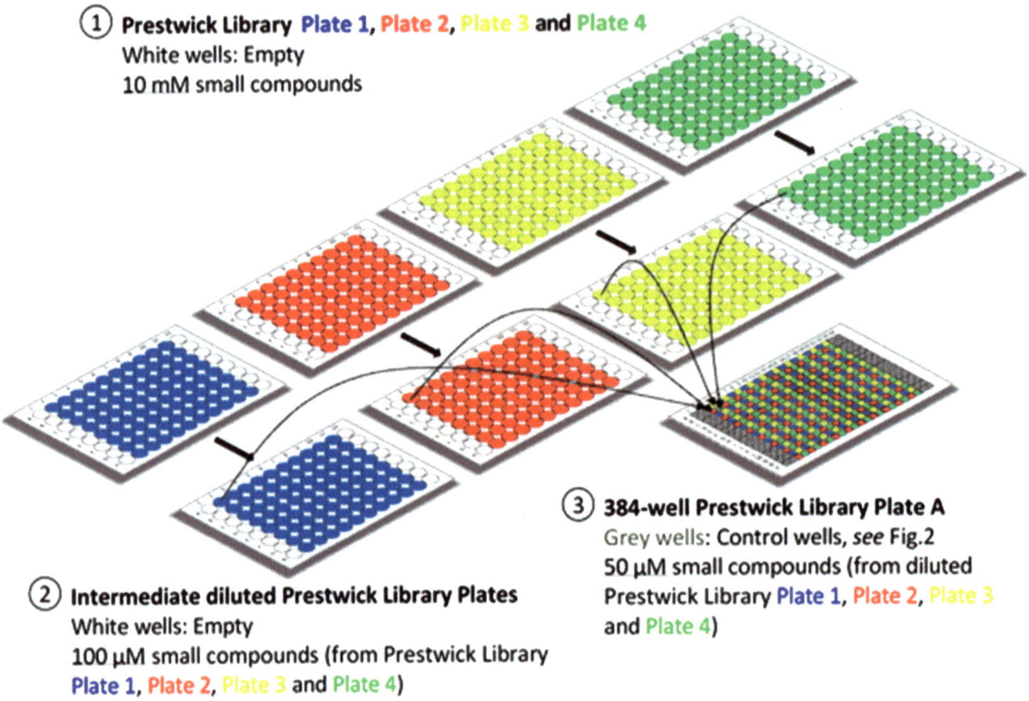

Fig. 5 Preparation of the final Prestwick library in a 384-well plate format. In total four 96-well intermediate diluted Prestwick library plates (plates 1, 2, 3, and 4) are distributed in the final Prestwick library 384-well plate A following the color-coded pipetting scheme. The 2 columns on both sides of the bank remain the control wells. In the resulting 384-well plate, each well contains 5 µL of one Prestwick library compound at 50 µM

4. Add 5 µL/well of purified GST-Nedd4 protein at 20 ng/µL (resulting in 125 ng/well) in each well containing 50 µM compounds and DMSO of the final Prestwick library 384-well plate A.
5. Cover the plate with a plastic 384-well plate cover.
6. Incubate the plate on the dark in the incubator at 25 °C during 30 min.
7. Add 5 µL/well of 0.1 µg/µL of nickel chelate AlphaLISA® acceptor beads solution in all wells (control and containing compound wells).
8. Add 5 µL/well of 0.1 µg/µL of AlphaScreen® glutathione donor beads solution in all wells (control and containing compound wells). Finally, each well contains 25 µL of a mix containing one Prestwick library compound at 10 µM, the partner proteins and beads (*see* **Note 20**).
9. Cover the plate with adhesive seals.
10. Transfer all drops to the bottom of the wells by gently tapping the edges of the plate with your hand.
11. Incubate the plate in the dark in the incubator at 25 °C during 1 h.

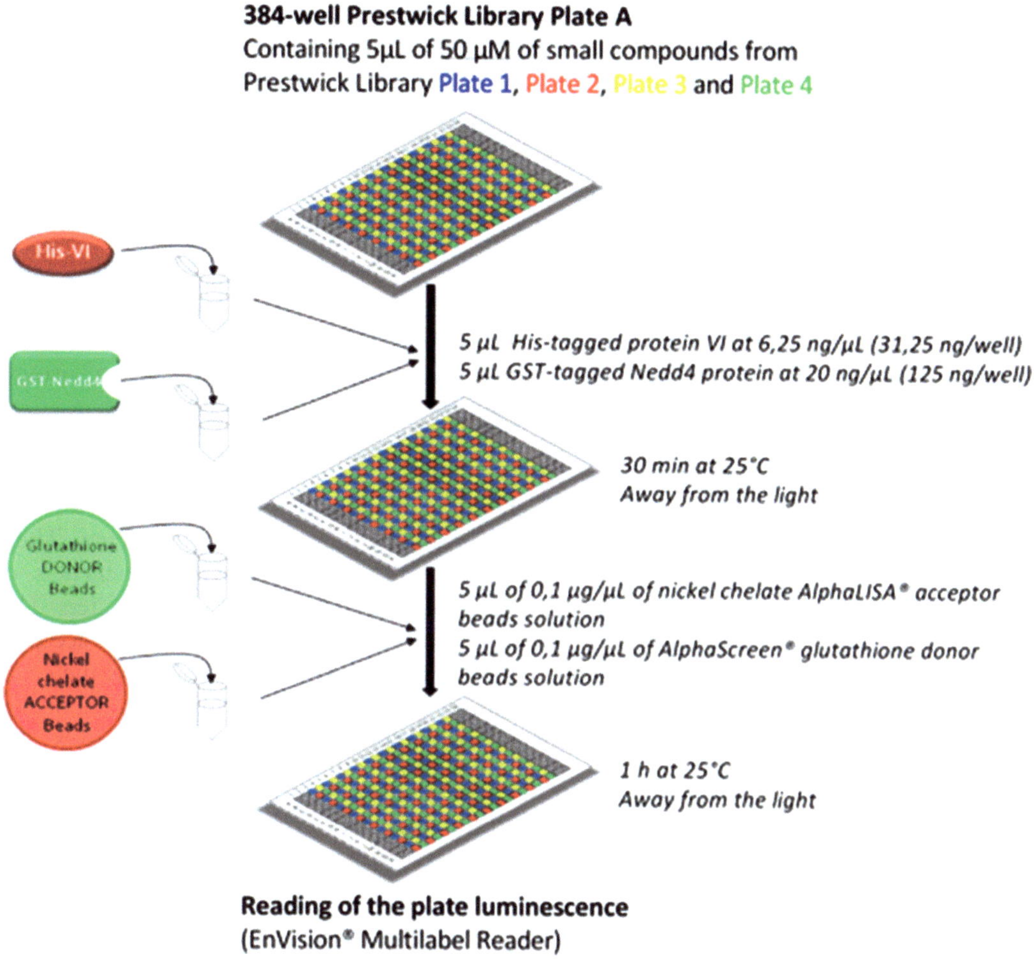

Fig. 6 Schematic distribution of reagents and protocol of AlphaScreen® assay using a part of the Prestwick chemical library (plate A, containing 4 of the 14 plates constituting the library) to find small-molecule inhibitors targeting VI⇔Nedd4 interaction

12. Read the luminescence of the plate at the EnVision® Multilabel Reader after 1, 4, and 24 h (*see* **Note 15**) of incubation to determinate the absolute signal intensity and the stability of the signal over time (*see* **Notes 16** and **17**) (Fig. 6).
13. Restart the protocol at **step 1** with the final Prestwick library 384-well plates B, C, and D to screen all compounds from the Prestwick Chemical library.

3.4 Data Analysis and Identification of Inhibitors Targeting a Conserved Host-Pathogen Interaction

1. For each plate, determine the background value (Fig. 2, *Controls 1, 2, 3, and 4*).
2. For each well with individual compound, subtract the mean background value from the luminescence value obtained with optimal conditions.

3. Determine the suitable cut off for a significant inhibition of the AlphaScreen® signal. Here, we choose [mean − (3 × Standard Deviation)] to select all compounds which inhibit 80 % of the interaction.
4. Because of the properties of AlphaScreen®, compounds that stimulate or stabilize the interaction can also be identified.
5. Remove all false-positive inhibitors using the AlphaScreen® TruHits kit (PerkinElmer) assay (*see* Subheading 3.5 for the protocol).
6. Determine IC_{50} (inhibitory concentration of inhibitors necessary to inhibit 50 % of the interaction) in AlphaScreen® using a variable concentration range of identified compound.

3.5 AlphaScreen® TruHits Assays

To exclude false-positive compounds, a counter screen was carried out using the AlphaScreen® TruHits kit (PerkinElmer). This kit assay permits to identify false-positive compounds of the screen. These compounds could be color quenchers, singlet oxygen quenchers, or light scatterers while blue compounds could act as inner filters. This assay is based on streptavidin donor beads and biotinylated acceptor beads resulting in a robust interaction between beads and the emission of strong constitutive AlphaScreen® signal. All compounds, which interfere with this signal, are considered as false positives.

1. According to the manufacturer indications (PerkinElmer), add 20 μL of the TruHits kit bead premix (pre-incubated 30 min at room temperature) into wells of a 384-well opaque plate.
2. Add 5 μL of the 50 μM selected compound solution take from each intermediate diluted Prestwick Library 96-well plates 1–14.
3. Add 5 μL of DMSO or interaction buffer as control in different wells.
4. Cover the plate with adhesive seals.
5. Incubate the plate 10 min at room temperature.
6. Read the luminescence of the plate at the EnVision® Multilabel Reader.

4 Notes

1. PerkinElmer offers different types of beads including streptavidin-coated beads, protein A-conjugated beads or antibody-conjugated beads which permit the analysis of a large range of interactors.
2. The protein VI M1 mutant is not able to interact with the Nedd4 [5] protein and is used here as our negative control for the interaction. An alternative for histidine-tagged proteins and nickel acceptor beads would be to use imidazole as negative

control to elute protein VI from the acceptor bead and inhibit the interaction. The amount of imidazole necessary to inhibit the interaction in our AlphaScreen® conditions was >10 mM.

3. AlphaScreen® assays are performed in white opaque plates to maximize signal intensity.
4. After purification, we dialyzed the proteins into glycerol supplemented buffer (110 mM potassium acetate, mM magnesium acetate, 2 mM DTT, 20 mM Hepes pH 7.3, 10 % glycerol) for storage and subsequent use. This buffer does not interfere with the assay signal. Alternative buffers should be tested for their compatibility with the AlphaScreen® assay.
5. It is recommended to resuspend beads by vortexing or pipetting before usage and briefly centrifuge them to avoid loss of reagent.
6. If no robot can be used, it is recommended to use an automatic monochannel pipette for sampling to minimize pipetting errors. This permits a quick, consistent, accurate, and homogeneous manual distribution of proteins and beads.
7. The manufacturer (PerkinElmer) recommends the use of the EnVision™ Multilabel Plate Reader to measure the AlphaScreen® signal. However, other readers can be used as well (e.g. EnSpire™ Alpha, AlphaPLUS).
8. Aliquots containing too small volume of protein solution (e.g. < 10 μL) tend to deteriorate due to rapid changes of concentration associated with evaporation.
9. We noticed that repeated freeze–thaw cycles for purified protein aliquots result in signal decrease (e.g. 3 cycles reduce the AlphaScreen® signal >10-fold compared to freshly thawed aliquots).
10. As recommended by PerkinElmer in "A Practical Guide to working with AlphaScreen®," the amount of blocking agent (BSA) or detergent (Tween20) in I-buffer can be optimized to avoid high background signals.
11. The ideal signal is in the range of 10^5 luminescence units to prevent masking weak compound effects on the interaction.
12. A signal for a true negative control of the interaction (here the GST-Nedd4 and the His-protein VI-M1 mutant) should be similar to the background signal obtained with each partner or beads alone.
13. The chemistry of AlphaScreen® reaction was designed to give best results at room temperature (ex.: 20–25 °C).
14. Brief plate centrifugation is an alternative to collect all solution drops at the bottom of the well.

15. As described by PerkinElmer in "A Practical Guide to working with AlphaScreen®," the beads should not be excited more than 2–3 times allowing to read the plate (e.g. at 1 and 24 h after the incubation) without changes in the signal intensity if the interaction is stable over time.
16. It is important to determinate the optimal incubation time to obtain a suitable signal. In our experiment the bead association reaction equilibrated at about 1 h of incubation. However, some compounds could be eliminated due to value fluctuations when read at 4 h post-incubation.
17. We could observe a signal decrease of the entire plate after 24 h incubation compared to 1 h or 4 h post-incubation suggesting that the best time point for readout has to be individually estimated.
18. As reported by the manufacturer, most AlphaScreen® assays will utilize partner concentration at 0.1–300 nM. We fixed the amount of one partner (the GST-Nedd4 purified protein) to 10 pmol and varied the amount of the second partner (His-protein VI-wt purified protein) starting with a 6:1 ratio and performing serial dilutions to reach a 0.5:1 ratio. This range should be individually adapted when using different proteins.
19. The Z' factor is a statistical parameter for evaluation of high-throughput assay [7]. Z' values are calculated from the measurement of >30 replicates of each positive and negative controls of interaction. A value equal to 1 is considered an ideal screening assay, when $1 > Z' > 0.5$ the assay is considered of excellent screening quality while $Z' < 0.5$ indicates strong sample and control signal variations not compatible with reliable screening conditions.
20. Prestwick library compounds are supplied at 10 mM in 14×96-well plates. It is recommended to test the compounds at 10 μM final concentration. Thus, we performed a first intermediate library plate at 100 μM and distribute 2.5 μL of this compound solution in the corresponding well of the 384-well plate. In each well, the final volume is 25 μL, resulting in compound concentration of 10 μM.

Acknowledgments

The work described in this protocol was supported by a PEPS IDEX grant of the Excellent initiative between the CNRS and Bordeaux University. S.A. received a fellowship from the Direction Générale de l'Armement (DGA). We acknowledge the support from the SFR TransBIOmed in buying the compound library. The authors would like to acknowledge networking support by the Proteostasis COST Action (BM1307).

References

1. Martin-Serrano J, Eastman SW, Chung W et al (2005) HECT ubiquitin ligases link viral and cellular PPXY motifs to the vacuolar protein-sorting pathway. J Cell Biol 168(1):89–101
2. Harty RN, Paragas J, Sudol M et al (1999) A proline-rich motif within the matrix protein of vesicular stomatitis virus and rabies virus interacts with WW domains of cellular proteins: implications for viral budding. J Virol 73(4):2921–2929
3. Bieniasz PD (2006) Late budding domains and host proteins in enveloped virus release. Virology 344(1):55–63
4. Zhadina M, Bieniasz PD (2010) Functional interchangeability of late domains, late domain cofactors and ubiquitin in viral budding. PLoS Pathog 6(10):e1001153
5. Wodrich H, Henaff D, Jammart B et al (2010) A capsid-encoded PPxY-motif facilitates adenovirus entry. PLoS Pathog 6(3):e1000808
6. Schreiner S, Martinez R, Groitl P et al (2012) Transcriptional activation of the adenoviral genome is mediated by capsid protein VI. PLoS Pathog 8(2):e1002549
7. Zhang JH, Chung TD, Oldenburg KR (1999) A simple statistical parameter for use in evaluation and validation of high throughput screening assays. J Biomol Screen 4(2):67–73

Chapter 31

Global MS-Based Proteomics Drug Profiling

Ana Sofia Carvalho and Rune Matthiesen

Abstract

DNA-based technologies such as RNAi, chemical-genetic profiling, or gene expression profiling by DNA microarrays combined with other biochemical methods are established strategies for surveying drug mechanisms. Such approaches can provide mechanistic information on how drugs act and affect cellular pathways. By studying how cancer cells compensate for the drug treatment, novel targets used in a combined treatment can be designed. Furthermore, toxicity effects on cells not targeted can be obtained on a molecular level. For example, drug companies are particularly interested in studying the molecular side effects of drugs in the liver. In addition, experiments with the purpose of elucidating liver toxicity can be studied using samples obtained from animal models exposed to different concentrations of a drug over time. More recently considerable advances in mass spectrometry (MS) technologies and bioinformatics tools allows informative global drug profiling experiments to be performed at a cost comparable to other large-scale technologies such as DNA-based technologies. Moreover, MS-based proteomics provides an additional layer of information on the dynamic regulation of proteins translation and particularly protein degradation. MS-based proteomics approaches combined with other biochemical methods delivers information on regulatory networks, signaling cascades, and metabolic pathways upon drug treatment. Furthermore, MS-based proteomics can provide additional information on single amino acid polymorphisms, protein isoform distribution, posttranslational modifications, and subcellular localization. In this chapter, we will share our experience using MS based proteomics as a pharmacoproteomics strategy to characterize drug mechanisms of action in single drug therapy or in multidrug combination. Finally, the emergence of integrated proteogenomics analysis, such as "The Cancer Genome Atlas" program, opened interesting perspectives to extend this approach to drug target discovery and validation.

Key words Pharmacoproteomics, MS-based proteomics, Posttranslational modifications, Subcellular localization, Single amino acid polymorphisms

1 Introduction

Interestingly most drugs used and approved by US Food and Drug Administration (FDA) are not fully characterized in terms of their targets and molecular effects and furthermore the molecular details of the disease itself are frequently poorly understood [1, 2]. There are many strategies developed using mass spectrometry-based proteomics that can be applied to define drug targets [3]; however, such targeted approaches do not necessarily reflect the effect of

Rune Matthiesen (ed.), *Proteostasis: Methods and Protocols*, Methods in Molecular Biology, vol. 1449,
DOI 10.1007/978-1-4939-3756-1_31, © Springer Science+Business Media New York 2016

drugs in vivo or in the context of all cellular pathways. Many drugs, small molecules as well as antibody therapies, target specific or a subset of proteins with similar three-dimensional structures. Although a narrow set of proteins are targeted, several other proteins are often affected if studied in a cellular context. This effect on other proteins can be explained by the fact that proteins frequently act in cellular pathways with other proteins and furthermore are part of large dynamic protein complexes. The composition, activity, and subcellular localization of protein complexes can be regulated by posttranslational modifications and other protein cofactors. Additionally, these activities can be affected by single amino acid polymorphisms and regulated by drugs. MS-based pharmacoproteomics is emerging as one of many novel technologies that hold potentials to elucidate the molecular details in a disease-therapy context [4, 5].

One relevant question when performing global MS-based drug profiling is: What is the most cost-effective way to obtained a global profiling? The balance between cost versus deepness of sampling demands consideration. A deeper insight into the proteome requires a more extensive fractionation of the complex mixture which will be at the expense of the number of LC-MS analysis needed and consequently timely and costly. In other words, it is important to address what is a reasonable number of fractions that can provide deep enough sampling to be informative in a global drug profiling framework. It is not possible to provide an exact answer to this question but it will depend on the quality of the MS instrument available as well as the complexity of the samples (e.g. bacteria versus human samples). We have considered several strategies for proteome fractionation such as: (1) 1D gel electrophoresis separation followed by cutting the lane into for example 10–20 bands [6], (2) trypsin digestion of complex protein mixture followed by strong cation exchange (SCX) fractionation [7, 8], and (3) subcellular fractionation (three to five crude fractions: mitochondrial, microsome, cytoplasmic, soluble nuclear, and insoluble nuclear [9, 10]).

Our approach consists in subcellular fractionation (into five crude fractions adapetd from Graham JM, 2001 [9]) followed by MS analysis using high resolution mass spectometry (Q-Exactive). Below are outlined the reasons for this implementation: (1) 20,000 proteins from all fractions can be identified before filtering protein isoforms, (2) more elaborate subcellular enrichment methods provides less cross-contamination but on the other hand in general cause loss of a large number of relevant protein factors from the targeted subcellular fractions, (3) this fractionation strategy provides additional information about subcellular localization, and (4) total protein concentration can be normalized before digestion of the subcellular fractionation and therefore provide more precise label-free quantitation compared to fractionation using SCX and 1D gel electrophoresis which in our hands are harder to perform reproducible for label-free quantitative studies. The schema in Fig. 1 below outlines how we perform global drug profiling.

There are several alternative protein digestion protocols that can be considered for the outline in Fig. 1 such as in solution digest [8], short SDS gel electrophoresis followed by in-gel digestion [11], and filter-aided sample preparation (FASP) [12, 13]. We decided to use FASP since some protein lysis buffers contained NP-40, sodium deoxycholate, and sodium dodecyl sulfate. Moreover, we were able to perform FASP with high reproducibility.

Applying label-free quantitation on drug-treated cell lines [14, 15] provides many significantly regulated proteins. Validation of all targets by Western blot is timely and imparticall for some proteins. We therefore recommend performing large-scale validation by using di or triple dimethyl labeling [14, 16–18] as outlined in Fig. 1 if a large number of proteins is significantly regulated.

2 Materials

2.1 Subcellular Fractionation

1. Confluent monolayer or suspension-cultured cells ($\sim 2 \times 10^8$).
2. Phosphate Buffered Saline (PBS): 10 mM Phosphate Buffer, pH 7.4, 0.14 M NaCl/KCl. To prepare 1 L of 1× PBS add 8.0 g of NaCl, 0.2 g of KCl, 1.44 g of Na_2HPO_4, 0.24 g of KH_2PO_4 to 800 mL of H_2O. Adjust the pH to 7.4 with HCl, and then add H_2O to 1 L. Alternatively, 1 L of 10× PBS solution can be prepared simply by weighing 10× the listed above reagents.
3. Cell Homogenization Medium (CHM): 150 mM $MgCl_2$, 10 mM KCl, 10 mM Tris–Cl, pH 6.7. To 100 mL H_2O add 30 μL 1 $MgCl_2$, 0.15 g KCl, 2.0 mL 1 M Tris–Cl, adjust pH to 6.7. Add H_2O to 200 mL.
4. CHM containing 1 M sucrose. For CHM containing 1 M sucrose add 68.4 g sucrose to 200 mL CHM.
5. Sucrose/Mg^{2+} medium: 0.15 M $MgCl_2$, 0.25 M sucrose, 10 mM Tris–Cl, pH 6.7. To 100 mL H_2O add: 30 μL 1 M $MgCl_2$, 17.1 g sucrose, 2 mL 1 M Tris–Cl, adjust pH to 6.7. Add H_2O to 200 mL.
6. cOmplete™, EDTA-free protease inhibitor cocktail (Roche).
7. PhosSTOP phosphatase inhibitor cocktail (Roche).
8. Low-speed centrifuge with swinging-bucket rotor and appropriate tubes.
9. High-speed centrifuge with fixed-angle rotor and appropriate tubes.
10. Ultracentrifuge with fixed-angle rotor and appropriate tubes.
11. Syringe and 20-gauge needle.

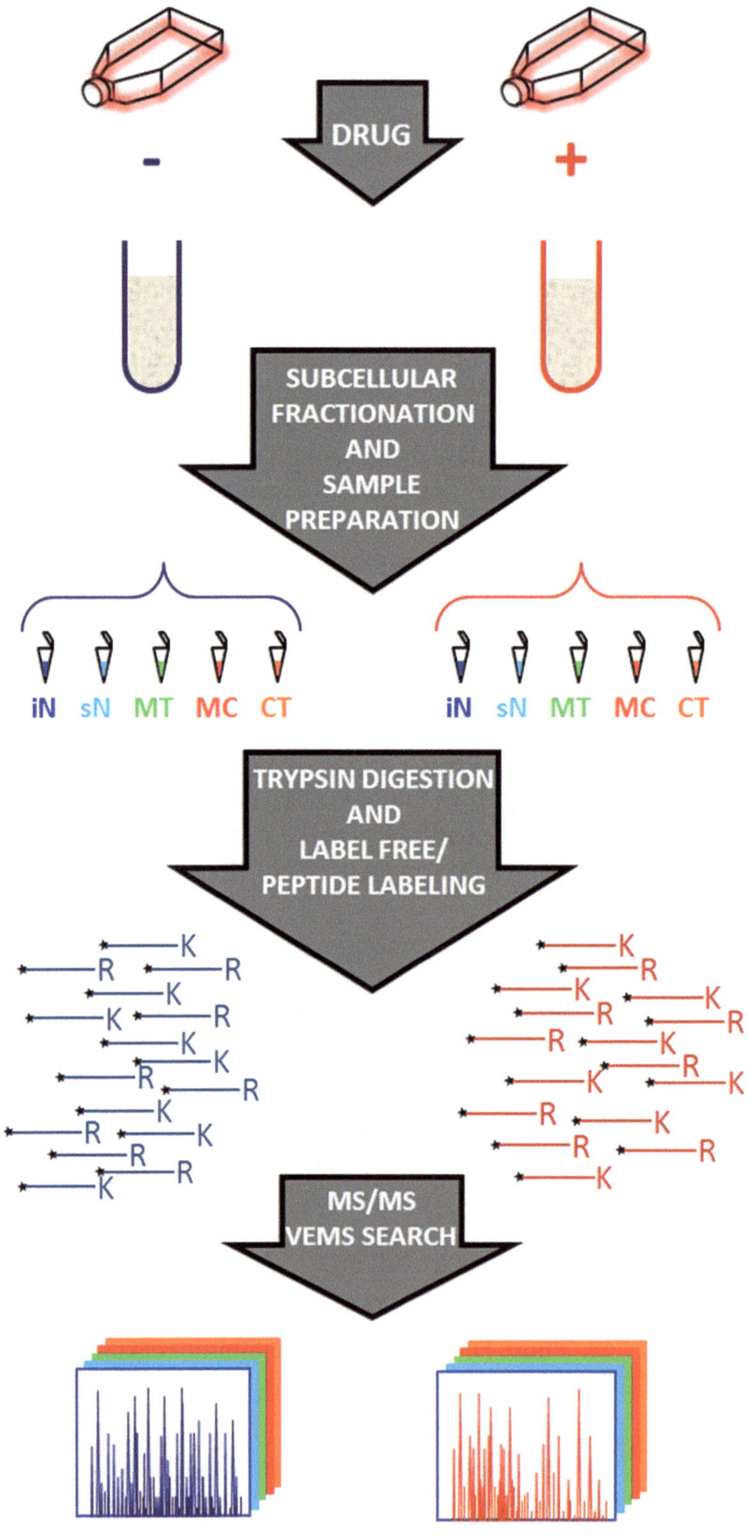

Fig. 1 Workflow for quantitative MS-based proteomics for global drug profiling to elucidate molecular drug effects. Step 1- Cell drug treatments. Cells are incubated in the presence and absence of drug(s) at different

2.2 Organelle Protein Extraction

1. Radio-Immunoprecipitation Assay (RIPA) Buffer: 50 mM Tris–HCl, pH 7.4, with 150 mM sodium chloride, 1.0% (v/v) Igepal (NP-40), 0.5% (w/v) sodium deoxycholate, and 0.1% (w/v) sodium dodecyl sulfate. 1 M Tris–HCl buffer stock solution pH 7.4: Dissolve 121.14 g Tris in 800 mL dH_2O. Adjust pH to 7.4 with the appropriate volume of concentrated HCl. Bring final volume to 1 L with deionized water. To 80 mL of H_2O add 5 mL 1 M Tris–HCl pH 7.4, 0.88 g NaCl, 1 mL Igepal (NP-40), 0.5 g sodium deoxycholate, and 0.5 mL of sodium dodecyl sulfate solution (20%). Measure the pH and if necessary adjust to 7.4. Add H_2O to a final volume of 100 mL.
2. cOmplete™, EDTA-free protease inhibitor cocktail (Roche).
3. PhosSTOP phosphatase inhibitor cocktail (Roche).
4. Refrigerated microcentrifuge with fixed-angle rotor.

2.3 Filter-Aided Sample Preparation

1. 1 M Dithiothreitol (DTT). Add 1.54 g DTT to 10 mL deionized H_2O.
2. 8 M Urea, 0.1 M HEPES, buffer pH 8.5. Add 48.0 g urea, 2.38 g HEPES to 80 mL deionized H_2O. Adjust pH to 8.5. Bring final volume to 0.1 L.
3. 0.05 M iodoacetamide in 8 M Urea, 0.1 M HEPES buffer, pH 8.5. To 25 mL 8 M Urea, 0.1 M HEPES, buffer pH 8.5 add 2.3 g iodoacetamide. Protect from light.
4. 40 mM NH_4HCO_3 and 10 mM NH_4HCO_3. To prepare 0.1 M NH_4HCO_3 stock solution, pH 8.5. Add 0.79 g NH_4HCO_3 to 80 mL deionized H_2O. Adjust pH to 8.5. Bring final volume to 0.1 L. For 40 mM NH_4HCO_3: Add 40 mL 0.1 M NH_4HCO_3, pH 8.5–60 mL deionized H_2O. For 10 mM NH_4HCO_3: Add 10 mL 0.1 M NH_4HCO_3, pH 8.5–90 mL deionized H_2O.
5. 0.05 μg/μL Trypsin in 40 mM NH_4HCO_3, pH 8.5. Reconstitute 100 μg of Trypsin Gold, Mass Spectrometry Grade (Promega) in 100 μL 50 mM acetic acid to 1 μg/μL. Add 50 μL 1 μg/μL of Trypsin Gold to 950 μL 40 mM NH_4HCO_3, pH 8.5.

Fig. 1 (continued) concentrations and at different intervals of time in order to select optimal drug(s) concentration(s)/treatment time; Step 2- Subcellular fractionation and sample preparation. Cells are harvested and disrupted followed by sequential differential centrifugation (method section 3.1). Protein extraction of nucleus, mitochondria and microsomal pellets results in a total of five fractions as described in method section 3.2, iN, insoluble nuclear fraction; sN, soluble nuclear fraction; MT, mitochondrial fraction; MC, microsomal fraction; CT, cytosolic fraction. Step 3- Tryptic digestion of protein samples and peptide labeling. Protein solutions are digested using FASP (method section 3.3) and peptides are dimethyl labeled as described in method section 3.4; Step 4- MS analysis and data quantitation analysis by VEMS or similar software tools (e.g. MaxQuant). "This workflow can be adapted using different quantitative MS methods based on label free or stabe isotope labelled peptides. The work flow can be further extended to include the study of different drugs, concentration and treatment intervals".

6. Thermomixer.
7. Microcentrifuge with fixed-angle rotor.

2.4 On-Column Stable Isotope Dimethyl Labeling

1. 50 mM sodium phosphate buffer pH 7.5. Add 50 mL of deionized H_2O to a 100 mL graduated glass beaker twice. Weigh 0.60 g NaH_2PO_4 to one beaker and 0.71 g of Na_2HPO_4 to the other beaker. Add deionized H_2O to a volume of 100 mL in each beaker.
2. 4% (vol/vol) formaldehyde in deionized H_2O (CH_2O, CD_2O or $^{13}CD_2O$). Add 40 μL of non-labeled and labeled formaldehyde (CH_2O, CD_2O or $^{13}CD_2O$) to 960 μL of ultrapure.
3. 0.6 M cyanoborohydride in water ($NaBH_3CN$). Into a 1.5 mL graduated tubes weigh 0.0377 g of sodium cyanoborohydride and add 1 mL of ultrapure H_2O.
4. 0.6 M cyanoborodeuteride in water ($NaBD_3CN$). Into a 1.5 mL graduated tubes weigh 0.0395 g of sodium cyanoborodeuteride and add 1 mL of ultrapure H_2O.
5. Reversed Phase (RP) solvent A, 0.6% (vol/vol) acetic acid. Add 0.6 mL of acetic acid and water to a volume of 100 mL.
6. RP solvent B, 0.6% (vol/vol) acetic acid and 80% (vol/vol) ACN. Add approximately 10 mL of ultrapure water to a beaker, 0.6 mL of acetic acid and 80 mL of acetonitrile. Mix and add water to a volume of 100 mL.
7. SepPak C18 cartridges (Waters).
8. Vaccum manifold system.
9. Vacuum Centrifuge.
10. Zip Tip columns.

3 Methods

Human cancer cell lines are a useful resource to study drug action and drug resistance mechanisms. The choice of cell line(s) is dependent on the drug(s) and vice versa. Cancer cell lines from a tumor type would respond differently to a specific drug and on the other hand different drugs may have distinct effects on a specific cell line. Taken this in consideration, preliminary results on the chosen model system testing drug concentration, incubation time on cell cycle progression, apoptosis, and protein response are desirable to be obtained prior to engage on obtaining 100–200 million cells for each condition studied.

3.1 Subcellular Fractionation

1. Pellet the cells (300 × *g*, 5 min, RT), decant the supernatant, wash twice with ~50 mL PBS per wash at room temperature.
2. Centrifuge the cells after the second wash at 1000 × *g*, room temperature for 15 min (*see* **Note 1**).

3. Aspirate or decant all of the supernatant and resuspend the cells in ice-cold cell homogenization medium containing protease and phosphatase inhibitors according to the manufacturer's instructions. Use a volume of medium equal to six times the volume of the pellet (*see* **Note 2**).
4. Leave on ice for 2 min.
5. Disrupt cells by passing the cell suspension ten times through a 20-gauge needle. Confirm that ≥90% cell breakage has occurred by examining the homogenate under a phase-contrast microscope (*see* **Note 3**).
6. Add one-third volume of ice-cold CHM containing 1 M sucrose (final 0.25 M) and mix gently by repeated inversion (*see* **Note 2**). Do not create foaming by rapid agitation.
7. Pellet nuclei by centrifuging for 5 min at 1000×*g*, 4 °C, in a swinging-bucket rotor using a low-speed centrifuge. Store nuclear pellet at −20 °C.
8. Decant or aspirate the supernatant and centrifuge for 10 min at 5000×*g*, 4 °C, in a fixed-angle rotor using a high-speed centrifuge. Keep supernatant on ice to use in **step 11**.
9. Resuspend the pellet in 10 mL ice-cold sucrose/Mg^{2+} medium by gently pipetting up and down.
10. Recentrifuge at 5000×*g*, 10 min, 4 °C. Discard supernatant. Store mitochondrial pellet at −20 °C.
11. Centrifuge the supernatant from **step 8** in an ultracentrifuge at 100,000×*g* for 60 min, 4 °C. Microsomes are pelleted and the supernatant constitutes the cytosol fraction.

3.2 Organelle Protein Extraction

To prepare the nuclear, mitochondrial and microsomal fractions pellets obtained by differential centrifugation, kept at −20 °C, were lysed.

1. Thaw nuclear, mitochondrial, and microsomal pellets on ice.
2. Add four times the volume of the pellet of RIPA buffer containing Mini protease inhibitor cocktail, PhosSTOP phosphatase inhibitor cocktail, to each pellet. Vortex briefly.
3. Leave on ice, vortex every 5 min for 20 min.
4. Transfer lysates to 1.5 mL graduated tubes.
5. Centrifuge lysates on a microcentrifuge at 15,000×*g* for 20 min, 4 °C.
6. Pipette supernatant to new 1.5 mL graduated tubes. The supernatants constitute soluble nuclear fraction, mitochondrial, and microsomal fractions and can be further used for proteome analysis or stored at −20 °C.
7. To the pellet of the nuclear lysate add four times the volume of the pellet of RIPA buffer containing inhibitors incubated 30 min on ice, vortexed every 5 min.

8. Centrifuge lysates on a microcentrifuge at 15,000×*g* for 20 min, 4 °C. The supernatant constitute the insoluble nuclear fraction. Stored at −20 °C.

3.3 Filter-Aided Sample Preparation (FASP)

1. Mix 100 μg of total protein with 10 μL of 1.0 M DTT (*see* **Note 4**).
2. Incubate in a thermomixer for 3 min at 40 °C.
3. Centrifuge at 14,000×*g*, 2 min and transfer the supernatant (a pellet was observed) to a Microcon YM-30 (Millipore).
4. Add 100 μL 8 M Urea, 0.1 M HEPES, pH 8.5 and centrifuge for 25 min at 14,000×*g*, 20 °C. Discard the flow-through from the collection tube. Repeat this step six times.
5. Add 100 μL of 0.05 M iodoacetamide in 8 M Urea, 0.1 M HEPES, pH 8.5 and mix in a thermo-mixer for 1 min. Incubate without mixing for 20 min in the dark.
6. Centrifuge for 35 min at 14,000×*g*, 20 °C. Discard the flow-through from the collection tube.
7. Add 100 μL of 8 M Urea, 0.1 M HEPES, pH 8.5. Centrifuge at 14,000×*g* for 25 min, 20 °C. Discard the flow-through from the collection tube. Repeat this step four times.
8. Add 100 μL of 40 mM NH_4HCO_3, pH 8.5. Centrifuge at 14,000×*g* for 25 min, 20 °C. Discard the flow-through from the collection tube. Repeat this step four times.
9. Add 40 μL of 40 mM NH_4HCO_3, pH 8.5 with trypsin (0.05 μg/μL) and mix for 1 min.
10. Incubate the filters at 37 °C O/N. Replace collection tube. Centrifuge filters for 20 min at 14,000×*g*, 20 °C.
11. Collect the flow-through.
12. Add 40 μL of 10 mM NH_4HCO_3 and centrifuge at 14,000×*g*, 20 °C and collect the flow-through. The total sample volume is approximately 80 μL in 25 mM NH_4HCO_3.

3.4 On-Column Stable Isotope Dimethyl Labeling

The peptide samples prepared with FASP were labeled according to the method described by Boersema et al. [17].

1. For each peptide sample mix 1 mL of 50 mM NaH_2PO_4 with 3.5 mL of 50 mM Na_2HPO_4 and with 250 μL of 4% (vol/vol) formaldehyde in water (CH_2O, CD_2O or $^{13}CD_2O$) and 250 μL of 0.6 M cyanoborohydride/cyanoborodeuteride in water ($NaBH_3CN$ or $NaBD_3CN$). For isotope peak deconvolution the mass difference to the previous light label should be at minimum 4 Da. Therefore, the formaldehyde/cyanoborohydride pairs for comparing three samples are: $CH_2O/NaBH_3CN$; $CD_2O/NaBH_3CN$; $^{13}CD_2O/NaBD_3CN$ (*see* **Note 5**). Prepare labeling reagent mixtures immediately prior to peptide reconstitution (*see* **Note 6**).

	CH_2O	CD_2O	$^{13}CD_2O$
$NaBH_3CN$	Light	Intermediate	
$NaBD_3CN$			Heavy

2. Reconstitute the peptides obtained in **step 12** of Subheading 3.3, after desalting (ZipTip or StageTip) and vacuum centrifugation, in 1 mL of 5 % formic acid.
3. Wash three SepPak columns with 2 mL of acetonitrile for each three different samples comparison (*see* **Note** 7).
4. Condition the SepPak columns twice with 2 mL of RP solvent A.
5. Load each of the three samples on a separate SepPak column.
6. Wash the SepPak columns with 2 mL of RP solvent A.
7. Flush each of the SepPak columns five times with 1 mL of the respective labeling reagent (light, intermediate, or heavy). This step should take at least 10 min to allow complete labeling.
8. Wash the SepPak columns with 2 mL of RP solvent A.
9. Elute and collect the labeled samples from the SepPak columns with 500 μL of RP solvent B.
10. Concentrate peptide samples by vacuum centrifugation to a final volume of 30 μL per sample.
11. Mix 10 μL of each of the three differentially labeled samples.

4 Notes

1. It is important to form a compact pellet to remove most of the PBS in **step 3**. Pellet the cells on a graduate tube to estimate the volume of CHM.
2. cOmplete™, EDTA-free protease inhibitor cocktail (Roche), and PhosSTOP phosphatase inhibitor cocktail (Roche) prevent protein degradation and maintain proteome phosphorylation status intact, respectively. In case of an extended analysis of other posttranslational modifications such as acetylation, glycosylation, or ubiquitylation, specific inhibitors can be purchased from life sciences suppliers such as, e.g. Roche, Sigma-Aldrich, or Selleckchem. Specificities of lysine deacetylase inhibitors can be found in Schölz et al. [19].
3. The needle (gauge number) must be adapted to the cell type. Prior to the experiment test cell homogenization in a small volume of cell suspension using needles with different gauge number. Confirm if cell breakage has occurred by examining the homogenate under a phase-contrast microscope.

4. Cell lysates obtained in Subheading 3.2 was accurately measured for protein concentration using BCA Assay kit (Pierce). 100 µg of total protein were prepared in RIPA buffer to a final concentration of 1 µg/µl. 100 µL were used in Subheading 3.3, **step 1**. Prepare three replicas per sample.
5. The labeling scheme presented gives low isotope peak overlap when combining three samples for MS/MS analysis. It is recommended that the labels are swapped between the three replicas per sample (e.g. sample A is labeled with light, intermediate, and heavy labels in three independent triple-sample sets). For comparison purposes, if the number of samples exceed 3 a common sample (e.g. sample A) can be included in each triple sample set.
6. Perform experiment in a fume hood, inclusive labeling reagents preparation.
7. The columns can be operated by using a manifold or by gravity.

Acknowledgement

This work was supported by EXPL/DTP-PIC/0616/2013. RM is sustained by the Fundação para a Ciência e a Tecnologia (FCT) investigator 2012 program. ASC is supported by grant SFRH/BPD/85569/2012 funded by Fundação para a Ciência e Tecnologia. The authors would like to acknowledge networking support by the Proteostasis COST Action (BM1307).

References

1. Drews J (2003) Stategic trends in the drug industry. Drug Discov Today 8(9):411–420
2. Yildirim MA et al (2007) Drug-target network. Nat Biotechnol 25(10):1119–1126
3. Schirle M, Bantscheff M, Kuster B (2012) Mass spectrometry-based proteomics in preclinical drug discovery. Chem Biol 19(1):72–84
4. Carvalho AS et al (2014) Global mass spectrometry and transcriptomics array based drug profiling provides novel insight into glucosamine induced ER stress. Mol Cell Proteomics 13(12):3294–3307
5. Hess S (2013) The emerging field of chemo- and pharmacoproteomics. Proteomics Clin Appl 7(1–2):171–180
6. Kim MS et al (2014) Heterogeneity of pancreatic cancer metastases in a single patient revealed by quantitative proteomics. Mol Cell Proteomics 13(11):2803–2811
7. Le Bihan T, Duewel HS, Figeys D (2003) On-line strong cation exchange micro-HPLC-ESI-MS/MS for protein identification and process optimization. J Am Soc Mass Spectrom 14(7):719–727
8. Schirmer EC, Yates JR 3rd, Gerace L (2003) MudPIT: a powerful proteomics tool for discovery. Discov Med 3(18):38–39
9. Graham JM (2001) Isolation of mitochondria from tissues and cells by differential centrifugation. Curr Protoc Cell Biol Chapter 3. Unit 3.3
10. Carvalho AS et al (2016) New insights into functional regulation in MS-based drug profiling, Scientific Reports 6:18826
11. Shevchenko A et al (2006) In-gel digestion for mass spectrometric characterization of proteins and proteomes. Nat Protoc 1(6):2856–2860
12. Manza LL et al (2005) Sample preparation and digestion for proteomic analyses using spin filters. Proteomics 5(7):1742–1745

13. Wisniewski JR et al (2009) Universal sample preparation method for proteome analysis. Nat Methods 6(5):359–362
14. Matthiesen R, Carvalho AS (2013) Methods and algorithms for quantitative proteomics by mass spectrometry. Methods Mol Biol 1007: 183–217
15. Matthiesen R (2013) LC-MS spectra processing. Methods Mol Biol 1007:47–63
16. Matthiesen R, Carvalho AS (2010) Methods and algorithms for relative quantitative proteomics by mass spectrometry. Methods Mol Biol 593:187–204
17. Boersema PJ et al (2009) Multiplex peptide stable isotope dimethyl labeling for quantitative proteomics. Nat Protoc 4(4):484–494
18. Hsu JL et al (2003) Stable-isotope dimethyl labeling for quantitative proteomics. Anal Chem 75(24):6843–6852
19. Scholz C et al (2015) Acetylation site specificities of lysine deacetylase inhibitors in human cells. Nat Biotechnol 33(4):415–423

Index

A

B

C

Rune Matthiesen (ed.), *Proteostasis: Methods and Protocols*, Methods in Molecular Biology, vol. 1449,
DOI 10.1007/978-1-4939-3756-1, © Springer Science+Business Media New York 2016

D

E

Q

R

S

W

X

Y

Z